FINITE ELEMENTS OF NONLINEAR CONTINUA

Advanced Engineering Series

Irving H. Shames, *Consulting editor*

HODGE: Continuum Mechanics
LEIGH: Nonlinear Continuum Mechanics
MEIROVITCH: Methods of Analytical Dynamics
ODEN: Finite Elements of Nonlinear Continua
THOMPSON: Compressible-fluid Dynamics

FINITE ELEMENTS OF NONLINEAR CONTINUA

J. T. ODEN
Professor of Engineering Mechanics
The University of Alabama in Huntsville

McGRAW-HILL BOOK COMPANY
NEW YORK ST. LOUIS SAN FRANCISCO DÜSSELDORF JOHANNESBURG
KUALA LUMPUR LONDON MEXICO MONTREAL NEW DELHI PANAMA
RIO DE JANEIRO SINGAPORE SYDNEY TORONTO

TO BARBARA

This book was set in Times Roman, and was printed and bound by The Maple Press Company. The designer was Edward Zytko; the drawings were done by John Cordes, J. & R. Technical Services, Inc. The editors were B. J. Clark and J. W. Maisel. John A. Sabella supervised production.

FINITE ELEMENTS OF NONLINEAR CONTINUA

Library of Congress Catalog Card Number 70-154237

07-047604-7

2 3 4 5 6 7 8 9 0 **MAMM** 7 9 8 7 6 5 4 3 2

CONTENTS

II. THEORY OF FINITE ELEMENTS 21

PREFACE

This book describes the finite element method and its application to a large class of nonlinear problems in structural and continuum mechanics. Special emphasis is given to the solution of problems in solid mechanics, but the general theory and methods of formulation are sufficiently general to be applied to nonlinear problems in, for example, fluid mechanics, electromagnetism, and partial differential equations. Various numerical methods for the solution of large systems of nonlinear equations are also examined.

My interest in the numerical analysis of nonlinear continua grew from a combination of a long interest in nonlinear mechanics, an appreciation of the great potential of modern digital computers for solving nonlinear problems, and a realization that much of the practical value of modern nonlinear theories of structural and material behavior will ultimately depend upon the availability of means to apply them to specific practical problems. Some years ago, I began to investigate the feasibility of applying the finite element method to the analysis of finite deformations of elastic solids. The surprising success of these early investigations, some of which form the basis for portions of this book, encouraged me to consider expanding the scope to nonlinear continua in general. In subsequent years, I developed and taught a graduate course on finite element applications in nonlinear mechanics at The University of Alabama in Huntsville, in which I attempted to draw together both the fundamentals of continuum mechanics and modern methods of numerical analysis. When these two subjects are brought together, each acquires new meaning and significance. The nonlinear field theories of mechanics then become valuable not only because they provide elegant generalizations of the classical theories, but also because, with the aid of electronic computing techniques, they provide a source for obtaining quantitative information on actual nonlinear phenomena encountered in nature. The finite element concept, with its simplicity and generality, provides the necessary ingredient for bringing these diverse subjects together in a manner which, in retrospect, may appear far more natural than many of the classical treatments of applied mechanics.

In selecting the topics to be covered in this book, it has not been my intention to provide an exhaustive collection of solutions to all kinds of nonlinear structural problems. Rather, the purpose here is to describe a general and physically appealing method for obtaining discrete models of continuous media, and to present a self-contained account of the application of this method to the analysis of representative nonlinear problems in solid mechanics. Once the basic notations are digested, applications to numerous nonlinear problems not examined herein should be straightforward.

So as to make the book self-contained, Chapter I contains an introductory discussion of the general concept of finite elements along with summary discussions of the kinematics of continuous media, the concept of stress, and the fundamental principles of conservation of mass and balance of momentum. Chapter II contains an account of the general theory of finite elements. Here the topological properties of finite element models of general fields are presented in forms valid for spaces of any finite dimension. Various types of finite element models are discussed as well as convergence criteria, and applications to linear and nonlinear differential equations, wave phenomena, and rarefied gas dynamics. This chapter also contains a detailed discussion of conjugate subspaces and the theory of conjugate approximations. Chapter III deals with the mechanics of a typical finite element of a continuous media. It begins with a discussion of appropriate thermodynamical concepts and principles, which is followed by derivations of local and global forms of the principle of conservation of energy for a continuum. These principles are used in conjunction with the theory developed in Chapter II to derive general kinematical equations and equations of motion and heat conduction for a finite element of arbitrary continuous media. A brief survey of the theory of constitutive equations is also included, and forms of constitutive equations cast in terms of discrete models of displacement and temperature fields are presented. In Chapter IV, applications of the finite element method to the analysis of nonlinear elasticity problems are presented. The chapter begins with an account of the theory of finite elastic deformations. Then nonlinear stiffness relations for elastic solids are derived, and solutions to a number of problems are presented. These include the problems of finite deformations of incompressible solids of revolution, stretching and inflation of elastic membranes, and finite plane strain of incompressible elastic solids. Also included in this chapter is a survey of various methods for the solution of large systems of nonlinear equations. Chapter V is devoted to inelastic behavior, with special emphasis on thermomechanically simple materials and materials with memory. General equations of motion and heat conduction for finite elements of such materials are derived. A number of applications of these equations to selected problems are examined, including problems in linear and nonlinear coupled thermoelasticity and nonlinear coupled thermoviscoelasticity.

I have discovered that writing a book is a nonlinear problem, the solution of which requires many iterations. Since the present form of this work varied very little in the last few iterations, I present it with the hope that it provides an approximate solution to the problem at hand. Nonlinear applied mechanics, however, is still in its infancy and is growing more rapidly with each passing day. Thus, the sequence is far from having converged. If this book provides a starting point for further iterations, it will have served a purpose for which it was intended.

I am grateful for the encouragement received from a number of colleagues and students during the preparation of this book. Of particular benefit were the comments and suggestions of Professors H. J. Brauchli and G. Aguirre-Ramirez. My discussions of related topics with Professors T. J. Chung and G. A. Wempner were also rewarding. Early versions of certain chapters of this book were distributed to students, and I sincerely appreciate their assistance and encouragement; among them, I am especially grateful to Messrs. J. E. Key, D. R. Bhandari, W. H. Armstrong, T. Sato, J. W. Poe, and D. A. Kross. For portions of my research which led to some of the ideas discussed in this book, I owe thanks to the support of the National Aeronautics and Space Administration and the U.S. Air Force Office of Scientific Research.

For their patience and diligency in keeping track of the multitude of subscripts, superscripts, and mathematical symbols in my manuscript, the staff at McGraw-Hill, particularly Jack Maisel, have my sincere thanks. It is also a pleasure to acknowledge the assistance of Mrs. D. Wigent, who, with enthusiasm and good humor, did an outstanding job of typing the entire manuscript.

Last, but far from least, I thank the one to whom this book is dedicated; for her continued encouragement, assistance in the proofreading and many personal sacrifices to help me finish this work, I shall always be grateful.

J. T. ODEN

I
Preliminary Discussions

1 INTRODUCTION

During the first half of the twentieth century, much of the literature on solid and structural mechanics was concerned with applications of long-standing linear theories to various boundary-value problems. There were notable exceptions, of course, such as the work which led to the rebirth and development of classical plasticity and viscoelasticity, the scattered attempts, some partially successful, at developing unified theories of material behavior, and the large number of studies of geometric nonlinearities by investigators who "retained nonlinear terms." To most of the engineering and scientific community, however, practical applications of solid mechanics meant the solution of linear problems.

The reason for this is easily understood, for the behavior of the majority of practical structures in the past could be adequately described by linear theories. The deformations of most structures under working loads, for example, were often scarcely detectable with the unaided eye, and for small deformations and steady uniform temperatures, the constitutive equations for such common materials as steel and aluminum can be treated as linear, without appreciable error.

This situation has drastically changed. Since 1950, many new materials have been introduced whose response cannot be described by classical linear theories. The thermoviscoelastic response of solid propellants, the postbuckling behavior of flexible structures, the use of highly deformable inflatable structures, and the nonlinear behavior of polymers and synthetic rubbers are only a few of the new problem areas that have encouraged the interest in nonlinear solid mechanics in recent times. The theory of elasticity has since been cast in general form, new nonlinear theories of viscoelasticity and thermoviscoelasticity have been proposed, and guiding principles for deriving constitutive equations for nonlinear materials are now generally accepted. The theme of modern research into nonlinear material behavior has been *generality*, and several theories have been proposed which span the gamut from elastic solids to thermoviscous fluids.

In spite of the advances in nonlinear theories of structural and material behavior, very little quantitative information is available to those who encounter nonlinear phenomena in practical applications. Nonlinear theories lead to nonlinear equations, which immediately render classical methods of analysis inapplicable. In all the work published on nonlinear behavior, only a handful of exact solutions to specific problems can be found; and these, without exception, deal with bodies of the most simple geometric shapes and boundary conditions. Often a "semi-inverse method" is employed, in which the shape of the deformed body is assumed to be known in advance (a situation that one seldom is so fortunate as to encounter in practice), and even in these cases numerical techniques must often be introduced in the final steps of the solution in order to obtain quantitative results.

This scarcity of quantitative information is, in some respects, quite ironic, for concurrent with the recent progress in nonlinear solid mechanics has been the development of the most powerful device for obtaining quantitative data that man has ever known—the digital computer. But, on the one hand, followers of the computational sciences have devoted full attention to new fields such as cybernetics and nonlinear programming, while, on the other hand, most researchers in continuum mechanics have been attracted to the purely theoretical aspects of the subject. In the middle ground lies a fertile and potentially important field: numerical analysis of nonlinear continua. It represents a marriage of modern theories of continuous media and modern methods of numerical analysis, so that, with the aid of electronic computation, quantitative information on the nonlinear behavior of solids and structures can be obtained. A systematic study of a portion of this middle ground is the subject of this book.

2 THE FINITE-ELEMENT CONCEPT

One must often resort to numerical procedures in order to obtain quantitative solutions to nonlinear problems in continuum mechanics. However, regardless of the initial assumptions and the methods used to formulate a problem, if numerical methods are employed in evaluating the results, the continuum is, in effect, approximated by a discrete model in the solution process. This observation suggests a logical alternative to the classical approach, namely, *represent the continuum by a discrete model at the onset.* Then further idealization in either the formulation or the solution may not be necessary. One such approach, based on the idea of piecewise approximating continuous fields, is referred to as the *finite-element method.* Its simplicity and generality make it an attractive candidate for applications to a wide range of nonlinear problems.

Classically, the analysis of continuous systems often began with investigations of the properties of small differential elements of the continuum under investigation. Relationships were established among mean values of various quantities associated with the infinitesimal elements, and partial differential equations or integral equations governing the behavior of the entire domain were obtained by allowing the dimensions of the elements to approach zero as the number of elements became infinitely large.

In contrast to this classical approach, the finite-element method begins with investigations of the properties of elements of finite dimensions. The equations describing the continuum may be employed in order to arrive at the properties of these elements, but the dimensions of the elements remain finite in the analysis, integrations are replaced by finite summations, and the partial differential equations of the continuous media are replaced, for example, by systems of algebraic or ordinary differential equations. The continuum with infinitely many degrees of freedom is thus represented by a discrete model which has finite degrees of freedom. Moreover, if certain completeness conditions are satisfied, then, as the number of finite elements is increased and their dimensions are decreased, the behavior of the discrete system converges to that of the continuous system. A significant feature of this procedure is that, in principle, it is applicable to the analysis of finite deformations of materially nonlinear, anisotropic, nonhomogeneous bodies of any geometrical shape with arbitrary boundary conditions.

2.1 HISTORICAL COMMENT

The idea of representing continuous functions by piecewise approximations is hardly a new one. Rudiments of the ideas of interpolation were supposedly used in ancient Babylonia and Egypt and, hence, preceded the calculus by over two thousand years. Much later, early Oriental mathematicians sought to evaluate the magical number π by determining the approximate area of the unit circle. This they accomplished to accuracies of almost forty significant figures by representing the circle as a collection of a large but finite number of rectangular or polygonal areas, the sum of which was taken as the area of the circle. It was left to Newton and Leibnitz to introduce the ideas of calculus, which have since made possible the formulation of most of the problems of mathematical physics in terms of partial differential and integral equations. Of course, the frequent failure of attempts to apply classical analytical methods to obtain solutions to many of these equations, plus the advent of the digital computer, has led an increasing number of investigators of modern times to consider approximate methods of analysis. It is interesting to note, however, that in many cases these investigators may unknowingly resort to concepts more primitive than those used to obtain the equations they wish to solve.

The practice of representing a structural system by a collection of discrete elements dates back to the early days of aircraft structural analysis, when wings and fuselages, for example, were treated as assemblages of stringers, skins, and shear panels. By representing a plane elastic solid as a collection of discrete elements composed of bars and beams, Hennikoff [1941] introduced his "framework method," a forerunner to the development of general discrete methods of structural mechanics. Topological properties of certain types of discrete systems were examined by Kron [1939]†, who developed systematic procedures for analyzing complex electrical networks and structural systems. Courant [1943]‡ presented an approximate solution to the St. Venant torsion problem in which he approximated the warping function linearly in each of an assemblage of triangular elements and proceeded to formulate the problem using the principle of minimum potential energy. Courant's piecewise application of the Ritz method involves all the basic concepts of the procedure now known as the finite-element method. Similar ideas were used later by Polya [1952]. The hypercircle method, presented in 1947 by Prager and Synge [1947] and discussed at length by Synge [1957]§, can be easily adapted to finite-element applications, and it provided further insight into the approximate solution of certain boundary-value problems in mathematical physics. In 1954, Argyris and his collaborators¶ began a series of papers in which they developed extensively certain generalizations of the linear theory of structures and presented procedures for analyzing complicated, discrete structural configurations in forms easily adapted to the digital computer.

† See also, for example, Kron [1944*a*, 1944*b*, 1953, 1954, 1955].

‡ See also Courant, Fredrichs, and Lewy [1928].

§ Synge [1957] speaks of linear interpolation over triangulated regions; his use of "polyhedral graphs" and "pyramid functions" is clearly in the spirit of the finite-element method.

¶ Argyris [1954, 1955, 1956, 1957], Argyris and Kelsey [1956, 1959, 1960, 1961, 1963], Argyris, Kelsey, and Kamel [1964].

The formal presentation of the finite-element method together with the direct stiffness method for assembling elements is attributed to Turner, Clough, Martin, and Topp [1956], who employed the equations of classical elasticity to obtain properties of a triangular element for use in the analysis of plane stress problems. It was Clough [1960] who first used the term "finite elements" in a later paper devoted to plane elasticity problems. In the intervening years, several hundred papers have appeared on the subject. Extensive references to previous work can be found in several books and survey papers on the subject.†

3 MECHANICS OF CONTINUA

In subsequent sections, it will be our principal aim to examine general methods for the formulation of finite-element models of continuous fields and to use these models in the analysis of nonlinear problems in structural and continuum mechanics. Equations describing the behavior of continuous media can generally be divided into four major categories: (1) kinematic, (2) kinetic (e.g., the mechanical-balance laws), (3) thermodynamic, and (4) constitutive. Thermodynamic principles are introduced in Chap. III as convenient means for obtaining general equations of motion and heat conduction for finite elements of continuous media. Constitutive equations establish relationships between kinematic, kinetic, and thermodynamic variables and, in so doing, characterize the material of which the continuum is composed. General notions of constitutive theory are discussed briefly in Chap. III, and constitutive equations for specific materials are examined in Chaps. IV and V. The remainder of this introductory chapter is devoted to a brief discussion of kinematics and kinetics of continuous media.

4 KINEMATIC PRELIMINARIES

Kinematics, in its traditional sense, is the study of the motion of bodies without consideration of the causes of motion. Its province is the description of changes in geometry with time, and the basic concepts it embodies

† See, for example, the conference proceedings and collections of papers edited by Fraeijs de Veubeke [1964*a*, 1971], Zienkiewicz and Holister [1965], Rydzewski [1965], Przemieniecki, Bader, Bozich, Johnson, and Mykytow [1966], Holland and Bell [1969], Rowan and Hackett [1969], Berke, Bader, Mykytow, Przemieniecki, and Shirk [1969], and Gallagher, Yamada, and Oden [1970] and the books of Pestel and Leckie [1963], Gallagher [1964], Martin [1966*a*], and Zienkiewicz and Cheung [1967]. Przemieniecki [1968] contains over 400 references on the subject. A number of survey articles and reports are available, e.g., Argyris [1958, 1966*a*, 1966*b*], Warren, Castle, and Gloria [1962], Parr [1964, 1967], and Felippa and Clough [1968]. See also Felippa [1966]. Singhal [1969] compiled a list of 775 references related to the finite-element method and matrix structural analysis. More recent surveys have been compiled by Zudans [1969] and Zienkiewicz [1970]. Survey articles devoted solely to applications to nonlinear problems were contributed by Marcal [1970], Martin [1970], and Oden [1969*c*, 1970*b*].

form an important part of the foundations of continuum mechanics. Our purpose here is to review and to record for future reference several of the more important kinematic relations.† Other relations are introduced, when convenient, in later sections.

4.1 GEOMETRY AND MOTION

A *body* is an infinite set of particles which can be brought into one-to-one correspondence with ordered triplets of real numbers called the *coordinates* of the particles. To each particle we assign a measure called *mass*, and we assume that the mass is absolutely continuous in the sense that as an arbitrary volume is shrunk to zero, so also is its associated mass. The simultaneous position of the set of particles comprising the body is called the *configuration* of the body. More rigorously, a configuration is a smooth mapping of a body onto a region of three-dimensional euclidean space. A sequence of mappings which defines the configurations at arbitrary times t (that is, a one-parameter family of configurations) is called the *motion* of the body.

The concepts of bodies and motions are, of course, closely related to our physical experience, and the rather formal tone implied in these definitions is not needed for most of our purposes. We cite them formally here only to bring added significance to certain approximations which are examined in subsequent chapters.

Consider, then, a continuous three-dimensional body in some reference configuration C_0. As a means of identification of particles, we associate with each particle $\mathbf{x}$ an ordered triple of real numbers $x_i = (x_1,x_2,x_3)$ called the *material coordinates* of $\mathbf{x}$. To give the particle labels x_i geometrical significance and to describe the motion of the body relative to C_0, we also establish a fixed rectangular cartesian coordinate system z_i in three-dimensional space called the *spatial coordinate system*. Then, while the body is in C_0, we set each triple of material coordinates x_i equal to the corresponding cartesian coordinates z_i of the place in space occupied by the particle. That is, the quantities x_i are geometrically the cartesian coordinates of the particle $\mathbf{x}$, relative to the frame z_i, while the body is in its reference configuration. The origin $x_i = (0,0,0)$ of the material frame is denoted o while that of the spatial frame is denoted 0. At time $\tau = t$ $(0 \leq \tau \leq t)$, the motion of the body has carried it from its initial (reference) configuration to a new configuration C, and the particle $\mathbf{x}$ has moved to a new place P, the spatial coordinates of which are denoted $z_i(\tau)$. Thus, the cartesian coordinates of a

† With a few exceptions, we follow generally the notations of Green and Zerna [1968, pp. 53–61]. A detailed account of the kinematics of continua, along with references to its historical development, is given in the treatise by Truesdell and Toupin [1960, pp. 241–463]. For a formal axiomatic treatment, see Truesdell and Noll [1965, pp. 48–56] and Truesdell [1966*a*, pp. 17–25].

particle at any time τ are $z_i(\tau)$, and at $\tau = 0$ the coordinates $z_i(\tau)$ and x_i coincide $[z_i(0) = x_i]$.†

We can now describe the motion of the body relative to C_0 by establishing a functional dependency of z_i on x_i and time. That is, we assume that the z_i are single-valued functions of x_i and τ which, except possibly at certain singular points, curves, and surfaces, are continuously differentiable as many times as desired. Then

$$z_i = z_i(x_1, x_2, x_3, \tau) \tag{4.1}$$

Further, we assume that a unique inverse to (4.1) exists and that

$$\left| \frac{\partial z_i(\mathbf{x}, \tau)}{\partial x_j} \right| > 0 \tag{4.2}$$

The functions indicated in (4.1) are said to define the motion of the body.

We emphasize that the triples of real numbers x_i are merely labels that we assign to material particles of the body. Thus the numerical values of x_i which define a particle in C_0 define the same particle in every configuration. We may consider these labels to be coordinates etched onto the body, so that they move continuously with the body as it passes from C_0 to some other configuration. It follows that while the coordinates x_i are cartesian in C_0, they are generally curvilinear in C; straight lines of particles parallel to x_i in C_0 become curved lines in C, and coordinate planes in C_0 become curved surfaces in C. Such coordinates are also called *convected* or *intrinsic* coordinates, and since they lead to rather simple kinematic relations, they are a natural choice in solid mechanics.

The position vector of a typical particle $\mathbf{x}$ at a place P_0 in C_0 is denoted $\mathbf{r}$, and the position vector of the *same* particle at a place P in C is denoted $\mathbf{R}$. If $\mathbf{i}_i$ denotes a set of orthonormal basis vectors tangent to the z_i axes, it follows that

$$\mathbf{r} = x_i \mathbf{i}_i \tag{4.3}$$

and

$$\mathbf{R} = z_i \mathbf{i}_i \tag{4.4}$$

The vector

$$\mathbf{u} = \mathbf{R} - \mathbf{r} = u_i \mathbf{i}_i \tag{4.5}$$

is called the *displacement vector*, and the functions $u_i = u_i(x_1, x_2, x_3, \tau)$ are

† Note that since the coordinates x_i and z_i are rectangular cartesian in C_0, the location of the index is immaterial. Thus, we use either x_i or x^i for the cartesian coordinates of a particle in C_0 and either z_i or z^i for the cartesian coordinates of a particle in C. However, in configuration C it is meaningful to write x^i instead of x_i.

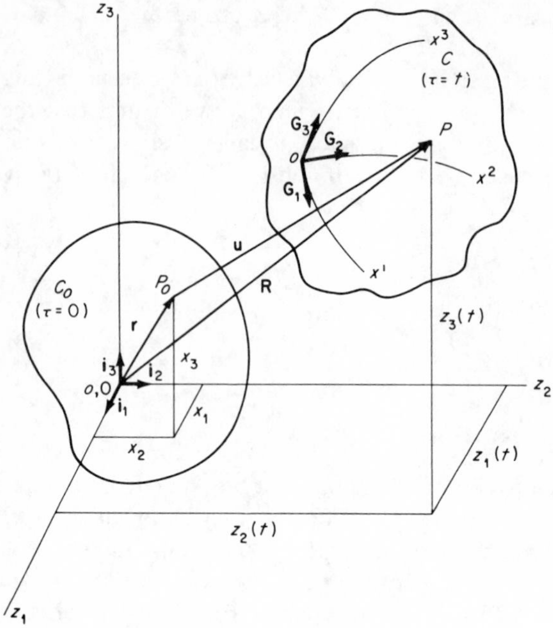

Fig. 4.1 Geometry of motion from C_0 to C.

called the *components of displacement*. The geometry of the motion from C_0 to C is indicated in Fig. 4.1.

4.2 DEFORMATION AND STRAIN

Let ds_0 denote a line element in C_0 and ds denote the *same* elemental line of particles in C. Then

$$ds_0^2 = d\mathbf{r} \cdot d\mathbf{r} = dx_i\, dx_i \tag{4.6}$$

$$ds^2 = d\mathbf{R} \cdot d\mathbf{R} = dz_i\, dz_i \tag{4.7}$$

The invariant

$$ds^2 - ds_0^2 \tag{4.8}$$

is a measure of the deformation of the body. If $ds^2 - ds_0^2 = 0$ everywhere, the motion of the body is referred to as a *rigid-body motion*. If $ds^2 - ds_0^2 \neq 0$ at $\mathbf{x}$, the body is said to be in a state of strain at $\mathbf{x}$.

According to (4.4) and (4.5),

$$d\mathbf{R} = \mathbf{R},_m\, dx^m = z_{i,m}\mathbf{i}_i\, dx^m = \mathbf{G}_m\, dx^m \tag{4.9}$$

where

$$\mathbf{G}_m = z_{i,m}\mathbf{i}_i = \mathbf{i}_m + \mathbf{u},_m = (\delta_{im} + u_{i,m})\mathbf{i}_i \tag{4.10}$$

Here the comma denotes partial differentiation with respect to the x_i (for example, $z_{i,m} \equiv \partial z_i/\partial x^m$; $u_{i,m} \equiv \partial u_i/\partial x^m$) and δ_{im} is the Kronecker delta. The vectors $\mathbf{G}_m$ are tangent to the deformed (convected) coordinate lines x^m. Note that the function $(\delta_{im} + u_{i,m})$ maps the initial tangent basis vectors $\mathbf{i}_i$ onto the tangent vectors $\mathbf{G}_m$ in C. The functions

$$z_{i,m} = \delta_{im} + u_{i,m} \tag{4.11}$$

are referred to as *deformation gradients.*

Introducing (4.9) into (4.7), we have

$$ds^2 = G_{ij}\, dx^i\, dx^j \tag{4.12}$$

in which

$$G_{ij} \equiv \mathbf{G}_i \cdot \mathbf{G}_j = G_{ji} \tag{4.13}$$

The functions G_{ij} are the covariant components of the metric tensor in C with respect to the convected system x^i; they are also referred to as components of *Green's deformation tensor.* In view of (4.10) and (4.13),

$$G_{ij} = z_{m,i} z_{m,j} = \delta_{ij} + u_{i,j} + u_{j,i} + u_{m,i} u_{m,j} \tag{4.14}$$

Returning to (4.8), we can now express the invariant deformation measure in the form

$$ds^2 - ds_0^2 = 2\gamma_{ij}\, dx^i\, dx^j \tag{4.15}$$

where

$$\gamma_{ij} = \tfrac{1}{2}(G_{ij} - \delta_{ij}) \tag{4.16}$$

Since $ds^2 - ds_0^2 \neq 0$ indicates that the body is in a state of strain at $\mathbf{x}$, the functions γ_{ij} serve as *strain measures.* They are the components of a symmetric second-order tensor called the *Green-Saint Venant strain tensor.* Since $ds = ds_0$ when $\gamma_{ij} = 0$ (and vice versa), a necessary and sufficient condition for the motion to be a rigid-body motion is that the strain components vanish throughout the body. Note also that, since $\gamma_{ij} = \gamma_{ji}$, we can always find an orthogonal coordinate system at a given point P in C such that $\gamma_{ij} = G_{ij} = 0$ for $i \neq j$. The directions of these coordinates are called *principal directions of strain*, and the strain components in the principal directions are called *principal strains.*

Substituting (4.14) into (4.16), we obtain the *strain-displacement relations*

$$\gamma_{ij} = \tfrac{1}{2}(u_{i,j} + u_{j,i} + u_{m,i} u_{m,j}) \tag{4.17}$$

Although the displacement gradients $u_{i,j}$ are not generally symmetrical (that is, $u_{i,j} \neq u_{j,i}$), they can be expressed as the sum of a symmetric tensor

e_{ij} and an antisymmetric tensor ω_{ij}, where

$$e_{ij} \equiv \tfrac{1}{2}(u_{i,j} + u_{j,i}) = e_{ji} \tag{4.18a}$$

$$\omega_{ij} \equiv \tfrac{1}{2}(u_{i,j} - u_{j,i}) = -\omega_{ji} \tag{4.18b}$$

The quantities e_{ij} and ω_{ij} are referred to as the components of the *infinitesimal strain* and *rotation tensors*, respectively. Clearly,

$$u_{i,j} = e_{ij} + \omega_{ij} \tag{4.19}$$

and

$$du_i = u_{i,j}\, dx^j = (e_{ij} + \omega_{ij})\, dx^j \tag{4.20}$$

We obtain the strain components in terms of e_{ij} and ω_{ij} by introducing (4.19) into (4.17):

$$\gamma_{ij} = e_{ij} + \tfrac{1}{2}(e_{mi} + \omega_{mi})(e_{mj} + \omega_{mj}) \tag{4.21}$$

In the case of deformations of very flexible bodies, such as plates and shells, the quantities e_{ij} may frequently be assumed to be infinitesimals of first order but the components ω_{ij} may be much larger.† Then, approximately,

$$\gamma_{ij} = e_{ij} + \tfrac{1}{2}\omega_{mi}\omega_{mj} \tag{4.22}$$

Further, if both e_{ij} and ω_{ij} (and, consequently, $u_{i,j}$) are infinitesimals of first order and if we therefore neglect their products and squares in comparison with their first powers, then (4.22) reduces to simply

$$\gamma_{ij} = e_{ij} \tag{4.23}$$

Notice that in this case if $e_{ij} = 0$ (rigid motion), (4.20) becomes

$$du_i = \omega_{ij}\, dx^j \tag{4.24}$$

Thus, we obtain an infinitesimal displacement without strain by a rigid rotation of the line elements dx^i.

4.3 STRAIN INVARIANTS

Three functions of the components of every second-order tensor can be formed which are the same in all coordinate systems. These are called the *invariants* of the tensor. The *principal invariants* of G_{ij}, denoted I_1, I_2, I_3, are called the *principal strain invariants* and are given by the formulas

$$I_1 = \delta^{ir}G_{ri} \tag{4.25a}$$

$$I_2 = \tfrac{1}{2}(\delta^{ir}\delta^{js}G_{ri}G_{sj} - \delta^{ir}\delta^{js}G_{ij}G_{rs}) \tag{4.25b}$$

$$I_3 = \det G_{ij} \equiv G \tag{4.25c}$$

† See Novozhilov [1953, p. 163].

or, in view of (4.16),

$$I_1 = 3 + 2\gamma_{ii} \tag{4.26a}$$

$$I_2 = 3 + 4\gamma_{ii} + 2(\gamma_{ii}\gamma_{jj} - \gamma_{ij}\gamma_{ji}) \tag{4.26b}$$

$$I_3 = 1 + 2\gamma_{ii} + 2(\gamma_{ii}\gamma_{jj} - \gamma_{ij}\gamma_{ji}) + \tfrac{4}{3}\epsilon^{ijk}\epsilon^{rst}\gamma_{ir}\gamma_{js}\gamma_{kt} \tag{4.26c}$$

where $\gamma_{ii} = \delta^{ir}\gamma_{ri}$ and ϵ^{ijk} and ϵ^{rst} are the permutation symbols. In later work, the partial derivatives of the invariants with respect to the strains are also of interest. Noting that

$$\frac{\partial \gamma_{ij}}{\partial \gamma_{mn}} = \delta_i^m \delta_j^n \tag{4.27}$$

we find

$$\frac{\partial I_1}{\partial \gamma_{mn}} = 2\delta^{mn} \tag{4.28a}$$

$$\frac{\partial I_2}{\partial \gamma_{mn}} = 4[\delta^{mn}(1 + \gamma_{ii}) - \delta^{nr}\delta^{ms}\gamma_{rs}] \tag{4.28b}$$

$$\frac{\partial I_3}{\partial \gamma_{mn}} = 2\delta^{mn}(1 + 2\gamma_{ii}) - 4\delta^{nr}\delta^{ms}\gamma_{rs} + 4\epsilon^{mjk}\epsilon^{nst}\gamma_{js}\gamma_{kt} \tag{4.28c}$$

4.4 CHANGES IN VOLUME AND AREA

The equation for an element of volume in C_0 is, simply,

$$dv_0 = dx_1\, dx_2\, dx_3 \tag{4.29}$$

After deformation, the vectors $\mathbf{i}_i\, dx_i$ (no sum) which formed the sides of dv_0 become $\mathbf{G}_i\, dx_i$. Thus, the same volume element acquires in the deformed body a new volume

$$dv = |\mathbf{G}_1 \cdot (\mathbf{G}_2 \times \mathbf{G}_3)|\, dx^1\, dx^2\, dx^3 \tag{4.30}$$

But

$$|\mathbf{G}_1 \cdot (\mathbf{G}_2 \times \mathbf{G}_3)| = |z_{i,j}| = \sqrt{|G_{ij}|} = \sqrt{I_3} = \sqrt{G} \tag{4.31}$$

Thus

$$dv = \sqrt{G}\, dx^1\, dx^2\, dx^3 = \sqrt{G}\, dv_0 = \sqrt{I_3}\, dv_0 \tag{4.32}$$

Deformations in which no changes in volume take place are called *isochoric deformations*. Clearly, in the case of isochoric deformations,

$$I_3 = 1 \tag{4.33}$$

In order to describe the deformation of elements of area, it is convenient to introduce a new set of vectors $\mathbf{G}^i$ which are normal to the respective x^i-coordinate surfaces in the deformed body:†

$$\mathbf{G}^i = \frac{\partial x^i}{\partial z_m} \mathbf{i}_m = \frac{1}{2\sqrt{G}} \epsilon^{ijk} \mathbf{G}_j \times \mathbf{G}_k \tag{4.34}$$

Here $\epsilon^{ijk} = \epsilon_{ijk}$ is the permutation symbol and $G = |G_{ij}| = I_3$. The vectors $\mathbf{G}^i$ are reciprocal to the tangent vectors $\mathbf{G}_i$; that is,

$$\mathbf{G}^i \cdot \mathbf{G}_j = \delta^i_j \tag{4.35}$$

Moreover, the functions

$$G^{ij} = \mathbf{G}^i \cdot \mathbf{G}^j \tag{4.36}$$

are the contravariant components of the metric tensor in C with respect to the convected system x^i, and G^{ij} is the inverse of the tensor G_{ij}. Now in the reference configuration C_0, vectors $\mathbf{i}_2\, dx_2$ and $\mathbf{i}_3\, dx_3$, for example, form the sides of an element of surface area dA_{10} on the x_1- material coordinate plane. The vector $\mathbf{i}_1 = \mathbf{i}_2 \times \mathbf{i}_3$ is normal to dA_{10}, and

$$dA_{10} = |\mathbf{i}_2 \times \mathbf{i}_3|\, dx_2\, dx_3 = dx_2\, dx_3 \tag{4.37}$$

After deformation, the originally plane area dA_{10} becomes a curved surface dA_1 with sides formed by the vectors $\mathbf{G}_2\, dx^2$ and $\mathbf{G}_3\, dx^3$. The vector $\sqrt{G}\,\mathbf{G}^1 = \mathbf{G}_2 \times \mathbf{G}_3$ is normal to dA_1. Thus

$$dA_1 = |\mathbf{G}_2 \times \mathbf{G}_3|\, dx^2\, dx^3 = \sqrt{GG^{11}}\, dx^2\, dx^3 \tag{4.38a}$$

where $G^{11} = \mathbf{G}^1 \cdot \mathbf{G}^1$ is the square of the magnitude of $\mathbf{G}^1$. Similarly, area elements dA_{20} and dA_{30} of the x_2- and x_3-material coordinate planes in C_0 become dA_2 and dA_3 in the deformed body, where

$$dA_2 = \sqrt{GG^{22}}\, dx^3\, dx^1 \tag{4.38b}$$

$$dA_3 = \sqrt{GG^{33}}\, dx^1\, dx^2 \tag{4.38c}$$

4.5 VELOCITY, ACCELERATION, AND DEFORMATION RATES

The velocity $\mathbf{v}$ of a particle is defined as its time rate of change of position. If we measure this change relative to the initial configuration C_0, then

$$\mathbf{v} = \dot{\mathbf{R}} = v_i \mathbf{i}_i \tag{4.39}$$

where the superimposed dot $(\cdot)$ indicates partial differentiation with respect

† The terms "undeformed body," "deformed body," etc., are, of course, not to be taken literally. By *deformation*, we mean here a deformation relative to the reference configuration, keeping in mind that it is not always possible or convenient to select a reference configuration corresponding to some natural, unstrained state of the body.

to time (that is, $\dot{\mathbf{R}} = \partial\mathbf{R}/\partial t$). Since, by definition, the convected coordinates x^i of a particle remain the same at all times (that is, $\dot{x}^i = 0$), it follows from (4.4) and (4.5) that

$$v_i = \dot{z}_i = \dot{u}_i \tag{4.40}$$

Similarly, the acceleration **a** of a particle is the time rate of change of velocity. Relative to C_0,

$$\mathbf{a} = a_i\mathbf{i}_i \equiv \dot{\mathbf{v}} = \ddot{\mathbf{R}} \tag{4.41}$$

and

$$a_i = \dot{v}_i = \ddot{z}_i = \ddot{u}_i \tag{4.42}$$

In studying the deformation of continuous bodies, it is also informative to examine the rates at which deformation takes place. The time rate of change of the invariant $ds^2 - ds_0^2$ of (4.8) is a natural measure of the *rate of deformation:*

$$\overline{ds^2 - ds_0^2}^{\,\boldsymbol{\cdot}} = \overline{ds^2}^{\,\boldsymbol{\cdot}} \tag{4.43}$$

In view of (4.15),

$$\overline{ds^2}^{\,\boldsymbol{\cdot}} = 2\dot{\gamma}_{ij}\,dx^i\,dx^j \tag{4.44}$$

where $\dot{\gamma}_{ij} = \partial\gamma_{ij}/\partial t$ are the strain rates. Introducing (4.17) into this result, we find

$$\overline{ds^2}^{\,\boldsymbol{\cdot}} = 2(\delta_{mi} + u_{m,i})\dot{u}_{m,j}\,dx^i\,dx^j = 2z_{m,i}\dot{u}_{m,j}\,dx^i\,dx^j \tag{4.45}$$

Moreover, in view of (4.19),

$$\overline{ds^2}^{\,\boldsymbol{\cdot}} = 2z_{m,i}(\dot{e}_{mj} + \dot{\omega}_{mj})\,dx^i\,dx^j \tag{4.46}$$

The functions $\dot{e}_{jm}$ and $\dot{\omega}_{jm}$ are referred to as the components of the *infinitesimal strain rate tensor* and *spin tensor*, respectively. Higher-order strain and deformation rates can be found by repeated differentiation of (4.44) with respect to time.

4.6 CURVILINEAR COORDINATES

It is not difficult to recast the kinematic relations derived previously in forms valid for general curvilinear coordinate systems. For example, let ξ^i denote a system of curvilinear coordinates in the reference configuration C_0 which are obtained through transformations of the form

$$\begin{aligned} \xi^i &= \xi^i(x_1,x_2,x_3) \\ x_i &= x_i(\xi^1,\xi^2,\xi^3) \end{aligned} \tag{4.47}$$

The coordinates ξ^i are also considered to be convected coordinates and merely serve as an alternative to x_i for describing the initial geometry of the body in C_0.

Basis vectors $\mathbf{g}_i$ tangent to the ξ^i-coordinate lines and $\mathbf{g}^i$ normal to the ξ^i-coordinate surfaces at $\tau = 0$ are given by

$$\mathbf{g}_i = \frac{\partial x_m}{\partial \xi^i}\mathbf{i}_m \qquad \mathbf{g}^i = \frac{\partial \xi^i}{\partial x_m}\mathbf{i}_m \tag{4.48a,b}$$

and covariant and contravariant components of the corresponding metric tensor are

$$g_{ij} = \mathbf{g}_i \cdot \mathbf{g}_j \qquad \text{and} \qquad g^{ij} = \mathbf{g}^i \cdot \mathbf{g}^j \tag{4.49a,b}$$

respectively.

The displacement vector $\mathbf{u}$, when referred to this curvilinear system, is of the form

$$\mathbf{u} = w_i\mathbf{g}^i = w^i\mathbf{g}_i \tag{4.50}$$

where w_i and w^i are the covariant and contravariant components of $\mathbf{u}$ with respect to the natural basis vectors of ξ^i. Similar formulas, of course, apply to other vectors referred to these coordinates.

Components of the strain tensor are now given by

$$\bar{\gamma}_{ij} = \tfrac{1}{2}(G_{ij} - g_{ij}) \tag{4.51}$$

and $\bar{\gamma}_{ij} = \gamma_{ij}$ of (4.17) only if $g_{ij} = \delta_{ij}$. Likewise, the strain-displacement relations are of the form

$$\bar{\gamma}_{ij} = \tfrac{1}{2}(w_{i:j} + w_{j:i} + w^m_{:i}w_{m:j}) \tag{4.52}$$

where the colon denotes covariant differentiation with respect to the coordinates ξ^i. That is,

$$w_{i:j} = \frac{\partial w_i}{\partial \xi^j} - \Gamma^m_{ij}w_m \tag{4.53a}$$

$$w^i_{:j} = \frac{\partial w^i}{\partial \xi^j} + \Gamma^i_{jm}w^m \tag{4.53b}$$

where Γ^i_{jk} are the Christoffel symbols of the second kind corresponding to the initial configuration C_0:

$$\Gamma^i_{jk} = \frac{\partial^2 z_m}{\partial \xi^j\,\partial \xi^k}\frac{\partial \xi^i}{\partial z_m} \tag{4.54}$$

The principal strain invariants are of the form

$$I_1 = g^{im}G_{im} = 3 + 2g^{im}\bar{\gamma}_{mi} \tag{4.55a}$$

$$I_2 = 3 + 4g^{im}\bar{\gamma}_{mi} + 2(g^{im}g^{nk}\bar{\gamma}_{mi}\bar{\gamma}_{nk} - g^{im}g^{nk}\bar{\gamma}_{mk}\bar{\gamma}_{ni}) \tag{4.55b}$$

$$I_3 = |\delta^i_j + 2g^{im}\bar{\gamma}_{mj}| = \frac{G}{g} \tag{4.55c}$$

where $g = |g_{ij}|$.

Other formulas derived previously for cartesian coordinates can be referred to the curvilinear system ξ^i by means of transformations similar to those used to obtain (4.48) to (4.55).

5 KINETIC PRELIMINARIES

Parallel to the kinematic relations, we may lay down certain kinetic relations which are assumed to hold for all continuous media. These follow from fundamental mechanical principles rather than from purely geometrical considerations, and they include local equations of motion, as derived from the momentum principles, and kinetic (traction) boundary conditions, as obtained from the concept of stress.† In finite-element formulations considered in subsequent chapters, these equations of motion and kinetic boundary conditions may be satisfied only in an average sense for a finite volume of the continuum. We shall then be concerned with the satisfaction of global equations of motion for finite material volumes and with only pointwise satisfaction of kinetic boundary conditions. For this reason, these kinetic relations often do not play as important a role in the development of discrete models of continuous bodies as do the kinematic relations presented in the previous section. They are, nevertheless, fundamental not only to mechanics per se but also to our approximate formulations, for to give meaning to any approximate theory it is essential that we identify carefully the nature of the approximation.

5.1 MASS AND MOMENTUM PRINCIPLES

We now cite several definitions and fundamental principles which lie at the foundations of continuum mechanics.

The principle of conservation of mass. Each body possesses a non-negative, additive, and unchanging property called mass. The mass of a body is conserved (invariant) during any motion of the body.

We assume that the mass is absolutely continuous and, therefore, that there exists a function ρ, called the *mass density*, which represents the mass per unit volume of the body. If ρ_0 and ρ denote mass densities of the body in configurations C_0 and C, respectively, then according to the above principle,

$$\int_{v_0} \rho_0 \, dv_0 = \int_v \rho \, dv \tag{5.1}$$

† For a more detailed account of these principles, consult, for example, Truesdell and Toupin [1960, pp. 465–491] or Eringen [1962, pp. 82–111], [1967, pp. 93–113].

In view of (4.32), we can also write

$$\int_{v_0} (\rho_0 - \sqrt{G}\rho)\, dv_0 = 0 \tag{5.2}$$

Since this must hold for arbitrary volumes, we conclude that

$$\rho = \frac{1}{\sqrt{G}}\rho_0 = \frac{1}{\sqrt{I_3}}\rho_0 \tag{5.3}$$

Equation (5.3) is referred to as the *equation of continuity.*

Definition 1 The linear momentum $\mathcal{T}$ of a body of mass $\mathcal{M}$ is defined by

$$\mathcal{T} \equiv \int_{\mathcal{M}} \mathbf{v}\, dm \tag{5.4}$$

where $\mathbf{v}$ is the velocity field and m is the mass.

Definition 2 The angular momentum $\mathcal{H}_0$ of a body of mass $\mathcal{M}$ with respect to a point 0 is defined by

$$\mathcal{H}_0 = \int_{\mathcal{M}} \mathbf{R} \times \mathbf{v}\, dm \tag{5.5}$$

where $\mathbf{R}$ is the position vector relative to 0, $\mathbf{v}$ is the velocity field, and m is the mass.

If the mass is absolutely continuous, dm in (5.4) and (5.5) can be replaced by $\rho\, dv$.

The principle of linear momentum. The time rate of change of linear momentum is equal to the resultant force $\mathbf{F}_{(R)}$ acting on a body:

$$\frac{d\mathcal{T}}{dt} = \mathbf{F}_{(R)} \tag{5.6}$$

The principle of angular momentum. The time rate of change of angular momentum about a fixed point 0 is equal to the resultant moment $\mathbf{M}_{(R)}$ about 0:

$$\frac{d\mathcal{H}_0}{dt} = \mathbf{M}_{(R)} \tag{5.7}$$

5.2 EXTERNAL FORCES AND STRESS

The external forces which act on a continuous body can be divided into two categories: (1) *body forces* and (2) *surface forces.* Body forces represent the force per unit mass (or unit volume) arising from external effects and are assumed to be given as continuous functions defined throughout the volume of the body. Surface forces (or surface tractions) act on the boundary surfaces of the body; they are defined in terms of external force per unit of boundary surface area and represent the contact of one body on another through the bounding surface. Surface forces are assumed to be sectionally continuous functions.

Within a body, internal forces are also developed due to the action of one particle on another. Since the force that one particle exerts on its neighbor is equal and opposite to that which the neighboring particle exerts on it, the resultant of these internal forces is zero. To measure these internal forces, we consider a particle **x** in the interior of the deformed body, through which we pass a surface of area A. Let **N** denote the net internal force on one side of A and $-\mathbf{N}$ the force (reaction) on the other side of A. Now consider a small element ΔA of A about **x**, and let $\Delta\mathbf{N}$ denote the net internal force on ΔA. Then the *stress vector* $\boldsymbol{\sigma}$ at **x** is defined by

$$\boldsymbol{\sigma} = \lim_{\Delta A \to 0} \frac{\Delta \mathbf{N}}{\Delta A} \tag{5.8}$$

The stress vector $\boldsymbol{\sigma}$ represents the forces per unit area in the deformed body. Clearly, the character of $\boldsymbol{\sigma}$ depends not only on **N** but also on the orientation of ΔA.

When referred to the tangent basis vectors $\mathbf{G}_j$ in the deformed body, the stress vector at **x** can be written in the form

$$\boldsymbol{\sigma} = \sigma^{ij} n_i \mathbf{G}_j \tag{5.9}$$

where σ^{ij} are the contravariant components of the *stress tensor*† *per unit area in the deformed body referred to the convected coordinates* x^i and where n_i are covariant components of a unit vector at **x** normal to the surface through **x** on which $\boldsymbol{\sigma}$ acts. The functions σ^{ij} are said to define the *state of stress* at **x**. Note the significance of the indices on σ^{ij}. The first index indicates the direction of the normal to the surface on which σ^{ij} acts, and the second index indicates the direction of the stress component.

Mixed and covariant components of stress can be computed using the metric tensor G_{ij} of (4.13):

$$\sigma_i^{\cdot j} = G_{im}\sigma^{mj} \qquad \text{and} \qquad \sigma_{ij} = G_{jm}\sigma_i^{\cdot m} \tag{5.10a,b}$$

Subsequently, we shall show that for monopolar media σ^{ij} is symmetric and that, therefore, $\sigma_i^{\cdot j} = \sigma_{\cdot i}^{j} = \sigma_i^j$. We also note that we can obtain *physical*

† The quantities σ^{ij} are sometimes referred to as the components of the second Piola-Kirchhoff stress tensor; see Truesdell and Noll [1965, pp. 124, 125]. The components $\pi_m^j = z_{m,i}\sigma^{ij}$ of the first Piola-Kirchhoff stress tensor are not, in general, symmetric.

components of stress s_{ij} referred to oblique unit vectors $\mathbf{G}_j/\sqrt{G_{(j)(j)}}$ by means of the formula

$$s_{ij} = \sigma^{ij} \frac{\sqrt{G_{jj}}}{\sqrt{G^{ii}}} \qquad \text{no sum on } i \text{ and } j \tag{5.11}$$

The quantities s_{ij} are not components of a tensor.

According to the definitions of $\boldsymbol{\sigma}$ and of the surface forces $\mathbf{S}$, the stress vector must be equal to the surface traction at points on the boundary surfaces of the deformed body. Thus, the condition

$$\boldsymbol{\sigma} = \mathbf{S} \tag{5.12}$$

which applies to boundary points is referred to as the *traction* or *kinetic boundary condition.* If the surface traction $\mathbf{S}$ is referred to the basis $\mathbf{G}_j$, we have

$$\sigma^{ij} n_i = S^j \tag{5.13}$$

where n_i are the covariant components of a unit normal to the boundary surface in the deformed body and S^j are the contravariant components of $\mathbf{S}$.

Since, in solid mechanics, we generally assume that the geometry in some reference configuration is known a priori, it is often convenient to define the state of stress at a point by a stress tensor t^{ij}, referred to the convected coordinates x^i in the deformed body but measured per unit area of the undeformed body.† When referred to the same coordinate system, the components t^{ij} and σ^{ij} are related by

$$t^{ij} = \sqrt{G}\, \sigma^{ij} \tag{5.14}$$

If $\mathbf{t}$ is the stress vector per unit area of undeformed body associated with a surface in the deformed body, then, instead of (5.9), we have

$$\mathbf{t} = t^{ij} \hat{n}_i \mathbf{G}_j \tag{5.15}$$

where $\hat{n}_i$ are the components of a unit normal $\hat{\mathbf{n}}$ to the undeformed area referred to the basis $\mathbf{i}_i$ in the reference configuration:

$$\hat{\mathbf{n}} = \hat{n}_i \mathbf{i}_i \tag{5.16}$$

5.3 CAUCHY'S LAWS OF MOTION

We can now use the momentum principles to obtain local equations of motion which apply to every particle $\mathbf{x}$ in C. Let $\mathbf{F}$ denote the body force per unit mass and $\mathbf{S}$ denote the surface force per unit surface area A. Then, according to (5.4) and (5.6),

$$\frac{d}{dt} \int_v \mathbf{v} \rho \, dv = \mathbf{F}_{(R)} = \int_v \mathbf{F} \rho \, dv + \int_A \mathbf{S} \, dA \tag{5.17}$$

† See Green and Adkins [1960, pp. 5–6].

Recalling (5.12) and (5.13) and using the Green-Gauss theorem to convert the surface integral into a volume integral, we have

$$\int_A \mathbf{S}\, dA = \int_A \sigma^{ij} n_i \mathbf{G}_j\, dA = \int_v \frac{1}{\sqrt{G}} \frac{\partial}{\partial x^i} (\sqrt{G}\, \sigma^{ij} \mathbf{G}_j)\, dv \tag{5.18}$$

Now $\mathbf{v}$, $\dot{\mathbf{v}}$, and ρ are assumed to be continuous throughout v and, because of the principle of conservation of mass, $d(\rho\, dv)/dt = 0$. Thus, we can rewrite (5.17) in the form

$$\int_v \left[\frac{1}{\sqrt{G}} (\sqrt{G}\, \sigma^{ij} \mathbf{G}_j)_{,i} + \rho \mathbf{F} - \rho \mathbf{a} \right] dv = 0 \tag{5.19}$$

where $\mathbf{a} = \dot{\mathbf{v}}$ is the acceleration. If (5.19) is to be valid for arbitrary volumes, the integrand must vanish. Thus

$$\frac{1}{\sqrt{G}} (\sqrt{G}\, \sigma^{ij} \mathbf{G}_j)_{,i} + \rho \mathbf{F} = \rho \mathbf{a} \tag{5.20}$$

Alternatively, we can express (5.20) in the component form

$$\sigma^{ij}_{;i} + \rho F^j = \rho a^j \tag{5.21}$$

where the semicolon denotes covariant differentiation with respect to the convected coordinates x^i in configuration C. That is,

$$\sigma^{ij}_{;i} = \sigma^{ij}_{,i} + \bar{\Gamma}^j_{im} \sigma^{im} + \bar{\Gamma}^m_{im} \sigma^{ij} \tag{5.22}$$

in which $\bar{\Gamma}^j_{im}$ are the Christoffel symbols of the second kind with respect to $\mathbf{x}^i$ in the deformed body:

$$\bar{\Gamma}^j_{im} = \frac{\partial^2 z_r}{\partial x^i\, \partial x^m} \frac{\partial x^j}{\partial z_r} \qquad \tau > 0 \tag{5.23}$$

Equation (5.21) is referred to as *Cauchy's first law of motion.*

It is often convenient to write the equation of motion in terms of quantities referred to the initial configuration C_0. Recalling (5.3) and (5.14) and noting that

$$\mathbf{F} = F^j \mathbf{G}_j = \hat{F}_j \mathbf{i}_j \tag{5.24}$$

$$\mathbf{a} = a^j \mathbf{G}_j = \ddot{u}_j \mathbf{i}_j \tag{5.25}$$

$$\hat{F}_j = z_{m,j} F^m \qquad \text{and} \qquad \ddot{u}_j = z_{m,j} a^m \tag{5.26a,b}$$

we observe that (5.21) can be recast in the form

$$(t^{ij} z_{m,j})_{,i} + \rho_0 \hat{F}_m = \rho_0 \ddot{u}_m \tag{5.27}$$

Turning now to the principle of angular momentum, we have†

$$\frac{d}{dt}\int_v \mathbf{R} \times \mathbf{v}\,\rho\,dv = \mathbf{M}_{(R)} = \int_v \mathbf{R} \times \mathbf{F}\,\rho\,dv + \int_A \mathbf{R} \times \mathbf{S}\,dA \tag{5.28}$$

Following the same procedure used previously, and using (5.20), we obtain for the local balance of angular momentum,

$$\mathbf{G}_i \times \sqrt{G}\,\sigma^{ij}\mathbf{G}_j = 0 \tag{5.29}$$

or, alternatively,

$$\epsilon_{ijk}\sigma^{ij} = 0 \tag{5.30}$$

from which we deduce that

$$\sigma^{ij} = \sigma^{ji} \tag{5.31}$$

Also, in view of (5.14),

$$t^{ij} = t^{ji} \tag{5.32}$$

Equation (5.30) is referred to as *Cauchy's second law of motion.* Equations (5.31) and (5.32) state that, as a consequence of this law, the stress tensors σ^{ij} and t^{ij} are symmetric. Thus, of the nine stress components, only six are independent.

† We confine our attention here to monopolar media. Thus, the possibility of couple stresses, body couples, etc., is not considered.

II
Theory of Finite Elements

6 INTRODUCTION TO THE CONCEPT OF FINITE ELEMENTS

The finite-element method may be described as a systematic procedure through which any continuous function is approximated by a discrete model which consists of a set of values of the given function at a finite number of points in its domain, together with piecewise approximations of the function over a finite number of subdomains. These subdomains are called *finite elements*, and the local approximation of the function over each finite element is uniquely defined in terms of the discrete values of the function at the finite number of preselected points in its domain. Thus, to construct a finite-element model of a given function, we proceed as follows:

1. A finite number of points are identified in the domain of the function, and the values of the function at these points are specified. These points are called *nodal points*, or simply *nodes*.
2. The domain of the function is represented approximately as a collection of a finite number of connected subdomains called finite elements. We

thereby regard the model of the actual domain as an assemblage of finite elements connected together appropriately at nodes on their boundaries.

3. The given function is approximated locally over each finite element by continuous functions which are uniquely defined in terms of the values of the function (and possibly of the values of its derivatives up to a certain order) at the nodal points belonging to each element.

An important aspect of the finite-element concept is that the finite elements may first be considered to be disjoint for the purpose of approximating a function locally over an element. That is, we can consider an individual element to be completely isolated from the collection and can proceed to approximate a function over the element in terms of its values at nodes of the element, independent of the ultimate location of the element in the connected model and independent of the behavior of the function in other finite elements. Thus, it is possible to develop a catalog of various finite elements in which nodal values of the local approximation are left arbitrary. We can then draw from this catalog whatever elements we need to approximate the domain and values of a given function.

6.1 ONE-DIMENSIONAL DOMAIN

To fix ideas, consider, for example, the real-valued continuous function $F(X)$, shown in Fig. 6.1*a*, whose domain is the closed interval $[A,B]$ of the real line. A finite-element model $\bar{F}(x)$ of this function is shown in Fig. 6.1*b*. We observe that the approximate function is designed so that its values coincide with those of the given function at a finite number (five, in this case) of nodal points in its domain and its local variation over a subinterval e approximates $F(X)$ by some assumed variation $f_{(e)}(x)$, where $x = x_{(e)}$ denotes the *local coordinates* of a point in subinterval e. We can construct such a model of $F(X)$ as follows:

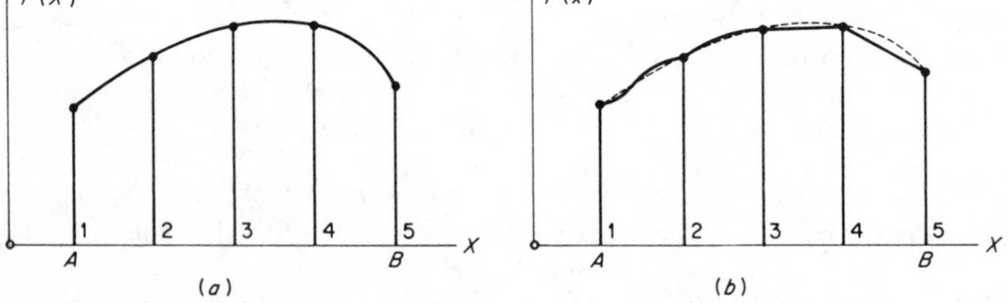

Fig. 6.1 A continuous function $F(X)$ and its finite-element representation $\bar{F}(x)$.

1. We identify a finite number (five) of nodal points in the domain $[A,B]$ of $F(X)$ and evaluate the function at each node (Fig. 6.2*a*):

$$F^{\Delta} = F(X^{\Delta}) \qquad \Delta = 1, 2, 3, 4, 5 \tag{6.1}$$

The five quantities (F^1,F^2,F^3,F^4,F^5) are called *global values* of the function and are elements of a set $\mathscr{G}$.

2. The interval $[A,B]$ is divided into a finite number (four) of subintervals called finite elements. Each element is temporarily considered to be distinct and disjoint from the other elements and, therefore, to have, for the moment, no points in common with other elements.

3. In a typical element e, a local coordinate system $x_{(e)}$ is established, and the function is approximated locally over e by a continuous function $f_{(e)}(x) \equiv f_{(e)}(x_{(e)})$ which takes on values $f^1_{(e)}$ and $f^2_{(e)}$ at nodes 1 and 2 at the end points of the element (Fig. 6.2*b*). The eight quantities $(f^1_{(1)}, f^2_{(1)}, f^1_{(2)}, \ldots, f^2_{(4)})$ may be considered to be elements of a set $\mathscr{L}$. The values $f^N_{(e)}$ $(N = 1, 2)$ are referred to as local values of the function corresponding to element e.

4. The interval $[A,B]$ is mapped onto the subintervals $[a_{(e)},b_{(e)}]$ $(e = 1, 2, 3, 4)$ by simple *incidence* relations, which relate the global coordinates X^{Δ}

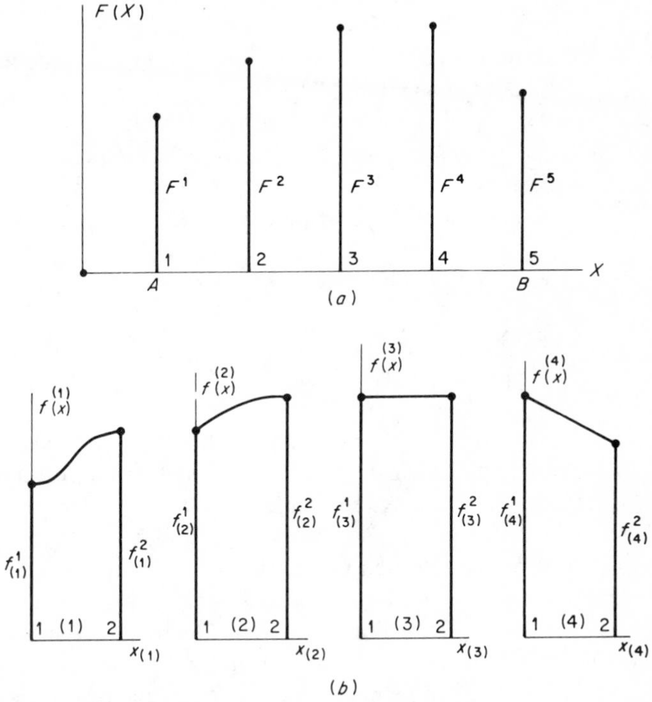

Fig. 6.2 Global values and local approximations of $F(X)$.

of various nodes to the local coordinates $x^N_{(e)}$:

$$
\begin{aligned}
&x^1_{(1)} = X^1 \qquad && x^2_{(1)} = x^1_{(2)} = X^2 \\
&x^2_{(2)} = x^1_{(3)} = X^3 \qquad && x^2_{(3)} = x^1_{(4)} = X^4 \\
&x^2_{(4)} = X^5
\end{aligned}
\tag{6.2}
$$

Though formally similar, these relations admit to two quite different interpretations: If one element is viewed at a time, relations of the type $x^1_{(2)} = X^2$ and $x^2_{(2)} = X^3$ describe an embedding of an element into the connected model of $[A,B]$. These establish the *connectivity* of the finite-element model. Alternatively, relations of the type $X^4 = x^2_{(3)} = x^1_{(4)}$ effectively describe a *decomposition* of the interval into finite elements. Mathematically, the two interpretations are quite different in character, as will be pointed out later.

5. The global values F^Δ ($\Delta = 1, 2, 3, 4, 5$) are related to the local values $f^N_{(e)}$ ($e = 1, 2, 3, 4$; $N = 1, 2$) by simple incidence relations of the form

$$
\begin{aligned}
&f^1_{(1)} = F^1 \qquad && f^2_{(1)} = f^1_{(2)} = F^2 \\
&f^2_{(2)} = f^1_{(3)} = F^3 \qquad && f^2_{(3)} = f^1_{(4)} = F^4 \\
&f^2_{(4)} = F^5
\end{aligned}
\tag{6.3}
$$

Equations (6.3) follow immediately from (6.2) due to the fact that the values of the functions $F(X)$ and $f_{(e)}(x)$ are in one-to-one correspondence with the points X and $x_{(e)}$ in their domains.

6. With the local approximations now given uniquely in terms of the global values F^Δ by (6.3), they are no longer considered disjoint and disconnected, and the final model of $F(X)$ is simply

$$
\bar{F}(x) = \sum_{e=1}^{4} \bar{f}_{(e)}(x, F^1, F^2, \ldots, F^5) \tag{6.4}
$$

where $\bar{f}_{(e)}(x, F^1, F^2, \ldots, F^5)$ are the local functions written in terms of the appropriate global values.

In constructing finite-element representations of other functions—$G(X)$, $H(X)$, etc.—defined on other intervals of the real line, there is little to be gained by developing separate and different local approximations for each subinterval that may be identified in a domain. Indeed, we can approach the problem, as indicated previously, by defining a function locally over an arbitrary interval in terms of arbitrary nodal values. This same local approximation can then be used repeatedly to construct a discrete model of a variety of different functions by assigning the arbitrary length of the interval and the locations of its nodes specific lengths and locations in the model of a given domain. Mathematically, this is accomplished by introducing incidence relations of the form in (6.2), which relate global coordinates X^Δ to local coordinates $x^N_{(e)}$. Similarly, the arbitrariness of the local nodal values $f^N_{(e)}$ is eliminated by relations of the form in (6.3), which specify the $f^N_{(e)}$ in terms of appropriate global values F^Δ.

The relationships described above between the connected model and individual finite elements and between the global nodes and the local nodes can be described mathematically by two rather simple mappings. First, we consider a typical member of a collection $\mathscr{R}^*$ of disjoint intervals, $\mathscr{R}^* = \{[x^1_{(1)},x^2_{(1)}], [x^1_{(2)},x^2_{(2)}], \ldots, [x^1_{(4)},x^2_{(4)}]\}$. Now let $\overset{(e)}{\Lambda}$ denote a mapping which embeds a *specific* interval e into its appropriate position in $[A,B]$. The mapping $\overset{(e)}{\Lambda}: [x^1_{(e)},x^2_{(e)}] \to [A,B]$ is accomplished by the transformation†

$$X^\Delta = \sum_{N=1}^{2} \overset{(e)}{\Lambda}{}^\Delta_N x^N_{(e)} \qquad e \text{ fixed} \tag{6.5}$$

where $\Delta = 1, 2, \ldots, 5$ and

$$\overset{(e)}{\Lambda}{}^\Delta_N = \begin{cases} 1 & \text{if node } \Delta \text{ of the connected model is incident on} \\ & \text{node } N \text{ of element } e \\ 0 & \text{if otherwise} \end{cases} \tag{6.6}$$

For example, for the fourth element in Eq. (6.2), $\overset{(4)}{\Lambda}{}^\Delta_N = 0$ if $\Delta \neq 4, 5$ and, in agreement with (6.2),

$$\begin{aligned} X^4 &= \overset{(4)}{\Lambda}{}^4_1 x^1_{(4)} + \overset{(4)}{\Lambda}{}^4_2 x^2_{(4)} = x^1_{(4)} \\ X^5 &= \overset{(4)}{\Lambda}{}^5_1 x^1_{(4)} + \overset{(4)}{\Lambda}{}^5_2 x^2_{(4)} = x^2_{(4)} \end{aligned} \tag{6.7}$$

since $\overset{(4)}{\Lambda}{}^4_1 = \overset{(4)}{\Lambda}{}^5_2 = 1$ and $\overset{(4)}{\Lambda}{}^4_2 = \overset{(4)}{\Lambda}{}^5_1 = 0$.

The mapping $\overset{(e)}{\Lambda}$ describes an embedding of an interval $[a_{(e)},b_{(e)}]$ into the interval $[A,B]$ and is referred to as *assembling* the element e into the connected model of the domain. Alternatively, we can consider the collection

$$\Lambda = \left\{\overset{(1)}{\Lambda},\overset{(2)}{\Lambda},\overset{(3)}{\Lambda},\overset{(4)}{\Lambda},\overset{(5)}{\Lambda}\right\}$$

of such mappings as a mapping of the collection $\mathscr{R}^*$ of disjoint elements

† Incidence relations such as this were introduced by Kron [1939], among others. Argyris [1954] employed similar transformations in matrix form in his "displacement method." The form given in (6.5) is similar to that proposed by Wissmann [1962, 1963, 1966] in connection with the finite-element analysis of large deformation of elastic bodies. Such transformations were also employed in nonlinear analyses by Oden [e.g., 1967*a*, 1967*b*, 1969*a*, 1969*b*]. Equation (6.5) is to be interpreted as simply a formal mathematical statement of connectivity requirements of the model. In applications, these relations can be established by inspection, after a judicious numbering of nodal points. The entire array $\overset{(e)}{\Lambda}{}^\Delta_N$, then, seldom need be evaluated.

into the "connected" interval $[A,B]$. For obvious reasons, the mappings Λ are said to establish the *connectivity* of the finite-element model.

The transpose of $\overset{(e)}{\Lambda}{}^{\Delta}_{N}$ is denoted $\overset{(e)}{\Omega}{}^{N}_{\Delta}$ and is interpreted as a mapping of the set $\mathscr{G}$ of global labels X^{Δ} into the set $\mathscr{L}$ of local labels $x^{N}_{(e)}$:

$$x^{N}_{(e)} = \sum_{\Delta=1}^{5} \overset{(e)}{\Omega}{}^{N}_{\Delta} X^{\Delta} \tag{6.8}$$

The elements of the array $\overset{(e)}{\Omega}{}^{N}_{\Delta}$ are defined in essentially the same manner as those of $\overset{(e)}{\Lambda}{}^{\Delta}_{N}$:

$$\overset{(e)}{\Omega}{}^{N}_{\Delta} = \begin{cases} 1 & \text{if node } N \text{ of element } e \text{ is incident on node } \Delta \text{ of the connected model} \\ 0 & \text{if otherwise} \end{cases} \tag{6.9}$$

For example, if we consider node $x^{2}_{(3)}$ in Fig. 6.2, we have

$$x^{2}_{(3)} = \overset{(3)}{\Omega}{}^{2}_{1} X^{1} + \overset{(3)}{\Omega}{}^{2}_{2} X^{2} + \overset{(3)}{\Omega}{}^{2}_{3} X^{3} + \overset{(3)}{\Omega}{}^{2}_{4} X^{4} + \overset{(3)}{\Omega}{}^{2}_{5} X^{5} \tag{6.10}$$

But $\overset{(3)}{\Omega}{}^{2}_{\Delta} = 0$ if $\Delta \neq 4$, by inspection, and $\overset{(3)}{\Omega}{}^{2}_{4} = 1$. Thus

$$x^{2}_{(3)} = X^{4} \tag{6.11}$$

The mapping $\overset{(e)}{\Omega}{}^{N}_{\Delta}$ amounts to merely a renumbering of integral labels assigned to various nodal points in the model. In a sense, $\overset{(e)}{\Omega}$ can be regarded as a mapping of a finite subset of $[A,B]$ into a subset of a finite subinterval. For this reason, we shall refer to the collection Ω of such mappings as the *decomposition* of $[A,B]$ into finite elements.

It is clear that the arrays $\overset{(e)}{\Lambda}{}^{\Delta}_{N}$ and $\overset{(e)}{\Omega}{}^{N}_{\Delta}$ can be used to form a composition $\overset{(e)(e)}{\Lambda\Omega} = I_e$, where I_e is the identity mapping of the collection of local nodes of element e into itself,

$$\sum_{\Delta=1}^{5} \overset{(e)}{\Lambda}{}^{\Delta}_{M} \overset{(e)}{\Omega}{}^{N}_{\Delta} = \delta^{N}_{M} \qquad N, M = 1, 2 \tag{6.12}$$

where δ^{N}_{M} is the Kronecker delta. A composition $\overset{(e)(e)}{\Omega\Lambda} = I^{(e)}_{G}$ can also be formed, where $I^{(e)}_{G}$ is a mapping of the collection of global nodes corresponding to element e into itself:

$$\sum_{N=1}^{2} \overset{(e)}{\Lambda}{}^{\Delta}_{N} \overset{(e)}{\Omega}{}^{N}_{\Gamma} = \begin{cases} \delta^{\Delta}_{\Gamma} & \text{if nodes } X^{\Delta}, X^{\Gamma} \text{ belong to element } e \\ 0 & \text{if nodes } X^{\Delta}, X^{\Gamma} \text{ do not belong to element } e \end{cases} \tag{6.13}$$
$$\Delta, \Gamma = 1, 2, 3, 4, 5$$

Returning now to $F(X)$, we recall that the given function $F(X)$ represents a one-to-one correspondence between the *values* $F = F(X)$ and the points X of its domain. Thus, for any point appearing in (6.7) or (6.8), local or global, there is a corresponding local or global value of the function. This correspondence is clearly indicated by the fact that the relations (6.3) are identical in form to (6.1). It follows that we can also write, for fixed e,

$$F^{\Delta} = \sum_{N=1}^{2\ (e)} \Lambda_N^{\Delta} f_{(e)}^N \tag{6.14}$$

and

$$f_{(e)}^N = \sum_{\Delta=1}^{5\ (e)} \Omega_{\Delta}^N F^{\Delta} \tag{6.15}$$

6.2 TWO-DIMENSIONAL DOMAIN

Practically the same procedure used in the one-dimensional case can be used to construct finite-element models of functions defined on two-dimensional domains. Consider, for example, the scalar-valued function $\Phi = \Phi(X, Y)$ defined on a closed region $\mathscr{R}$ of the X, Y plane (Fig. 6.3). Briefly, we identify a finite number G of points and label them consecutively $1, 2, \ldots, G$. The values Φ^{Δ} at these points are given by

$$\Phi^{\Delta} = \Phi(\mathbf{X}^{\Delta}) \tag{6.16}$$

where $\mathbf{X}^{\Delta}$ denotes the point Δ with coordinates (X^{Δ}, Y^{Δ}).

The region $\mathscr{R}$ is approximated by a collection of E subregions r_e, the points of which are given locally by $\mathbf{x}_{(e)} = (x_{(e)}, y_{(e)})$. The local approximations of the function of each element are denoted $\varphi_{(e)}(\mathbf{x})$, and their values at nodes $\mathbf{x}_{(e)}^N$ are denoted $\varphi_{(e)}^N$, where $N = 1, 2, \ldots, N_e$, N_e being the total number of nodes of element e. The connectivity and decomposition of the model are established by the incidence relations

$$\mathbf{X}^{\Delta} = \sum_{N=1}^{N_e\ (e)} \Lambda_N^{\Delta} \mathbf{x}_{(e)}^N \qquad \text{and} \qquad \mathbf{x}_{(e)}^N = \sum_{\Delta=1}^{G\ (e)} \Omega_{\Delta}^N \mathbf{X}^{\Delta} \tag{6.17a,b}$$

and the local values are related to the global values by the mappings

$$\Phi^{\Delta} = \sum_{N=1}^{N_e\ (e)} \Lambda_N^{\Delta} \varphi_{(e)}^N \qquad \text{and} \qquad \varphi_{(e)}^N = \sum_{\Delta=1}^{G\ (e)} \Omega_{\Delta}^N \Phi^{\Delta} \tag{6.18a,b}$$

The final model of $\Phi(\mathbf{X})$ is then of the form

$$\bar{\Phi}(\mathbf{x}) = \sum_{e=1}^{E} \varphi_{(e)}(\mathbf{x}, \Phi^1, \ldots, \Phi^G) \tag{6.19}$$

7 FINITE-ELEMENT MODELS OF GENERAL FUNCTIONS†

With the basic ideas of finite elements laid down in the previous section, we now direct our attention to generalizations of these concepts. These involve the construction of finite-element representations of general functions defined on spaces of any finite dimension k.

† This article is primarily based on the papers of Oden [1969*a*, 1969*b*]. See also Oden [1967*b*] and Oden and Aguirre-Ramirez [1969].

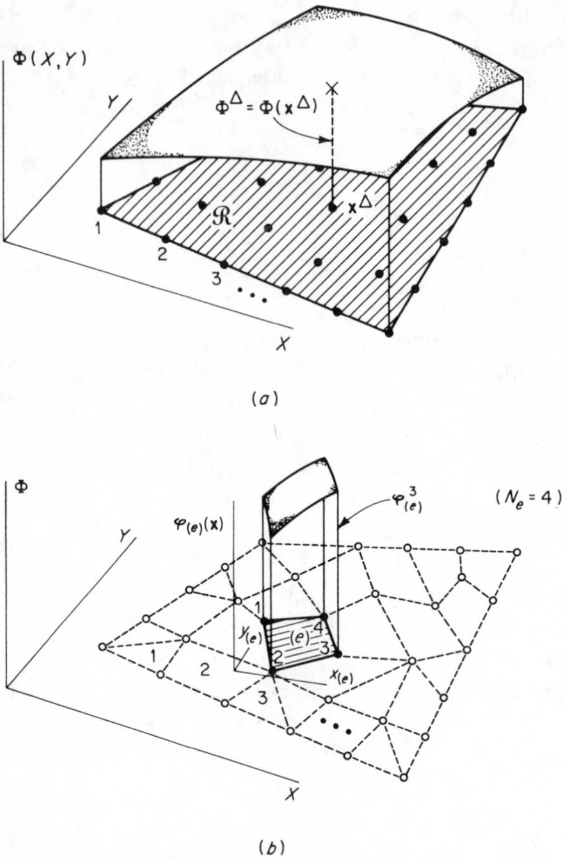

Fig. 6.3 Finite-element representation of a scalar-valued function defined on a two-dimensional domain.

7.1 SOME MATHEMATICAL PRELIMINARIES

We first review briefly a number of concepts and definitions† which are prerequisite to the study of properties of general finite-element models of continuous fields.

We adopt the usual set theoretics and notations.‡ For example, $a \in \mathscr{A}$ and is read "a belongs to the set $\mathscr{A}$"; $a \notin \mathscr{A}$ indicates that a is not a member of $\mathscr{A}$; $\mathscr{A} \subseteq \mathscr{B}$ indicates

† We elaborate further on certain of these concepts in Arts. 9 and 10.

‡ See, for example, Birkhoff and MacLane [1941], Moore [1962], or McCoy [1960]. For a readable yet comprehensive treatment of linear and abstract algebra, consult Mostow, Sampson, and Meyer [1963]. See also Finkbeiner [1966] and Greub [1963]. For introductory treatments on functional analysis, consult Kolmogorov and Fomin [1957, 1961], Liusternik and Sobolev [1961], or Taylor [1958].

that $\mathscr{A}$ is a subset of $\mathscr{B}$; and $\mathscr{A} \subset \mathscr{B}$ indicates that $\mathscr{A}$ is a proper subset of $\mathscr{B}$. The notation $\mathscr{A} = \{a \mid a \text{ has property } P\}$ is used to identify specific sets of quantities all having the defining property P. By the equality of two sets $\mathscr{A}$ and $\mathscr{B}$, written $\mathscr{A} = \mathscr{B}$, we mean that $\mathscr{A} \subseteq \mathscr{B}$ and $\mathscr{B} \subseteq \mathscr{A}$. The union, intersection, and difference of two sets $\mathscr{A}$ and $\mathscr{B}$ are denoted $\mathscr{A} \cup \mathscr{B}$, $\mathscr{A} \cap \mathscr{B}$, and $\mathscr{A} - \mathscr{B}$, respectively, and the cartesian product of two sets $\mathscr{A}$ and $\mathscr{B}$ is a set $\mathscr{A} \times \mathscr{B}$ of ordered pairs (a, b), that is, $\mathscr{A} \times \mathscr{B} = \{(a, b) \mid a \in \mathscr{A}, b \in \mathscr{B}\}$.

If $\mathscr{A}$ is a subset of $\mathscr{B}$, an element $b \in \mathscr{B}$ is called an *accumulation point* of $\mathscr{A}$ if every neighborhood of b (that is, every open set containing b) contains a point of $\mathscr{A}$ different from b. Equivalently, if b is an accumulation point of $\mathscr{A}$, then every neighborhood of b contains an infinite number of points of $\mathscr{A}$. The set consisting of $\mathscr{A}$ and all its accumulation points is referred to as the *closure* of $\mathscr{A}$. If $\mathscr{A}$ and $\mathscr{B}$ are two sets such that the closure of $\mathscr{A}$ contains the set $\mathscr{B}$, then $\mathscr{A}$ is said to be *dense* in $\mathscr{B}$. Further, if the closure of $\mathscr{A}$ equals $\mathscr{B}$, $\mathscr{A}$ is said to be *everywhere dense* in $\mathscr{B}$.

The term "function," used here synonymously with "mapping" and "transformation," carries its general meaning: Given two sets $\mathscr{A}$ and $\mathscr{B}$, a function F *on* $\mathscr{A}$ to $\mathscr{B}$, denoted $F: \mathscr{A} \to \mathscr{B}$, is a rule which assigns to each element $a \in \mathscr{A}$ one and only one element $F(a) \in \mathscr{B}$. Formally, $F \subseteq \mathscr{A} \times \mathscr{B}$ such that $(a, b) \in F$ and $(a, c) \in F \Rightarrow b = c$, where $\Rightarrow$ is read "implies." The set $\mathscr{A}$ is called the *domain* of F, and the element $F(a) \in \mathscr{B}$ is the *image* of a under the mapping F, or the *value* of the function F at a. The total set of images $F(\mathscr{A}) = \{f(a) \mid a \in \mathscr{A}\}$ is the *image space* of F, and F is said to be *onto* if $F(\mathscr{A}) = \mathscr{B}$ and *one to one* if $f(a) = f(b) \Rightarrow a = b$. For a one-to-one onto function $F: \mathscr{A} \to \mathscr{B}$, for every $a \in \mathscr{A}$ there is a unique $b = f(a) \in \mathscr{B}$ and an *inverse function* $F^{-1}: \mathscr{B} \to \mathscr{A}$ exists.

A *group* $\mathscr{G}$ is an algebraic system consisting of a set $\mathscr{A}$ together with an operation "$*$" (that is, a relation from $\mathscr{A} \times \mathscr{A}$ into $\mathscr{A}$) which associates each ordered pair (a, b), $a, b \in \mathscr{A}$, a third element $c \in \mathscr{A}$, denoted $c = a * b$, such that:

1. $(a * b) * c = a * (b * c)$.
2. There exists an element $e \in \mathscr{A}$ such that
   $e * a = a * e = a$ for every $a \in \mathscr{A}$.
3. For every $a \in \mathscr{A}$, there exists an element
   $a^{-1} \in \mathscr{A}$ such that $a^{-1} * a = a * a^{-1} = e$.

If $a * b = b * a$, $\mathscr{G}$ is said to be *abelian*.

Let R denote the real (or complex) number field [i.e., the set of real (or complex) numbers plus the usual operations of addition and multiplication of partially ordered real (or complex) numbers]. A *linear space* (or *vector space*) $\mathscr{V}$ is an abelian group, consisting of a set of elements **a**, **b**, . . . , sometimes called *vectors*, and an operation $+$ sometimes called *vector addition*, together with a function on $\text{R} \times \mathscr{V} \to \mathscr{V}$ called *scalar multiplication*, such that for all pairs $\alpha, \beta \in \text{R}$ and $\mathbf{a}, \mathbf{b} \in \mathscr{V}$, the following hold:

1. $(\alpha\beta)\mathbf{a} = \alpha(\beta\mathbf{a})$
2. $(\alpha + \beta)\mathbf{a} = \alpha\mathbf{a} + \beta\mathbf{a}$
   $\alpha(\mathbf{a} + \mathbf{b}) = \alpha\mathbf{a} + \alpha\mathbf{b}$
3. $1\mathbf{a} = \mathbf{a}$

Unless noted otherwise, we shall take R to be only the real number field in the following.

The maximum number k of linearly independent elements in a linear space $\mathscr{V}$ is called the *dimension* of $\mathscr{V}$. We use the notation $\mathscr{V}^k$ to denote spaces of dimension k. A set of k linearly independent elements is said to be a *basis* of $\mathscr{V}$ if the dimension of $\mathscr{V}$ is k; then every vector $\mathbf{a} \in \mathscr{V}$ is a linear combination of the elements comprising the basis.

A subset S of a linear space $\mathscr{V}$, together with the operations of vector addition and scalar multiplication, is a subspace of $\mathscr{V}$ if and only if it is also a linear space. If $\mathscr{S}$ and $\mathscr{T}$ are subspaces of a linear space $\mathscr{V}$, the sum of $\mathscr{S}$ and $\mathscr{T}$, denoted $\mathscr{S} + \mathscr{T}$, is the set of all elements $\mathbf{s} + \mathbf{t}$, where $\mathbf{s} \in \mathscr{S}$, $\mathbf{t} \in \mathscr{T}$. The intersection of $\mathscr{S}$ and $\mathscr{T}$, $\mathscr{S} \cap \mathscr{T}$, is the set of elements common to $\mathscr{S}$ and $\mathscr{T}$. Both $\mathscr{S} + \mathscr{T}$ and $\mathscr{S} \cap \mathscr{T}$ are subspaces of $\mathscr{V}$. If $\mathscr{S} + \mathscr{T} = \mathscr{V}$ and $\mathscr{S} \cap \mathscr{T} = \varnothing$, where $\varnothing$ is the null or empty space, then $\mathscr{V}$ is said to be the direct sum of $\mathscr{S}$ and $\mathscr{T}$, and is written $\mathscr{V} = \mathscr{S} \oplus \mathscr{T}$.

In a linear space $\mathscr{V}$ it is often possible to associate with each $\mathbf{a} \in \mathscr{V}$ a nonnegative real number, denoted $\|\mathbf{a}\|$ and called the *norm* of $\mathbf{a}$, such that

1. $\|\mathbf{a}\| \geq 0$, the equality holding if and only if $\mathbf{a} = \mathbf{0}$.
2. $\|\lambda\mathbf{a}\| = |\lambda| \, \|\mathbf{a}\|$, λ being a scalar.
3. $\|\mathbf{a} + \mathbf{b}\| \leq \|\mathbf{a}\| + \|\mathbf{b}\|$.

A linear space in which a norm has been defined is called a *normed linear space*. The concept of a norm is a generalization of the familiar notion of the length of a vector.

In a normed linear space, a sequence of elements $\mathbf{a}_1, \mathbf{a}_2, \ldots, \mathbf{a}_n, \ldots$, written $\{\mathbf{a}_n\}$, which has the property that $\lim_{m,n\to\infty} \|\mathbf{a}_m - \mathbf{a}_n\| = 0$ is called a *Cauchy sequence*. If there exists an $\mathbf{a}_0$ such that $\lim_{n\to\infty} \|\mathbf{a}_n - \mathbf{a}_0\| = 0$, we say that the sequence $\{\mathbf{a}_n\}$ converges (or converges in the norm $\|\cdot\|$) to $\mathbf{a}_0$. Clearly, if $\{\mathbf{a}_n\}$ converges, then it is a Cauchy sequence. It may happen, however, that the element to which a sequence converges does not belong to the linear space from which members of the sequence are obtained. If every Cauchy sequence in a normed linear space $\mathscr{V}$ converges to a point in the space, the space is said to be *complete*. If a normed linear space is not complete, it is always possible to add new elements so that the resulting collection, called the *completion* of the space, is a complete normed linear space. If every Cauchy sequence in a space has a subsequence that also converges, the space is said to be *compact*. A complete normed linear space is called a *Banach space*.

Let $\mathscr{B}_1$ and $\mathscr{B}_2$ be Banach spaces, and let $\mathscr{D}$ be an arbitrary subspace of $\mathscr{B}_1$. Let A be a mapping (i.e., a function) from $\mathscr{D}$ into $\mathscr{B}_2$. It is customary to refer to A as an operator. The domain of the operator A is then the set $\mathscr{D}$, and the range of A is the set of all elements in $\mathscr{B}_2$ of the form $A\mathbf{a}$, where $A\mathbf{a} \in \mathscr{B}_2$.

An operator A is additive if

$$A(\mathbf{a} + \mathbf{b}) = A\mathbf{a} + A\mathbf{b}$$

and it is homogeneous if

$$A(\lambda\mathbf{a}) = \lambda(A\mathbf{a})$$

where λ is a real number and $\mathbf{a}, \mathbf{b} \in \mathscr{D}$. If A is both additive and homogeneous, we say that it is a linear operator; otherwise, A is said to be nonlinear. An operator is bounded if there exists a positive real number M such that

$$\|A\mathbf{a}\| \leq M \, \|\mathbf{a}\|$$

where the symbols $\| \ \|$ denote the norms in either space $\mathscr{B}_1$ or space $\mathscr{B}_2$.

We define the sum of two operators A_1 and A_2 by

$$(A_1 + A_2)\mathbf{a} = A_1\mathbf{a} + A_2\mathbf{a}$$

Also, we define the multiplication of an operator by a scalar λ according to

$$(\lambda A)\mathbf{a} = \lambda(A\mathbf{a})$$

Using these definitions, it can be seen that the family of all linear operators on a Banach space $\mathscr{B}$ is itself a Banach space.

An operator of special importance is the functional Φ. An operator Φ is called a *functional* if its range space is a set of real numbers. A linear functional is an operator $\Phi(\mathbf{a}) = \lambda$, where $\mathbf{a} \in \mathscr{D}$ and λ is a real (or complex) number, such that $\Phi(\lambda\mathbf{a}) = \lambda\Phi(\mathbf{a})$ and $\Phi(\mathbf{a} + \mathbf{b}) = \Phi(\mathbf{a}) + \Phi(\mathbf{b})$. The collection of all linear functionals on a Banach space $\mathscr{B}$ is, itself, a Banach space. This space is denoted $\mathscr{B}^*$ and is called the *conjugate space* of $\mathscr{B}$.

An *inner-product space* is a linear space in which an inner product has been defined. By an inner product we mean a scalar-valued function from $\mathscr{V} \times \mathscr{V}$ into R such that to every pair of elements $(\mathbf{a},\mathbf{b}) \in \mathscr{V} \times \mathscr{V}$ we assign a nonnegative real number, denoted $\langle\mathbf{a},\mathbf{b}\rangle$, which satisfies the following axioms:

1. $\langle\mathbf{a},\mathbf{a}\rangle \geq 0$, the equality holding if and only if $\mathbf{a} = \mathbf{0}$.
2. $\langle\mathbf{a},\mathbf{b}\rangle = \langle\mathbf{b},\mathbf{a}\rangle$.
3. $\langle\lambda\mathbf{a} + \mu\mathbf{b}, \mathbf{c}\rangle = \lambda\langle\mathbf{a},\mathbf{c}\rangle + \mu\langle\mathbf{b},\mathbf{c}\rangle$.

The introduction of an inner product gives meaning to the idea of orthogonality ($\mathbf{a}$ is orthogonal to $\mathbf{b}$ if $\langle\mathbf{a},\mathbf{b}\rangle = 0$), as well as the idea of length, since we can take $\|\mathbf{a}\| = \sqrt{\langle\mathbf{a},\mathbf{a}\rangle}$. It follows from the last observation that every inner-product space is also a normed linear space. An infinite-dimensional Banach space in which an inner product is defined, and which is complete with respect to the norm $\|\mathbf{a}\| = \sqrt{\langle\mathbf{a},\mathbf{a}\rangle}$, is called a *Hilbert space*.

It is often possible to define an inner product between two spaces $\mathscr{V}$ and $\bar{\mathscr{V}}$ defined over the same field by mapping pairs $(\mathbf{a},\bar{\mathbf{a}})$, where $\mathbf{a} \in \mathscr{V}$ and $\bar{\mathbf{a}} \in \bar{\mathscr{V}}$, into the real numbers. Then, we require that for every $\mathbf{a}$ and $\bar{\mathbf{a}}$

1. $\langle\lambda\mathbf{a} + \mu\mathbf{b}, \bar{\mathbf{a}}\rangle = \lambda\langle\mathbf{a},\bar{\mathbf{a}}\rangle + \mu\langle\mathbf{b},\bar{\mathbf{a}}\rangle$.
2. $\langle\mathbf{a}, \lambda\bar{\mathbf{a}} + \mu\bar{\mathbf{b}}\rangle = \lambda\langle\mathbf{a},\bar{\mathbf{a}}\rangle + \mu\langle\mathbf{a},\bar{\mathbf{b}}\rangle$.
3. $\langle\mathbf{a},\bar{\mathbf{a}}\rangle = 0$ for fixed $\bar{\mathbf{a}} \in \bar{\mathscr{V}} \Rightarrow \bar{\mathbf{a}} = \mathbf{0}$, and $\langle\bar{\mathbf{a}},\mathbf{a}\rangle = 0$ for fixed $\mathbf{a} \in \mathscr{V} \Rightarrow \mathbf{a} = \mathbf{0}$.

The spaces $\mathscr{V}$ and $\bar{\mathscr{V}}$ are then called *dual* spaces. If $\mathbf{g}_i$ is a basis of $\mathscr{V}$ and $\mathbf{g}^i$ is a basis of $\bar{\mathscr{V}}$, they are said to provide a biorthogonal basis for $\mathscr{V}$ and $\bar{\mathscr{V}}$ if $\langle\mathbf{g}_i,\mathbf{g}^j\rangle = \delta_i^j$. Algebraically, there is essentially little difference between the dual $\bar{\mathscr{V}}$ of an inner-product space and its conjugate space $\mathscr{V}^*$; indeed, the space $\bar{\mathscr{V}}$ is isomorphic to $\mathscr{V}^*$. Consequently, the terms "dual space" and "conjugate space" are often used synonymously.

An inner-product space is also a *metric space*, though the converse is not necessarily true. A metric space consists of a collection of elements $\mathbf{X}, \mathbf{Y}, \mathbf{Z}, \ldots$ called *points*, together with a function $d(\mathbf{X},\mathbf{Y})$ which assigns to every pair of points $\mathbf{X}$, $\mathbf{Y}$ a nonnegative real number, called the "distance between" points $\mathbf{X}$ and $\mathbf{Y}$, which satisfies the following three metric axioms:

1. $d(\mathbf{X},\mathbf{X}) = 0$
2. $d(\mathbf{X},\mathbf{Y}) = d(\mathbf{Y},\mathbf{X})$
3. $d(\mathbf{X},\mathbf{Y}) \leq d(\mathbf{X},\mathbf{Z}) + d(\mathbf{Z},\mathbf{Y})$

By setting $d(\mathbf{X},\mathbf{Y}) = \|\mathbf{X} - \mathbf{Y}\|$ or $d(\mathbf{X},\mathbf{Y}) = \sqrt{\langle\mathbf{X} - \mathbf{Y}, \mathbf{X} - \mathbf{Y}\rangle}$, it is easily seen that every normed linear space—and, therefore, every inner-product space—is also a metric space. A metric space $\mathscr{M}$ is said to be separable if there exists in $\mathscr{M}$ a countable everywhere dense set.

In subsequent sections, it is convenient to speak of points in a k-dimensional space without introducing specific coordinate systems. This can be done by employing the idea of a *point space* $\mathscr{E}^k$. A nonempty set $\mathscr{R}^k$ is said to form a point space $\mathscr{E}^k$ associated with an inner-product space $\mathscr{V}^k$ if there exists a function $g: \mathscr{R}^k \times \mathscr{R}^k \to \mathscr{V}^k$ with the following properties:

1. For every $\mathbf{X} \in \mathscr{R}^k$ and $\mathbf{a} \in \mathscr{V}^k$, there exists a unique $\mathbf{Y} \in \mathscr{R}^k$ such that $g(\mathbf{X},\mathbf{Y}) = \mathbf{a}$.
2. If $g(\mathbf{X},\mathbf{Y}) = \mathbf{a}$ and $g(\mathbf{X},\mathbf{Z}) = \mathbf{b}$ and $g(\mathbf{Y},\mathbf{Z}) = \mathbf{c}$, for $\mathbf{X}$, $\mathbf{Y}$, $\mathbf{Z} \in \mathscr{R}^k$ and $\mathbf{a}$, $\mathbf{b}$, $\mathbf{c} \in \mathscr{V}^k$, then $\mathbf{a} + \mathbf{b} = \mathbf{c}$.

The elements $\mathbf{X}, \mathbf{Y}, \ldots \in \mathscr{R}^k$ are called points. The function $g(\mathbf{X},\mathbf{Y}) = \mathbf{a}$ is said to define a vector $\mathbf{a}$ as the difference between points $\mathbf{X}$ and $\mathbf{Y}$, and it is convenient to write

$$g(\mathbf{X},\mathbf{Y}) = \mathbf{X} - \mathbf{Y} = \mathbf{a}$$

The dimension of $\mathscr{E}^k$ is defined as the dimension of $\mathscr{V}^k$.

Thus, the difference between two points $\mathbf{Y}$, $\mathbf{X}$ is a vector $\overrightarrow{\mathbf{XY}} = \mathbf{Y} - \mathbf{X}$. However, when no confusion is likely, we shall take for granted that a fixed point $\mathbf{0}_{\mathscr{E}} \in \mathscr{E}^k$, called the *origin*, is identified so that $\mathbf{X}$ may, for simplicity, denote the vector $\mathbf{X} - \mathbf{0} = \overrightarrow{\mathbf{0X}}$, the *position vector* of point $\mathbf{X}$ relative to point $\mathbf{0}_{\mathscr{E}}$. Then we can associate with $\mathbf{X}$ (or $\mathbf{X} - \mathbf{0}_{\mathscr{E}}$) an ordered k-tuple of real numbers $\mathbf{X} = (X^1, X^2, \ldots, X^k)$ called the *coordinates* of point X. The nature of these coordinates depends upon the basis of $\mathscr{V}^k$. For example, if the orthonormal set of vectors $\{\mathbf{i}_1, \mathbf{i}_2, \ldots, \mathbf{i}_k\}$ is a basis of $\mathscr{V}^k(\langle \mathbf{i}_i, \mathbf{i}_j \rangle = \delta_{ij}, i,j = 1, 2 \ \ldots, k)$, then $\mathbf{X} - \mathbf{0}_{\mathscr{E}} = X^i \mathbf{i}_i$, with the repeated index summed from 1 to k, and the X^i are called *cartesian coordinates*. Then $\mathscr{E}^k$ is referred to as a *euclidean point space*. We note, however, that in general the quantities $X^1, X^2, \ldots, X^k$ need not be cartesian and $\mathscr{E}^k$ need not be euclidean.

By the neighborhood $\mathscr{N}(\mathbf{X}_0)$ of a point $\mathbf{X}_0 \in \mathscr{E}^k$, we shall mean the open sphere defined by the set of all $\mathbf{X} \in \mathscr{E}^k$ such that $d(\mathbf{X}_0,\mathbf{X}) < r$, r being a real number called the *radius* of the sphere. If $\mathscr{S}$ denotes a subset of $\mathscr{E}^k$, $\mathscr{S} \subset \mathscr{E}^k$, a point $\mathbf{X} \in \mathscr{E}^k$ is called an *exterior point* of $\mathscr{S}$ if there exists a neighborhood of $\mathbf{X}$ containing no points of $\mathscr{S}$. If every neighborhood of $\mathbf{X}$ contains at least one point of $\mathscr{S}$ and one point not belonging to $\mathscr{S}$, then $\mathbf{X}$ is called a *boundary point* of $\mathscr{S}$. The set of all boundary points of $\mathscr{S}$ is called the *boundary* of $\mathscr{S}$ and is denoted $\partial\mathscr{S}$.

Let $\mathscr{A}_1$ and $\mathscr{A}_2$ denote nonempty subsets of $\mathscr{E}^k$. If $\mathscr{A}_1$ and $\mathscr{A}_2$ have no elements in common, then the intersection $\mathscr{A}_1 \cap \mathscr{A}_2$ is the null or empty set $\varnothing$ and $\mathscr{A}_1$ and $\mathscr{A}_2$ are said to be *disjoint*. A set $\mathscr{S} \subset \mathscr{E}^k$ is said to be *disconnected* whenever $\mathscr{S} = \mathscr{A}_1 \cup \mathscr{A}_2$ and $\mathscr{A}_1$ and $\mathscr{A}_2$ are nonempty disjoint sets. If $\mathscr{S}$ is not disconnected, it is called a *connected* set.

We refer to certain connected subsets $\mathscr{R} \subset \mathscr{E}^k$ as *regions* of the k-dimensional space $\mathscr{E}^k$. Specifically, a region is the union of an open connected subset of $\mathscr{E}^k$ with some, none, or all of its boundary points. If $\mathscr{R}$ contains none of its boundary points, it is called an *open region*; if $\mathscr{R}$ contains all its boundary points, it is called a *closed region*. Clearly a region may be neither open nor closed. If the norm $\|\mathbf{X}\| = \|\mathbf{X} - \mathbf{0}_{\mathscr{E}}\| \leq M$, M being a positive constant, for every $\mathbf{X} \in \mathscr{R}$, we say that $\mathscr{R}$ is *bounded*, and if $\mathscr{R}$ is both closed and bounded, it is also *compact*.

Our main concern will be the approximation of certain types of continuous functions defined on compact subsets of a k-dimensional point space $\mathscr{E}^k$. More specifically, let $\mathscr{T}$ denote a set of elements $\mathbf{T}, \mathbf{U}, \mathbf{V}, \ldots$ which, to a great extent, are arbitrary. In almost all our applications, however, we shall consider the quantities $\mathbf{T}$ to be real or complex numbers, vectors or tensors of a certain order. We refer to a mapping $F: \mathscr{R} \to \mathscr{T}$ which associates with every $\mathbf{X}$ belonging to a compact subset $\mathscr{R}$ of $\mathscr{E}^k$ an element $\mathbf{T} \in \mathscr{T}$, a function on $\mathscr{R}$ to $\mathscr{T}$, and to indicate such functions the notation $\mathbf{T} = \mathbf{F}(\mathbf{X})$ is used, where $\mathbf{T}$ is the value of the function as point $\mathbf{X}$. The region $\mathscr{R}$ is the domain of the function $\mathbf{F}(\mathbf{X})$. We assume that $\mathbf{F}$ is continuous on $\mathscr{R}$; that is, at every point $\mathbf{X}_0 \in \mathscr{R}$, $\mathbf{F}(\mathbf{X}) \to \mathbf{F}(\mathbf{X}_0)$ as $d(\mathbf{X},\mathbf{X}_0) \to 0$. Thus, the image space $\mathbf{F}(\mathscr{R})$ is also compact; and since $\mathbf{F}$ is also one to one, $\mathbf{F}^{-1}$ exists and $\mathbf{F}$ may be described as a *topological mapping*, or *homeomorphism*, on $\mathscr{R}$.

7.2 COMMENT ON NOTATION

In all discussions to follow we adopt indicial notation and the summation convention: An indexed quantity indicates a typical element of an ordered array, and successive values are to be assigned to each index from 1 to its admissible range. All repeated indices are to be summed throughout their admissible range unless they are enclosed by parentheses. For example, Eqs. (6.5) and (6.14) are now written

$$X^\Delta = \overset{(e)}{\Lambda}{}^\Delta_N x^N_{(e)} \qquad \text{and} \qquad F^\Delta = \overset{(e)}{\Lambda}{}^\Delta_N f^N_{(e)}$$

where the repeated index N is summed from 1 to N_e, N_e being the total number of nodes of element e, and (6.8) and (6.15) are now written

$$x^N_{(e)} = \overset{(e)}{\Omega}{}^N_\Delta X^\Delta \qquad \text{and} \qquad f^N_{(e)} = \overset{(e)}{\Omega}{}^N_\Delta F^\Delta$$

where the repeated index Δ is summed from 1 to G, G being the total number of global nodes of the connected model.

To indicate associations with typical finite elements, we affix to certain symbols an element-identification label, usually represented by the letter e. This symbol is a label and not an index and is usually enclosed by parentheses when confusion is likely.

We shall frequently need to distinguish between global quantities (e.g., quantities associated with an assemblage of finite elements) and local quantities (e.g., quantities associated with a particular finite element). With few exceptions, uppercase letters are used to indicate global quantities. Lowercase Latin indices are used to indicate elements in an array or vector and tensor components. Uppercase Greek and Latin indices indicate quantities defined at specific nodal points in a discrete model, the Greek indices pertaining to nodes in the global, connected model and the Latin indices pertaining to local nodes in specific, disconnected finite elements.

7.3 DISCRETE MODEL OF THE REGION $\mathscr{R}$

The construction of a finite-element model of a region $\mathscr{R}$ of $\mathscr{E}^k$ is, of course, independent of the field $\mathbf{F}(\mathbf{X})$ for which $\mathscr{R}$ may be the domain. The discrete model of $\mathscr{R}$ is constructed as follows:

1. A finite number G of points are identified in $\mathscr{R}$ and are labeled consecutively $\mathbf{X}^1, \mathbf{X}^2, \ldots, \mathbf{X}^G$ or, more concisely, $\mathbf{X}^\Delta(\Delta = 1, 2, \ldots, G)$. These points are called *global nodal points* or *global nodes*. The finite set $\{\mathbf{X}^1, \mathbf{X}^2, \ldots, \mathbf{X}^G\}$ of such points is denoted $\bar{\mathscr{R}}_{\mathbf{X}}$.

2. The region $\mathscr{R}$ is represented approximately by another region $\bar{\mathscr{R}}$, which contains all the nodal points of $\mathscr{R}$. We refer to the difference $\mathscr{R} - \bar{\mathscr{R}}$ as the *error region*, denoted $\mathscr{R}_\epsilon$. If $\mathscr{R}_\epsilon$ is empty, we say that $\bar{\mathscr{R}}$ and $\mathscr{R}$ are equal. Naturally, $\bar{\mathscr{R}}_{\mathbf{X}} \subset \bar{\mathscr{R}}$.
3. We now consider a finite number E of disconnected subsets r_e of $\mathscr{E}^k$ called finite elements. Ultimately, we hope to connect these finite elements together to form $\bar{\mathscr{R}}$, but at this point we consider all the finite elements to be closed and disjoint: no overlapping of boundaries is permitted, and even though boundary points of two finite elements may eventually coincide in the connected model, the boundary of a typical element r_e is considered to be distinct from those of all other elements. Indeed, we refer to the disconnectedness of the elements as the *fundamental property of finite elements.* Mathematically,

$$r_i \cap r_j = \varnothing \qquad i \neq j;\, i, j = 1, 2, \ldots, E \tag{7.1}$$

where $\varnothing$ is the null set. The union

$$\mathscr{R}^* = \bigcup_{e=1}^{E} r_e \tag{7.2}$$

is referred to as the *disconnected* or *unassembled region.*
4. We now examine a typical finite element r_e which, for the moment, is considered to be completely isolated from the collection of elements. To identify points *locally* in r_e, we use the notation $\mathbf{x}^{(e)}$,† where in some cases it is convenient to regard $\mathbf{x}^{(e)}$ as the local position vector of a point in r_e relative to some fixed origin $\mathbf{o}_{ee}$. We may also associate with $\mathbf{x}^{(e)}$ an ordered k-tuple of real numbers $\mathbf{x}^{(e)} = (x^{1(e)}, x^{2(e)}, \ldots, x^{k(e)})$ called the *local coordinates* of the point relative to element e; but, for the moment, it is convenient to speak of local points without reference to specific origins or coordinates.

   A finite number N_e of points in r_e is now identified and these are labeled consecutively $\overset{1}{\mathbf{x}}_{(e)}, \overset{2}{\mathbf{x}}_{(e)}, \ldots, \overset{N_e}{\mathbf{x}}_{(e)}$ or, more concisely, $\overset{N}{\mathbf{x}}_{(e)}$ $(N = 1, 2, \ldots, N_e)$. These points are called *local nodal points* or simply *local nodes.* The finite set of local nodes corresponding to element r_e, $\{\overset{1}{\mathbf{x}}_{(e)}, \overset{2}{\mathbf{x}}_{(e)}, \ldots, \overset{N_e}{\mathbf{x}}_{(e)}\}$, is denoted $\mathscr{R}^*_{\mathbf{x}(e)}$.
5. Up to now, the subregions r_e are effectively unrelated to $\bar{\mathscr{R}}$. The connectivity of the discrete model of $\mathscr{R}$ is established by mapping $\mathscr{R}^*$ onto $\bar{\mathscr{R}}$ or each $\mathscr{R}^*_{\mathbf{x}(e)}$ onto $\bar{\mathscr{R}}_{\mathbf{X}}$. For obvious reasons, we refer to this mapping as *assembling the elements*, since it transforms the unassembled region $\mathscr{R}^*$ into the assembled region $\bar{\mathscr{R}}$. Assuming, for the moment,

† The letter e is an element-identification label, and no significance is to be attributed to its location. We use interchangeably $\mathbf{x}^{(e)}$ or $\mathbf{x}_{(e)}$ to denote points locally in r_e.

that the proper correspondence exists between nodes in a specific subregion r_e and those in $\bar{\mathscr{R}}$, the connectivity of the model is established by the mapping $\overset{(e)}{\Lambda}: \mathscr{R}^*_{\mathbf{x}(e)} \to \bar{\mathscr{R}}_{\mathbf{X}}$, for fixed e, by

$$\mathbf{X}^\Delta = \overset{(e)}{\Lambda}{}^\Delta_N \mathbf{x}^N_{(e)} \tag{7.3}$$

where $\Delta = 1, 2, \ldots, G$, N is summed from 1 to N_e, and $\overset{(e)}{\Lambda}{}^\Delta_N$ is defined in the same manner as the array in (6.6), that is,

$$\overset{(e)}{\Lambda}{}^\Delta_N = \begin{cases} 1 & \text{if node } \Delta \text{ of } \bar{\mathscr{R}} \text{ is coincident with node } N \text{ of} \\ & r_e \text{ in the connected model} \\ 0 & \text{if otherwise} \end{cases} \tag{7.4}$$

In (7.3) we prefer to think of $\mathbf{X}^\Delta$ and $\mathbf{x}^N_{(e)}$ as points in $\mathscr{E}^k$ rather than as position vectors (otherwise, we simply use instead of $\mathbf{X}^\Delta$ the vector $\mathbf{X}^\Delta - \mathbf{B}^{(e)}$, where $\mathbf{B}^{(e)}$ is the position vector of the origin of $\mathbf{x}^N_{(e)}$ relative to the origin of $\mathbf{X}^\Delta$). The question of "translating" origins $\mathbf{o}_{\epsilon e}$ to $\mathbf{0}_\epsilon$, etc., is, therefore, irrelevant, for the mapping merely indicates which points in $\mathscr{R}^*_{\mathbf{x}(e)}$ are incident on a given point in $\bar{\mathscr{R}}_{\mathbf{X}}$. For this reason, we say that the function $\overset{(e)}{\Lambda}{}^\Delta_N$ establishes an *incidence relation* between $\bar{\mathscr{R}}_{\mathbf{X}}$ and $\mathscr{R}^*_{\mathbf{x}(e)}$ (or $\bar{\mathscr{R}}$ and $\mathscr{R}^*$). Since the identification of points in $\bar{\mathscr{R}}$ and $\mathscr{R}^*_{\mathbf{x}(e)}$ as nodal points is arbitrary, we require that the proper correspondence for connectivity exists for *all* points in $\bar{\mathscr{R}}$ and $\mathscr{R}^*$, so that (7.3) also implies the mapping of $\bar{\mathscr{R}}$ into $\mathscr{R}^*$. Indeed, the collection

$$\Lambda = \left\{ \overset{(1)}{\Lambda}, \overset{(2)}{\Lambda}, \ldots, \overset{(E)}{\Lambda} \right\}$$

of element mappings effectively maps the entire collection $\mathscr{R}^*$ into $\bar{\mathscr{R}}$.

6. The reverse procedure of assembling elements into a connected model is called *decomposition*. The process of decomposing $\bar{\mathscr{R}}$ into finite elements often amounts to simply renumbering the global nodal points associated with an element so that they correspond to the numbering scheme adopted for local nodes. This is accomplished by a mapping $\overset{(e)}{\Omega}$ of $\bar{\mathscr{R}}_{\mathbf{X}}$ into $\mathscr{R}^*_{\mathbf{x}(e)}$ defined by

$$\mathbf{x}^N_{(e)} = \overset{(e)}{\Omega}{}^N_\Delta \mathbf{X}^\Delta \tag{7.5}$$

where the repeated index Δ is summed from 1 to G and $\overset{(e}{\Omega}{}^N_\Delta$ is defined by

$$\overset{(e)}{\Omega}{}^N_\Delta = \begin{cases} 1 & \text{if node } N \text{ of element } r_e \text{ is coincident with} \\ & \text{node } \Delta \text{ of the connected model } \bar{\mathscr{R}} \\ 0 & \text{if otherwise} \end{cases} \tag{7.6}$$

The array $\overset{(e)}{\Omega}{}^N_\Delta$ is simply the transpose of the array $\overset{(e)}{\Lambda}{}^\Delta_N$ defined in (7.4).

It is clear that for a fixed element $\mathscr{r}_e$, the composition $\overset{(e)(e)}{\Lambda\Omega}$ defines the identity mapping I_e of the set of local nodes $\mathscr{R}^*_{\mathbf{x}(e)}$ into itself:

$$\overset{(e)}{\Lambda}{}^\Delta_N \overset{(e)}{\Omega}{}^M_\Delta = \delta^M_N \tag{7.7}$$

Likewise, the composition $\overset{(e)(e)}{\Omega\Lambda}$ defines a mapping $I^{(e)}_G$ of the collection of global nodes corresponding to element e into itself:

$$\overset{(e)}{\Omega}{}^N_\Delta \overset{(e)}{\Lambda}{}^\Gamma_N = \begin{cases} \delta^\Gamma_\Delta & \text{if } \mathbf{X}^\Gamma, \mathbf{X}^\Delta \in \mathscr{r}_e \\ 0 & \text{if } \mathbf{X}^\Gamma, \mathbf{X}^\Delta \notin \mathscr{r}_e \end{cases} \tag{7.8}$$

Compatibility Clearly, we require that the boundaries and the locations of nodes in each element be such that the collection of disjoint elements can be connected together properly to form the discrete model of $\mathscr{R}$. The existence of the function $\overset{(e)}{\Lambda}{}^\Delta_N$ which assembles the elements into one unit depends upon the location of nodes in and on the boundaries of each finite element and on a certain correspondence of these nodes with the global nodes $\mathbf{X}^\Delta$ in $\bar{\mathscr{R}}$.

Consider two finite elements $\mathscr{r}_e$ and $\mathscr{r}_f$ which have boundaries $\partial\mathscr{r}_e$ and $\partial\mathscr{r}_f$, respectively. Suppose that these finite elements are to be adjacent to one another in the connected model $\bar{\mathscr{R}}$. Then portions $\partial\mathscr{r}'_e$ and $\partial\mathscr{r}'_f$ of these boundaries should be mapped into a common boundary in $\bar{\mathscr{R}}$, which we shall call an *interelement boundary* of elements $\mathscr{r}_e$ and $\mathscr{r}_f$. The following conditions must hold:

1. For every point $\mathbf{x}^{(e)} \in \partial\mathscr{r}'_e$ there is a corresponding point $\mathbf{x}^{(f)} \in \partial\mathscr{r}'_f$.
2. If assembling the elements carries $\mathbf{x}^{(e)} \in \partial\mathscr{r}'_e$ into a point $\mathbf{X} \in \bar{\mathscr{R}}$, it carries the point $\mathbf{x}^{(f)} \in \partial\mathscr{r}'_f$, which, by condition 1, corresponds to $\mathbf{x}^{(e)}$, into the *same* point $\mathbf{X}$.
3. For each local nodal point $\mathbf{x}^N_{(e)} \in \partial\mathscr{r}'_e$ there is a corresponding local nodal point $\mathbf{x}^M_{(e)} \in \partial\mathscr{r}'_f$.
4. For every local node $\mathbf{x}^N_{(e)} \in \mathscr{r}_e$, there corresponds a single node $\mathbf{X}^\Delta \in \bar{\mathscr{R}}$, and assembling the elements carries $\mathbf{x}^N_{(e)}$ into its corresponding node $\mathbf{X}^\Delta$.

We refer to the above requirements as the *compatibility conditions* for the discrete model. They are necessary and sufficient conditions for the existence of the function $\overset{(e)}{\Lambda}{}^\Delta_N$, and they ensure that the elements fit properly together to form the discrete model of $\mathscr{R}$. Obviously, the sets of points which constitute $\partial\mathscr{r}'_e$ and $\partial\mathscr{r}'_f$ in $\mathscr{R}^*$ are, according to condition 2, mapped into an interelement boundary in $\bar{\mathscr{R}}$. Conditions 1 to 3 ensure that under the mapping the elements are connected at their nodes and that no gaps or overlappings of finite-element boundaries occur in the connected model. Condition 4 covers the possibility of interior nodes (local nodes of r_e which do not fall on the boundary of the element) and provides that there be corresponding nodes in $\bar{\mathscr{R}}$. An interesting corollary follows immediately from conditions 1 to 3:

5. If (7.3) carries a local nodal point $\mathbf{x}_{(e)}^N \in \partial r'_e$ into a nodal point $\mathbf{X}^\Delta \in \bar{\mathscr{R}}$, it carries the local nodal point $\mathbf{x}_{(f)}^M \in \partial r'_f$ which corresponds to $\mathbf{x}_{(e)}^N$ into the *same* nodal point $\mathbf{X}^\Delta$.

Note that the nodal point $\mathbf{X}^\Delta$ in condition 5 is on an interelement boundary in $\bar{\mathscr{R}}$.

Multiplicity of boundary points Let $\mathbf{X}$ be a point in $\bar{\mathscr{R}}$ which lies on an interelement boundary. The point $\mathbf{X}$ is said to have *multiplicity m* if m finite elements meet at $\mathbf{X}$. If $\mathbf{X}$ does not belong to an interelement boundary, it has a multiplicity $m = 0$.

7.4 DISCRETE MODEL OF A FUNCTION

With the procedure for developing a finite-element model of a subset $\mathscr{R}$ of $\mathscr{E}^k$ now established, we return to the problem of constructing a discrete model of a continuous function $\mathbf{F}: \mathscr{R} \to \mathscr{T}$. We begin by representing the domain $\mathscr{R}$ of $\mathbf{F}$ by a finite-element model $\bar{\mathscr{R}}$ in the manner described previously. The model $\bar{\mathscr{R}}$ is the domain of another function $\bar{\mathbf{F}}$, which is to approximate $\mathbf{F}$. We proceed as follows:

1. The value $\mathbf{F}(\mathbf{X})$ of the function at a global node $\mathbf{X}^\Delta \in \mathscr{R}$ is denoted $\mathbf{F}^\Delta$:

$$\mathbf{F}^\Delta = \mathbf{F}(\mathbf{X}^\Delta) \tag{7.9}$$

The quantities $\mathbf{F}^\Delta$ are called the *global values* of the function $\mathbf{F}(\mathbf{X})$, and the set of global values $\{F^1, F^2, \ldots, F^G\}$ is denoted $\mathscr{G}$ and is called the *global* set corresponding to the function $\mathbf{F}(\mathbf{X})$.

2. The collection $\mathscr{R}^*$ of disconnected finite elements r_e is now considered. Each finite element r_e is considered to be the domain of a local function, denoted $\mathbf{f}^{(e)}$, of the same type as $\mathbf{F}$. The local function $\mathbf{f}^{(e)}$ and function values $\mathbf{f}^{(e)}(\mathbf{x})$ are defined *only for* $\mathbf{x} \in r_e$. Ultimately, we intend that the localized functions be pieced together to form the approximation of the original function $\mathbf{F}$, after the finite elements have been connected together to give $\bar{\mathscr{R}}$. However, at this point we consider all the local functions to be completely disconnected and independent of $\mathbf{F}(\mathbf{X})$ and of each other. The function $\mathbf{f}^{(e)}(\mathbf{x})$ is called the *restriction* of $\mathbf{F}(\mathbf{X})$ to r_e.

3. The value $\mathbf{f}^{(e)}(\mathbf{x})$ of the local function $\mathbf{f}^{(e)}$ at a local node $\mathbf{x}_{(e)}^N$ of element e

is denoted $\mathbf{f}^N_{(e)}$:

$$\mathbf{f}^N_{(e)} = \mathbf{f}^{(e)}(\mathbf{x}^N_{(e)}) \tag{7.10}$$

The quantities $\mathbf{f}^N_{(e)}$ are called the *local values* of the functions corresponding to element e. The ordered sets of local values $\{\mathbf{f}^1_{(1)}, \ldots, \mathbf{f}^{N_1}_{(1)}\}$, $\{\mathbf{f}^1_{(2)}, \ldots, \mathbf{f}^{N_2}_{(2)}\}, \ldots, \{\mathbf{f}^1_{(E)}, \ldots, \mathbf{f}^{N_E}_{(E)}\}$ are denoted $\mathscr{L}^e$ and are called *local sets*. Their union is denoted $\mathscr{L}$:

$$\mathscr{L} = \bigcup_{e=1}^{E} \mathscr{L}_e$$

4. The local functions are approximated over their respective finite elements by continuous functions of the form

$$\mathbf{f}^{(e)}(\mathbf{x}) \approx \bar{\mathbf{f}}^{(e)}(\mathbf{x}) = \boldsymbol{\psi}_{(e)}(\mathbf{x}, \mathbf{f}^N_{(e)}) \tag{7.11}$$

where the functions $\boldsymbol{\psi}_{(e)}(\mathbf{x}, \mathbf{f}^N_{(e)})$ are defined so as to have the property

$$\boldsymbol{\psi}_{(e)}(\mathbf{x}^M_{(e)}, \mathbf{f}^N_{(e)}) = \mathbf{f}^M_{(e)} \qquad \text{at node } \mathbf{x}^M \in \imath_e \tag{7.12}$$

Thus, the values of the functions $\boldsymbol{\psi}_{(e)}(\ )$ coincide with the values of the local functions at the nodal points of the element.†

In general, it is convenient to choose approximating functions $\boldsymbol{\psi}_{(e)}(\mathbf{x}, \mathbf{f}^N_{(e)})$ which are linear in the nodal values $\mathbf{f}^N_{(e)}$:

$$\boldsymbol{\psi}_{(e)}(\mathbf{x}, \mathbf{f}^N_{(e)}) = \mathbf{f}^N_{(e)}\psi^{(e)}_N(\mathbf{x}) \tag{7.13}$$

where the repeated index N is to be summed from 1 to N_e and $\psi^{(e)}_N(\mathbf{x})$ denotes N_e scalar-valued functions associated with each finite element. In view of (7.12),

$$\psi^{(e)}_N(\mathbf{x}^M) = \delta^M_N \qquad \mathbf{x}^M \equiv \mathbf{x}^M_{(e)} \in \imath_e \tag{7.14}$$

where δ^N_M $(M,N = 1, 2, \ldots, N_e)$ is the Kronecker delta.‡ The functions $\psi^{(e)}_N(\ )$ are similar to those of the familiar Lagrange interpolation polynomials, although in the present case they need not be polynomials.

With (7.13), (7.11) becomes

$$\bar{\mathbf{f}}^{(e)}(\mathbf{x}) = \mathbf{f}^N_{(e)}\psi^{(e)}_N(\mathbf{x}) \tag{7.15}$$

5. The finite-element model of $\mathbf{F}(\mathbf{X})$ is obtained by connecting all the disjoint elements together by applying (7.3) and applying successive

† We remark that if the localized functions $\mathbf{f}^{(e)}$ are of differentiability class C^r, $r \geq 1$, it is generally possible to construct a function $\mathbf{f}^{(e)}$ whose values not only match the values $\mathbf{f}^N_{(e)}$ at the nodes of each element but also the values of its first r partial derivatives match those of $\mathbf{f}^{(e)}(\mathbf{x})$ at these points. Such higher-order approximations are discussed in Art. 8.

‡ Additional properties of the local interpolation functions $\psi^{(e)}_N(\mathbf{x})$ are discussed in Arts. 9 and 10.

mappings of the local sets $\mathscr{L}_e$ into the global set $\mathscr{G}$. This is accomplished by the mapping $\overset{(e)}{\Lambda}: \mathscr{L}_e \to \mathscr{G}$ defined, for fixed e, by

$$\mathbf{F}^\Delta = \overset{(e)}{\Lambda}{}^\Delta_N \mathbf{f}^N_{(e)} \tag{7.16}$$

where $\overset{(e)}{\Lambda}{}^\Delta_N$ is defined in (7.4). Equivalently, the series of mappings

$$\Lambda = \sum_{e=1}^{E} \overset{(e)}{\Lambda} \qquad \text{maps} \qquad \mathscr{L} = \bigcup_{e=1}^{E} \mathscr{L}_e \qquad \text{into} \qquad \mathscr{G}$$

6. Alternatively, the array $\overset{(e)}{\Omega}{}^N_\Delta$ of (7.6) can be used to decompose $\mathscr{G}$ into $\mathscr{L}_e$:

$$\mathbf{f}^N_{(e)} = \overset{(e)}{\Omega}{}^N_\Delta \mathbf{F}^\Delta \tag{7.17}$$

7. The final discrete model of $\mathbf{F}(\mathbf{X})$ is given by

$$\mathbf{F}(\mathbf{X}) \approx \bar{\mathbf{F}}(\mathbf{x}) = \sum_{e=1}^{E} \bar{\mathbf{f}}^{(e)}(\mathbf{x}) \tag{7.18}$$

or, in view of (7.15) and (7.17),†

$$\bar{\mathbf{F}}(\mathbf{x}) = \sum_{e=1}^{E} \psi_N^{(e)}(\mathbf{x}) \overset{(e)}{\Omega}{}^N_\Delta \mathbf{F}^\Delta = \Phi_\Delta(\mathbf{x}) \mathbf{F}^\Delta \tag{7.19}$$

$\Phi_\Delta(\mathbf{x})$ are *global* approximation functions defined by

$$\Phi_\Delta(\mathbf{x}) = \sum_{e=1}^{E} \overset{(e)}{\Omega}{}^N_\Delta \psi_N^{(e)}(\mathbf{x}) \tag{7.20}$$

It is clear from (7.14) that for $\mathbf{x} \in \imath_e$,

$$\begin{aligned} \bar{\mathbf{F}}(\mathbf{x}) &= \mathbf{0} + \mathbf{0} + \cdots + \overset{(e)}{\Omega}{}^N_\Delta \psi_N^{(e)}(\mathbf{x}) \mathbf{F}^\Delta + \mathbf{0} + \cdots + \mathbf{0} \\ &= \mathbf{f}^N_{(e)} \psi_N^{(e)}(\mathbf{x}) \\ &= \bar{\mathbf{f}}^{(e)}(\mathbf{x}) \end{aligned}$$

† Technically, this relation holds everywhere in $\bar{\mathscr{R}}$ except on a set of measure zero. If, for example, $\bar{\mathbf{X}}$ is an interelement boundary point in $\bar{\mathscr{R}}$ of multiplicity m and if $\bar{\mathbf{T}}$ is the desired value of the discrete model of $\mathbf{F}(\mathbf{X})$ at $\bar{\mathbf{X}}$, then (7.19) gives $m\bar{\mathbf{T}}$ instead of $\bar{\mathbf{T}}$. To avoid such complications, we might simply redefine $\Phi_\Delta(\mathbf{x})$ so as to include a function $r(\mathbf{X})$ which equals unity if the multiplicity of $\mathbf{X}$ is zero and which equals the multiplicity m of $\mathbf{X}$ if $m \neq 0$ [e.g., instead of $\Phi_\Delta(\mathbf{x})$, use $\Phi_\Delta(\mathbf{x})$, $\Phi_\Delta(\mathbf{x})/r(\mathbf{x})$]. Henceforth we shall use the notation of (7.18) to (7.20) with the provision that the factor $1/r(\mathbf{x})$ be introduced at interelement boundaries. In metric spaces, for example, the point made here is largely trivial since any two functions are said to be equal if they are equal everywhere except on a set of measure zero. Consequently, Lebesque integrals of two functions equal everywhere except on a set of measure zero are identical. See Kolmogorov and Fomin [1961].

Moreover, for a particular nodal point $\mathbf{x}_{(e)}^N \in r_e$ which is coincident with $\mathbf{X}^\Delta$ in the connected model,

$$\begin{aligned}\bar{\mathbf{F}}(\mathbf{x}_{(e)}^N) &= \bar{\mathbf{f}}^{(e)}(\mathbf{x}_{(e)}^N) \equiv \mathbf{f}_{(e)}^N = \overset{(e)}{\Omega}{}_\Delta^N \mathbf{F}^\Delta \\ &= \mathbf{F}^\Delta \qquad \text{for } \mathbf{x}_{(e)}^N \text{ coincident with } \mathbf{X}^\Delta \text{ in } \bar{\mathscr{R}}\end{aligned}$$

7.5 EXAMPLES

The great utility of the finite-element concept arises from the fact that an approximation of a given function $\mathbf{F}(\mathbf{X})$ can be studied in detail within a small subregion of its domain, independent of the behavior of the function in other subregions. This means, for example, that in applying the concept to the study of the behavior of a solid body, we can isolate a typical finite element of the body, approximate various fields locally over the element, and completely define the behavior of an element in terms of these approximate fields without considering its ultimate location in the model, its mode of connection with adjacent elements, or the behavior of any other element in the model. Once the local approximation of a field over typical finite elements is defined, the complete model of the field is obtained by means of the mapping indicated in (7.17) and (7.19).

To fix ideas, we now consider some simple examples.

Example 7.1 Scaler-valued function on a two-dimensional domain Consider first a region $\mathscr{R}$ in two-dimensional space $\mathscr{E}^2$, shown in Fig. 7.1*a*, which is the domain of a scalar-valued function $\Phi(\mathbf{X})$, similar to that considered in Art. 6.2. A point $\mathbf{X} \in \mathscr{R}$ is given by the coordinates X^1 and X^2, so that we may write

$$\Phi = \Phi(X^1, X^2)$$

For simplicity, we consider a rather crude model $\bar{\mathscr{R}}$ of $\mathscr{R}$, which consists of only seven nodes and two finite elements, as shown in Fig. 7.1*b*. Thus, in this case $G = 7$ and $E = 2$. Note also that the boundaries of $\bar{\mathscr{R}}$ are drawn so as to approximate $\mathscr{R}$, the error region $\mathscr{R}_\epsilon$ being shaded in the figure. Whereas it is always desirable that $\bar{\mathscr{R}}$ closely approximate or even coincide with $\mathscr{R}$, it is often impossible or impractical to select as boundaries of $\bar{\mathscr{R}}$ the boundaries of $\mathscr{R}$. This is due, in part, to the fact that we prefer to define finite elements of very simple geometric shapes for ease in constructing the interpolating functions $\psi_N^{(e)}(\mathbf{x})$ for each element. Moreover, in a subsequent section we are to impose certain continuity requirements on the functions $\psi_N^{(e)}(\mathbf{x})$ in the connected model which, in turn, may impose severe restrictions on the shapes of finite-element boundaries. Hence, in general, a smooth boundary of $\mathscr{R}$ is approximated by a sectionally smooth boundary in $\bar{\mathscr{R}}$.

Returning to the global description of the discrete model of $\Phi(\mathbf{X})$, we label the nodes in $\bar{\mathscr{R}}$ consecutively from 1 to 7. Then, if $\Phi^\Delta(\Delta = 1, 2, \ldots, 7)$ denotes the values of the function at each of the respective nodes, we have for the global set $\mathscr{G}$

$$\{\Phi^\Delta\} = \{\Phi^1, \Phi^2, \Phi^3, \Phi^4, \Phi^5, \Phi^6, \Phi^7\} \tag{7.21}$$

We now identify two finite elements, shown in Fig. 7.1*c* and *d*, as elements (1) and (2) and establish in each element local coordinate systems $\mathbf{x}_{(1)} = (x_{(1)}^1, x_{(1)}^2)$ and

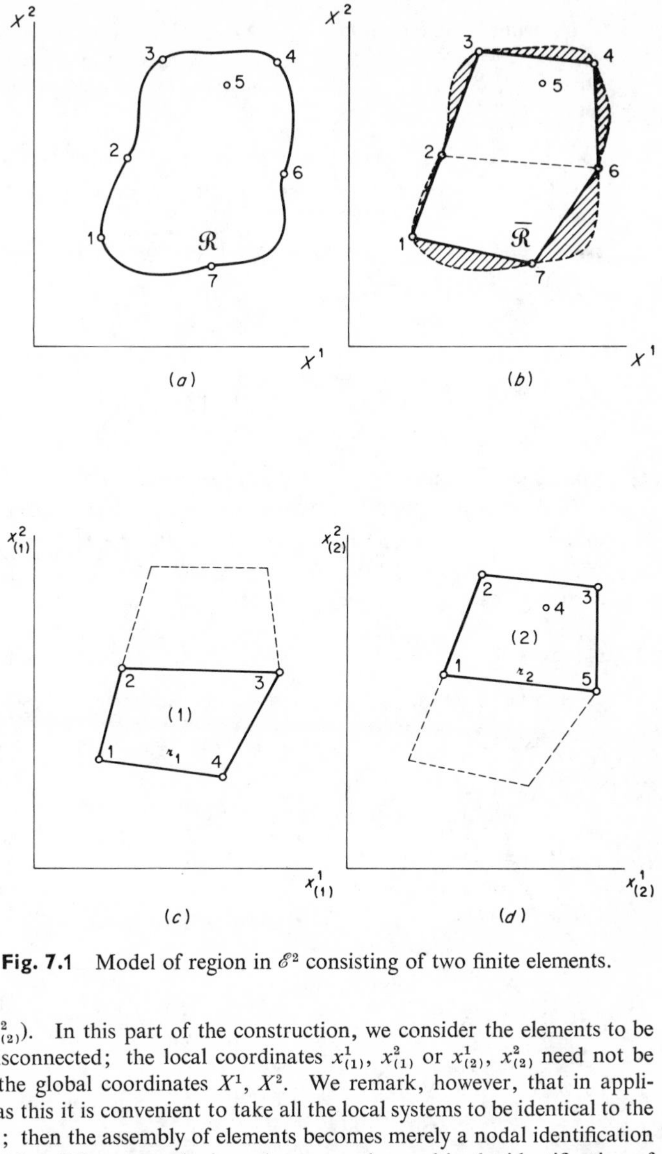

Fig. 7.1 Model of region in $\mathscr{E}^2$ consisting of two finite elements.

$\mathbf{x}_{(2)} = (x^1_{(2)}, x^2_{(2)})$. In this part of the construction, we consider the elements to be completely disconnected; the local coordinates $x^1_{(1)}$, $x^2_{(1)}$ or $x^1_{(2)}$, $x^2_{(2)}$ need not be the same as the global coordinates X^1, X^2. We remark, however, that in applications such as this it is convenient to take all the local systems to be identical to the global system; then the assembly of elements becomes merely a nodal identification process in which a different numbering scheme may be used in the identification of global nodes than in the identification of local nodes.

Proceeding further, we label the nodes belonging to element (1) as 1, 2, 3, and 4 ($N_{e=1} = N_1 = 4$) and those of element (2) as 1 to 5 ($N_{e=2} = N_2 = 5$). The portions of $\Phi(\mathbf{X})$ defined locally over elements (1) and (2) are denoted $\varphi_{(1)}(\mathbf{x}_{(1)})$ and $\varphi_{(2)}(\mathbf{x}_{(2)})$, respectively [these are not to be confused with the functions defined in (7.19) and (7.20)]. For simplicity, we take the local systems $\mathbf{x}_{(1)}$ and $\mathbf{x}_{(2)}$ to be coincident with the global system $\mathbf{X}$; we may then write for the local fields simply $\varphi_{(1)}(\mathbf{x})$ and $\varphi_{(2)}(\mathbf{x})$ and for the global field $\Phi(\mathbf{X})$ or $\Phi(\mathbf{x})$.

The values of the local field $\varphi_{(1)}(\mathbf{x})$ at the four nodes of element (1) are denoted $\varphi^1_{(1)}$, $\varphi^2_{(1)}$, $\varphi^3_{(1)}$, and $\varphi^4_{(1)}$, and the values of $\varphi_{(2)}(\mathbf{x})$ at the five nodes of element (2) are denoted $\varphi^1_{(2)}, \varphi^2_{(2)}, \ldots, \varphi^5_{(2)}$. Thus, we have for elements in the respective local sets $\mathscr{L}_1$ and $\mathscr{L}_2$,

$$\{\varphi^N_{(1)}\} = \{\varphi^1_{(1)},\varphi^2_{(1)},\varphi^3_{(1)},\varphi^4_{(1)}\} \qquad \text{and} \qquad \{\varphi^N_{(2)}\} = \{\varphi^1_{(2)},\varphi^2_{(2)},\varphi^3_{(2)},\varphi^4_{(2)},\varphi^5_{(2)}\} \tag{7.22}$$

Within each finite element, we construct approximations of the local fields of the form

$$\varphi_{(1)}(\mathbf{x}) \approx \bar{\varphi}_{(1)}(\mathbf{x}) = \varphi^1_{(1)}\psi^{(1)}_1(\mathbf{x}) + \varphi^2_{(1)}\psi^{(1)}_2(\mathbf{x}) + \varphi^3_{(1)}\psi^{(1)}_3(\mathbf{x}) + \varphi^4_{(1)}\psi^{(1)}_4(\mathbf{x}) \tag{7.23a}$$

$$\varphi_{(2)}(\mathbf{x}) \approx \bar{\varphi}_{(2)}(\mathbf{x}) = \varphi^1_{(2)}\psi^{(2)}_1(\mathbf{x}) + \varphi^2_{(2)}\psi^{(2)}_2(\mathbf{x}) + \varphi^3_{(2)}\psi^{(2)}_3(\mathbf{x}) + \varphi^4_{(2)}\psi^{(2)}_4(\mathbf{x}) + \varphi^5_{(2)}\psi^{(2)}_5(\mathbf{x}) \tag{7.23b}$$

or, more concisely,

$$\left.\begin{aligned} \varphi_{(1)}(\mathbf{x}) &\approx \varphi^N_{(1)}\psi^{(1)}_N(\mathbf{x}) \qquad N = 1, 2, 3, 4 \\ \varphi_{(2)}(\mathbf{x}) &\approx \varphi^N_{(2)}\psi^{(2)}_N(\mathbf{x}) \qquad N = 1, 2, 3, 4, 5 \end{aligned}\right\} \tag{7.24}$$

where the functions $\psi^{(e)}_N(\mathbf{x})$ have the property defined in (7.14).

We assume that points on the boundaries $\overline{23}$ and $\overline{15}$ of elements (1) and (2) are in one-to-one correspondence with points on the line $\overline{26}$ in the connected model $\bar{\mathscr{R}}$ (shown dashed in Fig. 7.1*b*) and that all nodal points in $\mathscr{R}^*$ and $\bar{\mathscr{R}}$ are in the proper correspondence so that the compatibility conditions of the model are satisfied. Obviously, if we start with a region $\bar{\mathscr{R}}$ and decompose it into elements, we shall always be able to select elements and nodal points so that the compatibility conditions are satisfied. Thus, we now introduce the array $\overset{(e)}{\Lambda}{}^\Delta_N$ of (7.4) and reconnect the elements into a single model of $\Phi(\mathbf{X})$ by the mappings $\underset{(e)}{\Lambda}: \mathscr{L}_e \to \mathscr{G}$:

$$\Phi^\Delta = \overset{(e)}{\Lambda}{}^\Delta_N \varphi^N_{(e)} \tag{7.25}$$

with $e = 1, 2$; $N = 1, 2, 3, 4$ for $e = 1$ and $N = 1, 2, 3, 4, 5$ for $e = 2$; and $\Delta = 1, 2, 3, 4, 5, 6, 7$. Thus,

$$\begin{aligned} e = 1: \quad \Phi^1 &= \overset{(1)}{\Lambda}{}^1_1\varphi^1_{(1)} + \cdots + \overset{(1)}{\Lambda}{}^1_4\varphi^4_{(1)} = \varphi^1_{(1)} \\ \Phi^2 &= \overset{(1)}{\Lambda}{}^2_1\varphi^1_{(1)} + \cdots + \overset{(1)}{\Lambda}{}^2_4\varphi^4_{(1)} = \varphi^2_{(1)} \\ \Phi^6 &= \overset{(1)}{\Lambda}{}^6_1\varphi^1_{(1)} + \cdots + \overset{(1)}{\Lambda}{}^6_4\varphi^4_{(1)} = \varphi^3_{(1)} \\ \Phi^7 &= \overset{(1)}{\Lambda}{}^7_1\varphi^1_{(1)} + \cdots + \overset{(1)}{\Lambda}{}^7_4\varphi^4_{(1)} = \varphi^4_{(1)} \end{aligned} \tag{7.26a}$$

$$\begin{aligned} e = 2: \quad \Phi^2 &= \overset{(2)}{\Lambda}{}^2_1\varphi^1_{(2)} + \cdots + \overset{(2)}{\Lambda}{}^2_5\varphi^5_{(2)} = \varphi^1_{(2)} \\ \Phi^3 &= \overset{(2)}{\Lambda}{}^3_1\varphi^1_{(2)} + \cdots + \overset{(2)}{\Lambda}{}^3_5\varphi^5_{(2)} = \varphi^2_{(2)} \\ \Phi^4 &= \overset{(2)}{\Lambda}{}^4_1\varphi^1_{(2)} + \cdots + \overset{(2)}{\Lambda}{}^4_5\varphi^5_{(2)} = \varphi^3_{(2)} \\ \Phi^5 &= \overset{(2)}{\Lambda}{}^5_1\varphi^1_{(2)} + \cdots + \overset{(2)}{\Lambda}{}^5_5\varphi^5_{(2)} = \varphi^4_{(2)} \\ \Phi^6 &= \overset{(2)}{\Lambda}{}^6_1\varphi^1_{(2)} + \cdots + \overset{(2)}{\Lambda}{}^6_5\varphi^5_{(2)} = \varphi^5_{(2)} \end{aligned} \tag{7.26b}$$

Similarly,

$$\varphi^N_{(e)} = \overset{(e)}{\Omega^N_\Delta}\Phi^\Delta \tag{7.27}$$

and

$$\begin{aligned}
\varphi^1_{(1)} &= \overset{(1)}{\Omega^1_1}\Phi^1 + \overset{(1)}{\Omega^1_2}\Phi^2 + \cdots + \overset{(1)}{\Omega^1_7}\Phi^7 = \Phi^1 \\
\varphi^2_{(1)} &= \overset{(1)}{\Omega^2_1}\Phi^1 + \overset{(1)}{\Omega^2_2}\Phi^2 + \cdots + \overset{(1)}{\Omega^2_7}\Phi^7 = \Phi^2 \\
&\cdots\cdots\cdots\cdots\cdots\cdots \\
\varphi^5_{(2)} &= \overset{(2)}{\Omega^5_1}\Phi^1 + \overset{(2)}{\Omega^5_2}\Phi^2 + \cdots + \overset{(2)}{\Omega^5_7}\Phi^7 = \Phi^6
\end{aligned} \tag{7.28}$$

In summary,

$$\begin{aligned}
&\Phi^1 = \varphi^1_{(1)} \qquad \Phi^2 = \varphi^2_{(1)} = \varphi^1_{(2)} \qquad \Phi^3 = \varphi^2_{(2)} \\
&\Phi^4 = \varphi^3_{(2)} \qquad \Phi^5 = \varphi^4_{(2)} \qquad \Phi^6 = \varphi^3_{(1)} = \varphi^5_{(2)} \\
&\Phi^7 = \varphi^4_{(1)}
\end{aligned} \tag{7.29}$$

Finally, we have the discrete model of $\Phi(\mathbf{X})$,

$$\bar{\Phi}(\mathbf{x}) = \sum_{e=1}^{2} \bar{\varphi}_{(e)}(\mathbf{x}) = \bar{\varphi}_{(1)}(\mathbf{x}) + \bar{\varphi}_{(2)}(\mathbf{x}) \tag{7.30}$$

or, from (7.23),

$$\begin{aligned}
\bar{\Phi}(\mathbf{x}) = {}& \varphi^1_{(1)}\psi^{(1)}_1(\mathbf{x}) + \varphi^2_{(1)}\psi^{(1)}_2(\mathbf{x}) + \varphi^3_{(1)}\psi^{(1)}_3(\mathbf{x}) + \varphi^4_{(1)}\psi^{(1)}_4(\mathbf{x}) + \varphi^1_{(2)}\psi^{(2)}_1(\mathbf{x}) \\
&+ \varphi^2_{(2)}\psi^{(2)}_2(\mathbf{x}) + \varphi^3_{(2)}\psi^{(2)}_3(\mathbf{x}) + \varphi^4_{(2)}\psi^{(2)}_4(\mathbf{x}) + \varphi^5_{(2)}\psi^{(2)}_5(\mathbf{x})
\end{aligned} \tag{7.31}$$

Introducing (7.29), we have

$$\begin{aligned}
\bar{\Phi}(\mathbf{x}) = {}& \Phi^1\psi^{(1)}_1(\mathbf{x}) + \Phi^2[\psi^{(1)}_2(\mathbf{x}) + \psi^{(2)}_1(\mathbf{x})] + \Phi^3\psi^{(2)}_2(\mathbf{x}) + \Phi^4\psi^{(2)}_3(\mathbf{x}) \\
&+ \Phi^5\psi^{(2)}_4(\mathbf{x}) + \Phi^6[\psi^{(1)}_3(\mathbf{x}) + \psi^{(2)}_5(\mathbf{x})] + \Phi^7\psi^{(1)}_4(\mathbf{x})
\end{aligned} \tag{7.32}$$

which completes the construction of the finite-element model of $\Phi(\mathbf{X})$.

It is easily verified that $\bar{\Phi}(\mathbf{x})$ coincides with $\Phi(\mathbf{X})$ at the seven nodal points in $\bar{\mathscr{R}}$ and that elsewhere $\Phi(\mathbf{X})$ is represented approximately by $\bar{\Phi}(\mathbf{x})$. By definition, the functions $\psi^{(1)}_N(\mathbf{x})$ are identically zero in element (2) and $\psi^{(2)}_N(\mathbf{x})$ are identically zero in element (1). For example, in the case of a point $\mathbf{x}$ in element (1) which is in the neighborhood of, say, node $\mathbf{X}^6$ of $\bar{\mathscr{R}}$, we may set $\psi^{(2)}_N(\mathbf{x}) = 0$ and observe that $\psi^{(1)}_1(\mathbf{x})$, $\psi^{(1)}_2(\mathbf{x})$, and $\psi^{(1)}_4(\mathbf{x}) \to 0$ and $\psi^{(1)}_3(\mathbf{x}) \to 1$ as $\mathbf{x} \to \mathbf{X}^6$, $\psi^{(1)}_N(\mathbf{x}) = 0$ and $\psi^{(2)}_1(\mathbf{x})$, $\psi^{(2)}_2(\mathbf{x})$, $\psi^{(2)}_3(\mathbf{x})$, and $\psi^{(2)}_4(\mathbf{x}) \to 0$ while $\psi^{(2)}_5(\mathbf{x}) \to 1$ as $\mathbf{x} \to \mathbf{X}^6$.

Nodal point $\mathbf{X}^6$ is of multiplicity 2 since it lies on the boundary of two elements. Consequently,† from (7.32), $\bar{\Phi}(\mathbf{X}^6) = 0 + \Phi^6[\psi^{(1)}_3(\mathbf{x}^3_{(1)}) + \psi^{(2)}_5(\mathbf{x}^5_{(2)})]/m(\mathbf{X}^6) + 0 = \Phi^6(1 + 1)/2 = \Phi^6$. Similar relations hold for other boundary points.

It follows that the approximation $\bar{\Phi}(\mathbf{x})$ is continuous at nodes 2 and 6 of the interelement boundary $\overline{26}$. Later we shall require that the geometries of the elements and the properties of the functions $\psi^{(e)}_N(\mathbf{x})$ be such that $\bar{\Phi}(\mathbf{x})$ is also continuous at every point of such interelement boundaries.

† See the footnote on page 39.

Example 7.2 Local coordinate transformations As implied in the previous example, the notion of local coordinate systems is often just a formality introduced to emphasize the fact that individual finite elements are to be considered disjoint for the purpose of constructing the local approximations $\bar{f}_{(e)}(\mathbf{x})$. In the actual construction of the model, the same coordinates can often be used for the global and all local systems, the only difference being the formal indexing of local versus global nodal points. There are cases, however, in which the careful identification of independent local coordinate systems is essential. A common characteristic of such cases is that the final finite-element model consists of an assembly of finite elements of a certain dimension embedded in a space of higher dimension; that is, the local fields $\bar{f}_{(e)}(\mathbf{x})$ are defined in a space of dimension n, but the final model is in a space of dimension $k > n$. One example of such a case is a three-dimensional framework composed of bars whose local behavior can be described by one-dimensional functions. Local systems are also used when the orientation or location of an element in the connected model or the geometry of the element makes it convenient to establish local systems for the purpose of calculating the functions $\psi_N^{(e)}(\mathbf{x})$.

We consider here a simple example in which the region $\mathscr{R}$ consists of a number of planar segments connected together to make up a piecewise smooth surface in three-dimensional space. Each planar segment is assumed to be of some relatively simple geometric shape, each typified by sectionally smooth straight-line boundaries and vertices (e.g., triangles, rectangles, hexagons). It is thus convenient to regard each segment as a finite element, to take $\bar{\mathscr{R}} = \mathscr{R}$, and to select the vertices of each segment as the nodal points of the element. Although in many applications a smooth surface might be represented by a suitable collection of flat finite elements, such an approximation is not necessarily implied here.

To describe the geometry of the model, we establish a global coordinate system X^i $(i = 1, 2, 3)$ which, for simplicity, is assumed to be rectangular cartesian. In addition, we establish in each finite element a special local coordinate system $x^i_{(e)}$ $(i = 1, 2, 3)$ which is also rectangular cartesian but for which $x^1_{(e)}$ and $x^2_{(e)}$ are in the plane of element e and $x^3_{(e)}$ is normal to the plane of element e. This finite-element model is illustrated in Fig. 7.2.

We are concerned with the problem of representing a sectionally continuous scalar field $\Phi(\mathbf{x})$ which is locally a two-dimensional field for each finite element. In other words, if $\varphi_{(e)}(\mathbf{x})$ is the local field corresponding to element $\imath_e$, then $\varphi_{(e)}(\mathbf{x})$ is a function of only the local coordinates $x^1_{(e)}$, $x^2_{(e)}$ in the plane of element e:

$$\varphi_{(e)} = \varphi_{(e)}(x^1_{(e)}, x^2_{(e)}) \qquad e = 1, 2, \ldots, E \tag{7.33}$$

Following the usual procedure, we denote by $\varphi^N_{(e)}$ the values of the local fields at the local nodes of each element and we approximate the local fields by functions of the form

$$\bar{\varphi}_{(e)} = \psi_N^{(e)}(x^1_{(e)}, x^2_{(e)})\varphi^N_{(e)} \qquad N = 1, 2, \ldots, N_e \tag{7.34}$$

The final model is then

$$\bar{\Phi}(\mathbf{x}) = \sum_e \bar{\varphi}_{(e)}(\mathbf{x}) = \sum_e \overset{(e)}{\Omega}{}^N_\Delta \Phi^\Delta \psi_N^{(e)}(x^1, x^2) \tag{7.35}$$

According to (7.35), to obtain the approximate value of the field at a point which is not a node, it is necessary to identify only the particular element to which the point belongs and the local coordinates of the point relative to the local reference frame $x^i_{(e)}$. The global coordinates X^i provide a means for identifying the location

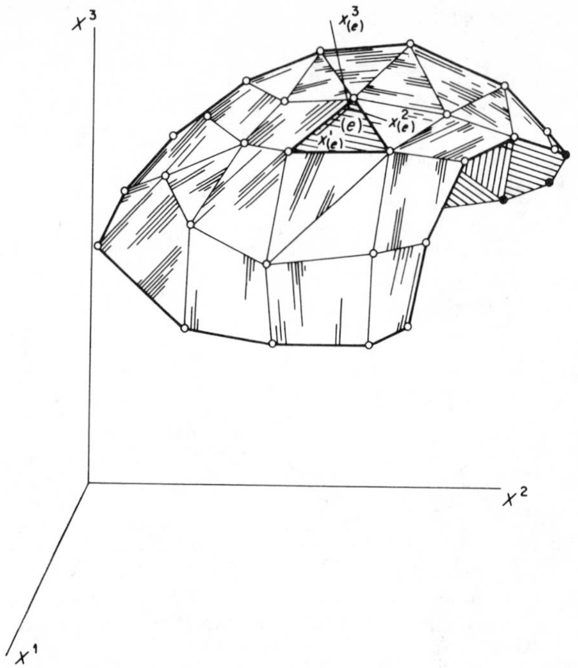

Fig. 7.2 Collection of flat two-dimensional elements embedded in a three-dimensional space.

of elements and nodes in the model and are used in establishing the connectivity of the model, but they do not appear in the final description of the approximate field $\bar{\Phi}(\mathbf{x})$ in (7.35). In some instances, however, it is convenient to refer all local descriptions to a common coordinate system, the global system X^i usually being the most logical choice.

To establish the connectivity of the model and to refer the approximate field to the global reference frame, we introduce other systems of local coordinates $\hat{x}^i_{(e)}$ ($i = 1, 2, 3$) which coincide with the X^i; that is, $\hat{x}^i_{(e)}$ is parallel to X^i for every element, and the origins of each $\hat{x}^i_{(e)}$ coincide with the origin of the system X^i. The local systems $\hat{x}^i_{(e)}$ and $x^i_{(e)}$ are related through the transformations

$$\begin{aligned} \hat{x}^i_{(e)} &= \overset{(e)}{\alpha^i_j} x^j_{(e)} + B^i_{(e)} \\ x^i_{(e)} &= \overset{(e)}{\alpha^j_i}(\hat{x}^j_{(e)} - B^j_{(e)}) \end{aligned} \tag{7.36}$$

where $\overset{(e)}{\alpha^i_j}$ is the direction cosine of X^i (or $\hat{x}^i_{(e)}$) relative to $x^j_{(e)}$ and $B^i_{(e)}$ are the global coordinates of the origin of $x^i_{(e)}$ relative to the origin 0 of X^i. The quantities $\overset{(e)}{\alpha^i_j}$ have the customary properties

$$\overset{(e)}{\alpha^i_m}\overset{(e)}{\alpha^j_m} = \delta^{ij} \qquad \det \overset{(e)}{\alpha^i_j} = 1 \tag{7.37}$$

The formal connectivity and decomposition of the model are established by the mappings

$$X^{\Delta i} = \overset{(e)}{\Lambda^{\Delta}_{N}} \hat{x}^{Ni}_{(e)} \qquad \hat{x}^{Ni}_{(e)} = \overset{(e)}{\Omega^{N}_{\Delta}} X^{\Delta i} \tag{7.38}$$

where

$$\hat{x}^{Ni}_{(e)} = \overset{(e)}{\alpha^{i}_{j}} x^{Nj}_{(e)} + B^{Ni}_{(e)} \tag{7.39}$$

Here $x^{Ni}_{(e)}$ are the local coordinates of node N of element e ($x^{N3}_{(e)} = 0$).

The local interpolating functions $\psi^{(e)}_{N}(\mathbf{x})$ are expressed in terms of the auxiliary coordinates $\hat{x}^{i}_{(e)}$ with the aid of (7.36), and (7.35) acquires the form

$$\Phi(\hat{\mathbf{x}}) = \sum_{e} \overset{(e)}{\Omega^{N}_{\Delta}} \Phi^{\Delta} \psi^{(e)}_{N}(\hat{\mathbf{x}}) \tag{7.40}$$

where

$$\psi^{(e)}_{N}(\hat{\mathbf{x}}) = \psi^{(e)}_{N}[\overset{(e)}{\alpha^{j}_{1}}(\hat{x}^{j}_{(e)} - B^{j}_{(e)}), \overset{(e)}{\alpha^{j}_{2}}(\hat{x}^{j}_{(e)} - B^{j}_{(e)})] \tag{7.41}$$

In (7.41), the global coordinates X^{i} may be written instead of $\hat{x}^{i}_{(e)}$ provided it is recognized that the functions $\psi^{(e)}_{N}(\hat{\mathbf{x}})$ are nonzero only within their associated finite element.

Example 7.3 Vector fields The previous example dealt with a scalar field given in a euclidean space in rectangular cartesian coordinates. However, the general concept of finite element is applicable to virtually any type of continuous vector or tensor field, defined on euclidean or non-euclidean spaces, which are given functions of any type of curvilinear coordinates.

To emphasize this point, consider a vector field $\mathbf{V}(\mathbf{X})$ defined over a region $\mathscr{R}$ of a non-euclidean space $\mathscr{E}^{k}$ whose points are given by the curvilinear coordinates X^{i} ($i = 1, 2, \ldots, k$). One such region in a two-dimensional space is shown represented by an assembly of curvilinear finite elements in Fig. 7.3. Let $\mathbf{v}_{(e)}(\mathbf{x})$ denote the local vector field corresponding to element e, and let $\mathbf{g}_{i(e)}$ and $\mathbf{g}^{i}_{(e)}$ denote the natural covariant and contravariant basis vectors associated with the frame $x^{i}_{(e)}$. Then

$$\mathbf{v}_{(e)} = v^{i}_{(e)}\mathbf{g}_{i(e)} = v_{i(e)}\mathbf{g}^{i}_{(e)} \tag{7.42}$$

where $v^{i}_{(e)}$ and $v_{i(e)}$ are, respectively, the contravariant and covariant components of the local vector field.

Now the vector functions $\mathbf{g}_{i(e)}$ and $\mathbf{g}^{i}_{(e)}$ are assumed to be given a priori, and each component $v^{i}_{(e)}$ or $v_{i(e)}$ can be regarded as a scalar field over $\imath_{e}$. Thus, we may introduce the approximations†

$$\bar{v}^{i}_{(e)} = v^{Ni}_{(e)}\psi^{(e)}_{N}(\mathbf{x}) \tag{7.43a}$$

$$\bar{v}^{(e)}_{i} = v^{N}_{i(e)}\psi^{(e)}_{N}(\mathbf{x}) \tag{7.43b}$$

† Use of orthogonal curvilinear coordinates to devise finite elements of cylindrical shells was made by Bogner, Fox, and Schmit [1966, 1967], among others. General finite-element representations of covariant and contravariant components of vectors defined on non-euclidean spaces, such as is represented by (7.35), were used by Oden [1968*a*], in the analysis of thin shells. See also Wempner, Oden, and Kross [1968], Wempner [1969], and Oden and Wempner [1967]. Ergatoudis, Irons, and Zienkiewicz [1968*a*, 1968*b*] and Ahmad, Irons, and Zienkiewicz [1969] developed families of "isoparametric" finite elements which have curvilinear boundaries. By fitting polynomials through prescribed boundary points, these authors succeeded in developing finite elements which approximate quite general boundaries. A more detailed discussion of such elements is given in Sec. 10.4.

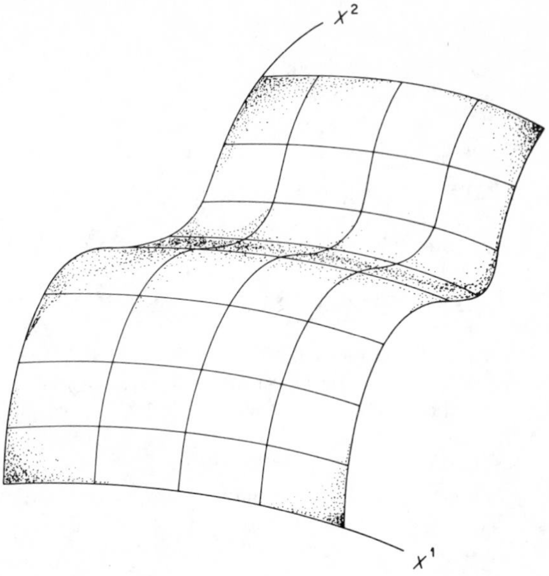

Fig. 7.3 Domain of a vector field $\mathbf{V}(\mathbf{X})$ defined on a two-dimensional non-euclidean space.

in which $v_{(e)}^{Ni}$ and $v_{i(e)}^{N}$ are the contravariant and covariant components, respectively, of the local field at node N of element e. In view of (7.42), we have

$$\bar{\mathbf{v}}_{(e)} = \mathbf{v}_{(e)}^{N}\psi_N^{(e)}(\mathbf{x}) \tag{7.44}$$

where $\mathbf{v}_{(e)}^{N}$ is the value of the local vector field at node N of $\mathscr{r}_e$:

$$\mathbf{v}_{(e)}^{N} = v_{(e)}^{Ni}\mathbf{g}_i^{(e)} = v_{i(e)}^{N}\mathbf{g}_{(e)}^{i} \tag{7.45}$$

Likewise,

$$\mathbf{v}_{(e)}^{N} = \overset{(e)}{\Omega}{}_{\Delta}^{N}\mathbf{V}^{\Delta} \tag{7.46}$$

Thus, the final finite-element representation of the vector field is given by

$$\bar{\mathbf{v}}(\mathbf{x}) = \sum_e \overset{(e)}{\Omega}{}_{\Delta}^{N}\mathbf{V}^{\Delta}\psi_N^{(e)}(\mathbf{x}) \tag{7.47}$$

It is interesting to note that an alternate approximation to $\mathbf{V}(\mathbf{X})$ is possible within the context of the theory presented earlier. Instead of (7.43), suppose that we set

$$\bar{\mathbf{v}}_{(e)} = \mathbf{v}_{(e)}(\mathbf{x}^N)\psi_N^{(e)}(\mathbf{x}) \tag{7.48}$$

and write

$$\hat{\mathbf{v}}_{(e)} = \mathbf{v}_{(e)}^{N}\psi_N^{(e)}(\mathbf{x}) \tag{7.49}$$

Then approximations (7.44) and (7.49) coincide at the nodes but differ elsewhere in the element. Indeed,

$$\hat{\mathbf{v}}_{(e)}(\mathbf{x}) = \psi_N^{(e)}(\mathbf{x})(v_{(e)}^{Ni}\mathbf{g}_{(N)_i}^{(e)}) = \psi_N^{(e)}(\mathbf{x})(v_{Ni}^{(e)}\mathbf{g}_{(N)(e)}^{i}) \tag{7.50}$$

while

$$\bar{\mathbf{v}}_{(e)}(\mathbf{x}) = \psi_N^{(e)}(\mathbf{x})v_{(e)}^{Ni}\mathbf{g}_i^{(e)}(\mathbf{x}) = \psi_N^{(e)}(\mathbf{x})v_{i(e)}^N\mathbf{g}_{(e)}^i(\mathbf{x}) \tag{7.51}$$

where

$$\mathbf{g}_{(N)i}^{(e)} = \mathbf{g}_{(e)i}(\mathbf{x}^N) \qquad \mathbf{g}_{(N)(e)}^i = \mathbf{g}_{(e)}^i(\mathbf{x}^N) \tag{7.52a,b}$$

Thus, the use of the approximation (7.50) amounts to referring $\mathbf{v}_{(e)}(\mathbf{x})$ to constant-basis vectors $\mathbf{g}_{(N)i}^{(e)}$, $\mathbf{g}_{(N)(e)}^i$, which represent some average of the values of the vectors $\mathbf{g}_i$ and $\mathbf{g}^i$ at each node. Approximation (7.51) refers $\mathbf{v}_{(e)}(\mathbf{x})$ to the actual basis at $\mathbf{x}$ and, in general, is a better approximation. The general form (7.47) of the global approximation is the same in either case.

Example 7.4 Tensor fields The same procedure can be used for tensor fields as well. For example, let $\mathbf{T}(\mathbf{X})$ denote a second-order tensor field defined on $\mathscr{R}$, and let $\mathbf{t}_{(e)}(\mathbf{x})$ denote the associated local field corresponding to element e. Then†

$$\mathbf{t}_{(e)}(\mathbf{x}) = t_{(e)}^{ij}\mathbf{g}_{i(e)} \otimes \mathbf{g}_{j(e)} = t_{(e)j}^{i\cdot}\mathbf{g}_{i(e)} \otimes \mathbf{g}^{j(e)} = t_{i(e)}^{\cdot j}\mathbf{g}_{(e)}^i \otimes \mathbf{g}_{j(e)} = t_{ij(e)}\mathbf{g}_{(e)}^i \otimes \mathbf{g}_{(e)}^j \tag{7.53}$$

where $t_{(e)}^{ij}$ and $t_{ij(e)}$ are the contravariant and covariant components and $t_{(e)j}^{i\cdot}$ and $t_{i(e)}^{\cdot j}$ are the mixed components of $\mathbf{t}_{(e)}(\mathbf{x})$. Thus

$$\bar{t}_{(e)}^{ij} = t_{(e)}^{Nij}\psi_N^{(e)}(\mathbf{x}) \qquad \bar{t}_{(e)j}^{i\cdot} = t_{(e)j}^{Ni\cdot}\psi_N^{(e)}(\mathbf{x}) \tag{7.54a,b}$$

$$\bar{t}_i^{(e)j} = t_{(e)i}^{N\cdot j}\psi_N^{(e)}(\mathbf{x}) \qquad \bar{t}_{ij}^{(e)} = t_{(e)ij}^N\psi_N^{(e)}(\mathbf{x}) \tag{7.54c,d}$$

where $t_{(e)}^{Nij}, \ldots, t_{(e)ij}^N$ denote the values of the respective components at node N of element e. It follows that

$$\mathbf{t}_{(e)}(\mathbf{x}) = \mathbf{t}_{(e)}^N\psi_N^{(e)}(\mathbf{x}) \tag{7.55}$$

where

$$\begin{aligned}\mathbf{t}_{(e)}^N &= t_{(e)}^{Nij}\mathbf{g}_i^{(e)} \otimes \mathbf{g}_j^{(e)} = t_{(e)\cdot j}^{Ni}\mathbf{g}_i^{(e)} \otimes \mathbf{g}_{(e)}^j \\ &= t_{(e)i}^{Nj}\mathbf{g}_{(e)}^i \otimes \mathbf{g}_j^{(e)} = t_{(e)ij}^N\mathbf{g}_{(e)}^i \otimes \mathbf{g}_{(e)}^j\end{aligned} \tag{7.56}$$

If $\mathbf{T}^\Delta$ denotes the global values of $\mathbf{T}(\mathbf{X})$, then

$$\mathbf{t}_{(e)}^N = \overset{(e)}{\Omega}{}_\Delta^N\mathbf{T}^\Delta \tag{7.57}$$

and the final finite-element representation of the tensor field is given by

$$\bar{\mathbf{T}}(\mathbf{x}) = \sum_e \overset{(e)}{\Omega}{}_\Delta^N\mathbf{T}^\Delta\psi_N^{(e)}(\mathbf{x}) \tag{7.58}$$

Note that alternate approximations of the type in (7.50) can also be made in the present case by using, instead of $\mathbf{g}_{(e)i}(\mathbf{x})$, $\mathbf{g}_{(e)}^i(\mathbf{x})$, the approximate-basis vectors $\mathbf{g}_{(N)i}^{(e)}$ and $\mathbf{g}_{(N)(e)}^i$ of (7.52).

Example 7.5 Linear interpolation For completeness and for future reference, we now return to the one-dimensional example discussed briefly in Art. 6.1 and cite one of the simplest examples of a finite-element model of a continuous function.

† We use $\mathbf{g}_i \otimes \mathbf{g}_j$, $\mathbf{g}_i \otimes \mathbf{g}^j, \ldots, \mathbf{g}^i \otimes \mathbf{g}^j$ to denote tensor products of the basis vectors $\mathbf{g}_i$ and $\mathbf{g}^j (i, j = 1, 2, \ldots, k)$. Each of the sets of dyadics $\mathbf{g}_i \otimes \mathbf{g}_j$, $\mathbf{g}_i \otimes \mathbf{g}^j$, $\mathbf{g}^i \otimes \mathbf{g}_j$, $\mathbf{g}^i \otimes \mathbf{g}^j$ provides bases for the k^2-dimensional vector space of second-order tensors $\mathbf{T}$.

Consider once again the real-valued function $F(X)$ defined on a closed interval of the real line $A \leq X \leq B$ (Fig. 6.1). We begin, as usual, by identifying a number of nodal points on the domain $[A,B]$, with $A = 1$ and $B = G$. The case in which $G = 5$ is indicated in Fig. 6.1*a*. The values $F^1 = F(X^1), \ldots, F^G = F(X^G)$ of the function are then computed at these points. We then consider a number of disjoint finite elements (Fig. 6.2*b*) which, when connected together, form subintervals of $[A,B]$. For compatibility with the global representation, each finite element has two local nodes, one at each end of the element. Locally, the function corresponding to element e is denoted $f_{(e)}(x)$, and the local values of these functions at local nodes 1 and 2 of the element are denoted $f^1_{(e)}$ and $f^2_{(e)}$ (Fig. 6.2*b*).

Up to now, the analysis is merely a repetition of that outlined previously. However, we can now examine the local functions in more detail. Accordingly, each local field is approximated by functions of the form

$$\bar{f}_{(e)}(x) = f^N_{(e)}\psi^{(e)}_N(x) = f^1_{(e)}\psi^{(e)}_1(x) + f^2_{(e)}\psi^{(e)}_2(x) \tag{7.59}$$

We take for illustration purposes the *linear* interpolating functions

$$\psi^{(e)}_1(x) = \frac{x^2_{(e)} - x_{(e)}}{L_{(e)}} \qquad \psi^{(e)}_2(x) = \frac{x_{(e)} - x^1_{(e)}}{L_{(e)}} \tag{7.60a,b}$$

where $L_{(e)} = x^2_{(e)} - x^1_{(e)}$ is the length of the element. It is emphasized, however, that we have not as yet placed any general restrictions on the functions $\psi^{(e)}_N(x)$ other than that they be continuous within the element and that they satisfy (7.14). Thus, these functions need not be linear, although in the case considered now they are linear. Indeed, by adding to the functions in (7.60) any multiple of any continuous function that vanishes at both nodes $x^1_{(e)}$ and $x^2_{(e)}$, we obtain new functions $\psi^{(e)}_N(x)$ which satisfy (7.14).†

Noting, as usual, that

$$f^N_{(e)} = \overset{(e)}{\Omega}{}^N_\Delta F^\Delta \tag{7.61}$$

with $N = 1, 2$ and $\Delta = 1, 2, \ldots, G$, we have for the final finite-element model of $F(X)$ the piecewise linear approximation

$$\bar{F}(x) = \sum_e \bar{f}_{(e)}(x) = \sum_e \overset{(e)}{\Omega}{}^N_\Delta F^\Delta \psi^{(e)}_N(x) \tag{7.62}$$

8 HIGHER-ORDER REPRESENTATIONS

With the basic properties of finite-element models of continuous functions established, we now examine the refinements necessary to construct higher-order finite-element representations.

8.1 GENERAL PROPERTIES OF HIGHER-ORDER MODELS

Consider once again a continuous function $\mathbf{F}(\mathbf{X}) = \mathbf{F}(X^1, X^2, \ldots, X^k)$ defined on a region $\mathscr{R}$ of $\mathscr{E}^k$. If $\mathbf{F}(\mathbf{X})$ and its first p partial derivatives exist

† For example, add $C(x^{(e)} - x^{(e)}_1)^m(x^{(e)} - x^{(e)}_2)^n$, m, $n > 0$, to the right sides of (7.60*a*) and (7.60*b*) and note that still $\psi^1_{(e)}(x_1) = 1$, $\psi^1_{(e)}(x_2) = 0$ and $\psi^2_{(e)}(x_1) = 0$, $\psi^2_{(e)}(x_2) = 1$. We examine further restrictions on the functions $\psi^N_{(e)}(x)$ in Art. 10.

and are continuous in $\mathscr{R}$, it is possible to construct higher-order representations which not only coincide with the local functions $\mathbf{f}_{(e)}(\mathbf{x})$ at each node of the finite element but also possess partial derivatives of order $r \le p$ which coincide with those of the given function at each node of the element.

It is convenient now to regard points in $\mathscr{E}^k$ as being given by specific k-tuples of global coordinates X^i $(i = 1, 2, \ldots, k)$ which, when all the elements are connected together, can be related to local coordinates $x^i_{(e)}$ $(i = 1, 2, \ldots, k; e = 1, 2, \ldots, E)$ by coordinate transformations of the form

$$X^i = X^i(x^1_{(e)}, x^2_{(e)}, \ldots, x^k_{(e)}) \tag{8.1}$$

which, for each e, is assumed to have unique inverse transformations

$$x^i_{(e)} = x^i_{(e)}(X^1, X^2, \ldots, X^k) \tag{8.2}$$

For higher-order representations, we use locally, instead of (7.11), the more general formula

$$\bar{\mathbf{f}}^{(e)}(\mathbf{x}) = \mathbf{H}_{(e)}(\mathbf{x}; \mathbf{f}^N_{(e)}; \mathbf{f}^N_{(e),1}; \mathbf{f}^N_{(e),2}; \ldots) \tag{8.3}$$

where we use the notation

$$\mathbf{f}^N_{(e),i_1 i_2 \cdots i_r} = \frac{\partial^r \mathbf{f}^{(e)}(\mathbf{x}^N)}{\partial x^{i_1}_{(e)}\, \partial x^{i_2}_{(e)} \cdots \partial x^{i_r}_{(e)}} \tag{8.4}$$

in which $i_1, i_2, \ldots, i_r = 1, 2, \ldots, k$. The functions $\mathbf{H}_{(e)}(\;)$ in (8.3) are defined so as to have the following properties:

$$\begin{aligned}
&1.\ \mathbf{H}_{(e)}(\mathbf{x}; \cdots) \text{ and all partial derivatives with respect to } x^i \text{ vanish for } \mathbf{x} \notin \imath_e.\\
&2.\ \mathbf{H}_{(e)}(\mathbf{x}^M; \cdots) = \mathbf{f}^M_{(e)} \qquad \mathbf{x}^M \in \imath_e\\
&3.\ \frac{\partial \mathbf{H}_{(e)}(\mathbf{x}^M; \cdots)}{\partial x^i} = \mathbf{f}^M_{(e),i} \qquad \mathbf{x}^M \in \imath_e\\
&4.\ \frac{\partial^r \mathbf{H}_{(e)}(\mathbf{x}^M; \cdots)}{\partial x^{i_1}\, \partial x^{i_2} \cdots \partial x^{i_r}} = \mathbf{f}^M_{(e),i_1 i_2 \cdots i_r} \qquad \mathbf{x}^M \in \imath_e
\end{aligned} \tag{8.5}$$

Within the element, the functions $\mathbf{H}_{(e)}(\;)$ have properties similar to Hermite interpolation polynomials, although these functions need not be polynomials.† For this reason, we refer to the $\mathbf{H}_{(e)}(\;)$ as *generalized Hermite*

† Hermite interpolation formulas and higher-order finite-element representations have been used by numerous investigators in the development of discrete models of one- and two-dimensional scalar fields. Among these are, for example, Bogner, Fox, and Schmit [1966], Pestel [1966], Felippa [1966], Deak and Pian [1967], and Tocher and Hartz [1967]. The elementary notions of Hermite interpolation polynomials can be found in many books on numerical analysis (e.g., Hildebrand [1956]). Related work on piecewise interpolation and higher-order representations has been reported by Courant [1943], Fort [1948], Birkhoff and de Boor [1965], and Langhaar and Chu [1968], among others.

interpolation functions. In general, it is convenient to choose the functions $\mathbf{H}_{(e)}(\ )$ so that their rth partial derivatives are linear in the nodal values of the corresponding partial derivatives of the local field $\mathbf{f}_{(e)}(\mathbf{x})$:

$$\frac{\partial^r \mathbf{H}_{(e)}(\mathbf{x}, \ldots)}{\partial x^{i_1}\, \partial x^{i_2} \cdots \partial x^{i_r}} = \mathbf{f}^N_{(e),i_1 i_2 \cdots i_r} \psi_N^{(e)i_1 i_2 \cdots i_r}(\mathbf{x}) \tag{8.6}$$

with $i_1, i_2, \ldots, i_r = 1, 2, \ldots, k$ and $N = 1, 2, \ldots, N_e$. Then (8.3) becomes

$$\bar{\mathbf{f}}^{(e)}(\mathbf{x}) = \mathbf{f}^N_{(e)} \psi_N^{(e)0}(\mathbf{x}) + \sum_{i=1}^{k} \mathbf{f}^N_{(e),i} \psi_N^{(e)i}(\mathbf{x}) + \sum_{i=1}^{k} \left(\sum_{j=i}^{k} \mathbf{f}^N_{(e),ij} \psi_N^{(e)ij}(\mathbf{x}) \right)$$
$$+ \cdots + \sum_{i_1=1}^{k} \left[\sum_{i_2=i_1}^{k} \left(\cdots \sum_{i_r=i_{r-1}}^{k} \mathbf{f}^N_{(e),i_1 i_2 \cdots i_r} \psi_N^{(e)i_1 i_2 \cdots i_r}(\mathbf{x}) \right) \cdots \right] \tag{8.7}$$

where k is the dimension of the space and the repeated nodal indices are to be summed from 1 to N_e. The scalar functions $\psi_N^{(e)0}(\mathbf{x})$, $\psi_N^{(e)i}(\mathbf{x}), \ldots,$ $\psi_N^{(e)i_1 i_2 \cdots i_r}(\mathbf{x})$ have the following properties:

1. $\psi_N^{(e)0}(\mathbf{x})$, $\psi_N^{(e)i}(\mathbf{x}), \ldots, \psi_N^{(e)i_1 i_2 \cdots i_r}(\mathbf{x})$ and all partial derivatives of these functions vanish for $\mathbf{x} \notin \imath_e$.
2. For arbitrary $\mathbf{x} \in \imath_e$, the only nonzero functions in the set $\psi_N^{(e)i_1 i_2 \cdots i_r}(\mathbf{x})$ are those for which $i_1 \leq i_2 \leq i_3 \leq \cdots \leq i_{r-1} \leq i_r$
3. At node $\mathbf{x}^M \in \imath_e$—with $s = 0, 1, 2, \ldots, m$; $i_1 \leq i_2 \leq \cdots \leq i_r$; $j_1 \leq j_2 \leq \cdots \leq j_r$; $i_1, i_2, \ldots, i_r, j_1, j_2, \ldots, j_r = 1, 2, \ldots, k$, k being the dimension of the space; and M, $N = 1, 2, \ldots, N_e$—the functions $\psi_N^{(e)i_1 i_2 \cdots i_r}(\mathbf{x})$ with $r \leq m$ are such that

$$\frac{\partial^s \psi_N^{(e)i_1 i_2 \cdots i_r}(\mathbf{x}^M)}{\partial x^{j_1}\, \partial x^{j_2} \cdots \partial x^{j_s}} = \begin{cases} 0 & \text{if } s \neq r \\ \delta_M^N\, \delta_{j_1}^{i_1}\, \delta_{j_2}^{i_2} \cdots \delta_{j_s}^{i_r} & \text{if } s = r \end{cases} \tag{8.8}$$

where $\delta_M^N, \delta_{j_1}^{i_1}, \ldots, \delta_{j_s}^{i_r}$ are Kronecker deltas.

The above properties ensure that $\bar{\mathbf{f}}^{(e)}(\mathbf{x})$ and its first m partial derivatives with respect to x^i coincide with $\bar{\mathbf{f}}^{(e)}(\mathbf{x})$ and its corresponding derivatives at each node of element e.

If a higher-order representation $\bar{\mathbf{f}}^{(e)}(\mathbf{x})$ and its first m partial derivatives contain at least one member of the set $\psi_N^{(e)i_1 i_2 \cdots i_m}(\mathbf{x})$ but none of order higher than m, we refer to $\bar{\mathbf{f}}^{(e)}(\mathbf{x})$ as an $(m + 1)$st-order representation of $\bar{\mathbf{f}}^{(e)}(\mathbf{x})$. Thus (7.15) represents a first-order representation. Note, however, that for an $(m + 1)$st-order representation, the normalized Hermite functions $\psi_N^{(e)0}(\mathbf{x})$ are *not* the Lagrange functions $\psi_N^{(e)}(\mathbf{x})$ of (7.15); for in (7.15) we place no restrictions on the derivatives of $\psi_N^{(e)}(\mathbf{x})$ of any order, whereas the derivatives of $\psi_N^{(e)0}(\mathbf{x})$ of order $r \leq m$ must vanish at all the nodal points of the finite element. Indeed, in the first-order representation (7.15) we require only that the function $\bar{\mathbf{f}}^{(e)}(\mathbf{x})$ match $\mathbf{f}^{(e)}(\mathbf{x})$ at the local nodes. Thus, only for a first-order representation need $\psi_N^{(e)}(\mathbf{x})$ and $\psi_N^{(e)0}(\mathbf{x})$ be the same.

When establishing the connectivity of finite-element models based on higher-order representations, it is necessary to introduce additional global and local sets corresponding to the nodal values of $\mathbf{F}(\mathbf{X})$ and $\mathbf{f}^{(e)}(\mathbf{x})$ and their partial derivatives. For example, we may have

$$\mathscr{G}^0 = \{\mathbf{F}^\Delta\}; \mathscr{G}^i = \{\mathbf{F}^\Delta_{,i}\}; \ldots ; \mathscr{G}^{i_1 i_2 \cdots i_r} = \{\mathbf{F}^\Delta_{,i_1 i_2 \cdots i_r}\} \tag{8.9}$$

and

$$\mathscr{L}^0_e = \{\mathbf{f}^N_{(e)}\}; \mathscr{L}^i_e = \{\mathbf{f}^N_{(e),i}\}; \ldots ; \mathscr{L}^{i_1 i_2 \cdots i_r} = \{\mathbf{f}^N_{(e),i_1 i_2 \cdots i_r}\} \tag{8.10}$$

in which $\mathbf{F}^\Delta_{,i}, \ldots, \mathbf{F}^\Delta_{,i_1 i_2 \cdots i_r}$ are the nodal values of various partial derivatives of $\mathbf{F}(\mathbf{X})$, corresponding to those of $\mathbf{f}^{(e)}(\mathbf{x})$ appearing in (8.7), at node Δ of $\overline{\mathscr{R}}$. We note, however, that by definition

$$\mathbf{F}^\Delta_{,i} = \frac{\partial \mathbf{F}(\mathbf{X}^\Delta)}{\partial X^i} \qquad \mathbf{f}^N_{(e),i} = \frac{\partial \mathbf{f}^{(e)}(\mathbf{x}^N)}{\partial x^i_{(e)}} \tag{8.11}$$

etc. Thus, to establish the connectivity and decomposition of the system, we use, instead of (7.16) and (7.17),

$$\mathbf{F}^\Delta = \overset{(e)}{\Lambda^\Delta_N} \mathbf{f}^N_{(e)}, \ \mathbf{F}^\Delta_{,i} = \frac{\partial x^j_{(e)}}{\partial X^i} \mathbf{f}^N_{(e),j} \overset{(e)}{\Lambda^\Delta_N}$$

$$\mathbf{F}^\Delta_{,ij} = \frac{\partial x^r_{(e)}}{\partial X^i} \frac{\partial x^s_{(e)}}{\partial X^j} \mathbf{f}^N_{(e),rs} \overset{(e)}{\Lambda^\Delta_N}, \ldots \tag{8.12}$$

and

$$\mathbf{f}^N_{(e)} = \overset{(e)}{\Omega^N_\Delta} \mathbf{F}^\Delta, \ \mathbf{f}^N_{(e),i} = \frac{\partial X^j}{\partial x^i_{(e)}} \mathbf{F}^\Delta_{,j} \overset{(e)}{\Omega^N_\Delta}$$

$$\mathbf{f}^N_{(e),ij} = \frac{\partial X^r}{\partial x^i_{(e)}} \frac{\partial X^s}{\partial x^j_{(e)}} \mathbf{F}^\Delta_{,rs} \overset{(e)}{\Omega^N_\Delta}, \ldots \tag{8.13}$$

or, in general,

$$\mathbf{f}^N_{(e),i_1 i_2 \cdots i_r} = \frac{\partial X^{j_1}}{\partial x^{i_1}_{(e)}} \frac{\partial X^{j_2}}{\partial x^{i_2}_{(e)}} \cdots \frac{\partial X^{j_r}}{\partial x^{i_r}_{(e)}} \mathbf{F}^\Delta_{,j_1 j_2 \cdots j_r} \overset{(e)}{\Omega^N_\Delta} \tag{8.14}$$

It is often possible to select the local coordinates $x^i_{(e)}$ to be coincident with the global coordinates X^i. Then $\partial X^i / \partial x^j_{(e)} = \delta^i_j$ and (8.14) reduces to

$$\mathbf{f}^N_{(e),i_1 i_2 \cdots i_r} = \overset{(e)}{\Omega^N_\Delta} \mathbf{F}^\Delta_{,i_1 i_2 \cdots i_r} \tag{8.15}$$

It is clear that the construction and manipulation of higher-order representations are generally a great deal more involved than that of the first-order representations. It is an unsettled question as to whether the improved accuracy that higher-order representations usually provide will outweigh the additional difficulties involved in their construction. We remark that since the derivatives $\mathbf{F}_{,i_1 i_2 \ldots i_r}$ are often continuous functions themselves, it is often possible to use first-order representations of not only $\mathbf{F}(\mathbf{X})$ but also its derivatives.

8.2 EXAMPLES

We examine briefly examples of higher-order finite-element representations.

Example 8.1 A second-order representation A second-order representation of a local field defined over an element in a two-dimensional space is of the form

$$\bar{\mathbf{f}}^{(e)}(\mathbf{x}) = \mathbf{f}^N_{(e)}\psi^{(e)0}_N(\mathbf{x}) + \mathbf{f}^N_{(e),1}\psi^{(e)1}_N(\mathbf{x}) + \mathbf{f}^N_{(e),2}\psi^{(e)2}_N(\mathbf{x}) \tag{8.16}$$

where $N = 1, 2, \ldots, N_e$. The first partial derivatives of $\bar{\mathbf{f}}^{(e)}(\mathbf{x})$ are

$$\frac{\partial \bar{\mathbf{f}}^{(e)}(\mathbf{x})}{\partial x^1} = \mathbf{f}^N_{(e)}\psi^0_{N,1} + \mathbf{f}^N_{(e),1}\psi^1_{N,1} + \mathbf{f}^N_{(e),2}\psi^2_{N,1}$$
$$\frac{\partial \bar{\mathbf{f}}^{(e)}(\mathbf{x})}{\partial x^2} = \mathbf{f}^N_{(e)}\psi^0_{N,2} + \mathbf{f}^N_{(e),1}\psi^1_{N,2} + \mathbf{f}^N_{(e),2}\psi^2_{N,2} \tag{8.17}$$

where, for clarity, we have temporarily dropped the element identification label (e) and where the dependence of $\psi_{N,i}$, etc., on $\mathbf{x}$ is understood (that is, $\psi^0_{N,1} \equiv \psi^{(e)0}_{N,1}(\mathbf{x})$, etc.).

We require that at a node $\mathbf{x}^M \in \imath_e$,

$$\bar{\mathbf{f}}^{(e)}(\mathbf{x}^M) = \mathbf{f}^M_{(e)} \qquad \bar{\mathbf{f}}^{(e)}_{,1}(\mathbf{x}^M) = \mathbf{f}^M_{(e),1} \qquad \bar{\mathbf{f}}^{(e)}_{,2}(\mathbf{x}^M) = \mathbf{f}^M_{(e),2} \tag{8.18a,b,c}$$

Thus, for every node $\mathbf{x}^M \in \imath_e$,

$$\psi^0_{N,1}(\mathbf{x}^M) = \psi^0_{N,2}(\mathbf{x}^M) = \psi^1_{N,2}(\mathbf{x}^M) = \psi^2_{N,1}(\mathbf{x}^M) = \psi^1_N(\mathbf{x}^M) = \psi^2_N(\mathbf{x}^M) = 0 \tag{8.19}$$

and for a specific node $\mathbf{x}^M$,

$$\psi^0_N(\mathbf{x}^M) = \psi^1_{N,1}(\mathbf{x}^M) = \psi^2_{N,2}(\mathbf{x}^M) = \delta^M_N = \begin{cases} 1 & \text{if } \mathbf{x}^N = \mathbf{x}^M \\ 0 & \text{if } \mathbf{x}^N \neq \mathbf{x}^M \end{cases} \tag{8.20}$$

As noted previously, the local nodal values of derivatives $\mathbf{f}^N_{(e),1}$ and $\mathbf{f}^N_{(e),2}$ and the global values $\mathbf{F}^\Delta_{,1}$ and $\mathbf{F}^\Delta_{,2}$ should be referred to a common system before assembling the elements. In the present case, we have

$$\mathbf{f}^N_{(e)} = \overset{(e)}{\Omega^N_\Delta}\mathbf{F}^\Delta \qquad \mathbf{F}^\Delta = \overset{(e)}{\Lambda^\Delta_N}\mathbf{f}^N_{(e)} \tag{8.21a,b}$$

$$\mathbf{f}^N_{(e),1} = \frac{\partial X^\alpha}{\partial x^1_{(e)}}\overset{(e)}{\Omega^N_\Delta}\mathbf{F}^\Delta_{,\alpha} \qquad \mathbf{F}^\Delta_{,1} = \frac{\partial x^\alpha_{(e)}}{\partial X^1}\overset{(e)}{\Lambda^\Delta_N}\mathbf{f}^N_{(e),\alpha} \tag{8.21c,d}$$

$$\mathbf{f}^N_{(e),2} = \frac{\partial X^\alpha}{\partial x^2_{(e)}}\overset{(e)}{\Omega^N_\Delta}\mathbf{F}^\Delta_{,\alpha} \qquad \mathbf{F}^\Delta_{,2} = \frac{\partial x^\alpha_{(e)}}{\partial X^2}\overset{(e)}{\Lambda^\Delta_N}\mathbf{f}^N_{(e),\alpha} \tag{8.21e,f}$$

wherein $\alpha = 1, 2$. The final model is then

$$\mathbf{F}(\mathbf{x}) = \sum_{e=1}^{E}\overset{(e)}{\Omega^N_\Delta}\left[\psi^{(e)0}_N(\mathbf{x})\mathbf{F}^\Delta + \frac{\partial X^\alpha}{\partial x^\beta_{(e)}}\psi^{(e)\beta}_N(\mathbf{x})\mathbf{F}^\Delta_{,\alpha}\right] \tag{8.22}$$

with $N = 1, 2, \ldots, N_e$; $\Delta = 1, 2, \ldots, G$; and $\alpha, \beta = 1, 2$.

Example 8.2 Hermite interpolation We comment briefly here on a higher-order representation of the real-valued function $F(X)$ defined on $A \leq X \leq B$ (see Fig. 6.1). After identifying global nodes and finite elements, we isolate a typical finite element for the purpose of describing the local approximations $\bar{f}_{(e)}(x)$. In the present example, we select the functions $\bar{f}_{(e)}(x)$ so as to coincide with $f_{(e)}(x)$ and $df_{(e)}(x)/dx$ at the nodal points which are at the end of each element. The local and global coordinates may be taken to be coincident.

One such approximation that meets this requirement is the cubic polynomial

$$\bar{f}_{(e)}(x) = \psi_1^0(x) f^1_{(e)} + \psi_2^0(x) f^2_{(e)} + \psi_1^1(x) f^1_{(e),1} + \psi_2^1(x) f^2_{(e),1} \tag{8.23}$$

where

$$\psi_1^0(x) = 1 - 3\xi^2 + 2\xi^3 \tag{8.24a}$$

$$\psi_2^0(x) = 3\xi^2 - 2\xi^3 \tag{8.24b}$$

$$\psi_1^1(x) = L_{(e)}(\xi - 2\xi^2 + \xi^3) \tag{8.24c}$$

$$\psi_2^1(x) = -L_{(e)}(\xi^2 - \xi^3) \tag{8.24d}$$

in which

$$\xi = \frac{x_{(e)} - x^1_{(e)}}{L_{(e)}} \qquad L_{(e)} = x^2_{(e)} - x^1_{(e)} \tag{8.25a,b}$$

It is easily verified that

$$\begin{aligned} \bar{f}_{(e)}(x^1) &= f^1_{(e)} \qquad & \bar{f}_{(e)}(x^2) &= f^2_{(e)} \\ \frac{d\bar{f}_{(e)}(x^1)}{dx} &= f^1_{(e),1} \qquad & \frac{d\bar{f}_{(e)}(x^2)}{dx} &= f^2_{(e),1} \end{aligned} \tag{8.26}$$

The functions in (8.24) are the well-known Hermite interpolation polynomials, and (8.23) represents the unique cubic which satisfies (8.26). In addition to (8.26), we can match $d^2f_{(e)}(x)/dx^2$ at each node with a complete polynomial of fifth degree, and in general, we can specify $d^nf_{(e)}(x)/dx^n$ at each node with a complete polynomial of degree $2n + 1$. However, we emphasize that (8.23) is only an example of one of an infinity of functions that satisfy (8.26).

The final finite-element model of $F(X)$ is given by

$$\bar{F}(x) = \sum_e \bar{f}_{(e)}(x) = \sum_e \overset{(e)}{\Omega^N_\Delta}[\psi_N^{(e)0}(x)F^\Delta + \psi_N^{(e)1}(x)F^\Delta_{,1}] \tag{8.27}$$

in which $N = 1, 2$ and $\Delta = 1, 2, \ldots, G$.

9 THEORY OF CONJUGATE APPROXIMATIONS†

Among the most fundamental properties of finite-element approximations is that the interpolation functions $\psi_N^{(e)}(\mathbf{x})$ discussed previously are assumed to provide a basis for a finite-dimensional subspace of the space $\mathscr{H}$ to which the function $\mathbf{F}(\mathbf{X})$ to be approximated belongs. Assuming that an inner product

† This article is based on the paper by Brauchli and Oden [1971]. See also Oden and Brauchli [1971*a*].

is defined on $\mathscr{H}$, the functions $\psi_N^{(e)}(\mathbf{x})$ are generally nonorthogonal, and this suggests the construction of a new system of functions called *conjugate-approximation functions.* In this article, we explore the concept of conjugate-approximation functions in some detail and show that these functions possess certain properties that are fundamental to methods of approximation in general and to the finite-element method in particular.

9.1 CONJUGATE SPACES AND BIORTHOGONAL BASES

We recall from Sec. 7.1 that the class of all linear functionals on a linear space $\mathscr{V}$ is itself a linear space, denoted $\mathscr{V}^*$ and called the conjugate of $\mathscr{V}$. In the case of certain inner-product spaces, the conjugate space can be shown to be isomorphic to the dual space $\overline{\mathscr{V}}$ by using the inner product to construct linear functionals on $\mathscr{V}$.

In particular, let $\mathscr{B}$ denote a Banach space (a complete, normed linear space) and $\mathscr{B}^*$ denote its conjugate. Let both $\mathscr{B}$ and $\mathscr{B}^*$ be separable; i.e., there exists in $\mathscr{B}$ and $\mathscr{B}^*$ countable everywhere dense sets of elements whose linear combinations are dense in $\mathscr{B}$ and $\mathscr{B}^*$. Then the sequences $\{x_n\} \in \mathscr{B}$ and $\{y^n\} \in \mathscr{B}^*$ are said to form a *countable biorthogonal basis* of $\mathscr{B}$ and $\mathscr{B}^*$ if

$$y^i(x_j) = \delta^i_j \tag{9.1}$$

and if, for any $x \in \mathscr{B}$ and $y \in \mathscr{B}^*$, the following decompositions hold:

$$x = \sum_{i=1}^{\infty} a^i x_i \qquad y = \sum_{i=1}^{\infty} a_i y^i \tag{9.2}$$

Here $y^i(x)$ are linear functionals defined on $\mathscr{B}$ and convergence of the series (9.2) is in the norm topology in $\mathscr{B}$ and $\mathscr{B}^*$:

$$\left\| x - \sum_{i=1}^{\infty} a^i x_i \right\| \to 0 \qquad \left\| y - \sum_{i=1}^{\infty} a_i y^i \right\| \to 0 \tag{9.3}$$

It follows from (9.1) and (9.2) that the coefficients a^i and a_i are given by

$$a^i = y^i(x) \qquad a_i = y(x_i) \tag{9.4}$$

An important property of such spaces is that, given a countable biorthogonal basis, it is possible to represent both $\mathscr{B}$ and $\mathscr{B}^*$ as the direct sum of a finite-dimensional and an infinite-dimensional space:

$$x = \sum_{i=1}^{n} a^i x_i + \sum_{i=n+1}^{\infty} a^i x_i \tag{9.5a}$$

$$y = \sum_{i=1}^{n} a_i y^i + \sum_{i=n+1}^{\infty} a_i y^i \tag{9.5b}$$

The decompositions indicated in (9.5) define projection operators $\Pi: \mathscr{B} \to \mathscr{B}'$ and $\Pi^*: \mathscr{B}^* \to \mathscr{B}^{*\prime}$ of $\mathscr{B}$ and $\mathscr{B}^*$ into finite-dimensional subspaces $\mathscr{B}'$ and

$\mathscr{B}^{*\prime}$, respectively; that is,

$$\Pi x = \sum_{i=1}^{n} a^i x_i \qquad \Pi^* y = \sum_{i=1}^{n} a_i y^i \tag{9.6}$$

It follows that each element $x \in \mathscr{B}$ has a unique representation as the sum of elements $\Pi x \in \mathscr{B}'$ and of elements of the form

$$\sum_{i=n+1}^{\infty} a^i x_i$$

belonging to a subspace $\mathscr{B}''$ of $\mathscr{B}$ (that is, $\mathscr{B} = \mathscr{B}' \oplus \mathscr{B}''$). The same applies to every element $y \in \mathscr{B}^*$; that is, $\mathscr{B} = \mathscr{B}^{*\prime} \oplus \mathscr{B}^{*\prime\prime}$ and $y = \Pi^* y + y''$, where $\Pi^* y \in \mathscr{B}^{*\prime}$ and

$$y'' = \sum_{i=n+1}^{\infty} a_i y^i \in \mathscr{B}^{*\prime\prime}$$

The subspaces $\mathscr{B}^{*\prime\prime}$ and $\mathscr{B}^{*\prime}$ are the orthogonal complements of $\mathscr{B}'$ and $\mathscr{B}''$, respectively, and $\mathscr{B}'$, $\mathscr{B}^{*\prime}$ and $\mathscr{B}''$, $\mathscr{B}^{*\prime\prime}$ constitute dual pairs.

9.2 CONJUGATE SUBSPACES

For simplicity, we shall generally confine our attention to Hilbert spaces, i.e., complete, infinite-dimensional inner-product spaces. Hilbert spaces have the convenient property of being self-conjugate in the sense that the topological conjugate of a Hilbert space is isomorphic to the given space. Moreover, a Hilbert space $\mathscr{H}$ is also a Banach space if $\mathscr{H}$ is complete with respect to the norm $\|x\| = \sqrt{\langle x,x \rangle}$, where $\langle x,x \rangle$ is the inner product defined on $\mathscr{H}$.

Consider, then, a Hilbert space $\mathscr{H}$, the elements of which are functions defined on a bounded domain $\mathscr{R}$ of an n-dimensional euclidean space. Elements of $\mathscr{H}$ are denoted $F(\mathbf{X})$, $G(\mathbf{X})$, . . . , where $\mathbf{X}$ is a point in $\mathscr{R}$. The inner product defined on $\mathscr{H} \times \mathscr{H}$ is denoted $\langle F,G \rangle$, and we shall use the natural norm $\|F\| = \sqrt{\langle F,F \rangle}$.

In many applications we are concerned with the special case in which $\mathscr{H}$ is the space of square integrable functions on $\mathscr{R}$. Then the inner product of two elements $F(\mathbf{X})$ and $G(\mathbf{X}) \in \mathscr{H}$ is given by

$$\langle F,G \rangle = \int_{\mathscr{R}} F(\mathbf{X}) G(\mathbf{X}) \, d\mathscr{R} \tag{9.7}$$

where integration in the Lebesque sense is implied; $F(\mathbf{X})$ and $G(\mathbf{X})$ are orthogonal if $\langle F,G \rangle = 0$, and

$$\langle F,F \rangle < +\infty \tag{9.8}$$

for every $F(\mathbf{X}) \in \mathscr{H}$.

We now decompose $\mathscr{H}$ into the sum of two subspaces Φ and $\mathscr{H}'$, the subspace Φ being of finite dimension G. To define this decomposition, we select from $\mathscr{H}$ a set of G linearly independent functions, denoted $\{\Phi_\Delta(\mathbf{X})\}$,† $\Delta = 1, 2, \ldots, G$, which forms a basis of the subspace Φ. Then every $F(\mathbf{X}) \in \mathscr{H}$ can be represented as the sum of an element of Φ and an element of $\mathscr{H}'$; that is,‡

$$F(\mathbf{X}) = \bar{F}(\mathbf{X}) + \hat{F}(\mathbf{X}) \tag{9.9}$$

where $\bar{F}(\mathbf{X}) \in \Phi$ and $\hat{F}(\mathbf{X}) \in \mathscr{H}'$. Since $\{\Phi_\Delta(\mathbf{X})\}$ provides a basis for Φ, every $\bar{F}(\mathbf{X}) \in \Phi$ can be represented in a unique manner as a linear combination of the functions $\Phi_\Delta(\mathbf{X})$:

$$\bar{F}(\mathbf{X}) = F^\Delta \Phi_\Delta(\mathbf{X}) \tag{9.10}$$

where the repeated index Δ is summed from 1 to G. The G coefficients F^Δ are called the *components* of $\bar{F}(\mathbf{X})$ with respect to the basis $\{\Phi_\Delta(\mathbf{X})\}$.

By identifying the set of functions $\Phi_\Delta(\mathbf{X})$, we have effectively defined a projection operator Π which maps $\mathscr{H}$ into a finite-dimensional subspace, $\Pi: \mathscr{H} \to \Phi$. Thus, we may write

$$\bar{F}(\mathbf{X}) = \Pi F(\mathbf{X}) \tag{9.11}$$

It is important to realize that the functions $\Phi_\Delta(\mathbf{X})$ are not necessarily orthogonal. Indeed, we shall introduce the symmetric $G \times G$ matrix

$$C_{\Delta\Gamma} \equiv \langle \Phi_\Delta, \Phi_\Gamma \rangle \qquad \Delta, \Gamma = 1, 2, \ldots, G \tag{9.12}$$

which is referred to as the *fundamental matrix* § of the subspace Φ. Then the $\Phi_\Delta(\mathbf{X})$ are orthogonal if and only if

$$C_{\Delta\Gamma} = c_{(\Delta)} \delta_{\Delta\Gamma} \tag{9.13}$$

where $\delta_{\Delta\Gamma}$ is the Kronecker delta; that is, the $\Phi_\Delta(\mathbf{X})$ are orthogonal if $C_{\Delta\Gamma}$ is diagonal.

Since the set $\{\Phi_\Delta(\mathbf{X})\}$ is linearly independent and is prescribed in the definition of Φ, the matrix $C_{\Delta\Gamma}$ is regular and can be generated directly by

† The use of the symbol Δ as an index in the same manner as that used previously for the identification of global nodes is not accidental; indeed, we show subsequently that the functions $\Phi_\Delta(\mathbf{X})$ provide a basis for finite-element approximations over $\bar{\mathscr{R}}$. See Sec. 9.6.

‡ In more general spaces, the approach followed here does not rely on the separability of $\mathscr{H}$. If $\mathscr{H}$ is separable, we often seek an *infinite* set of basis functions $\Phi_\Delta(\mathbf{X})$ so as to set

$$\hat{F}(X) = \sum_{\Delta=G+1}^{\infty} F^\Delta \Phi_\Delta(\mathbf{X})$$

Here we simply identify a subspace Φ by constructing a set of linearly independent functions. The method used to construct such a set of basis functions for finite-element approximations was discussed in Sec. 7.4 and is examined more fully in Sec. 9.6.

§ The matrix $C_{\Delta\Gamma}$ is sometimes referred to as the Gram matrix of the basis functions $\{\Phi_\Delta\}$. If $\mathscr{H}$ is complex, the transpose of $C_{\Delta\Gamma}$ is its complex conjugate.

means of (9.12). We denote the inverse of $C_{\Delta\Gamma}$ by $C^{\Delta\Gamma}$:

$$(C_{\Delta\Gamma})^{-1} \equiv C^{\Delta\Gamma} \tag{9.14}$$

or

$$C_{\Delta\Lambda}C^{\Lambda\Gamma} = \delta_{\Delta}^{\Gamma} \tag{9.15}$$

where $\Delta,\Lambda,\Gamma = 1, 2, \ldots, G$. The matrices $C_{\Delta\Gamma}$ and $C^{\Delta\Gamma}$ play an important role in the theory of conjugate approximations.

We can now proceed to define the conjugate space Φ^* of Φ by introducing a finite sequence $\{\Phi^{\Delta}\}$ of linear functionals on Φ. Toward this end, let $\{\Phi^{\Delta}(\mathbf{X})\}$ denote another set of G linearly independent functions on $\mathscr{R}$ which are required to have the property

$$\langle \Phi^{\Delta},\Phi_{\Gamma} \rangle = \delta_{\Gamma}^{\Delta} \tag{9.16}$$

We recall that the conjugate space $\mathscr{H}^*$ of $\mathscr{H}$ is the space of all linear functionals defined on $\mathscr{H}$. Since (9.16) defines a set of G linear functionals on $\mathscr{H}$ and since the functions $\Phi^{\Delta}(\mathbf{X})$ are linearly independent, it follows that $\{\Phi^{\Delta}(\mathbf{X})\}$ defines a projection Π^* of $\mathscr{H}^*$ into a finite-dimensional subspace Φ^* which is conjugate to Φ. According to (9.16), the two sets $\{\Phi_{\Delta}(\mathbf{X})\}$ and $\{\Phi^{\Delta}(\mathbf{X})\}$ form a countable biorthogonal basis of Φ. We shall refer to the set $\{\Phi^{\Delta}(\mathbf{X})\}$ as the *conjugate basis* of Φ.

Now (9.16) describes a property that the functions $\Phi^{\Delta}(\mathbf{X})$ must possess in order to qualify as conjugate basis functions, but this property is not sufficiently restrictive to define a unique collection of independent functions $\Phi^{\Delta}(\mathbf{X})$. To overcome this difficulty, we shall construct the functions $\Phi^{\Delta}(\mathbf{X})$ by means of the formula

$$\Phi^{\Delta}(\mathbf{X}) = C^{\Delta\Gamma}\Phi_{\Gamma}(\mathbf{X}) \tag{9.17}$$

where $C^{\Delta\Gamma}$ is the inverse of the fundamental matrix as defined in (9.14). Then also

$$\Phi_{\Delta}(\mathbf{X}) = C_{\Delta\Gamma}\Phi^{\Gamma}(\mathbf{X}) \tag{9.18}$$

It is easily shown that the G functions $\Phi^{\Delta}(\mathbf{X})$ of (9.17) are linearly independent, for since

$$K^{\Delta}\Phi_{\Delta}(\mathbf{X}) = 0 \Rightarrow K^{\Delta} = 0 \qquad \Delta = 1, 2, \ldots, G$$

then, if

$$K_{\Delta}\Phi^{\Delta}(\mathbf{X}) = K_{\Delta}C^{\Delta\Gamma}\Phi_{\Gamma}(\mathbf{X}) = 0$$

it is clear that $K_{\Delta}C^{\Delta\Gamma} = K^{\Gamma} = 0$. But this implies that $K_{\Delta} = 0$ because $C^{\Delta\Gamma}$ is positive definite; thus $K_{\Delta}\Phi^{\Delta}(\mathbf{X}) = 0 \Rightarrow K_{\Delta} = 0$ ($\Delta = 1, 2, \ldots, G$), and therefore, the functions $\Phi^{\Delta}(\mathbf{X})$ are linearly independent. Moreover, taking the inner product of (9.17) with $\Phi_{\Gamma}(\mathbf{X})$, we find that

$$\langle \Phi^{\Delta},\Phi_{\Gamma} \rangle = \langle C^{\Delta\Lambda}\Phi_{\Lambda},\Phi_{\Gamma} \rangle = C^{\Delta\Lambda}\langle \Phi_{\Lambda},\Phi_{\Gamma} \rangle = C^{\Delta\Lambda}C_{\Lambda\Gamma} \tag{9.19}$$

or, in view of (9.15),

$$\langle \Phi^\Delta, \Phi_\Gamma \rangle = \delta^\Delta_\Gamma \tag{9.20}$$

which is precisely the condition (9.16). Therefore, (9.17) defines a unique biorthogonal basis for the subspaces Φ and Φ^*.

Two important properties of Φ^* follow immediately from (9.17). First, notice that by virtue of (9.17) each member of the set $\{\Phi^\Delta(\mathbf{X})\}$ is a linear combination of the original basis functions $\Phi_\Delta(\mathbf{X})$. Therefore, the functions $\Phi^\Delta(\mathbf{X})$ of Φ^* belong to Φ, which means that Φ and its conjugate Φ^* coincide; that is, Φ is *self-conjugate*. Consequently, every $\bar{F}(\mathbf{X}) \in \Phi$ can be represented as a linear combination of either the functions $\Phi_\Delta(\mathbf{X})$ or the functions $\Phi^\Delta(\mathbf{X})$. Indeed, we may write, instead of (9.10),

$$\bar{F}(\mathbf{X}) = F^\Delta \Phi_\Delta(\mathbf{X}) = F_\Delta \Phi^\Delta(\mathbf{X}) \tag{9.21}$$

where the coefficients F_Δ may be referred to as the components of $\bar{F}(\mathbf{X})$ with respect to $\Phi^\Delta(\mathbf{X})$. The form of (9.21) suggests that F_Δ and F^Δ be referred to as the covariant and contravariant components of $\bar{F}(\mathbf{X})$, respectively. A second important property of Φ^* follows from (9.17). Observe that

$$\langle \Phi^\Delta, \Phi^\Gamma \rangle = C^{\Delta\Lambda} \langle \Phi_\Lambda, \Phi^\Gamma \rangle = C^{\Delta\Lambda} \delta^\Gamma_\Lambda \tag{9.22}$$

or

$$\langle \Phi^\Delta, \Phi^\Gamma \rangle = C^{\Delta\Gamma} \tag{9.23}$$

Therefore, *the inverse of the fundamental matrix* $C_{\Delta\Gamma}$ *of the subspace* Φ *is the fundamental matrix of the conjugate space* Φ^*.

Returning to (9.21), we observe that†

$$\langle \bar{F}, \Phi^\Delta \rangle = \langle F^\Gamma \Phi_\Gamma, \Phi^\Delta \rangle = F^\Gamma \langle \Phi_\Gamma, \Phi^\Delta \rangle = F^\Delta \tag{9.24}$$

$$\langle \bar{F}, \Phi_\Delta \rangle = \langle F_\Gamma \Phi^\Gamma, \Phi_\Delta \rangle = F_\Gamma \langle \Phi^\Gamma, \Phi_\Delta \rangle = F_\Delta \tag{9.25}$$

Moreover, in view of (9.17) and (9.18), the components F^Δ and F_Δ are related to each other according to

$$F^\Delta = C^{\Delta\Gamma} F_\Gamma \tag{9.26}$$

$$F_\Delta = C_{\Delta\Gamma} F^\Gamma \tag{9.27}$$

so that the fundamental matrices provide a means for "raising and lowering" indices for components of functions in Φ.

Inner products and norms Let $\bar{F}(\mathbf{X})$ and $\bar{G}(\mathbf{X})$ denote two functions belonging to the subspace Φ:

$$\bar{F}(\mathbf{X}) = F^\Delta \Phi_\Delta(\mathbf{X}) = F_\Delta \Phi^\Delta(\mathbf{X}) \quad \text{and} \quad \bar{G}(\mathbf{X}) = G^\Delta \Phi_\Delta(\mathbf{X}) = G_\Delta \Phi^\Delta(\mathbf{X}) \tag{9.28}$$

† Alternatively, if $F(\mathbf{X})$ is known, a better choice for the components of $\bar{F}(\mathbf{X})$ would be $F^\Delta = \langle F, \Phi^\Delta \rangle$ instead of (9.24) and $F_\Delta = \langle F, \Phi_\Delta \rangle$ instead of (9.25). See Sec. 9.3.

Then the inner product of $\bar{F}$ and $\bar{G}$ is given by

$$\langle \bar{F},\bar{G}\rangle = \langle F^{\Delta}\Phi_{\Delta},G^{\Gamma}\Phi_{\Gamma}\rangle = \langle F^{\Delta}\Phi_{\Delta},G_{\Gamma}\Phi^{\Gamma}\rangle$$
$$= \langle F_{\Delta}\Phi^{\Delta},G^{\Gamma}\Phi_{\Gamma}\rangle = \langle F_{\Delta}\Phi^{\Delta},G_{\Gamma}\Phi^{\Gamma}\rangle \tag{9.29}$$

or

$$\langle \bar{F},\bar{G}\rangle = C_{\Delta\Gamma}F^{\Delta}G^{\Gamma} = F^{\Delta}G_{\Delta} = F_{\Delta}G^{\Delta} = C^{\Delta\Gamma}F_{\Delta}G_{\Gamma} \tag{9.30}$$

In particular,

$$\|\bar{F}\|^2 = \langle \bar{F},\bar{F}\rangle = C_{\Delta\Gamma}F^{\Delta}F^{\Gamma} = F^{\Delta}F_{\Delta} = C^{\Delta\Gamma}F_{\Delta}F_{\Gamma} \geq 0 \tag{9.31}$$

which demonstrates the positive-definite character of $C_{\Delta\Gamma}$ mentioned previously.

Affine transformations Let $\{\Upsilon_{\Delta}(\mathbf{X})\}$ be another set of G linearly independent functions. If this set also belongs to Φ, it is always possible to find a matrix A_{Δ}^{Γ} which describes an affine transformation of the set $\{\Upsilon_{\Delta}\}$ into $\{\Phi_{\Delta}\}$. Moreover, if A_{Δ}^{Γ} is nonsingular, its inverse $\hat{A}_{\Delta}^{\Gamma}$ transforms $\{\Phi_{\Delta}\}$ into $\{\Upsilon_{\Delta}\}$:

$$\Phi_{\Delta}(\mathbf{X}) = A_{\Delta}^{\Gamma}\Upsilon_{\Gamma}(\mathbf{X}) \qquad \Upsilon_{\Delta}(\mathbf{X}) = \hat{A}_{\Gamma}^{\Delta}\Phi_{\Gamma}(\mathbf{X}) \tag{9.32}$$

Similarly, it can be shown that

$$\Phi^{\Delta}(\mathbf{X}) = \hat{A}_{\Gamma}^{\Delta}\Upsilon^{\Gamma}(\mathbf{X}) \qquad \Upsilon^{\Delta}(\mathbf{X}) = A_{\Delta}^{\Gamma}\Phi^{\Gamma}(\mathbf{X}) \tag{9.33}$$

Let F_{*}^{Δ} and F_{Δ}^{*} denote the components of an element $\bar{F}(\mathbf{X})$ of Φ with respect to the new biorthogonal basis $\{\Upsilon_{\Delta}\}$, $\{\Upsilon^{\Delta}\}$. Then, if $\bar{F}(\mathbf{X})$ is to be invariant under transformations of the type in (9.32) and (9.33), it is easily shown that

$$F_{*}^{\Delta} = A_{\Gamma}^{\Delta}F^{\Gamma} \qquad F_{\Delta}^{*} = \hat{A}_{\Delta}^{\Gamma}F_{\Gamma} \tag{9.34}$$

Then

$$\bar{F}(\mathbf{X}) = F^{\Delta}\Phi_{\Delta}(\mathbf{X}) = F_{\Delta}\Phi^{\Delta}(\mathbf{X}) = F_{*}^{\Delta}\Upsilon_{\Delta}(\mathbf{X}) = F_{\Delta}^{*}\Upsilon^{\Delta}(\mathbf{X}) \tag{9.35}$$

Also note that

$$C_{\Delta\Gamma} = A_{\Delta}^{\Lambda}\bar{C}_{\Lambda\Omega}A_{\Gamma}^{\Omega} \qquad \bar{C}_{\Delta\Gamma} = \hat{A}_{\Delta}^{\Lambda}C_{\Lambda\Omega}\hat{A}_{\Gamma}^{\Omega} \tag{9.36}$$
$$C^{\Delta\Gamma} = \hat{A}_{\Lambda}^{\Delta}\bar{C}^{\Lambda\Omega}\hat{A}_{\Omega}^{\Gamma} \qquad \bar{C}^{\Delta\Gamma} = A_{\Lambda}^{\Delta}C^{\Lambda\Omega}A_{\Omega}^{\Gamma} \tag{9.37}$$

where

$$\langle \Upsilon_{\Delta},\Upsilon_{\Gamma}\rangle = \bar{C}_{\Delta\Gamma} \qquad \langle \Upsilon^{\Delta},\Upsilon^{\Gamma}\rangle = \bar{C}^{\Delta\Gamma} \qquad \langle \Upsilon_{\Delta},\Upsilon^{\Gamma}\rangle = \delta_{\Delta}^{\Gamma} \tag{9.38}$$

We remark that (9.32) and (9.33) describe rather special changes in basis because of the assumption that the functions $\Upsilon_{\Delta}(\mathbf{X})$ belong to the subspace Φ. It is not difficult to introduce a set of basis functions which do not belong to Φ, in which case transformations of the type in (9.32) and (9.33) do not exist.

Summary By way of a brief summary of the ideas established thus far, we reiterate here some of the basic notions. First of all, to approximate a function $F(\mathbf{X})$, we define a finite-dimensional subspace Φ of the given Hilbert space $\mathscr{H}$ to which $F(\mathbf{X})$ belongs by identifying a collection of G linearly independent elements $\Phi_\Delta(\mathbf{X})$ of $\mathscr{H}$. The projection of an element $F(\mathbf{X}) \in \mathscr{H}$ in Φ is denoted $\bar{F}(\mathbf{X})$, and $\bar{F}(\mathbf{X})$ can be expressed as a linear combination of the basis functions $\Phi_\Delta(\mathbf{X})$ [that is, $\bar{F}(\mathbf{X}) = F^\Delta \Phi_\Delta(\mathbf{X})$; $\Delta = 1, 2, \ldots, G$]. We then proceed to construct another set of conjugate basis functions $\Phi^\Delta(\mathbf{X})$ by means of the formula

$$\Phi^\Delta(\mathbf{X}) = C^{\Delta\Gamma}\Phi_\Gamma(\mathbf{X}) \tag{9.39}$$

where $C^{\Delta\Gamma}$ is the inverse of a regular, positive-definite matrix $C_{\Delta\Gamma}$ defined by

$$C_{\Delta\Gamma} = \langle \Phi_\Delta, \Phi_\Gamma \rangle \tag{9.40}$$

where $\langle\ ,\ \rangle$ is the inner product defined on $\mathscr{H}$. Clearly,

$$\Phi_\Delta(\mathbf{X}) = C_{\Delta\Gamma}\Phi^\Gamma(\mathbf{X}) \tag{9.41}$$

Since (9.39) uniquely defines a set of conjugate basis functions satisfying

$$\langle \Phi^\Delta, \Phi_\Gamma \rangle = \delta^\Delta_\Gamma \tag{9.42}$$

the two sets of functions $\{\Phi_\Delta(\mathbf{X})\}$ and $\{\Phi^\Delta(\mathbf{X})\}$ provide a countable biorthogonal basis for Φ. Moreover, since (9.42) effectively defines a set of G linear functionals on Φ, the functions $\{\Phi^\Delta(\mathbf{X})\}$ provide a basis for a G-dimensional space Φ^* which is conjugate to Φ. Thus in Φ^*, elements are expressed as linear combinations of $\Phi^\Delta(\mathbf{X})$ [that is, $\bar{F}^*(\mathbf{X}) = F_\Delta \Phi^\Delta(\mathbf{X})$; $\Delta = 1, 2, \ldots, G$]. However, (9.17) shows that each $\Phi^\Delta(\mathbf{X})$ is a linear combination of the original basis functions $\Phi_\Delta(\mathbf{X})$; thus, $\Phi^\Delta(\mathbf{X}) \in \Phi$, which means that Φ and Φ^* coincide. It follows that every $\bar{F}(\mathbf{X}) \in \Phi$ can be expressed in a unique way as a linear combination of either the functions $\Phi_\Delta(\mathbf{X})$ or the functions $\Phi^\Delta(\mathbf{X})$. Indeed,

$$\bar{F}(\mathbf{X}) = F^\Delta \Phi_\Delta(\mathbf{X}) = F_\Delta \Phi^\Delta(\mathbf{X}) \tag{9.43}$$

where

$$F^\Delta = \langle \bar{F}, \Phi^\Delta \rangle \qquad F_\Delta = \langle \bar{F}, \Phi_\Delta \rangle \tag{9.44}$$

and

$$F_\Delta = C_{\Delta\Gamma}F^\Gamma \qquad F^\Delta = C^{\Delta\Gamma}F_\Gamma \tag{9.45}$$

The inner product of two functions and the norm of a function in Φ are given by (9.30) and (9.31), respectively. Finally, affine transformations of a biorthogonal basis in the space Φ are described by (9.32) and (9.33).

9.3 BEST APPROXIMATION

We shall now investigate the problem of determining, for a given element $F(\mathbf{X}) \in \mathscr{H}$ and a given projection defined by a prescribed set of functions

$\Phi_\Delta(\mathbf{X})$, the best approximation of $F(\mathbf{X})$ in Φ. By "best approximation," we shall mean the element in Φ closest to $F(\mathbf{X})$ in the sense of a natural metric $d[F(\mathbf{X}),G(\mathbf{X})] = \|F(\mathbf{X}) - G(\mathbf{X})\|$ defined on $\mathscr{H}$.

Let Λ^Δ denote an arbitrary collection of G quantities such that $\Lambda^\Delta\Phi_\Delta(\mathbf{X}) \in \Phi$. We are concerned with determining the Λ^Δ so as to minimize the functional

$$J(\Lambda^\Delta) = \langle F - \Lambda^\Delta\Phi_\Delta, F - \Lambda^\Gamma\Phi_\Gamma\rangle \tag{9.46}$$

which, since $J(\Lambda^\Delta) = d^2[F(\mathbf{X}),\Lambda^\Delta\Phi_\Delta(\mathbf{X})]$, is a measure of the distance between $F(\mathbf{X})$ and $\Lambda^\Delta\Phi_\Delta(\mathbf{X})$. Recalling (9.9) and (9.25), we note that

$$\langle F,\Phi_\Delta\rangle = \langle \bar{F} + \hat{F}, \Phi_\Delta\rangle = F_\Delta + \hat{F}_\Delta \tag{9.47}$$

where we have used the notation $\langle \hat{F},\Phi_\Delta\rangle = \hat{F}_\Delta$. Using this result and (9.30), we see that (9.46) can be rewritten in the form

$$J(\Lambda^\Delta) = C_{\Delta\Gamma}F^\Delta F^\Gamma + \|\hat{F}\|^2 - 2F^\Delta\hat{F}_\Delta - 2\Lambda^\Delta(F_\Delta + \hat{F}_\Delta) + C_{\Delta\Gamma}\Lambda^\Delta\Lambda^\Gamma \tag{9.48}$$

To find Λ^Δ so as to minimize $J(\Lambda^\Delta)$, we set

$$\frac{\partial J(\Lambda^\Gamma)}{\partial \Lambda^\Delta} = 0 \tag{9.49}$$

and solve the resulting system of linear equations to obtain

$$\Lambda^\Delta_{\min} = F^\Delta + C^{\Delta\Gamma}\hat{F}_\Gamma \tag{9.50}$$

To interpret this result, notice that, like (9.47),

$$\langle F,\Phi^\Delta\rangle = \langle \bar{F} + \hat{F}, \Phi^\Delta\rangle = F^\Delta + C^{\Delta\Gamma}\hat{F}_\Gamma = \Lambda^\Delta_{\min} \tag{9.51}$$

Therefore, *the best approximation of a given function* $F(\mathbf{X})$ *in the subspace* Φ *is the function* $\bar{F}(\mathbf{X}) \in \Phi$ *whose components with respect to the basis* $\{\Phi_\Delta\}$ *are equal to the inner products of* $F(\mathbf{X})$ *with the conjugate basis functions* $\Phi^\Delta(\mathbf{X})$.

We remark that in applications the function $F(\mathbf{X})$ is generally unknown, and the best that we can do is to use the components F^Δ and F_Δ defined by (9.24) and (9.25). However, if $F(\mathbf{X})$ is given, we can determine immediately the best set of coefficients for its approximation in Φ by means of (9.51). If $F^\Delta_{(b)}$ and $F^{(b)}_\Delta$ denote the best coefficients, we see that

$$F^\Delta_{(b)} = F^\Delta + \langle \hat{F},\Phi^\Delta\rangle \qquad F^{(b)}_\Delta = F_\Delta + \langle \hat{F},\Phi_\Delta\rangle \tag{9.52}$$

Another interesting property of the subspaces Φ and Φ^* follows if it is possible to obtain the best coefficients $F^\Delta_{(b)}$ and $F^{(b)}_\Delta$ as defined in (9.52). In this case, we use, instead of (9.43),

$$\bar{F}(\mathbf{X}) = F^\Delta_{(b)}\Phi_\Delta(\mathbf{X}) = F^{(b)}_\Delta\Phi^\Delta(\mathbf{X}) \tag{9.53}$$

Then†

$$\langle \hat{F},\Phi_\Delta\rangle = \langle F - \bar{F}, \Phi_\Delta\rangle = 0$$
$$\langle \hat{F},\Phi^\Delta\rangle = \langle F - \bar{F}, \Phi^\Delta\rangle = 0 \tag{9.54}$$

that is, the space Φ (or Φ^*) is the orthogonal complement of $\mathscr{H}'$ (or $\mathscr{H}^{*\prime}$). Furthermore, suppose that (9.53) and (9.54) hold; then a measure E of the error induced by representing $F(\mathbf{X})$ by its projection $\bar{F}(\mathbf{X}) = \Pi F(\mathbf{X})$ is

$$E = \|F - \bar{F}\|^2 = \|\hat{F}\|^2 \tag{9.55}$$

since $\bar{F}$ and $\hat{F}$ are now orthogonal. Since, in general, $\|\hat{F}\|$ may be unbounded, E is small only if $F(\mathbf{X})$ is close to the subspace Φ.

Generalization It is not difficult to generalize the above results to dual spaces which are not self-conjugate. Briefly, let $\mathscr{F}$ and $\mathscr{G}$ denote dual linear spaces, the elements of which are functions denoted f and g, respectively, and let $\langle f,g\rangle$ denote the inner product on $\mathscr{F}$ and $\mathscr{G}$. Further, let Φ denote a finite-dimensional subspace of $\mathscr{F}$ and Ψ its dual in $\mathscr{G}$. The spaces Φ and Ψ are spanned by the biorthogonal basis $\Phi_\Delta \in \Phi$, $\Psi^\Delta \in \Psi$ such that

$$\langle \Phi_\Delta,\Psi^\Gamma\rangle = \delta_\Delta^\Gamma \qquad \Delta,\Gamma = 1, 2, \ldots, G \tag{9.56}$$

Then, if F and G are the projections of f and g into Φ and Ψ,

$$F = F^\Delta\Phi_\Delta \qquad G = G_\Delta\Psi^\Delta \tag{9.57}$$

where the repeated indices are summed from one to G and

$$F^\Delta = \langle f,\Psi^\Delta\rangle \qquad G_\Delta = \langle g,\Phi_\Delta\rangle \tag{9.58}$$

Let $\Lambda^\Delta\Phi_\Delta$ and $M_\Delta\Psi^\Delta$ denote arbitrary elements in Φ and Ψ, and consider the bilinear functional

$$J(\Lambda^\Delta,M_\Gamma) = \langle f - \Lambda^\Delta\Phi_\Delta, g - M_\Gamma\Psi^\Gamma\rangle \tag{9.59}$$

which, upon simplification, can be written

$$J(\Lambda^\Delta,M_\Gamma) = \langle f - F, g - G\rangle + \langle F^\Delta - \Lambda^\Delta\rangle\langle G_\Gamma - M_\Gamma\rangle \tag{9.60}$$

Obviously, either of the projections defined by (9.57) makes the second term in (9.60) vanish. If we now introduce a linear, positive-definite, regular mapping κ from $\mathscr{F}$ to $\mathscr{G}$, such that $\kappa(\Phi) \subset \Psi$, then κ defines a matrix $\kappa_{\Gamma\Delta} = \langle \Phi_\Delta,\kappa(\Phi_\Gamma)\rangle$ such that

$$\kappa_{\Gamma\Delta}\Lambda^\Delta = M_\Gamma \tag{9.61}$$

† The selection of $\bar{F}$ such that $F - \bar{F}$ is orthogonal to Φ is motivated by the classical projection theorem for Hilbert spaces. See Theorem 10.8.

Then $J(\Lambda^\Delta, M_\Gamma) = J(\Lambda^\Delta, \kappa_{\Gamma\Omega}\Lambda^\Omega)$ becomes a positive-definite quadratic form.

$$J(\Lambda^\Delta) = \langle f - F, \kappa(f - F)\rangle + \kappa_{\Delta\Gamma}(\Lambda^\Delta - F^\Delta)(\Lambda^\Gamma - F^\Gamma) \tag{9.62}$$

Consequently, (9.62) assumes a minimum value when Λ^Δ is chosen according to (9.58).

9.4 SOME PROPERTIES OF CONJUGATE APPROXIMATIONS

Suppose that we enlarge the space $\mathscr{H}$ so that it includes distributions such as the Dirac delta function† $\delta(\mathbf{X} - \mathbf{A})$, $\mathbf{A} \in \mathscr{R}$, defined by

$$\delta(\mathbf{X} - \mathbf{A}) = \begin{cases} 0 & \mathbf{X} \neq \mathbf{A} \\ \infty & \mathbf{X} = \mathbf{A} \end{cases} \qquad \int_{\mathscr{R}} \delta(\mathbf{X} - \mathbf{A})\, d\mathscr{R} = 1 \tag{9.63}$$

Here $\mathbf{A}$ is a specified point in the domain $\mathscr{R}$.

Let $\Delta(\mathbf{X} - \mathbf{A})$ denote the projection of $\delta(\mathbf{X} - \mathbf{A})$ on the subspace Φ:

$$\Delta(\mathbf{X} - \mathbf{A}) = \Pi\delta(\mathbf{X} - \mathbf{A}) = \Delta^\Delta\Phi_\Delta(\mathbf{X}) = \Delta_\Delta\Phi^\Delta(\mathbf{X}) \tag{9.64}$$

where, according to (9.24) and (9.25) [or (9.44)],

$$\Delta^\Delta = \langle\delta(\mathbf{X} - \mathbf{A}), \Phi^\Delta(\mathbf{X})\rangle = \Phi^\Delta(\mathbf{A}) \tag{9.65}$$

$$\Delta_\Delta = \langle\delta(\mathbf{X} - \mathbf{A}), \Phi_\Delta(\mathbf{X})\rangle = \Phi_\Delta(\mathbf{A}) \tag{9.66}$$

Hence

$$\Delta(\mathbf{X} - \mathbf{A}) = \Phi^\Delta(\mathbf{A})\Phi_\Delta(\mathbf{X}) = \Phi_\Delta(\mathbf{A})\Phi^\Delta(\mathbf{X}) \tag{9.67}$$

Thus, *the values of the functions* $\Phi^\Delta(\mathbf{X})$ [or $\Phi_\Delta(\mathbf{X})$] *at an arbitrary point* $\mathbf{A} \in \mathscr{R}$ *are the components* Δ^Δ (or Δ_Δ) *of the projection of the delta function* $\delta(\mathbf{X} - \mathbf{A})$ *in* Φ.

We observe that while the function $\delta(\mathbf{X} - \mathbf{A})$ assumes a nonzero value only at $\mathbf{A}$, the projection $\Delta(\mathbf{X} - \mathbf{A})$ may take on nonzero values almost everywhere in the domain of functions in Φ. However, the essential properties of $\delta(\mathbf{X} - \mathbf{A})$ are preserved under the projection $\Pi: \mathscr{H} \to \Phi$. For example, note that

$$\begin{aligned} \langle\Delta(\mathbf{X} - \mathbf{A}), \bar{F}(\mathbf{X})\rangle &= \langle\Phi_\Delta(\mathbf{A})\Phi^\Delta(\mathbf{X}), F^\Gamma\Phi_\Gamma(\mathbf{X})\rangle \\ &= \Phi_\Delta(\mathbf{A})F^\Gamma\langle\Phi^\Delta, \Phi_\Gamma\rangle \\ &= F^\Delta\Phi_\Delta(\mathbf{A}) \\ &= \bar{F}(\mathbf{A}) \end{aligned} \tag{9.68}$$

Also, if $\bar{F}(\mathbf{X}) = 1$ is an element of Φ, then

$$\langle\Delta(\mathbf{X} - \mathbf{A}), 1\rangle = 1 \tag{9.69}$$

† We recognize that $\delta(\mathbf{X} - \mathbf{A})$ is not square integrable. Thus, our conclusions concerning the best approximation of elements of $\mathscr{H}$ do not hold. However, the projection $\Pi\delta(\mathbf{X} - \mathbf{A})$ is well defined.

To demonstrate another property of the functions $\Phi_\Delta(\mathbf{X})$, suppose that we identify in $\mathscr{R}$ a finite number G of *nodal points* $\mathbf{X}^\Delta$ ($\Delta = 1, 2, \ldots, G$). Further, suppose the function $\Phi_\Delta(\mathbf{X})$ has a value of unity at node $\mathbf{X}^\Gamma$ but is zero at all other nodes:

$$\Phi_\Delta(\mathbf{X}^\Gamma) = \delta_\Delta^\Gamma \tag{9.70}$$

Then the projection of the delta function in (9.64) becomes

$$\Delta(\mathbf{X} - \mathbf{X}^\Gamma) = \Phi_\Delta(\mathbf{X}^\Gamma)\Phi^\Delta(\mathbf{X}) = \Phi^\Gamma(\mathbf{X}) \tag{9.71}$$

We shall refer to basis functions $\Phi_\Delta(\mathbf{X})$ with property (9.70) as being *normalized* with respect to the G nodes $\mathbf{X}^\Delta$. In Sec. 9.6 we discuss the procedure for computing such normalized basis functions for finite-element approximations. Equation (9.71) shows that the conjugate functions $\Phi^\Delta(\mathbf{X})$ represent the projection of the delta function $\delta(\mathbf{X} - \mathbf{X}^\Delta)$ at node $\mathbf{X}^\Delta$.

Moments and volumes We remark that in certain applications in which $\mathscr{H}$ is a Hilbert space with an inner product given by (9.7), it is convenient to introduce moments M_Δ and M^Δ defined by

$$M_\Delta = \int_{\mathscr{R}} \Phi_\Delta(\mathbf{X})\, d\mathscr{R} \tag{9.72}$$

$$M^\Delta = \int_{\mathscr{R}} \Phi^\Delta(\mathbf{X})\, d\mathscr{R} = C^{\Delta\Gamma} M_\Gamma \tag{9.73}$$

Then

$$\int_{\mathscr{R}} \bar{F}(\mathbf{X})\, d\mathscr{R} = F^\Delta M_\Delta = F_\Delta M^\Delta \tag{9.74}$$

Assuming that $\bar{F}(\mathbf{X}) = 1 \in \Phi$, then in addition to being normalized with respect to G nodal points by means of (9.70), the functions $\Phi_\Delta(\mathbf{X})$ satisfy the condition

$$\sum_{\Delta=1}^{G} \Phi_\Delta(\mathbf{X}) = 1 \tag{9.75}$$

Then the moments M^Δ of (9.73) become

$$M^\Delta = C^{\Delta\Gamma} M_\Gamma = C^{\Delta\Gamma} \int_{\mathscr{R}} \Phi_\Gamma(\mathbf{X}) \sum_{\Lambda=1}^{G} \Phi_\Lambda(\mathbf{X})\, d\mathscr{R} \tag{9.76}$$

or

$$M^\Delta = C^{\Delta\Gamma} \sum_{\Lambda=1}^{G} \langle \Phi_\Gamma, \Phi_\Lambda \rangle \tag{9.77}$$

But $\langle \Phi_\Gamma, \Phi_\Lambda \rangle = C_{\Gamma\Lambda}$, so that

$$M^\Delta = \sum_{\Lambda=1}^{G} C^{\Delta\Gamma} C_{\Gamma\Lambda} = \sum_{\Lambda=1}^{G} \delta_\Lambda^\Delta = 1 \tag{9.78}$$

Similarly,

$$M_\Delta = \sum_{\Lambda=1}^{G} \langle \Phi_\Delta, \Phi_\Lambda \rangle = \sum_{\Lambda=1}^{G} C_{\Delta\Lambda} \tag{9.79}$$

Observe that

$$\sum_{\Delta=1}^{G} M_\Delta = \int_{\mathscr{R}} \sum_{\Delta=1}^{G} \Phi_\Delta(\mathbf{X})\, d\mathscr{R} = \int_{\mathscr{R}} d\mathscr{R} = \mathscr{V} \tag{9.80}$$

where $\mathscr{V}$ is the volume of $\mathscr{R}$. Note also that

$$\mathscr{V} = \int_{\mathscr{R}} d\mathscr{R} = \sum_{\Delta=1}^{G} \sum_{\Gamma=1}^{G} C_{\Delta\Gamma} \tag{9.81}$$

Thus, *if* (9.75) *is satisfied, the sum of all elements of the fundamental matrix* $C_{\Delta\Gamma}$ *is the volume of the domain* $\mathscr{R}$ *over which elements of* Φ *are defined.* Moreover, in view of (9.74) and (9.78),

$$\int_{\mathscr{R}} \bar{F}(\mathbf{X})\, d\mathscr{R} = \sum_{\Delta=1}^{G} F_\Delta \tag{9.82}$$

We also observe that if the basis functions $\Phi_\Delta(\mathbf{X})$ satisfy (9.70) and (9.75), then

$$\bar{F}(\mathbf{X}^\Delta) = F^\Delta = C^{\Delta\Gamma} F_\Gamma \tag{9.83}$$

Thus, *the average value of* $\bar{F}(\mathbf{X})$ *over* $\mathscr{R}$ *as given by the integral in* (9.82) *is not the sum of the values of* $\bar{F}(\mathbf{X})$ *at the nodes* $\mathbf{X}^\Delta$*; rather, it is the sum of the conjugate values* $F_\Delta = C_{\Delta\Gamma}F^\Gamma$. *We see that* F^Δ *is the value of* $\bar{F}(\mathbf{X})$ *at node* $\mathbf{X}^\Delta$, *but* F_Δ *represents an average value of* $\bar{F}(\mathbf{X})$ *in the neighborhood of* $\mathbf{X}^\Delta$.

9.5 LINEAR OPERATORS

We now investigate briefly certain properties of approximations of a linear operator $\mathscr{L}$. In general, we are concerned with equations of the type

$$\mathscr{L}F(\mathbf{X}) = G(\mathbf{X}) \tag{9.84}$$

where $\mathscr{L}$ is a linear operator mapping an element $F(\mathbf{X})$ belonging to $\mathscr{H}$ into an element $G(\mathbf{X})$ of another Hilbert space $\hat{\mathscr{H}}$, the elements of which are assumed to be functions defined over the same domain $\mathscr{R}$ as that of $\mathscr{H}$. For simplicity, suppose for the moment that $\mathscr{H} = \hat{\mathscr{H}}$; that is, $\mathscr{L}: \mathscr{H} \to \mathscr{H}$. Then $\mathscr{L}F(\mathbf{X}) \in \mathscr{H}$, and it is meaningful to speak of its projection $\Pi\mathscr{L}F(\mathbf{X})$ in the subspace Φ:

$$\Pi\mathscr{L}F(\mathbf{X}) = L^\Delta \Phi_\Delta(\mathbf{X}) = L_\Delta \Phi^\Delta(\mathbf{X}) \tag{9.85}$$

Here L^Δ and L_Δ are coefficients characteristic of the operator $\mathscr{L}$:

$$L^\Delta = \langle \Phi^\Delta, \Pi\mathscr{L}F \rangle \qquad L_\Delta = \langle \Phi_\Delta, \Pi\mathscr{L}F \rangle \tag{9.86}$$

Now, in general,

$$\Pi \mathscr{L} F(\mathbf{X}) = \Pi \mathscr{L} \bar{F}(\mathbf{X}) + \Pi \mathscr{L} \hat{F}(\mathbf{X}) \tag{9.87}$$

Thus, to simplify the analysis further, we shall assume that the operators Π and $\mathscr{L}$ are commutative:

$$\Pi \mathscr{L} = \mathscr{L} \Pi \tag{9.88}$$

We recognize, of course, that (9.88) holds only approximately in most cases, the difference $\mathscr{L}\Pi - \Pi\mathscr{L}$ being a measure of how well the subspace Φ corresponds to the problem at hand. Clearly, (9.88) implies that $\mathscr{L}\Phi_\Delta \subset \Phi$. With the commutativity (9.88) in effect, we can set

$$\Pi \mathscr{L} \hat{F}(\mathbf{X}) = \mathscr{L} \Pi \hat{F}(\mathbf{X}) = 0 \tag{9.89}$$

and

$$\Pi \mathscr{L} F(\mathbf{X}) = \mathscr{L} \Pi F(\mathbf{X}) = \mathscr{L} \bar{F}(\mathbf{X}) = F^\Delta \mathscr{L} \Phi_\Delta(\mathbf{X}) \tag{9.90}$$

Thus, from (9.86),

$$L_\Delta = L_{\Delta\Gamma} F^\Gamma = L_{\Delta}^{.\Gamma} F_\Gamma \qquad \text{and} \qquad L^\Delta = L^{\Delta\Gamma} F_\Gamma = L_{.\Gamma}^{\Delta} F^\Gamma \tag{9.91}$$

where

$$L_{\Delta\Gamma} = \langle \Phi_\Delta, \mathscr{L}\Phi_\Gamma \rangle \qquad L_{\Delta}^{.\Gamma} = \langle \Phi_\Delta, \mathscr{L}\Phi^\Gamma \rangle \tag{9.92a,b}$$

$$L^{\Delta\Gamma} = \langle \Phi^\Delta, \mathscr{L}\Phi^\Gamma \rangle \qquad L_{.\Gamma}^{\Delta} = \langle \Phi^\Delta, \mathscr{L}\Phi_\Gamma \rangle \tag{9.92c,d}$$

Clearly, the action of the operator $\mathscr{L}$ on Φ is described by the matrices $L_{\Delta\Gamma}, \ldots, L_{.\Gamma}^{\Delta}$. It follows that

$$\begin{aligned} \Pi \mathscr{L} F(\mathbf{X}) &= L_{\Delta\Gamma} F^\Gamma \Phi^\Delta(\mathbf{X}) = L_{\Delta}^{.\Gamma} F_\Gamma \Phi^\Delta(\mathbf{X}) \\ &= L^{\Delta\Gamma} F_\Gamma \Phi_\Delta(\mathbf{X}) = L_{.\Gamma}^{\Delta} F^\Gamma \Phi_\Delta(\mathbf{X}) \end{aligned} \tag{9.93}$$

The approximation of (9.84) in Φ is then

$$L_{.\Gamma}^{\Delta} F^\Gamma \Phi_\Delta(\mathbf{X}) = G^\Delta \Phi_\Delta(\mathbf{X}) \tag{9.94}$$

or, equating corresponding components,

$$L_{.\Gamma}^{\Delta} F^\Gamma = G^\Delta \tag{9.95}$$

Assuming $L_{.\Gamma}^{\Delta}$ is nonsingular,

$$F^\Delta = N_{.\Gamma}^{\Delta} G^\Gamma \tag{9.96}$$

$N_{.\Gamma}^{\Delta}$ being the inverse of $L_{.\Gamma}^{\Delta}$, and the approximate solution of (9.84) is

$$\bar{F}(\mathbf{X}) = N_{.\Gamma}^{\Delta} G^\Gamma \Phi_\Delta(\mathbf{X}) \tag{9.97}$$

For a Hilbert space $\mathscr{H}$, the adjoint operator $\hat{\mathscr{L}}$ of $\mathscr{L}$ has the property

$$\langle \mathscr{L}F, G \rangle = \langle \hat{\mathscr{L}}G, F \rangle \tag{9.98}$$

In our approximation,

$$\langle \mathscr{L} F^{\Delta}\Phi_{\Delta}, G^{\Gamma}\Phi_{\Gamma}\rangle = \langle \hat{\mathscr{L}} G^{\Delta}\Phi_{\Delta}, F^{\Gamma}\Phi_{\Gamma}\rangle$$

or

$$L_{\Gamma\Delta} G^{\Gamma} F^{\Delta} = \hat{L}_{\Gamma\Delta} F^{\Gamma} G^{\Delta} \tag{9.99}$$

where

$$\hat{L}_{\Delta\Gamma} = \langle \Phi_{\Gamma}, \hat{\mathscr{L}} \Phi_{\Delta}\rangle \tag{9.100}$$

It follows that

$$\hat{L}_{\Gamma\Delta} = L_{\Delta\Gamma} \tag{9.101}$$

Thus, as expected, the matrix of the adjoint operator $\hat{\mathscr{L}}$ is the transpose of the matrix of $\mathscr{L}$.

We examine approximations of specific linear operators as examples in Art. 11.

Eigenvalue problems Consider the two eigenvalue problems

$$\mathscr{L} F = \lambda F \qquad \text{and} \qquad \hat{\mathscr{L}} G = \lambda G \tag{9.102}$$

For our approximation, these equations can be represented by

$$(L^{\Delta\Gamma} - \lambda C^{\Delta\Gamma}) F_{\Gamma} = 0 \tag{9.103}$$

and

$$(\hat{L}^{\Delta\Gamma} - \lambda C^{\Delta\Gamma}) F_{\Gamma} = 0 \tag{9.104}$$

But it is well known that a matrix and its transpose have the same sets of eigenvalues. It is further possible to choose biorthogonal sets of eigenfunctions, these being uniquely defined when all eigenvalues are distinct. Thus, if $F_{\Delta}^{(\alpha)}$ and $G_{\Delta}^{(\beta)}$ denote eigenvectors corresponding to eigenvalues λ_{α} and λ_{β} of $L^{\Delta\Gamma}$ and $\hat{L}^{\Delta\Gamma}$, then

$$L^{\Delta\Gamma} F_{\Gamma}^{(\alpha)} = \lambda_{\alpha} C^{\Delta\Gamma} F_{\Gamma}^{(\alpha)} \tag{9.105}$$

$$\hat{L}^{\Delta\Gamma} G_{\Gamma}^{(\beta)} = \lambda_{\beta} C^{\Delta\Gamma} G_{\Gamma}^{(\beta)} \tag{9.106}$$

and we have

$$F_{\Delta}^{(\alpha)} G_{(\beta)}^{\Delta} = C^{\Delta\Gamma} F_{\Gamma}^{(\alpha)} G_{\Delta}^{(\beta)} = \delta_{\alpha\beta} \tag{9.107}$$

In particular, *if the functions $\Phi_{\Delta}(\mathbf{X})$ are eigenfunctions of a linear operator $\mathscr{L}$ on $\mathscr{H}$, the conjugate functions $\Phi^{\Delta}(\mathbf{X})$ are eigenfunctions of the adjoint operator $\hat{\mathscr{L}}$.* That is,

$$\mathscr{L}\Phi_{\Delta}(\mathbf{X}) = \lambda_{(\Delta)}\Phi_{\Delta}(\mathbf{X}) \Rightarrow \hat{\mathscr{L}}\Phi^{\Delta}(\mathbf{X}) = \lambda_{(\Delta)}\Phi^{\Delta}(\mathbf{X}) \tag{9.108}$$

If $\mathscr{L}$ is self-adjoint, the eigenfunctions coincide: $\Phi_\Delta(\mathbf{X}) = \Phi^\Delta(\mathbf{X})$ (within a constant).

Derivatives of conjugate approximations One of the most familiar linear operators is the partial differential operator. We now examine certain properties of derivatives of approximate functions. Let $\partial_\mu F(\mathbf{X})$ denote the partial derivatives of a function $F(\mathbf{X}) \in \mathscr{H}$. We shall assume that $\partial_\mu F(\mathbf{X})$ exists and that the derivatives of the basis functions $\Phi_\Delta(\mathbf{X})$ also belong to the subspace Φ: $\partial_\mu \Phi_\Delta \in \Phi$. (The latter assumption, of course, is not always valid.) We begin by introducing the array

$$D_\mu^{\Delta\Gamma} = \langle \partial_\mu \Phi^\Delta, \Phi^\Gamma \rangle \tag{9.109}$$

Then, according to (9.11),

$$\Pi[\partial_\mu \Phi^\Delta(\mathbf{X})] = D_\mu^{\Delta\Gamma} \Phi_\Gamma(\mathbf{X}) \tag{9.110}$$

A fundamental question again arises: Under what conditions can the projection (9.110) be used as an approximation of the derivatives of the conjugate functions $\Phi^\Delta(\mathbf{X})$? In other words, when is it legitimate to set

$$\partial_\mu \Phi^\Delta(\mathbf{X}) = D_\mu^{\Delta\Gamma} \Phi_\Gamma(\mathbf{X}) \tag{9.111}$$

The answer to this question is provided by the following theorem.

Theorem 9.1 *The following conditions are equivalent to* (9.111)

a. $\partial_\mu \Pi = \Pi \partial_\mu$ *on the space* Φ.

b. $\partial_\mu \Phi \subset \Phi$.

c. Φ *is the solution space of a system of linear differential equations in* Υ_Δ *with constant coefficients.*

Proof Briefly, the proof follows by showing that (9.111) implies conditions a and c, that a implies b, and that b implies (9.111). To wit, (9.111) $\Rightarrow$ a:

$$\Pi \partial_\mu \Phi^\Delta(\mathbf{X}) = D_\mu^{\Delta\Gamma} \Phi_\Gamma(\mathbf{X}) = \partial_\mu \Phi^\Delta(\mathbf{X}) = \partial_\mu[\Pi \Phi^\Delta(\mathbf{X})]$$

for each $\Phi^\Delta(\mathbf{X}) \in \Phi$.

a $\Rightarrow$ b: For $\partial_\mu(\Phi) = \partial_\mu \Pi \Phi = \Pi(\partial_\mu \Phi)$ and $\Pi(\partial_\mu \Phi) \subset \Phi$, by definition.

b $\Rightarrow$ (9.111): The fact that $\partial_\mu(\Phi) \subset \Phi$ implies that $\partial_\mu \Phi^\Delta(\mathbf{X})$ is a linear combination of the functions $\Phi_\Delta(\mathbf{X})$, as indicated by (9.111).

(9.111) $\Rightarrow$ c: This follows immediately from the fact that the set $\{\Phi_\Delta\}$ is a complete soluton of the equation

$$\partial_\mu \Upsilon_\Delta = C_{\Delta\Gamma} D_\mu^{\Gamma\Lambda} \Upsilon_\Lambda$$

Finally, c ⇒ (9.111) because a system of linear differential equations of order n can always be transformed into a system of first-order equations. Hence $\partial_\mu \Phi_\Delta(\mathbf{X})$ is contained in Φ. This completes the proof.

Assuming that the conditions of the theorem are met, we can use (9.111) to obtain the derivative of an arbitrary element $\bar{F}(\mathbf{X}) = F^\Delta \Phi_\Delta(\mathbf{X}) \in \Phi$:

$$\partial_\mu \bar{F}(\mathbf{X}) = D_\mu^{\Delta\Gamma} F_\Delta \Phi_\Gamma(\mathbf{X}) \tag{9.112}$$

However, the relation

$$\Pi \partial_\mu F(\mathbf{X}) = D_\mu^{\Delta\Gamma} F_\Delta \Phi_\Gamma(\mathbf{X}) \tag{9.113}$$

need not hold unless the following stronger conditions are imposed on Φ:

a′. $\partial_\mu \Pi = \Pi \partial_\mu$ on $\mathscr{H}$.
b′. $\partial_\mu(\Phi) = \Phi$.

It can be shown that conditions (9.113), a′ and b′ are equivalent.

Under the assumption that these stronger conditions are also met, the action of a differentiation ∂_μ in $\mathscr{H}$ can be completely described in our approximation by the array $D_\mu^{\Delta\Gamma}$ of (9.109), and higher derivatives may be represented as well using powers of $D_\mu^{\Delta\Gamma}$. Indeed, introducing the discrete operators

$$D_{\mu\lambda}^{\Delta\Gamma} = D_\mu^{\Delta\Lambda} C_{\Lambda\Omega} D_\lambda^{\Omega\Gamma} \tag{9.114}$$

and

$$B_{\mu\lambda}^{\Delta\Gamma} = D_\mu^{\Delta\Lambda} C_{\Lambda\Omega} D_\lambda^{\Omega\Gamma} \tag{9.115}$$

it is easily shown, for example, that

$$\partial_\mu \partial_\lambda F(\mathbf{X}) = D_{\mu\lambda}^{\Delta\Gamma} F_\Delta \Phi_\Gamma(\mathbf{X}) \tag{9.116}$$

$$\operatorname{div} \mathbf{a} = \partial_\mu a^\mu = D_\mu^{\Delta\Gamma} A_\Delta^\mu \Phi_\Gamma(\mathbf{X}) \tag{9.117}$$

$$\int_{\mathscr{R}} \operatorname{div} \mathbf{a} \, d\mathscr{R} = D_\mu^{\Delta\Gamma} A_\Delta^\mu M_\Gamma \tag{9.118}$$

$$\langle F, \partial_\mu G \rangle = D_\mu^{\Gamma\Delta} F_\Delta G_\Gamma \tag{9.119}$$

$$\langle \partial_\mu F, \partial_\lambda G \rangle = B_{\mu\lambda}^{\Delta\Gamma} F_\Delta G_\Gamma \tag{9.120}$$

$$\int_{\mathscr{R}} \operatorname{grad} \mathbf{F} \cdot \operatorname{grad} \mathbf{G} \, d\mathscr{R} = B_{\mu\mu}^{\Gamma\Delta} F_\Delta G_\Gamma \tag{9.121}$$

etc. We also note that it is possible to use other types of discrete operators such as

$$D_{\mu,\Delta\Gamma} = \langle \partial_\mu \Phi_\Delta, \Phi_\Gamma \rangle = C_{\Delta\Lambda} C_{\Gamma\Omega} D_\mu^{\Omega\Lambda} \tag{9.122}$$

$$D_{\mu\cdot\Gamma}^{\Delta\cdot} = C_{\Gamma\Lambda} D_\mu^{\Lambda\Delta} = C^{\Delta\Lambda} D_{\mu,\Lambda\Gamma}, \quad D_{\mu\cdot\Gamma}^{\cdot\Delta} = C_{\Gamma\Lambda} D_\mu^{\Delta\Lambda} = C^{\Delta\Lambda} D_{\mu,\Gamma\Lambda} \tag{9.123a,b}$$

$$B_{\lambda\mu,\Delta\Gamma} = \langle \partial_\lambda \Phi_\Delta, \partial_\mu \Phi_\Gamma \rangle \tag{9.124}$$

etc. Then

$$\partial_\mu \Phi_\Delta(\mathbf{X}) = D_{\mu,\Delta\Gamma}\Phi^\Gamma(\mathbf{X}) = D_{\mu,\Delta}^{\cdot\Gamma}\Phi_\Gamma(\mathbf{X}) \tag{9.125}$$

$$\partial_\mu \Phi^\Delta(\mathbf{X}) = D_{\mu,\Gamma}^{\Delta\cdot}\Phi^\Gamma(\mathbf{X}) = D_\mu^{\Delta\Gamma}\Phi_\Gamma(\mathbf{X}) \tag{9.126}$$

$$\partial_\mu \bar{F}(\mathbf{X}) = F^\Delta D_{\mu,\Delta\Gamma}\Phi^\Gamma(\mathbf{X}) = D_\mu^{\Delta\Gamma}F_\Delta\Phi_\Gamma(\mathbf{X}) \tag{9.127}$$

Clearly, (9.122) to (9.124) can be used to obtain a variety of alternate forms of the examples (9.116) to (9.121).

Derivatives of order higher than the second can be computed, in general, by the formula

$$\partial_{\mu_1}\partial_{\mu_2}\cdots\partial_{\mu_{r-1}}\partial_{\mu_r}\bar{F}(\mathbf{X}) = C_{\Delta_1\Gamma_1}C_{\Delta_2\Gamma_2}\cdots C_{\Delta_{r-1}\Gamma_{r-1}}D_{\mu_1}^{\Delta_1\Lambda}D_{\mu_2}^{\Delta_2\Gamma_1} \cdots D_{\mu_{r-1}}^{\Delta_{r-1}\Gamma_{r-2}}D_{\mu_r}^{\Gamma_{r-1}\Omega}F_\Lambda\Phi_\Omega(\mathbf{X}) \tag{9.128}$$

It should be noted that the stronger conditions a′ and b′ (and even the weaker conditions a, b, c) are rarely satisfied in applications. However, we assume that the dimension G can be taken sufficiently large so that these conditions are satisfied in some approximate sense. If these conditions are not met, it is interesting to note that the derivative of the projection of a constant need not be zero! We examine other consequences subsequently by means of specific examples.

9.6 APPLICATIONS TO FINITE-ELEMENT APPROXIMATIONS

In order to apply the theory of conjugate approximations developed thus far to finite-element approximations, it is necessary that we identify the character of the basis functions $\Phi_\Delta(\mathbf{X})$ for general finite-element representations of the functions $F(\mathbf{X})$. Toward this end, we follow precisely the procedure discussed in Arts. 6 and 7 and represent the domain $\mathscr{R}$ by a domain $\bar{\mathscr{R}}$ which consists of a collection of E finite elements $\imath_e$ connected appropriately together at G global nodal points $\mathbf{X}^\Delta$, $\Delta = 1, 2, \ldots, G$. As usual, we select within each element N_e local nodal points $\mathbf{x}_{(e)}^N$ such that $\mathbf{x}_{(e)}^N = \overset{(e)}{\Omega}{}_\Delta^N \mathbf{X}^\Delta$ and, for fixed e, $\mathbf{X}^\Delta = \overset{(e)}{\Lambda}{}_N^\Delta \mathbf{x}_{(e)}^N$, where $\overset{(e)}{\Lambda}{}_N^\Delta$ and $\overset{(e)}{\Omega}{}_\Delta^N$ are the arrays defined in (7.4) and (7.6). To simplify notation, we shall assume here that the local and global coordinate systems $\mathbf{x}_{(e)}$ and $\mathbf{X}$ coincide, so that either $\mathbf{x}$ or $\mathbf{X}$ may denote a point in $\bar{\mathscr{R}}$. Then, if we need to emphasize that a point belongs to a specific element, we affix the element identification label (e), as before.

Considering now a Hilbert space $\mathscr{H}$ whose elements are defined over $\mathscr{R}$, we recall that the projection $\Pi: \mathscr{H} \to \Phi$ describes a G-dimensional subspace of $\mathscr{H}$, spanned by functions $\Phi_\Delta(\mathbf{X})$. For finite-element approximations, the domain of functions in the subspace Φ is the collection $\bar{\mathscr{R}}$ of finite elements. Then, on $\bar{\mathscr{R}}$,

$$F(\mathbf{X}) \approx \bar{F}(\mathbf{X}) = F^\Delta\Phi_\Delta(\mathbf{X}) \tag{9.129}$$

where, assuming that the functions $\Phi_\Delta(\mathbf{X})$ are normalized with respect to the nodes $\mathbf{X}^\Delta$ in accordance with (9.70),

$$\bar{F}(\mathbf{X}^\Delta) = F^\Gamma \Phi_\Gamma(\mathbf{X}^\Delta) = F^\Gamma \delta_\Gamma^\Delta = F^\Delta \tag{9.130}$$

The finite-element concept suggests that we follow a procedure similar to that used to obtain (9.129) on a local scale. Thus, let $P^{(e)}$ denote a projection of $\mathscr{H}$ into an N_e-dimensional subspace $\Psi^{(e)}$ of functions with domain r_e; $P^{(e)}: \mathscr{H} \to \Psi^{(e)}$. We define such a projection by introducing a system of N_e linearly independent *local basis functions* $\psi_N^{(e)}(\mathbf{x})$ which are normalized with respect to the local nodal points:

$$\psi_N^{(e)}(\mathbf{x}^M) = \delta_N^M \tag{9.131}$$

with $M, N = 1, 2, \ldots, N_e$. Then, locally, the projection of $F(\mathbf{X})$ is given by

$$P^{(e)}F(\mathbf{X}) = \bar{f}^{(e)}(\mathbf{x}) \tag{9.132}$$

where the local approximation $\bar{f}^{(e)}(\mathbf{x})$ is uniquely defined as a linear combination of the local basis functions [see (7.15)]:

$$\bar{f}^{(e)}(\mathbf{x}) = f_{(e)}^N \psi_N^{(e)}(\mathbf{x}) \tag{9.133}$$

In view of (9.131), the components $f_{(e)}^N$ are the values of the local approximation at the local nodes:

$$\bar{f}^{(e)}(\mathbf{x}^M) = f_{(e)}^N \psi_N^{(e)}(\mathbf{x}^M) = f_{(e)}^N \delta_N^M = f_{(e)}^M \tag{9.134}$$

Thus, as usual,

$$f_{(e)}^N = \overset{(e)}{\Omega}{}_\Delta^N F^\Delta \qquad \text{and} \qquad F^\Delta = \overset{(e)}{\Lambda}{}_N^\Delta f_{(e)}^N \tag{9.135}$$

To relate the local functions $\psi_N^{(e)}(\mathbf{x})$ to the global functions $\Phi_\Delta(\mathbf{X})$, *we require that* $\psi_N^{(e)}(\mathbf{x})$ *be the restriction of* $\Phi_\Delta(\mathbf{X})$ *to element* r_e *described by*

$$\psi_N^{(e)}(\mathbf{x}) = \overset{(e)}{\Lambda}{}_N^\Delta \Phi_\Delta(\mathbf{x}) \tag{9.136}$$

where by $\Phi_\Delta(\mathbf{x})$ we mean $\Phi_\Delta(\mathbf{X})$ restricted to the domain $\mathbf{X} \in r_e$. This relation implies an embedding $\overset{(e)}{\Lambda}: \Phi \to \Psi^{(e)}$ and follows from the fact that the values of $\psi_N^{(e)}(\mathbf{x})$ and $\Phi_\Delta(\mathbf{X})$ are in one-to-one correspondence with points in their domains r_e and $\bar{\mathscr{R}}$. It follows that for $\mathbf{x} \in r_e$, we also have the relation

$$\Phi_\Delta(\mathbf{X}) = \overset{(e)}{\Omega}{}_\Delta^N \psi_N^{(e)}(\mathbf{x}) \qquad \text{for fixed } e \tag{9.137}$$

and almost everywhere† in $\bar{\mathscr{R}}$,

$$\Phi_\Delta(\mathbf{X}) = \sum_{e=1}^{E} \overset{(e)}{\Omega}{}_\Delta^N \psi_N^{(e)}(\mathbf{x}) \tag{9.138}$$

† See the footnote on page 39. Note also that (9.138) defines the basis functions $\Phi_\Delta(\mathbf{X})$ as those presented previously in (7.20).

From the definitions of $P^{(e)}$ and Π, it is clear that

$$P^{(e)} = \overset{(e)}{\Lambda}\Pi \tag{9.139}$$

Thus, with the same provisions as those that apply to (9.138), we see that (9.130) can now be written

$$\bar{F}(\mathbf{X}) = F^{\Delta}\Phi_{\Delta}(\mathbf{X}) = \sum_{e=1}^{E} F^{\Delta}\overset{(e)}{\Omega}{}_{\Delta}^{N}\psi_N^{(e)}(\mathbf{x}) = \sum_{e=1}^{E} f_{(e)}^{N}\psi_N^{(e)}(\mathbf{x}) \tag{9.140}$$

or, from (9.133),

$$\bar{F}(\mathbf{X}) = \sum_{e=1}^{E} \bar{f}^{(e)}(\mathbf{x}) \tag{9.141}$$

Fundamental properties With the basis $\Phi_{\Delta}(\mathbf{X})$ now described for finite-element approximations, we can proceed to apply the theory developed previously for general conjugate approximations. Introducing (9.138) into (9.12), we obtain for the fundamental matrix of Φ for finite-element models:

$$C_{\Delta\Gamma} = \langle \Phi_{\Delta}, \Phi_{\Gamma} \rangle = \sum_{e=1}^{E}\sum_{f=1}^{E} \overset{(e)}{\Omega}{}_{\Delta}^{N}\overset{(f)}{\Omega}{}_{\Gamma}^{M}\langle \psi_N^{(e)}, \psi_M^{(f)} \rangle \tag{9.142}$$

However, the local approximation functions $\psi_N^{(e)}(\mathbf{x})$ are defined so as to have *almost disjoint support;* that is,

$$\langle \psi_N^{(e)}, \psi_M^{(f)} \rangle = 0 \qquad e \neq f \tag{9.143}$$

Thus (9.142) can be written

$$C_{\Delta\Gamma} = \sum_{e=1}^{E} \overset{(e)}{\Omega}{}_{\Delta}^{N}\overset{(e)}{\Omega}{}_{\Gamma}^{M} c_{NM}^{(e)} \tag{9.144}$$

where $c_{NM}^{(e)}$ is the local fundamental matrix of $\Psi^{(e)}$:

$$c_{NM}^{(e)} = \langle \psi_N^{(e)}, \psi_M^{(e)} \rangle \tag{9.145}$$

We are tempted to proceed on a local level to use $c_{NM}^{(e)}$ and its inverse to construct a set of local conjugate basis functions in a manner analogous to that used earlier to compute $\Phi^{\Delta}(\mathbf{X})$. However, we have limited the spaces Φ and Φ^* to include only continuous functions, and while we generally design the functions $\psi_N^{(e)}(\mathbf{x})$ so as to give continuous basis functions $\Phi_{\Delta}(\mathbf{X})$ by (9.138), we have no reason to expect a rather arbitrary linear combination $\sum_{e} (c_{NM}^{(e)})^{-1}\psi_M^{(e)}(\mathbf{x})$ to be also continuous on $\bar{\mathscr{R}}$. Therefore, an alternate procedure must be sought.

With this in mind, we now direct our attention to determining local conjugate approximations for finite-element representations. We begin by considering a linear functional obtained by forming the inner product of the function

$$\bar{F}(\mathbf{X}) = F^{\Delta}\Phi_{\Delta}(\mathbf{X}) = \sum_{e} \overset{(e)}{\Omega}{}^{N}_{\Delta} F^{\Delta} \psi^{(e)}_{N}(\mathbf{x}) = F_{\Delta}\Phi^{\Delta}(\mathbf{X}) \tag{9.146}$$

and an arbitrary function $\bar{G}(\mathbf{X}) \in \Phi^*$ which is also the sum of E local approximations,

$$\bar{G}(\mathbf{X}) = \sum_{e} \bar{g}^{(e)}(\mathbf{x}) = G_{\Delta}\Phi^{\Delta}(\mathbf{X}) \tag{9.147}$$

where

$$\Phi^{\Delta}(\mathbf{X}) = C^{\Delta\Gamma}\Phi_{\Gamma}(\mathbf{X}) = C^{\Delta\Gamma} \sum_{e} \overset{(e)}{\Omega}{}^{N}_{\Gamma} \psi^{(e)}_{N}(\mathbf{x}) \tag{9.148}$$

We have

$$\langle \bar{F},\bar{G} \rangle = F^{\Delta}G_{\Delta} = F_{\Delta}G^{\Delta} = F^{\Delta}\langle \sum_{e} \overset{(e)}{\Omega}{}^{N}_{\Delta} \psi^{(e)}_{N}(\mathbf{x}),\bar{G} \rangle \tag{9.149}$$

where

$$G_{\Delta} = \langle \bar{G},\Phi_{\Delta} \rangle = \sum_{e} \overset{(e)}{\Omega}{}^{N}_{\Delta} g^{(e)}_{N} \tag{9.150}$$

and we have defined

$$g^{(e)}_{N} = \langle \bar{G},\psi^{(e)}_{N} \rangle \tag{9.151}$$

We see that, unlike (9.135), the *global values* G_{Δ} *at a global node* Δ *are obtained by summing all the local values* $g^{(e)}_{N}$ *at local nodes incident on* Δ. We elaborate on this point later.

Returning to (9.147), we can now define systems of local conjugate basis functions, for

$$\bar{G}(\mathbf{X}) = \sum_{e} \overset{(e)}{\Omega}{}^{N}_{\Delta} g^{(e)}_{N} \Phi^{\Delta}(\mathbf{X}) = \sum_{e} \bar{g}^{(e)}(\mathbf{x}) \tag{9.152}$$

which suggests that we write

$$\bar{g}^{(e)}(\mathbf{x}) = g^{(e)}_{N} \psi^{N}_{(e)}(\mathbf{x}) \qquad N = 1, 2, \ldots, N_e \tag{9.153}$$

where, from (9.152),

$$\psi^{N}_{(e)}(\mathbf{x}) = \overset{(e)}{\Omega}{}^{N}_{\Delta} \Phi^{\Delta}(\mathbf{X}) \tag{9.154}$$

This equation defines the local conjugate basis functions for element e of the finite-element approximation. In view of (9.138),

$$\psi^{N}_{(e)}(\mathbf{x}) = \overset{(e)}{\Omega}{}^{N}_{\Delta} C^{\Delta\Gamma} \sum_{f=1}^{E} \overset{(f)}{\Omega}{}^{M}_{\Gamma} \psi^{(f)}_{M}(\mathbf{x}) \tag{9.155}$$

The form of (9.155) is significant; it shows that *the "local" conjugate basis functions for element e are linear combinations of the basis functions*

$\psi_N^{(f)}(\mathbf{x})$ *of all E finite elements. Thus, the functions* $\psi_{(e)}^N(\mathbf{x})$ *need not have local support; indeed, the support of each local function* $\psi_{(e)}^N(\mathbf{x})$ *is the entire connected domain* $\bar{\mathscr{R}}$. *This means that the usual procedure of calculating local values of conjugate approximations by taking local averages of the nodal values* $g_N^{(e)}$ (for example, computing element stresses from a displacement approximation) *is not strictly correct. In order that the local conjugate approximation be consistent with the linear functional defined on* Φ (for example, energy), *it is necessary that it be referred to a basis which has as its domain the entire collection of finite elements.* We shall demonstrate these properties of local conjugate approximations by means of examples in Sec. 9.8.

Observe that

$$\langle \psi_{(e)}^N, \psi_R^{(e)} \rangle = \overset{(e)}{\Omega_\Delta^N} C^{\Delta\Gamma} \sum_{f=1}^{E} \overset{(f)}{\Omega_\Gamma^M} \langle \psi_M^{(f)}, \psi_R^{(e)} \rangle \tag{9.156}$$

thus, in view of (9.145),

$$\langle \psi_{(e)}^N, \psi_R^{(e)} \rangle = \overset{(e)}{\Omega_\Delta^N} C^{\Delta\Gamma} \overset{(e)}{\Omega_\Gamma^M} c_{MR}^{(e)} \tag{9.157}$$

with $M, N, R = 1, 2, \ldots, N_e$ and $\Delta, \Gamma = 1, 2, \ldots, G$. Clearly, the sets of functions $\psi_N^{(e)}(\mathbf{x})$ and $\psi_{(e)}^N(\mathbf{x})$ do not form a local biorthogonal basis for functions in $\Psi^{(e)}$.

With the basis functions $\Phi_\Delta(\mathbf{X})$ and $\Phi^\Delta(\mathbf{X})$ determined for finite-element representations by (9.138) and (9.148), it is now a simple matter to generate other quantities needed in finite-element approximations. For example, from (9.72) and (9.73),

$$M_\Delta = \int_{\mathscr{R}} \Phi_\Delta(\mathbf{X})\, d\mathscr{R} = \sum_{e=1}^{E} \overset{(e)}{\Omega_\Delta^N} m_N^{(e)} = \sum_{\Gamma=1}^{G} C_{\Delta\Gamma} \tag{9.158}$$

where

$$m_N^{(e)} = \int_{r_e} \psi_N^{(e)}(\mathbf{x})\, dr_e \tag{9.159}$$

$$M^\Delta = C^{\Delta\Gamma} \sum_{e=1}^{E} \overset{(e)}{\Omega_\Gamma^N} m_N^{(e)} = 1 \tag{9.160}$$

Also, if we define a local discrete operator by

$$d_{\mu,NM}^{(e)} = \langle \partial_\mu \psi_N^{(e)}, \psi_M^{(e)} \rangle \tag{9.161}$$

then, according to (9.109) and (9.122),

$$D_{\mu,\Delta\Gamma} = \sum_{e=1}^{E} \overset{(e)}{\Omega_\Delta^N} \overset{(e)}{\Omega_\Gamma^M}\, d_{\mu,NM}^{(e)} \tag{9.162}$$

$$D_\mu^{\Delta\Gamma} = C^{\Delta\Lambda} C^{\Gamma\Omega} \sum_{e=1}^{E} \overset{(e)}{\Omega_\Omega^N} \overset{(e)}{\Omega_\Lambda^M}\, d_{\mu,NM}^{(e)} \tag{9.163}$$

For example,

$$\partial_\mu F(\mathbf{X}) = \sum_e \overset{(e)}{\Omega}{}^N_\Delta \overset{(e)}{\Omega}{}^M_\Gamma \, d^{(e)}_{\mu,NM} \, F^\Delta \Phi^\Gamma(\mathbf{X}) \tag{9.164}$$

The discrete operators $D^{\Delta\Gamma}_{\mu\lambda}$ and $B^{\Delta\Gamma}_{\mu\lambda}$ of (9.114) and (9.115) for finite-element models are now

$$D^{\Delta\Gamma}_{\mu\lambda} = C^{\Delta\Sigma} C^{\Gamma\Theta} C^{\Omega\Upsilon} \left(\sum_e \overset{(e)}{\Omega}{}^N_\Sigma \overset{(e)}{\Omega}{}^M_\Omega \, d^{(e)}_{\mu,NM} \right) \left(\sum_f \overset{(f)}{\Omega}{}^R_\Upsilon \overset{(f)}{\Omega}{}^S_\Theta \, d^{(f)}_{\lambda,RS} \right) \tag{9.165}$$

$$B^{\Delta\Gamma}_{\mu\lambda} = C^{\Delta\Lambda} C^{\Gamma\Sigma} C^{\Theta\Omega} \left(\sum_e \overset{(e)}{\Omega}{}^N_\Lambda \overset{(e)}{\Omega}{}^M_\Theta \, d^{(e)}_{\mu,NM} \right) \left(\sum_f \overset{(f)}{\Omega}{}^R_\Sigma \overset{(f)}{\Omega}{}^S_\Omega \, d^{(f)}_{\lambda\cdot RS} \right) \tag{9.166}$$

It is now a simple matter to compute relations such as (9.116) to (9.121) or (9.122) to (9.128) in a form appropriate for finite-element approximations by substituting (9.162), (9.163), (9.164), or (9.165) into these formulas. Since the procedure is straightforward, we omit these details here.

In the case of a linear-operator equation of the type in (9.84), we examine a local form

$$P^{(e)} \mathscr{L} F(\mathbf{X}) = P^{(e)} G(\mathbf{X}) \tag{9.167}$$

which we assume to be approximated by

$$P^{(e)} \mathscr{L} F(\mathbf{X}) \approx \mathscr{L} P^{(e)} F(\mathbf{X}) = \mathscr{L} \bar{f}^{(e)}(\mathbf{x}) = f^N_{(e)} \mathscr{L} \psi^{(e)}_N(\mathbf{x}) \tag{9.168}$$

and

$$P^{(e)} G(\mathbf{X}) = g^N_{(e)} \psi^{(e)}_N(\mathbf{x}) \tag{9.169}$$

Thus, within a finite element,

$$f^N_{(e)} \mathscr{L} \psi^{(e)}_N(\mathbf{x}) = g^N_{(e)} \psi^{(e)}_N(\mathbf{x}) \tag{9.170}$$

Denoting

$$l^{(e)}_{MN} = \langle \psi^{(e)}_M, \mathscr{L} \psi^{(e)}_N \rangle \tag{9.171}$$

we have

$$l^{(e)}_{MN} f^N_{(e)} = c^{(e)}_{NM} g^N_{(e)} \tag{9.172}$$

Local matrices of the type $l^{(e)}_{MN}$ are often singular because the local functions $\bar{f}^{(e)}(\mathbf{x})$ are allowed, at this point, to assume arbitrary values on the local element boundaries ∂r_e. Since

$$f^N_{(e)} = \overset{(e)}{\Omega}{}^N_\Gamma F^\Gamma \qquad \text{and} \qquad g^N_{(e)} = \overset{(e)}{\Omega}{}^N_\Gamma G^\Gamma \tag{9.173}$$

we multiply (9.172) by $\overset{(e)}{\Omega}{}^M_\Delta$ and sum on e:

$$F^\Gamma \sum_e \overset{(e)}{\Omega}{}^M_\Delta l^{(e)}_{MN} \overset{(e)}{\Omega}{}^N_\Gamma = G^\Gamma \sum_e \overset{(e)}{\Omega}{}^M_\Delta c^{(e)}_{MN} \overset{(e)}{\Omega}{}^N_\Gamma \tag{9.174}$$

Denoting

$$L_{\Delta\Gamma} = \sum_e \overset{(e)}{\Omega}{}_{\Delta}^{M} l_{MN}^{(e)} \overset{(e)}{\Omega}{}_{\Gamma}^{N} \tag{9.175}$$

and observing that

$$G^{\Gamma} \sum_e \overset{(e)}{\Omega}{}_{\Delta}^{M} c_{MN}^{(e)} \overset{(e)}{\Omega}{}_{\Gamma}^{N} = G^{\Gamma} C_{\Delta\Gamma} = G_{\Delta} \tag{9.176}$$

we get

$$L_{\Delta\Gamma} F^{\Gamma} = G_{\Delta} \tag{9.177}$$

which is equivalent to (9.95). Note that the above procedure is based on the assumption of commutativity of $\mathscr{L}$ and $P^{(e)}$. We consider alternate procedures, involving both linear and nonlinear operators, in Art. 10 and specific examples in Art. 11 (and in Sec. 9.8).

9.7 GENERALIZED CONJUGATE VARIABLES†

As a slight generalization of the theory of conjugate approximations developed thus far, we now consider the procedure for calculating generalized conjugate variables for cases in which other linear functionals appear in the analysis.

Let $\mathscr{S}$ and $\mathscr{T}$ denote two spaces, the elements of which are, respectively, functions $\mathbf{S}(\mathbf{X})$ and $\mathbf{T}(\mathbf{X})$, the domain of each being a region $\mathscr{R} \subset \mathscr{E}^k$. In the present case, the values of the functions $\mathbf{S}(\mathbf{X})$ and $\mathbf{T}(\mathbf{X})$ may be vectors, tensors, etc. Further, let G denote a continuous mapping from $\mathscr{T} \times \mathscr{S}$ into R, R being the real-number field, which defines a scalar-valued function $G(\mathbf{X})$, $\mathbf{X} \in \mathscr{R}$. We shall denote the binary operation which, for a given $\mathbf{X}$, maps the pair $[\mathbf{T}(\mathbf{X}),\mathbf{S}(\mathbf{X})]$ into $G(\mathbf{X})$ by the symbol $*$; that is,

$$G(\mathbf{X}) = \mathbf{T}(\mathbf{X}) * \mathbf{S}(\mathbf{X}) \tag{9.178}$$

In addition, let $Q[G(\mathbf{X})]$ denote a functional on $\mathscr{T} \times \mathscr{S}$ defined by

$$Q[\mathbf{T}(\mathbf{X}),\mathbf{S}(\mathbf{X})] = \int_{\mathscr{R}} \mathbf{T}(\mathbf{X}) * \mathbf{S}(\mathbf{X})\, d\mathscr{R} \tag{9.179}$$

where integration in the Lebesque sense is implied. We shall require that $Q[\mathbf{T}(\mathbf{X}),\mathbf{S}(\mathbf{X})] = Q[\mathbf{T},\mathbf{S}]$ be linear in that

$$\begin{aligned} Q[\alpha\mathbf{T}_1 + \beta\mathbf{T}_2, \mathbf{S}] &= \alpha Q[\mathbf{T}_1,\mathbf{S}] + \beta Q[\mathbf{T}_2,\mathbf{S}] \\ Q[\mathbf{T}, \alpha\mathbf{S}_1 + \beta\mathbf{S}_2] &= \alpha Q[\mathbf{T},\mathbf{S}_1] + \beta Q[\mathbf{T},\mathbf{S}_2] \end{aligned} \tag{9.180}$$

where α and β are scalars. Thus, for fixed $\mathbf{S}(\mathbf{X})$, (9.179) defines a linear functional on $\mathscr{T}$. The collection of all such linear functionals is the conjugate space of $\mathscr{T}$. We shall refer to $\mathbf{S}(\mathbf{X})$ as being *conjugate* to $\mathbf{T}(\mathbf{X})$ with respect to Q.

† See Oden [1969*a*, 1971*a*]; for further generalization, see Oden and Reddy [1971].

We are concerned with the following problem: Given a finite-element approximation $\bar{\mathbf{T}}(\mathbf{X})$ of $\mathbf{T}(\mathbf{X})$, find the corresponding discrete model of $\mathbf{S}(\mathbf{X})$ which is consistent with (9.179). We shall describe the meaning of the term "consistent" shortly.

To resolve this problem, we begin by considering a typical finite element r_e of the approximation $\bar{\mathscr{R}}$ of $\mathscr{R}$, over which the restrictions of the fields $\mathbf{T}(\mathbf{X})$ and $\mathbf{S}(\mathbf{X})$ are $\mathbf{t}_{(e)}(\mathbf{x})$ and $\mathbf{s}_{(e)}(\mathbf{x})$, respectively. Locally, we introduce

$$q^{(e)}[\mathbf{t},\mathbf{s}] = \int_{r_e} \mathbf{t}_{(e)}(\mathbf{x}) * \mathbf{s}_{(e)}(\mathbf{x})\, dr_e \tag{9.181}$$

Now, for a finite-element model of $\mathbf{T}(\mathbf{X})$, we have the local approximations

$$\mathbf{t}^{(e)}(\mathbf{x}) \approx \bar{\mathbf{t}}_{(e)}(\mathbf{x}) = \mathbf{t}^N_{(e)}\psi^{(e)}_N(\mathbf{x}) \qquad \mathbf{x} \in r_e \tag{9.182}$$

so that instead of (9.181) we have

$$q^{(e)} \approx \bar{q}^{(e)} = \mathbf{t}^N_{(e)} * \mathbf{s}^{(e)}_N \qquad N = 1, 2, \ldots, N_e \tag{9.183}$$

where

$$\mathbf{s}^{(e)}_N = \int_{r_e} \psi^{(e)}_N(\mathbf{x})\mathbf{s}_{(e)}(\mathbf{x})\, dr_e \tag{9.184}$$

The quantities $\mathbf{s}^{(e)}_N$ $(e = 1, 2, \ldots, E;\ N = 1, 2, \ldots, N_e)$ are the *local generalized* values of $\mathbf{S}(\mathbf{X})$ at node N of element e consistent† with the approximation $\bar{\mathbf{t}}_{(e)}(\mathbf{x})$ and the functional Q.

Globally, we introduce

$$Q \approx \bar{Q} = \int_{\bar{\mathscr{R}}} \bar{\mathbf{T}}(\mathbf{X}) * \mathbf{S}(\mathbf{X})\, d\mathscr{R} = \sum_{e=1}^{E} \bar{q}^{(e)} \tag{9.185}$$

where

$$\bar{\mathbf{T}}(\mathbf{X}) = \mathbf{T}^\Delta\Phi_\Delta(\mathbf{X}) = \mathbf{T}^\Delta \sum_{e=1}^{E} \overset{(e)}{\Omega}{}^N_\Delta \psi^{(e)}_N(\mathbf{x}) \tag{9.186}$$

Thus, we may write

$$\bar{Q} = \mathbf{T}^\Delta * \mathbf{S}_\Delta \qquad \Delta = 1, 2, \ldots, G \tag{9.187}$$

in which $\mathbf{S}_\Delta$ is the *global generalized value of* $\mathbf{S}(\mathbf{X})$ *at node* $\mathbf{X}^\Delta$:

$$\mathbf{S}_\Delta = \int_{\bar{\mathscr{R}}} \Phi_\Delta(\mathbf{X})\mathbf{S}(\mathbf{X})\, d\mathscr{R} = \sum_{e=1}^{E} \overset{(e)}{\Omega}{}^N_\Delta \int_{r_e} \psi^{(e)}_N(\mathbf{x})\mathbf{s}_{(e)}(\mathbf{x})\, dr_e \tag{9.188}$$

† The term "consistent" in regard to finite-element approximations was introduced by Archer [1963, 1965] in connection with consistent approximation of forces and mass distributions in finite-element analyses.

From (9.184) and (9.188) it is clear that

$$\mathbf{S}_\Delta = \sum_e \overset{(e)}{\Omega}{}_\Delta^N \mathbf{s}_N^{(e)} \tag{9.189}$$

Thus, unlike the simple incidence relation $\mathbf{t}_{(e)}^N = \overset{(e)}{\Omega}{}_\Delta^N \mathbf{T}^\Delta$ between local and global values of $\overline{\mathbf{T}}(\mathbf{x})$, *the global value* $\mathbf{S}_\Delta$ *of the conjugate field at a node* Δ *of the assembled model is equal to the sum of the local values* $\mathbf{s}_N^{(e)}$ *of all elements connected at* Δ, *the nodes N of each element that contributes to the sum being incident on node* Δ.

For example, at global node 1 of the two-dimensional network of finite elements in Fig. 9.1, we have

$$\mathbf{T}^1 = \mathbf{t}_{(1)}^3 = \mathbf{t}_{(2)}^2 = \mathbf{t}_{(3)}^5 = \mathbf{t}_{(4)}^2 \tag{9.190}$$

However, for the conjugate field,

$$\mathbf{S}_1 = \mathbf{s}_3^{(1)} + \mathbf{s}_2^{(2)} + \mathbf{s}_5^{(3)} + \mathbf{s}_2^{(4)} \tag{9.191}$$

Notice that

$$\bar{Q} = \mathbf{T}^\Delta * \mathbf{S}_\Delta = \sum_e \mathbf{t}_{(e)}^N * \mathbf{s}_N^{e} \tag{9.192}$$

It is important to note that

$$\mathbf{T}^\Delta = \overline{\mathbf{T}}(\mathbf{X}^\Delta) \qquad \text{but} \qquad \mathbf{S}_\Delta \neq \mathbf{S}(\mathbf{X}^\Delta) \tag{9.193}$$

Indeed, the variation of the conjugate approximation over $\bar{\mathscr{R}}$ is given by the

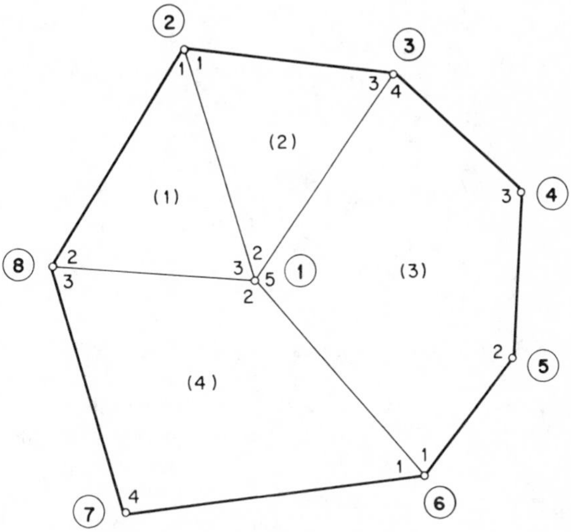

Fig. 9.1 Collection of finite elements.

more complicated formulas

$$\mathbf{S}(\mathbf{X}) \approx \mathbf{S}_\Delta \Phi^\Delta(\mathbf{X}) = \sum_{e=1}^{E} \bar{\mathbf{s}}^{(e)}(\mathbf{x}) \tag{9.194}$$

$$\bar{\mathbf{s}}^{(e)}(\mathbf{x}) = \overset{(e)}{\mathbf{s}_N} \psi^N_{(e)}(\mathbf{x}) \qquad \psi^N_{(e)}(\mathbf{x}) = \overset{(e)}{\Omega^N_\Delta} C^{\Delta\Gamma} \sum_f \overset{(f)}{\Omega^M_\Gamma} \psi^{(f)}_M(\mathbf{x}) \tag{9.195}$$

9.8 EXAMPLES

To fix ideas, we cite in this section several examples involving conjugate approximations.

Example 9.1 Mechanical work We begin with a simple example involving the elementary concept of work. Consider two vector fields $\mathbf{U}(\mathbf{X})$ and $\mathbf{F}(\mathbf{X})$ defined on a region $\mathscr{R}$ of $\mathscr{E}^3$. If $\mathbf{U}(\mathbf{X})$ is the primary field to be approximated, then

$$\bar{\mathbf{U}}(\mathbf{x}) = \sum_e \bar{\mathbf{u}}_{(e)}(\mathbf{x}) = \sum_e^{E} \overset{(e)}{\Omega^N_\Delta} \psi^{(e)}_N(\mathbf{x}) \mathbf{U}^\Delta \tag{9.196}$$

Let us assume that the vector-valued function $\mathbf{U}(\mathbf{X})$ represents the displacement field in a continuous body and that $\mathbf{F}(\mathbf{X})$ represents a force field acting on the body. Then $\bar{q}^{(e)}$ is the work done by the local force field $\mathbf{f}^{(e)}(\mathbf{x})$ in moving through the displacements $\bar{\mathbf{u}}_{(e)}(\mathbf{x})$ and $*$ denotes the ordinary scalar product of vectors $(* = \cdot)$:

$$\bar{q}^{(e)} = \int_{\imath_e} \bar{\mathbf{u}}_{(e)} \cdot \mathbf{f}^{(e)} \, d\imath_e = \mathbf{u}^N_{(e)} \cdot \mathbf{f}^{(e)}_N \tag{9.197}$$

Here $\mathbf{f}^{(e)}_N$ is the consistent generalized force at node N:

$$\mathbf{f}^{(e)}_N = \int_{\imath_e} \mathbf{f}^{(e)}(\mathbf{x}) \psi^{(e)}_N(\mathbf{x}) \, d\imath_e \tag{9.198}$$

Finally, the global values of the nodal forces are obtained by adding vectorially all local nodal forces common to a node in the connected model:

$$\mathbf{F}_\Delta = \sum_e \overset{(e)}{\Omega^N_\Delta} \mathbf{f}^{(e)}_N \tag{9.199}$$

The functional $\bar{Q}$ represents, in this case, the total work done by the nodal forces $\mathbf{F}_\Delta$ in moving through the corresponding nodal displacements $\mathbf{U}^\Delta$:

$$\bar{Q} = \mathbf{U}^\Delta \cdot \mathbf{F}_\Delta = \sum_e \mathbf{u}^N_{(e)} \cdot \mathbf{f}^{(e)}_N \tag{9.200}$$

Example 9.2 Stress calculations To demonstrate the significance of the conjugate functions $\psi^N_{(e)}(x)$, we present a simple example† involving the computation of stresses in a model based on approximate displacement fields.

Consider a nonhomogeneous bar for which the stress $\sigma(x)$ is given by the formula

$$\sigma(x) = k(x) \frac{du(x)}{dx} \tag{9.201}$$

† See Oden and Brauchli [1971*a*].

Here $u(x)$ is the displacement field, and the modulus $k(x)$ is assumed to vary linearly according to

$$k(x) = k_0(1 + x) \tag{9.202}$$

k_0 being a material constant.

For simplicity, we shall employ a rather crude finite-element representation consisting of only three one-dimensional elements, each of unit length. To further simplify matters, we take for the local basis functions $\psi_N^{(e)}(x)$ corresponding to a typical element e the linear forms

$$\psi_1^{(e)}(\xi) = 1 - \xi \qquad \psi_2^{(e)}(\xi) = \xi \tag{9.203}$$

ξ being a local coordinate, so that the local fundamental matrices are

$$c_{NM}^{(e)} = \langle \psi_N^{(e)}, \psi_M^{(e)} \rangle = \frac{1}{6}\begin{bmatrix} 2 & 1 \\ 1 & 2 \end{bmatrix} \tag{9.204}$$

Determining, by inspection, the incidence operators $\overset{(e)}{\Omega}{}_{\Delta}^{N}$ and introducing (9.204) into (9.142), we get

$$C_{\Delta\Gamma} = \frac{1}{6}\begin{bmatrix} 2 & 1 & 0 & 0 \\ 1 & 4 & 1 & 0 \\ 0 & 1 & 4 & 1 \\ 0 & 0 & 1 & 2 \end{bmatrix} \quad \text{and} \quad C^{\Delta\Gamma} = \frac{6}{45}\begin{bmatrix} 26 & -7 & 2 & -1 \\ -7 & 14 & -4 & 2 \\ 2 & -4 & 14 & -7 \\ -1 & 2 & -7 & 26 \end{bmatrix} \tag{9.205}$$

The conjugate basis functions can now be computed with the aid of (9.155):

$$\begin{aligned}
\psi_{(1)}^{1}(x) &= \tfrac{6}{45}\{26\psi_1^{(1)}(x) - 7[\psi_2^{(1)}(x) + \psi_1^{(2)}(x)] + 2[\psi_2^{(2)}(x) + \psi_1^{(1)}(x)] - \psi_2^{(3)}(x)\} \\
\psi_1^{(2)}(x) &= \psi_{(2)}^{1}(x) \\
&= \tfrac{6}{45}\{-7\psi_1^{(1)}(x) + 14[\psi_2^{(1)}(x) + \psi_1^{(2)}(x)] - 4[\psi_2^{(2)}(x) + \psi_1^{(3)}(x)] + 2\psi_2^{(3)}(x)\} \\
\psi_{(2)}^{2}(x) &= \psi_{(3)}^{1}(x) \\
&= \tfrac{6}{45}\{2\psi_1^{(1)}(x) - 4[\psi_2^{(2)}(x) + \psi_1^{(2)}(x)] + 14[\psi_2^{(2)}(x) + \psi_1^{(3)}(x)] - 7\psi_2^{(3)}(x)\} \\
\psi_{(3)}^{2}(x) &= \tfrac{6}{45}\{-\psi_1^{(1)}(x) + 2[\psi_2^{(1)}(x) + \psi_1^{(2)}(x)] - 7[\psi_2^{(2)}(x) + \psi_1^{(3)}(x)] + 26\psi_2^{(3)}(x)\}
\end{aligned} \tag{9.206}$$

The functions $\psi_N^{(e)}(x)$ and $\psi_{(e)}^{N}(x)$ are shown in Fig. 9.2.

We shall now assume that the bar is given a prescribed quadratic displacement field of the form

$$u(x) = \alpha\left[1 - \left(\frac{x}{3}\right)^2\right] \tag{9.207}$$

where α is a small constant. Clearly, the exact stress distribution is

$$\sigma(x) = \frac{-2\alpha k_0}{9}\, x(1 + x) \tag{9.208}$$

However, the displacement field, as represented by the finite-element model, is piecewise linear:

$$U(x) = \frac{\alpha}{9}\,[9\Phi_1(x) + 8\Phi_2(x) + 5\Phi_3(x)] \tag{9.209}$$

$$\Phi_1(x) = \psi_1^{(1)}(x) \qquad \Phi_2(x) = \psi_2^{(1)}(x) + \psi_1^{(2)}(x) \qquad \Phi_3(x) = \psi_2^{(2)}(x) + \psi_1^{(3)}(x) \tag{9.210}$$

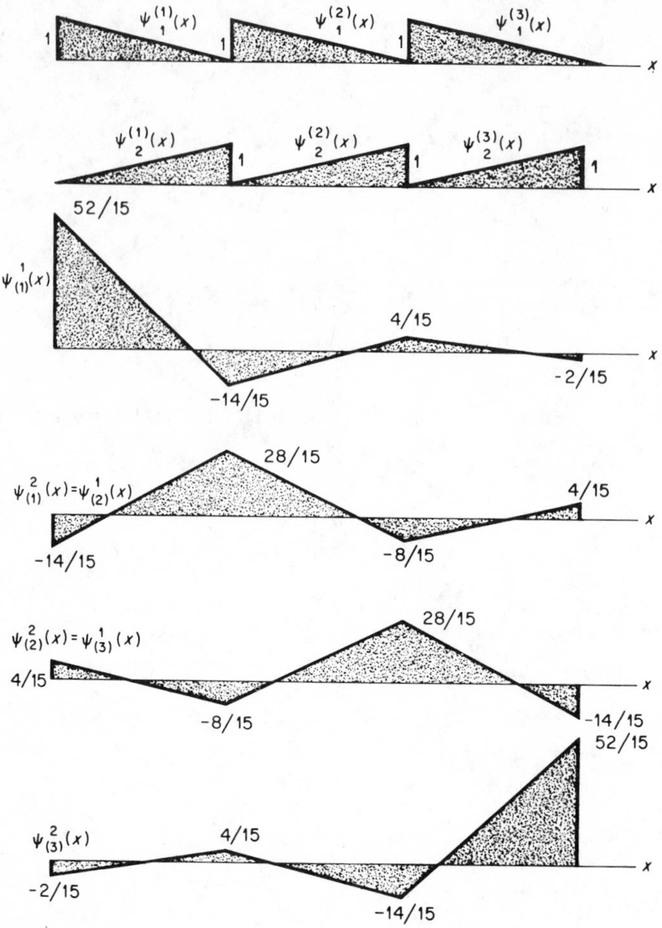

Fig. 9.2 Local basis functions $\psi_N^{(e)}(x)$ and their conjugate basis functions $\psi_{(e)}^N(x)$.

If the usual procedure for computing stresses in finite elements is used, we simply introduce (9.209) into the constitutive equation (9.201) for each element. This results in a discontinuous stress distribution which exhibits a finite discontinuity at the juncture of each element (Fig. 9.3). Further, the maximum stress computed in this manner is 16.7 percent in error.

A quite different profile is obtained if conjugate approximations are used. Assuming that $k(x)\, du/dx \in \Phi$, at least approximately, the strain energy defines a linear functional on $\overline{\mathscr{R}}$ and identifies $\sigma(x)$ as an element in the conjugate space Φ^*. Thus, it should be represented as a linear combination of the conjugate basis functions. Introducing (9.209) into (9.201) gives, as before, a local stress field $\sigma^{(e)}(x)$ for each element. The conjugate (nodal) components $\sigma_N^{(e)}$ are then obtained with the aid

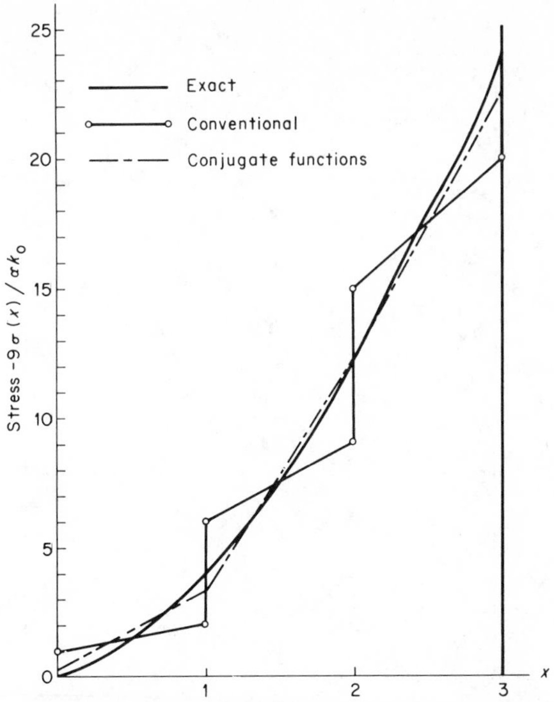

Fig. 9.3 Comparison of stress distributions computed conventionally with those obtained by conjugate approximation.

of (9.151):

$$\sigma_N^{(e)} = \langle \sigma(x), \psi_N^{(e)}(x) \rangle \tag{9.211}$$

Therefore, the conjugate-function representation of stress is given by

$$\sigma(x) = \sum_{e=1}^{3} \sigma_N^{(e)} \psi_{(e)}^N(x) \qquad N = 1, 2 \tag{9.212}$$

This conjugate stress profile is shown in Fig. 9.3 together with the exact solution and the discontinuous distribution obtained using conventional procedures. We see that the distribution obtained using conjugate-approximation functions is continuous at the junction of adjacent elements and that it indicates a maximum stress which is less than 6.5 percent in error.

Figure 9.4 indicates the results of a similar calculation using six elements and $u(x) = \alpha(1 - x^2/36)$. While averaged "conventional" stresses appear to be more accurate at interior nodes, the conjugate stresses lead to less mean-square error and, as can be seen in the figure, generally depict peaks in the stress variation more accurately.

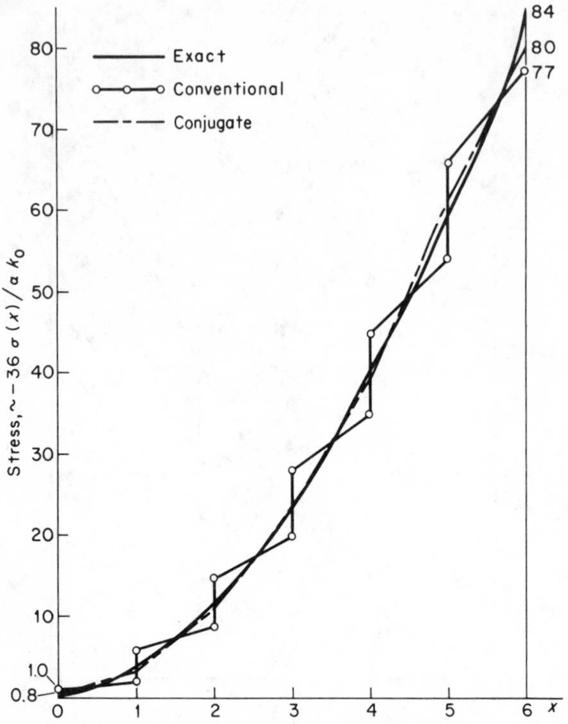

Fig. 9.4 Six-element stress approximation.

Example 9.3 Two-dimensional conjugate-approximation functions Essentially the same procedure outlined previously can be used for two- and three-dimensional finite elements. As a third example, we outline briefly the construction of the conjugate-approximation functions corresponding to a two-dimensional network of triangular elements.

Consider a triangular element in the x_1, x_2 plane, the vertices of which are the local nodal points. The local interpolation functions $\psi_N^{(e)}(\mathbf{x})$, where $\mathbf{x} = (x_1,x_2)$, are linear functions of x_1 and x_2 and satisfy $\psi_N^{(e)}(\mathbf{x}^M) = \delta_N^M$; $M, N = 1, 2, 3$. Introducing these functions into (9.145), we obtain for the local component of the fundamental matrix $C_{\Gamma\Delta}$,

$$c_{NM}^{(e)} = \int_A \psi_N^{(e)}(\mathbf{x})\psi_M^{(e)}(\mathbf{x})\,dA = \frac{A}{12}\begin{bmatrix} 2 & 1 & 1 \\ 1 & 2 & 1 \\ 1 & 1 & 2 \end{bmatrix} \tag{9.213}$$

where A is the area of the triangle. We observe that (9.213) is independent of the included angles α, β, γ formed by sides of the triangle. However, discrete models of various differential operators may depend on these angles—for example, for the triangle shown in Fig. 9.5a,

$$d_{NM}^{(e)} = \int_A \operatorname{grad}\psi_N^{(e)}(\mathbf{x})\operatorname{grad}\psi_M^{(e)}(\mathbf{x})\,dA$$

$$= \frac{1}{2}\begin{pmatrix} \cot\beta + \cot\gamma & -\cot\gamma & -\cot\beta \\ -\cot\gamma & \cot\gamma + \cot\alpha & -\cot\alpha \\ -\cot\beta & -\cot\alpha & \cot\alpha + \cot\beta \end{pmatrix} \tag{9.214}$$

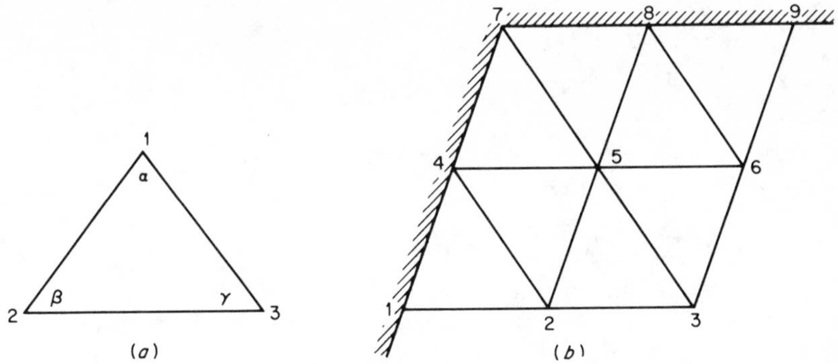

Fig. 9.5 Two-dimensional network of triangular elements.

To demonstrate the character of the conjugate-approximation functions for a specific finite-element representation, consider the network shown in Fig. 9.5*b*. In this case, we have from (9.142)

$$C_{\Gamma\Delta} = \langle \Phi_\Delta, \Phi_\Gamma \rangle = \frac{A}{12} \begin{bmatrix} 2 & 1 & 0 & 1 & 0 & 0 & 0 & 0 & 0 \\ & 6 & 1 & 2 & 2 & 0 & 0 & 0 & 0 \\ & & 4 & 0 & 2 & 1 & 0 & 0 & 0 \\ & & & 6 & 2 & 0 & 1 & 0 & 0 \\ & & & & 12 & 2 & 2 & 2 & 0 \\ & & & & & 6 & 0 & 2 & 1 \\ & \text{symmetric} & & & & & 4 & 1 & 0 \\ & & & & & & & 6 & 1 \\ & & & & & & & & 2 \end{bmatrix} \tag{9.215}$$

A being the area of an element. Inverting this matrix and making use of (9.148), we obtain the conjugate-approximation functions $\Phi^\Delta(\mathbf{X})$. Since, in the present example, the functions $\Phi_\Delta(\mathbf{X})$ are piecewise linear, $\Phi^\Delta(\mathbf{X})$ are also piecewise linear, and it is sufficient to merely calculate the values of the conjugate functions at each node. Rather than write out the entire collection of functions, we cite as representative examples the nodal values

$$\Phi^1(\mathbf{X}^\Delta) = \frac{12}{A}(0.580, -0.080, 0.009, -0.080, 0.027, -0.009, 0.009, -0.009, 0.009)$$

$$\Phi^5(\mathbf{X}^\Delta) = \frac{12}{A}(0.027, -0.027, -0.045, -0.027, 0.116, -0.027, -0.045, -0.027, 0.027) \tag{9.216}$$

$$\Phi^8(\mathbf{X}^\Delta) = \frac{12}{A}(-0.009, 0.000, 0.027, 0.018, -0.027, -0.054, -0.045, 0.214, -0.080)$$

These functions are illustrated in Fig. 9.6.

Observe that, in general, for the piecewise-linear approximation in two dimensions, the global approximation function $\Phi_\Delta(\mathbf{X})$ takes the shape of the "pyramid" function, as indicated in Fig. 9.7*a*, reaching a peak at node Δ and being zero in elements

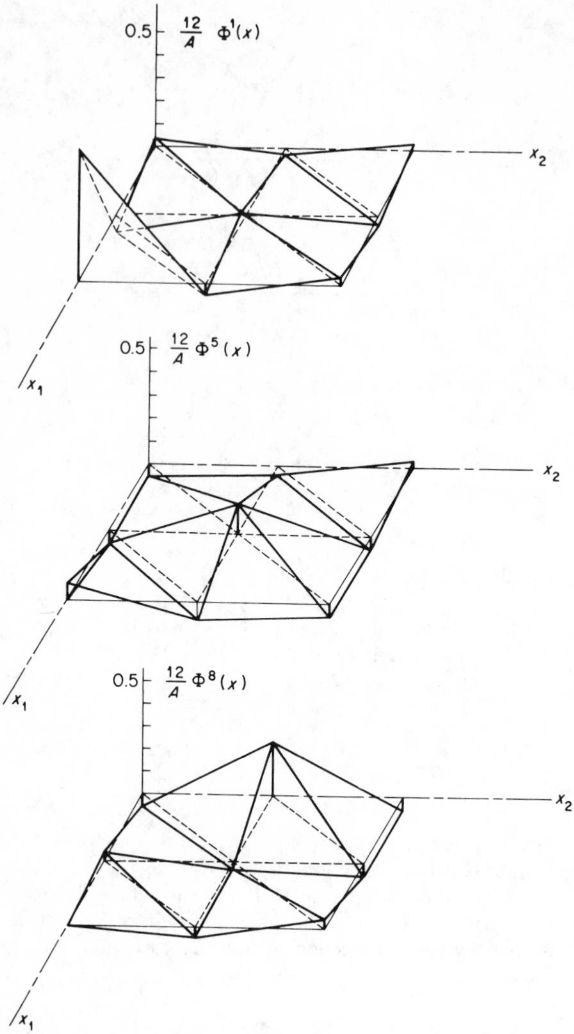

Fig. 9.6 Representative conjugate functions.

not containing node Δ. The conjugate approximation function $\Phi^{\Delta}(\mathbf{X})$ also peaks at node Δ but assumes nonzero values in all elements of the domain (Fig. 9.7*b*).

Example 9.4 Piecewise-linear approximation functions of one variable As an explicit but simple example to illustrate the foregoing discussions, consider the piecewise-linear approximation functions arising by dividing an interval I into $G - 1$ equal subintervals of length h and requiring $\Phi_{\Delta}(x)$ to vanish at all nodes except node $x_{\Delta} = h\Delta$, as shown in Fig. 9.8. Then

$$M_1 = M_G = \int_{\mathscr{R}} \Phi_1 \, d\mathscr{R} = \frac{h}{2} \qquad M_{\Delta} = \int_{\mathscr{R}} \Phi_{\Delta} \, d\mathscr{R} = h$$

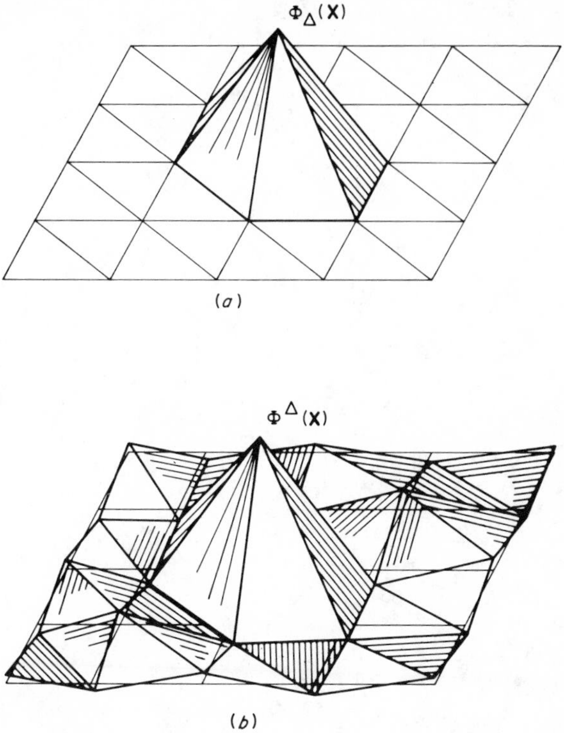

Fig. 9.7 Piecewise-linear two-dimensional global approximation function $\Phi_\Delta(\mathbf{X})$ and its conjugate function $\Phi^\Delta(\mathbf{X})$.

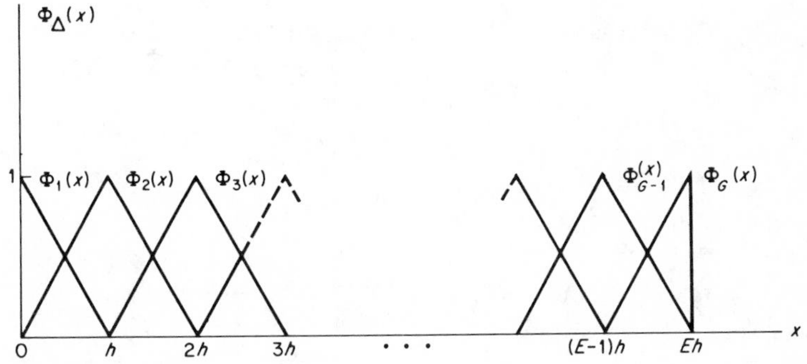

Fig. 9.8 Basis functions $\Phi_\Delta(x)$ for finite-element representation of interval on real axis.

for $\Delta = 1, \ldots, G-1$, $d\mathscr{R} = dx$, and the fundamental matrix is

$$C_{\Delta\Gamma} = \int_{\mathscr{R}} \Phi_\Delta \Phi_\Gamma \, dx = \frac{h}{6} \begin{bmatrix} 2 & 1 & 0 & \cdots & 0 \\ 1 & 4 & 1 & \cdots & 0 \\ 0 & 1 & 4 & \cdots & 0 \\ \cdot & \cdot & \cdot & \cdot & \cdot \\ 0 & \cdots & 4 & 1 & 0 \\ 0 & \cdots & 1 & 4 & 1 \\ 0 & \cdots & 0 & 1 & 2 \end{bmatrix} \tag{9.217}$$

Let A_n denote the $n \times n$ determinants,

$$A_n = \begin{vmatrix} 4 & 1 & 0 & \cdots & 0 & 0 & 0 \\ 1 & 4 & 1 & \cdots & 0 & 0 & 0 \\ 0 & 1 & 4 & \cdots & 0 & 0 & 0 \\ \cdot & \cdot & \cdot & \cdot & \cdot & \cdot & \cdot \\ 0 & 0 & 0 & \cdots & 4 & 1 & 0 \\ 0 & 0 & 0 & \cdots & 1 & 4 & 1 \\ 0 & 0 & 0 & \cdots & 0 & 1 & 4 \end{vmatrix} \tag{9.218}$$

satisfying the recurrence relation

$$A_{n+1} - 4A_n + A_{n-1} = 0 \tag{9.219}$$

Since

$$B_n = \begin{vmatrix} 2 & 1 & 0 & \cdots & 0 & 0 \\ 1 & 4 & 1 & \cdots & 0 & 0 \\ 0 & 1 & 4 & \cdots & 0 & 0 \\ \cdot & \cdot & \cdot & \cdot & \cdot & \cdot \\ 0 & 0 & 0 & \cdots & 4 & 1 \\ 0 & 0 & 0 & \cdots & 1 & 4 \end{vmatrix} \qquad C_n = \begin{vmatrix} 2 & 1 & 0 & \cdots & 0 & 0 \\ 1 & 4 & 1 & \cdots & 0 & 0 \\ 0 & 1 & 4 & \cdots & 0 & 0 \\ \cdot & \cdot & \cdot & \cdot & \cdot & \cdot \\ 0 & 0 & 0 & \cdots & 4 & 1 \\ 0 & 0 & 0 & \cdots & 1 & 2 \end{vmatrix} \tag{9.220}$$

are linear combinations of A_k, that is,

$$B_n = 2A_{n-1} - A_{n-2} = A_n - 2A_{n-1} \tag{9.221}$$

$$C_n = B_n - 2B_{n-1} = 3A_{n-2} \tag{9.222}$$

they also satisfy (9.219). Now (9.219) admits two independent solutions in geometric series $a_k = \alpha^k$, $b_k = \alpha^{-k}$, where $\alpha = 2 + \sqrt{3} = 3.732050808$ and $\alpha^{-1} = 2 - \sqrt{3} = 0.267949191$ satisfy the quadratic equation

$$\alpha^2 - 4\alpha + 1 = 0 \tag{9.223}$$

A_k and B_k are then linear combinations of α^k and α^{-k}:

$$A_k = \gamma(\alpha^k - \alpha^{-k-2}) \qquad B_k = \tfrac{1}{2}(\alpha^k + \alpha^{-k}) \tag{9.224}$$

where $\gamma = \alpha/(\alpha - \alpha^{-1}) = 1.077350269$. The beginnings of the series A_k, B_k are represented in Table 9.1, in which the relation

$$\Delta^2 A_k = A_{k+1} - 2A_k + A_{k-1} = 2A_k \tag{9.225}$$

has been used.

Table 9.1

k	A_k	ΔA_k	$\Delta^2 A_k$	B_k	ΔB_k	$\Delta^2 B_k$
−2	−1			7		
		1			−5	
−1	0		0	2		4
		1			−1	
0	1		2	1		2
		3			1	
1	4		8	2		4
		11			5	
2	15		30	7		14
		41			19	
3	56		112	26		52
		153			71	
4	209			97		

Now the inverse of the fundamental matrix is

$$C^{\Delta\Gamma} = (-1)^{\Delta+\Gamma}\frac{2}{h}B_{\Delta-1}\frac{B_{G-\Gamma}}{A_{G-2}} \qquad \Delta \le \Gamma \tag{9.226}$$

the elements for $\Delta > \Gamma$ being given by symmetry $C^{\Delta\Gamma} = C^{\Gamma\Delta}$. As an example, for $G = 7$,

$$C^{\Delta\Gamma} = \frac{1}{390h}\begin{bmatrix} 1{,}351 & -362 & -7 & -26 & 7 & -2 & 1 \\ -362 & 724 & -194 & 52 & -14 & 4 & -2 \\ 97 & -194 & 679 & -182 & 49 & -14 & 7 \\ -26 & 52 & -182 & 676 & -182 & 52 & -26 \\ 7 & -14 & 49 & -182 & 679 & -194 & 97 \\ -2 & 4 & -14 & 52 & -194 & 724 & -362 \\ 1 & -2 & 7 & -26 & 97 & -362 & 1{,}351 \end{bmatrix} \tag{9.227}$$

The corresponding conjugate functions are shown in Fig. 9.9. As a check, we can verify

$$M^{\Delta} = M_{\Gamma}C^{\Delta\Gamma} = 1 = \int_{\mathscr{R}} \Phi^{\Delta}(x)\,dx$$

Using (9.224), (9.226) becomes

$$C^{\Delta\Gamma} = (-1)^{\Delta+\Gamma}\frac{\sqrt{3}}{h}\frac{(\alpha^{\Delta} + \alpha^{-\Delta+2})(\alpha^{G-\Gamma} + \alpha^{-G+\Gamma})}{\alpha^{G} - \alpha^{-G+2}} \qquad \Delta \le \Gamma \tag{9.228}$$

or in the limit $G \to \infty$, $\Delta \approx \Gamma \approx G/2$,

$$C^{\Delta\Gamma} \approx (-1)^{|\Delta-\Gamma|}\frac{\sqrt{3}}{h}\alpha^{-|\Delta-\Gamma|} \qquad \Delta, \Gamma \text{ arbitrary} \tag{9.229}$$

or

$$C^{\Delta\Gamma} \approx \frac{\sqrt{3}}{h}\begin{bmatrix} \cdots & \cdots & \cdots & \cdots & \cdots \\ 1 & -\alpha^{-1} & \alpha^{-2} & -\alpha^{-3} & \cdots \\ -\alpha^{-1} & 1 & -\alpha^{-1} & \alpha^{-2} & -\alpha^{-3} \\ \cdots & -\alpha^{-1} & 1 & -\alpha^{-1} & \alpha^{-2} \\ \cdots & \cdots & -\alpha^{-1} & 1 & -\alpha^{-1} \\ \cdots & \cdots & \cdots & \cdots & \cdots \end{bmatrix} \tag{9.230}$$

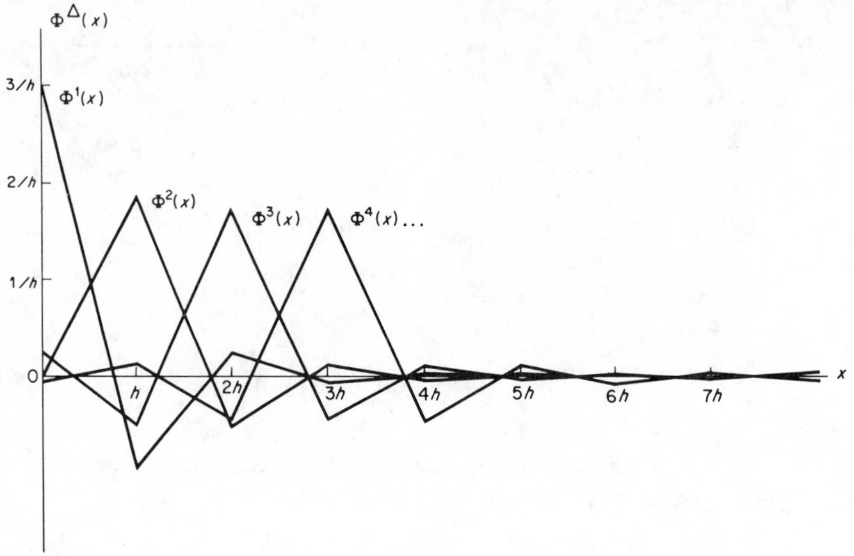

Fig. 9.9 Conjugate functions corresponding to basis functions in Fig. 9.8.

We now examine approximately the derivatives of functions corresponding to the piecewise linear basis functions of Fig. 9.8. Since the derivatives $\partial\Phi_\Delta(x)$ are discontinuous, the commutativity condition $\partial\Pi = \Pi\partial$ (see Theorem 9.1) is clearly violated. Yet it makes sense to introduce matrices $D_{\Delta\Gamma}$ and $B_{\Delta\Gamma}$ corresponding to (9.122) and (9.124),

$$D_{\Delta\Gamma} = \int_{\mathscr{R}} \Phi_\Delta \, \partial\Phi_\Gamma \, dx \qquad B_{\Delta\Gamma} = \int_{\mathscr{R}} \partial\Phi_\Delta \, \partial\Phi_\Gamma \, dx \tag{9.231}$$

While $B_{\Delta\Gamma}$ is symmetric, $D_{\Delta\Gamma}$ is almost antisymmetric, because

$$\int_{\mathscr{R}} \Phi_\Delta \, \partial\Phi_\Gamma \, dx = -\int_{\mathscr{R}} \Phi_\Gamma \, \partial\Phi_\Delta \, dx \qquad \text{for } 1 < \Delta, \Gamma < G \tag{9.232}$$

Numerically,

$$D_{\Delta\Gamma} = \frac{1}{2}\begin{vmatrix} -1 & 1 & 0 & 0 & \cdots & 0 & 0 & 0 \\ -1 & 0 & 1 & 0 & \cdots & 0 & 0 & 0 \\ 0 & -1 & 0 & 1 & \cdots & 0 & 0 & 0 \\ \cdots & \cdots & \cdots & \cdots & \cdots & \cdots & \cdots & \cdots \\ 0 & 0 & 0 & 0 & \cdots & -1 & 0 & 1 \\ 0 & 0 & 0 & 0 & \cdots & 0 & -1 & 1 \end{vmatrix}$$

$$B_{\Delta\Gamma} = \frac{1}{h}\begin{vmatrix} 1 & -1 & 0 & 0 & \cdots & 0 & 0 & 0 \\ -1 & 2 & -1 & 0 & \cdots & 0 & 0 & 0 \\ 0 & -1 & 2 & -1 & \cdots & 0 & 0 & 0 \\ \cdots & \cdots & \cdots & \cdots & \cdots & \cdots & \cdots & \cdots \\ 0 & 0 & 0 & 0 & \cdots & -1 & 2 & -1 \\ 0 & 0 & 0 & 0 & \cdots & 0 & -1 & 1 \end{vmatrix} \tag{9.233}$$

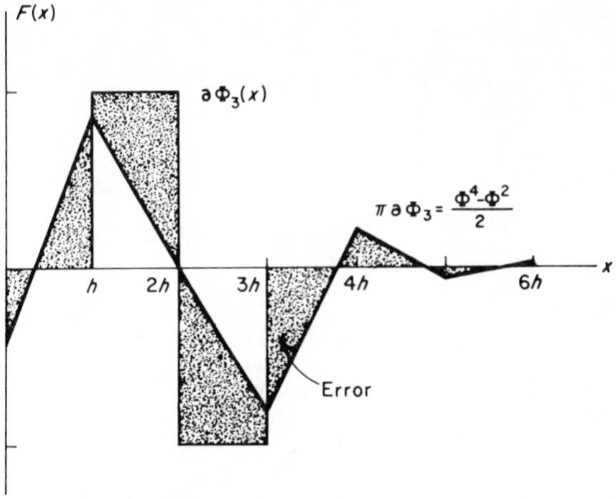

Fig. 9.10 Projection of derivative.

Now

$$\int_{\mathscr{R}} \partial F \, \partial G \, dx \approx B_{\Delta\Gamma} F^{\Delta} G^{\Delta} \tag{9.234}$$

is a reasonable approximation, while

$$\partial F \approx D_{\Delta\Gamma} F^{\Delta} \Phi^{\Gamma}(x) \tag{9.235}$$

will be rather crude. This is illustrated by Fig. 9.10, where $D_{\Gamma\Delta}$ is applied to a basis function $\Phi_3(x)$. This explains why computing $B_{\Delta\Gamma}$ out of $D_{\Delta\Gamma}$ by (9.116) would give a poor approximation. On the other hand, (9.231) suggests that $-B_{\Delta\Gamma}$ may be used to compute second derivatives:

$$\partial^2 F(x) \approx -B_{\Delta\Gamma} F^{\Delta} \Phi^{\Gamma}(x) \tag{9.236}$$

instead of (9.115), (9.117). In fact,

$$\begin{aligned} \Pi \, \partial^2 \Phi_{\Delta} &= \Pi \left[\frac{1}{h} \, \delta(x - \Delta h - 2h) - \frac{2}{h} \, \delta(x - \Delta h + h) + \frac{1}{h} \, \delta(x - \Delta h) \right] \\ &= \frac{1}{h} \, (\Phi^{\Delta-2} - 2\Phi^{\Delta-1} + \Phi^{\Delta}) \end{aligned} \tag{9.237}$$

which is $1/h$ times the central difference operator at $x = (\Delta - 1)h$; this yields

$$\int_{\mathscr{R}} \Phi_{\Delta} \, \partial^2 \Phi_{\Gamma} \, dx = -B_{\Delta\Gamma} \tag{9.238}$$

Example 9.5 Polynomials, analytic functions Due to the simplicity of the functions $\Phi_{\Delta}(x)$ in Fig. 9.8, it is possible to find explicit formulas for the components of an analytic function $F(x) = \sum a_n x^n$. To simplify the forms of these formulas, we shall shift the node numbers so that $\Delta = 0, 1, 2, \ldots, N$, $N = G - 1$. Now consider the integral,

$$\int_0^h x^n \Phi_{\Delta}(x) \, dx = \int_0^h (h\Delta + x)^n \left(1 - \frac{x}{h}\right) dx + \int_0^h (\Delta h - x)^n \left(1 - \frac{x}{h}\right) dx \tag{9.239}$$

Using

$$1 - \frac{x}{h} = \Delta + 1 - \frac{\Delta h + x}{h} = 1 - \Delta + \frac{\Delta h - x}{h} \tag{9.240}$$

we find

$$\int_0^h x^n \Phi_\Delta(x)\, dx = \frac{h^{n+1}}{(n+1)(n+1)} \hat{\Delta}^2(\Delta)^{n+2} \qquad 0 < \Delta < N \tag{9.241}$$

where

$$\hat{\Delta}^2(\Delta)^{n+2} = (\Delta + 1)^{n+2} - 2(\Delta)^{n+2} + (\Delta - 1)^{n+2} \tag{9.242}$$

For the boundaries, we have

$$\int_0^h x^n \Phi_0(x)\, dx = \begin{cases} 0 & n \text{ odd} \\ \dfrac{2h^{n+1}}{(n+1)(n+2)} & n \text{ even} \end{cases} \tag{9.243}$$

and

$$\int_0^h x^n \Phi_N(x)\, dx = \frac{h^{n+1}}{(n+1)(n+2)} [(N-1)^{n+2} - N^{n+2} + (n+2)N^{n+1}] \tag{9.244}$$

or, for large N,

$$\int_0^h x^n \Phi_N(x)\, dx \approx \frac{h^{n+1}}{(n+1)(n+2)} \binom{n+2}{2} N^n \tag{9.245}$$

Returning now to $F(x) = \sum a_n x^n$, it is clear that for finite-element approximations,

$$\sum_n a_n x^n \approx F^\Delta \Phi_\Delta(x) = F_\Delta \Phi^\Delta(x) \tag{9.246}$$

where, for the best approximation,

$$F_\Delta = C_{\Delta\Gamma} F^\Gamma = \langle \Phi_\Delta, \sum_n a_n x^n \rangle = \sum_n a_n \langle \Phi_\Delta, x^n \rangle = \sum_n a_n \int_{\mathscr{R}} \Phi_\Delta(x) x^n\, dx \tag{9.247}$$

or

$$F_\Delta = \sum_{e=1}^{N} \sum \left[a_n \overset{(e)}{\Omega}{}_\Delta^N \int_0^h \psi_N^{(e)}(x) x^n\, dx \right] \tag{9.248}$$

We obtain

$$F_\Delta = \frac{1}{h} \hat{\Delta}^2 F^{-2}(h\Delta) \tag{9.249}$$

where $F^{-2}(x) = \sum_0^\infty a_n x^{n+2}/(n+1)(n+2)$ is a second antiderivative of $F(x)$ and the operator $\hat{\Delta}^2$ is defined in (9.242). At the boundaries

$$F_0 = \frac{1}{h} F^{-2}(h) \qquad F_N = F^{-1}(Nh) - \frac{1}{h} \hat{\Delta} F^{-2}(Nh) \tag{9.250}$$

F^{-1} being a first antiderivative and $\hat{\Delta}$ the first backward difference operator. Obviously, for very large N, $F_\Delta = hF(h\Delta)$.

10 FINITE ELEMENTS AND INTERPOLATION FUNCTIONS

The theory of finite elements, as developed to this point, has largely dealt with elements of arbitrary shape, containing an arbitrary number of nodal points. The local approximations $\bar{f}^{(e)}(\mathbf{x})$ over each element have been described in terms of general interpolation functions $\psi_N^{(e)}(\mathbf{x})$, the precise forms of which have, with a few exceptions, not been specified. In applications of the theory, however, this generality must be abandoned and specific types of elements and their corresponding interpolation functions must be identified. In devising various finite-element models, we generally require that the individual elements be of reasonably simple geometric shapes and that the functions $\psi_N^{(e)}(\mathbf{x})$ be of correspondingly simple forms, usually polynomials. If to this requirement we add the condition that the functions $\psi_N^{(e)}(\mathbf{x})$ be such that the approximation $\bar{F}(\mathbf{x})$ is continuous at interelement boundaries and that $\bar{F}(\mathbf{x})$ converges to $F(\mathbf{x})$ as the network of elements is refined,† we effectively impose rather severe restrictions on both the geometry of acceptable finite elements and on the forms of the interpolation functions. In this article, we examine finite-element models of operator equations as well as the question of convergence for a wide class of elements. We also present examples of a number of acceptable finite elements and their corresponding interpolation functions.

† Synge [1957, pp. 207–213] investigated convergence of polyhedral functions defined by linear interpolations over triangular regions. His analysis is directly applicable to similar finite-element models, but applications, as in the earlier work of Courant [1943], are restricted to first-order representations. The studies of spline functions and bicubic interpolation of Birkhoff and Garabedian [1960] and Birkhoff and de Boor [1965] are applicable to finite-element approximations of smooth surfaces; see also Ahlberg, Nilson, and Walsh [1967] in this connection. Melosh [1963*a*] presented sufficient criteria for monotone convergence of finite-element analyses of linear-plate problems using potential energy concepts and Fraeijs de Veubeke [1964*b*, 1965, 1966] investigated bounds on various finite-element solutions to linear elasticity problems. Discussions of convergence in applications of finite elements to certain boundary-value problems were presented by Friedrichs and Keller [1967], Varga [1967], and Ciarlet, Schultz, and Varga [1967]. Detailed studies of convergence of finite-element approximations of a class of linear-operator equations were presented by Key [1966] and Arantes e Oliveira [1968, 1969, 1971]. Hybrid models were considered by Tong and Pian [1967] and completeness and convergence properties of certain types of finite-element approximations were examined by McLay [1963, 1967], Johnson and McLay [1968], Dunne [1968], Zlamal [1968], Göel [1968], Walz, Fulton, and Cyrus [1969], Tabandeh [1970], and Bramble and Zlamal [1970]. The generation of "hill functions" for finite-element approximations was discussed by Babuška [1970], and stability, error, and convergence of certain finite-element approximations were examined by Fix and Strang [1969], Strang [1970] and Fried [1971]. Several details of convergence and completeness criteria are summarized in the paper by Felippa and Clough [1968]. See also Clough and Tocher [1966], Greene, Jones, McLay, and Strome [1969], Oden and Brauchli [1971*b*], and Oden [1970*c*, 1971*b*.].

10.1 OPERATORS ON NORMED SPACES

In general, we are concerned with solving problems characterized by equations of the form

$$\mathscr{P}(u) = f \tag{10.1}$$

wherein $u = u(\mathbf{x})$ is an element of a normed linear space $\mathscr{V}$, generally consisting of functions defined over some open region $\mathscr{R}$ and its boundary $\partial\mathscr{R}$ of k-dimensional euclidean space, $f = f(\mathbf{x})$ is an element of another normed space $\mathfrak{U}$ (often $\mathfrak{U} = \mathscr{V}$), and $\mathscr{P}: \mathscr{V} \to \mathfrak{U}$ is an operator from $\mathscr{V}$ into $\mathfrak{U}$ or, more specifically, $\mathscr{P}$ carries some set $\mathscr{D}$ in $\mathscr{V}$, called the domain of $\mathscr{P}$, into $\mathfrak{U}$. In addition to (10.1), the solution u must also satisfy certain boundary conditions on $\partial\mathscr{R}$.

Most methods of approximation amount to constructing sequences $\{\bar{u}^n\}$ of approximations of u in linear subspaces of $\mathscr{V}$ that converge in some sense to u so as to give meaning to the symbolism $\lim_{n\to\infty} \mathscr{P}(\bar{u}^n) = \mathscr{P}(u)$. It may happen, however, that the domain $\mathscr{D}$ of $\mathscr{P}$ does not include the elements $\bar{u}^n$; then it is necessary to extend the domain of the operator $\mathscr{P}$ in such a way that it belongs to (is dense in) a complete normed space. Before examining these ideas in connection with finite-element approximations, we review a number of basic concepts.

Cauchy sequences and completeness Consider a real normed linear space $\mathscr{V}$; that is, a normed linear space whose elements can be multiplied by real numbers. Let u denote an element of $\mathscr{V}$ and $\|u\|$ its corresponding norm. If to every positive integer n we assign an element $u^n \in \mathscr{V}$, then the collection $u^1, u^2, \ldots, u^n, \ldots$, written $\{u^n\}$, forms a sequence in $\mathscr{V}$. A sequence $\{u^n\}$ in $\mathscr{V}$ is said to have a limit $u^* \in \mathscr{V}$ if for every $\epsilon > 0$ there is an integer m such that $\|u^n - u^*\| < \epsilon$ for all $n > m$. If such is the case, we say that the sequence $\{u^n\}$ converges to the limit u^* in the norm $\|u\|$ (i.e., mean convergence), and we write $\lim_{n\to\infty} \{u^n\} = u^*$ or, more precisely, $\lim_{n\to\infty} \|u^n - u^*\| = 0$.

A sequence $\{u^n\}$ is called a Cauchy sequence if $\lim_{m,n\to\infty} \|u^m - u^n\| \to 0$; that is, for every $\epsilon > 0$, there is an integer N such that $\|u^m - u^n\| < \epsilon$ for $m, n > N$. It is clear that every convergent sequence is a Cauchy sequence, for if u^* is the limit of $\{u^n\}$, then $\|u^m - u^n\| = \|(u^m - u^*) - (u^n - u^*)\| \leq \|u^m - u^*\| + \|u^n - u^*\| < \epsilon$, where $\epsilon = \epsilon_1 + \epsilon_2$ and $\|u^m - u^*\| < \epsilon_1$ for $m > N$ and $\|u^n - u^*\| < \epsilon_2$ for $n > N$. Thus $\|u^m - u^n\| < \epsilon$ for m, $n > N$. The converse is *not* true, however, for it may happen that a sequence $\{u^n\} \in \mathscr{V}$ converges to a limit $u^* \notin \mathscr{V}$. A well-known example of an incomplete metric space is the space of rational numbers; e.g., the sequence $\{1, \frac{3}{4}, \frac{13}{15}, \ldots\}$ converges to the irrational number $\pi/4$ and, hence, does not converge in the space of rational numbers.

If every Cauchy sequence of elements of a normed linear space $\mathscr{V}$

converges to an element of $\mathscr{V}$, then the space $\mathscr{V}$ is said to be *complete*. A complete normed linear space is called a Banach space. If $\mathscr{V}$ is not complete, it is always possible to embed it in a complete space $\overline{\mathscr{V}}$ called the completion of $\mathscr{V}$. Indeed, $\overline{\mathscr{V}}$ is a completion of $\mathscr{V}$ if $\mathscr{V}$ is a subspace of $\overline{\mathscr{V}}$ and if $\mathscr{V}$ is everywhere dense in $\overline{\mathscr{V}}$. By "everywhere dense," we mean that $\overline{\mathscr{V}}$ is the closure of $\mathscr{V}$; that is, every element $u \in \mathscr{V}$ is arbitrarily close to an element $\bar{u}$ of $\overline{\mathscr{V}}$ in the sense that for every $\epsilon > 0$, there exists an element $\bar{u} \in \overline{\mathscr{V}}$ such that $\|u - \bar{u}\| < \epsilon$. For normed spaces, the notion of completeness has meaning only in connection with the type of norm used to characterize the space.

Technically, the process of completing an incomplete space amounts to adding to the space the limit points of Cauchy sequences in the space (more specifically, we may complete a space by constructing equivalence classes of Cauchy sequences and their limit points so that if $\bar{u}$ is an equivalence class, $\|\bar{u}\| = \lim_{n\to\infty} \|u^n\|$, where $\{u^n\} \in \mathscr{V}$). As an important example, consider the case in which $\mathscr{V} = \mathscr{H}$ is an inner-product space, in which the inner product is given by

$$\langle u,v\rangle = \int_{\mathscr{R}} uv \, d\mathscr{R} \tag{10.2}$$

where integration in the Lebesque sense is implied. Here u and v are functions defined on $\mathscr{R}$ and $\|u\| = \langle u,u\rangle^{\frac{1}{2}} < \infty$. Suppose that $\mathscr{H}$ consists of the functions $u \in C^0(\mathscr{R})$; that is, the class of functions continuous on $\mathscr{R}$. Then $\mathscr{H}$ is incomplete, because we can construct a sequence of continuous functions that converges to a discontinuous function. A common example of such a sequence is

$$u^n(x) = \begin{cases} 0 & 0 \le x \le \dfrac{1}{2} - \dfrac{1}{n} \\[2ex] nx + 1 - \dfrac{n}{2} & \dfrac{1}{2} - \dfrac{1}{n} \le x \le \dfrac{1}{2} \\[2ex] 1 & \dfrac{1}{2} \le x \le 1 \end{cases}$$

each component being continuous for $x \in [0,1]$ but which converges in the norm $\|u\| = \langle u,u\rangle^{\frac{1}{2}}$ to the discontinuous unit-step function $u^*(x) = h(x - \frac{1}{2})$. Moreover, it is impossible to make $u^*(x)$ continuous by changing its values by a set of measure zero. To complete $\mathscr{H}$, we associate with $\{u^n\}, \{v^n\} \in \mathscr{H}$ the limit points u^*, v^* so that by $u^* = v^*$ we mean $\lim_{m,n\to\infty} \|u^n - v^n\| = 0$. Then $\langle u^*,v^*\rangle = \lim_{n\to\infty} \langle u^n,v^n\rangle$ and $\|u^*\| = \lim_{n\to\infty} \|u^n\|$. The success of such a completion procedure is guaranteed by the famous Hahn-Banach theorem.† *Every linear functional* [such as $g(u) = \langle u,v \rangle$] *defined on a linear subspace* $\mathscr{V}'$ *of a normed linear space* $\mathscr{V}$ *can be extended to the entire space with preservation of the norm.* Hence, in defining $\|u^*\|$, we may use the same norm defined for the incomplete space $\mathscr{H}$. In our example, the completion of $\mathscr{H}$ is $L_2(\mathscr{R})$, the space of square integrable functions on $\mathscr{R}$. Thus $C^0(\mathscr{R})$ is dense in $L_2(\mathscr{R})$; we can generate a sequence of continuous functions $u^n \in C^0(\mathscr{R})$ such that for every $\epsilon > 0$, there is an N such that $\|u^* - u^n\| < \epsilon$ for $n > N$. In other words, any square integrable function on $\mathscr{R}$ can be approximated in the mean by a continuous function. Since in accordance with the Weierstrass approximation theorem every continuous function $u \in C^0(\mathscr{R})$ can be uniformly approximated by a polynomial p, then for every n, there is a polynomial p^n such that $\|u^n - p^n\| \le \epsilon$, $\epsilon > 0$, for all $\mathbf{x} \in \mathscr{R}$. This implies also that for bounded $\mathscr{R}$, $\|u^n - p^n\| < \epsilon'$, where ϵ' depends on $\mathscr{R}$, so that $\|u^* - u^n\| \le \|u^* - p^n\| + \|u^n - p^n\| < \epsilon''$ and $\|u^* - p^n\| < \epsilon'' - \epsilon' = \epsilon$ for $n > N$. Thus, we may also approximate in the mean any square integrable function by a polynomial. More generally, we may consider the space $\mathscr{H}$ as that of functions $u \in C^r(\mathscr{R})$; that is, functions belonging to the class of all functions continuous on $\mathscr{R}$ (or $\mathscr{R} + \partial\mathscr{R}$) together with their first r partial derivatives. Then, if, instead of (10.2), we define

$$\langle u,v \rangle = \int_{\mathscr{R}} \left(uv + \frac{\partial u}{\partial x_i}\frac{\partial v}{\partial x_i} + \frac{\partial^2 u}{\partial x_i\, \partial x_j} \cdot \frac{\partial^2 v}{\partial x_i\, \partial x_j} + \cdots + \frac{\partial^r u}{\partial x_{i_1}\, \partial x_{i_2} \cdots \partial x_{i_r}} \cdot \frac{\partial^r v}{\partial x_{i_1}\, \partial x_{i_2} \cdots \partial x_{i_r}} \right) d\mathscr{R} \tag{10.3}$$

wherein $(x_1,x_2, \ldots ,x_k) = \mathbf{x}$ is a point in $\mathscr{R}$, $i_1 + i_2 + \cdots + i_r = r$, $i \le j$, $i_1 \le i_2 \le \cdots \le i_r$, and $i, j, i_1, i_2, \ldots, i_r = 1, 2, \ldots, k$, we find that $\mathscr{H}$ is still incomplete in the norm $\|u\| = \langle u,u \rangle^{\frac{1}{2}}$. By completing $\mathscr{H}$ in a manner similar to that described above,‡ we obtain a space denoted $L_2^r(\mathscr{R})$.

† Cf. Kolmogorov and Fomin [1957, p. 86].

‡ In order that $L_2^r(\mathscr{R})$ be complete, it is necessary to consider u and v in (10.3) to be not only square integrable on $\mathscr{R}$ but also to possess *generalized derivatives* of all orders up to and including r. By a generalized derivative of u of order r in the L_2 sense, we mean an integrable function w such that

$$\int_{\mathscr{R}} u \frac{\partial^r \varphi}{\partial x_{i_1} \cdots \partial x_{i_r}} d\mathscr{R} = (-1)^r \int_{\mathscr{R}} w\varphi \, d\mathscr{R}$$

where φ is a r-times differentiable function that together with its derivatives vanishes on a boundary layer of $\mathscr{R}$. Thus u may not be continuously differentiable r times in $\mathscr{R}$, but it must be square integrable on $\mathscr{R}$. [The space which consists of functions on $\mathscr{R}$ which have all generalized derivatives of order r and which are integrable in powers of p is called the *Sobolev space* $W_p^r(\mathscr{R})$.] For a more detailed discussion, consult Mikhlin [1965, pp. 45–49] or Mikhlin and Smolitskiy [1967, pp. 151–155].

In view of our previous definition of a dense set, it is clear that a set $\mathscr{A}$ is dense in a space $\mathscr{V}$ if each element $u \in \mathscr{V}$ is the limit of a sequence of elements $u^n \in \mathscr{A}$. If $\mathscr{V}$ is obtained by the completion of an incomplete space $\mathscr{A}$ [as $L_2(\mathscr{R})$ was obtained by completing $C^0(\mathscr{R})$], then obviously $\mathscr{A}$ is dense in $\mathscr{V}$. We are often faced with cases in which a subspace $\mathscr{A}^n$ of finite dimension n is spanned by a collection of n linearly independent elements $\varphi_1, \varphi_2, \ldots, \varphi_n$. Then each element $u^n \in \mathscr{A}^n$ is of the form

$$u^n = \sum_{i=1}^{n} \alpha_i \varphi_i$$

α_i being real or complex numbers. Since $\mathscr{A}^n$ is dense in a complete normed linear space $\mathscr{V}$ if for some $\epsilon > 0$ and every $u \in \mathscr{V}$, $\|u^n - u\| < \epsilon$ for $n > N$, it is meaningful to speak of *completeness of the set of elements* $\varphi_1, \varphi_2, \ldots, \varphi_n$ in the same sense. Specifically, a set of elements $\varphi_1, \varphi_2, \ldots, \varphi_n$ is said to be complete in the norm $\|\cdot\|$ with respect to a complete normed space $\mathscr{V}$, if for an arbitrary element $u \in \mathscr{V}$ with finite norm, it is possible to find scalars $\alpha_1, \alpha_2, \ldots, \alpha_n$ such that for every $\epsilon > 0$ there exists an integer N such that

$$\left\| u - \sum_{i=1}^{n} \alpha_i \varphi_i \right\| < \epsilon \qquad \text{for } n > N \tag{10.4}$$

A space $\mathscr{V}$ is *separable* if there exists a countable set of elements $(\varphi_1, \varphi_2, \ldots, \varphi_n)$ whose finite linear combinations are dense in $\mathscr{V}$. Hence, if (10.4) holds for all $u \in \mathscr{V}$, then $\mathscr{V}$ is separable.

Linear operators† Consider a complex Hilbert space $\mathscr{H}$, the elements $u, v, w, \ldots$ of which are functions defined on a region $\mathscr{R}$ of $\mathscr{E}^k$. The inner product on $\mathscr{H}$ is again denoted $\langle u,v\rangle$ and the norm $\|u\| = \langle u,u\rangle^{\frac{1}{2}}$, but now $\langle u,v\rangle = \overline{\langle v,u\rangle}$, where $\overline{\langle v,u\rangle}$ denotes the complex conjugate of $\langle v,u\rangle$. In most applications, the elements $u, v, \ldots \in \mathscr{H}$ are taken to be the square-integrable functions [that is, $u \in L_2(\mathscr{R})$], so that $\langle u,v\rangle$ is as in (10.2) with v replaced by $\bar{v}$.

Now consider an operator $\mathscr{P} \equiv \mathscr{L}$ defined on a dense (linear) set $\mathscr{D}$ of $\mathscr{H}$. We will make use of the following definitions concerning $\mathscr{L}$.

$\mathscr{L}$ is *linear* if it is additive and homogeneous; that is, if for any two elements $u, v \in \mathscr{D}$ and any two scalars α_1, α_2, $\mathscr{L}(\alpha_1 u + \alpha_2 v) = \alpha_1 \mathscr{L}(u) + \alpha_2 \mathscr{L}(v)$. Likewise, if $0 \in \mathscr{D}$ and $\mathscr{L}$ is linear, it follows that $\mathscr{L}(0) = 0$.

$\mathscr{L}$ is *symmetric* if for any two elements $u, v \in \mathscr{D}$, $\langle \mathscr{L}u,v\rangle = \langle u,\mathscr{L}v\rangle$. If $\mathscr{H}$ is complex, a necessary and sufficient condition that $\mathscr{L}$ be symmetric is that $\langle \mathscr{L}u,u\rangle$ be real. If $\mathscr{H}$ is real, additional conditions must be imposed to insure that $\mathscr{L}$ is symmetric.

† In this section, we draw heavily from the works of Mikhlin [1964, 1965].

$\mathscr{L}$ is *positive definite* if for every $u \in \mathscr{D}$, $\langle u, \mathscr{L}u \rangle \geq 0$ and $\langle u, \mathscr{L}u \rangle = 0$ if and only if $u = 0$, $\mathscr{L}$ being symmetric.

$\mathscr{L}$ is *bounded* (above and below) if for every $u \in \mathscr{D}$ there exist constants c_1 and c_2 such that $c_1 \|u\| \leq \|\mathscr{L}u\| \leq c_2 \|u\|$.

Let $\mathscr{L}$ be bounded; then the real number c providing the least upper bound is called the *norm* of $\mathscr{L}$ and is denoted $\|\mathscr{L}\|$. Hence $\|\mathscr{L}u\| \leq \|\mathscr{L}\| \, \|u\|$.

$\mathscr{L}$ is *positive-bounded below* if for every $u \in \mathscr{D}$ there exists a constant $\gamma \neq 0$ such that $\langle u, \mathscr{L}u \rangle \geq \gamma^2 \|u\|^2$.

If for every sequence $\{u^n\}$ of elements in $\mathscr{D}$ such that $\lim_{n \to \infty} \|u^n - u^*\| = 0$, $\lim_{n \to \infty} \|\mathscr{L}u^n - \mathscr{L}u^*\| = 0$, then $\mathscr{L}$ is continuous at u^*.

An operator $\mathscr{M}$ with domain $\mathscr{D}_m$ is called an *extension* of an operator $\mathscr{L}$ with domain $\mathscr{D}_l$ if $\mathscr{D}_l \subset \mathscr{D}_m$ and if, for every $u \in \mathscr{D}_l$, $\mathscr{M}u = \mathscr{L}u$.

A linear operator $\mathscr{L}$ from $\mathscr{H}$ into the real or complex number field is called a *linear functional.* Every bounded linear functional l on $\mathscr{H}$ has the form $lu = \langle u, v \rangle$, where v is a unique, fixed element of $\mathscr{H}$ (technically, v belongs to the conjugate space $\mathscr{H}^*$; that is, the space of all linear functionals on $\mathscr{H}$; but in the present context, $\mathscr{H} = \mathscr{H}^*$).

Let us now consider the problem of solving the equation

$$\mathscr{L}u = f \tag{10.5}$$

where $\mathscr{L}$ is a linear operator defined on a set $\mathscr{D}$ dense in $\mathscr{H}$ and f is a prescribed element of $\mathscr{H}$. In addition to (10.5), we also have a set of boundary conditions

$$\mathscr{B}u = g \tag{10.6}$$

on $\partial\mathscr{R}$, where $\mathscr{B}$ is also a linear operator. However, we shall first consider the case of homogeneous boundary conditions; that is, $g = 0$; so that $\mathscr{D}$ is temporarily restricted to those elements satisfying $\mathscr{B}u = 0$ on $\partial\mathscr{R}$.

We now cite three important theorems† concerning the solution of (10.5).

Theorem 10.1 (Uniqueness) *If the operator* $\mathscr{L}$ *in* (10.5) *is positive-definite, then* (10.5) *can have no more than one solution.*

Proof The proof is simple; suppose $\mathscr{L}u_1 = \mathscr{L}u_2 = f$. Then $u = u_1 - u_2$ is such that $\mathscr{L}u = 0$. Since $\mathscr{L}$ is positive definite, $\langle u, \mathscr{L}u \rangle = 0 \Rightarrow u = 0$. Hence $u_1 = u_2$.

Theorem 10.2 *Let* $\mathscr{L}$ *be a positive-definite linear operator and suppose that* (10.5) *has a solution* u^*. *Then the functional*

$$J(u) = \langle \mathscr{L}u, u \rangle - \langle u, f \rangle - \langle f, u \rangle \tag{10.7}$$

† Cf. Mikhlin [1964, pp. 74, 318].

assumes its minimal value when $u = u^*$. *Conversely, if there exists in the domain* $\mathscr{D}$ *of* $\mathscr{L}$ *a function* u^* *that minimizes* $J(u)$, *then* u^* *satisfies* (10.5). *Here* $\mathscr{H}$ *is complex.*

Proof Let u^* be such that $\mathscr{L}u^* = f$ and let η be an arbitrary element of $\mathscr{D}$. Then, since $\mathscr{L}$ is symmetric,

$$J(u^* + \eta) = J(u^*) + \langle \mathscr{L}u^* - f, \eta\rangle + \langle \eta, \mathscr{L}u^* - f\rangle + \langle \mathscr{L}\eta,\eta\rangle$$

$$= J(u^*) + \langle \mathscr{L}\eta,\eta\rangle$$

Since $\mathscr{L}$ is positive definite, $\langle \mathscr{L}\eta,\eta\rangle \geq 0$. Hence, if v is an arbitrary element in $\mathscr{D}$ of the form $v = u^* + \eta$, $\eta \neq 0$, then

$$J(v) = J(u^*) + \langle \mathscr{L}\eta,\eta\rangle > J(u^*)$$

Therefore, of all $v \in \mathscr{D}$, $J(v)$ assumes its minimum value at $v = u^*$. This completes the proof of the first part of the theorem.

The proof of the second part rests on the fact that *any element* $u \in \mathscr{H}$ *orthogonal to every element* η *of a set* $\mathscr{D}$ *dense in* $\mathscr{H}$ *must be zero.* Let $\epsilon > 0$ and $\eta \in \mathscr{D}$. Then

$$J(u^* + \epsilon\eta) - J(u^*) \geq 0$$

if u^* is the element that minimizes $J(u)$. Making use of (10.7), we can obtain

$$\epsilon^2\langle \mathscr{L}\eta,\eta\rangle + 2\epsilon \operatorname{Re}\,[\langle \mathscr{L}u^* - f, \eta\rangle] \geq 0$$

which, since ϵ can be chosen to be real and positive, is possible only if $\operatorname{Re}\,[\langle \mathscr{L}u^* - f, \eta\rangle] = 0$. Further, $\operatorname{Im}\,[\langle \mathscr{L}u^* - f, \eta\rangle] = 0$, as can be seen by replacing η by $i\eta$. Hence

$$\langle \mathscr{L}u^* - f, \eta\rangle = 0$$

That is, $\mathscr{L}u^* - f$ is orthogonal to an arbitrary element η of a set $\mathscr{D}$ dense in $\mathscr{H}$. This can happen only if $\mathscr{L}u^* = f$, which proves the theorem.

Unfortunately, the above theorem rests on several important assumptions that may not be valid in many cases. First, we assumed that there exists in the linear manifold $\mathscr{D}$ an element u^* that either satisfies (10.5) or minimizes the functional in (10.7). The question of existence of minimal elements u^* for $J(u)$ is answered, in part, by a theorem, presented below, that involves the concept of *energy norms* and *convergence in energy*. Briefly, a new inner product $[u,v] \equiv \langle \mathscr{L}u,v\rangle$ leads to another definition of a norm on $\mathscr{D}$, $\|u\|_{\mathscr{L}} = [u,u]^{\frac{1}{2}}$, called the energy norm or the energy of the element u. If given an $\epsilon > 0$ and a sequence $\{u^n\} \in \mathscr{D}$ such that $\|u^n - u\|_{\mathscr{L}} < \epsilon$ for $n > N$, then $\{u^n\}$ is said to *converge in energy* to u. Likewise, a set of elements $\varphi_1, \varphi_2, \ldots, \varphi_n$ is said to be *complete in energy* if given $\epsilon > 0$ and

$u^* \in \mathscr{V}$, there exist constants $\alpha_1, \alpha_2, \ldots, \alpha_n$ such that

$$\left\| u^* - \sum_{i=1}^{n} \alpha_i \varphi_i \right\|_{\mathscr{L}} < \epsilon$$

for n greater than some integer N. If $\mathscr{L}$ is positive-bounded-below, it is easily shown that convergence in energy of a sequence $\{u^n\}$ to u^* also implies mean convergence of $\{u^n\}$ to u^*. In seeking elements u^* that minimize $J(u)$, we require that (10.5) have a solution possessing finite energy; that is, if $\mathscr{L}u^* = f$, then $\|u^*\|_{\mathscr{L}} < \infty$. Conditions on $\mathscr{L}$ that make possible the existence of an element u^* that minimizes $J(u)$ are now provided by the following theorem.†

Theorem 10.3 (Existence) *Let $J(u)$ be a functional of the form* (10.7). *Then there exists an element u^* with finite energy that minimizes $J(u)$ only if $\mathscr{L}$ is a positive-bounded-below operator.*

However, even if $\mathscr{L}$ is positive-bounded-below, it may happen that $\mathscr{D}$ does not contain an element that minimizes $J(u)$. Then $\mathscr{D}$ does not constitute a complete space with respect to the norm $\|u\|_{\mathscr{L}}$, and it is necessary to extend $\mathscr{L}$ (and $\mathscr{D}$) to obtain a complete normed linear space. The completion process is outlined as follows.

1. Let $\mathscr{H}$ be a complete normed linear space and $\mathscr{L}$ be a linear, positive-bounded-below operator on a set $\mathscr{D} \subset \mathscr{H}$. Construct a new space $\mathscr{H}_{\mathscr{L}}$ of functions with finite energy in which the inner product $[u,v]$ is given by

   $$[u,v] = \langle \mathscr{L}u,v \rangle$$

   and the norm by

   $$\|u\|_{\mathscr{L}} = [u,u]^{\frac{1}{2}}$$

2. If $\mathscr{H}_{\mathscr{L}}$ is incomplete, complete it by the addition of limit points, as described previously. Then every $u \in \mathscr{H}_{\mathscr{L}}$ either belongs to $\mathscr{D}$ or there exists a sequence $\{u^n\} \in \mathscr{D}$ such that $\lim_{n\to\infty} \|u - u^n\|_{\mathscr{L}} \to 0$; that is, $\mathscr{D}$ is dense in $\mathscr{H}_{\mathscr{L}}$ with respect to the energy norm $\|\cdot\|_{\mathscr{L}}$. It can then be shown [Mikhlin, p. 323] that every element $u \in \mathscr{H}_{\mathscr{L}}$ also $\in \mathscr{H}$; moreover, if $\mathscr{H}$ is separable, so also is $\mathscr{H}_{\mathscr{L}}$. Thus, $\mathscr{D}$ is now a normed space (indeed, an inner-product space) dense in a complete space $\mathscr{H}_{\mathscr{L}}$.
3. Extend the domain of the functional $J(u)$ to the entire space $\mathscr{H}_{\mathscr{L}}$. This can be accomplished by setting $\langle \mathscr{L}u,u \rangle = \|u\|^2_{\mathscr{L}}$ in the definition of

† This theorem is proved in Mikhlin [1964, p. 122]. See also Friedrichs [1934], wherein the first treatment of this problem and the proof of this theorem are presented.

$J(u)$ and observing that

$$|\langle u,f \rangle| \leq \|u\| \, \|f\| \leq \frac{1}{\gamma} \|f\| \, \|u\|_{\mathscr{L}}$$

Consequently, the linear functional $lu \equiv \langle u,f \rangle$ is bounded in $\mathscr{H}_{\mathscr{L}}$ and there exists a unique element u^* of $\mathscr{H}_{\mathscr{L}}$ (and, hence, of $\mathscr{H}$) such that $\langle u,f \rangle = [u,u^*]$. Thus, if $u \in \mathscr{H}_{\mathscr{L}}$,

$$J(u) = \|u - u^*\|^2_{\mathscr{L}} - \|u^*\|^2_{\mathscr{L}}$$

and $\min J(u) = J(u^*) = -\|u^*\|^2_{\mathscr{L}}$. Thus, we write for the extended functional

$$J(u) = [u,u] - [u,u^*] - \overline{[u,u^*]}$$

or

$$J(u) = [u,u] - 2\text{Re}\,[u,u^*] \tag{10.8}$$

4. Since $\langle u,f \rangle = [u,u^*]$, there exists an operator $\mathscr{M}$ on $\mathscr{H}$, which can be shown to be bounded and positive-definite, such that $u^* = \mathscr{M}f$. If $\mathscr{M}f = 0$, then $\langle u,f \rangle = 0$ and f is orthogonal to every element of the set $\mathscr{D}$ dense in $\mathscr{H}_{\mathscr{L}}$. Consequently $\mathscr{M}f = 0 \Rightarrow f = 0$, which means that $\mathscr{M}^{-1}$ exists. If $u^* \in \mathscr{D}$, $\mathscr{L}u^* = f$, and $u^* = \mathscr{M}f$, then $\mathscr{M}^{-1}u^* = \mathscr{L}u^*$. Hence, for every $u \in \mathscr{D}$, $\mathscr{M}^{-1}u = \mathscr{L}u$ and $\mathscr{D}_{\mathscr{M}^{-1}} \supseteq \mathscr{D}_{\mathscr{L}}$. It follows that $\mathscr{M}^{-1}$ is the extension of $\mathscr{L}$ in $\mathscr{H}_{\mathscr{L}}$.

In the case of nonhomogeneous boundary conditions, the functional $J(u)$ must be expanded. Suppose, for example, that $\mathscr{H} = L_2(\mathscr{R})$, $\mathscr{H}$ is real, $\mathscr{L}$ of (10.5) is a positive-definite operator, and that a solution to (10.5) exists. Consider also the case in which the solution must satisfy the nonhomogeneous boundary conditions (10.6) on $\partial\mathscr{R}$, where the functions g are differentiable as many times as desired and $\partial\mathscr{R}$ is at least piecewise smooth. We then consider u to be the sum of two parts, v and w, where v satisfies the homogeneous boundary conditions $\mathscr{B}v = 0$. Clearly, v minimizes the functional

$$J(v) = \langle \mathscr{L}v,v \rangle - 2\langle v,\bar{f} \rangle \tag{10.9a}$$

where

$$\bar{f} = f - \mathscr{L}w \tag{10.9b}$$

Since $v = u - w$,

$$J(v) = J(u) + Q_1(u,w) - J(w) \tag{10.10}$$

where

$$Q_1(u,w) = \langle u,\mathscr{L}w \rangle - \langle \mathscr{L}u,w \rangle \tag{10.11}$$

In some cases $Q_1(u,w) = 0$. Otherwise, $Q_1(u,w)$ is assumed to be separable in the sense that

$$Q_1(u,w) = Q(u) + P(w) \tag{10.12}$$

Then, if $R(w) \equiv P(w) - J(w)$, (10.10) can be put in the form

$$J(v) = J(u) + Q(u) + R(w) \tag{10.13}$$

The quantity $R(w)$ is a scalar depending only on w and does not contribute to the gradient of $J(v)$, while the functional $Q(u)$ is generally representative of boundary conditions and is of the form of surface integrals over the hypersurface $\partial\mathscr{R}$. Thus, the solution of (10.5) subject to the boundary conditions (10.6) is obtained by minimizing the new functional

$$I(u) = \langle \mathscr{L}u,u \rangle - 2\langle u,f \rangle + Q(u) \tag{10.14}$$

where, clearly, $I(u) = J(v) - R(w)$.

For an alternate interpretation, suppose that $\mathscr{L}^*$ is the *formal adjoint* of $\mathscr{L}$; that is,

$$u\mathscr{L}v - v\mathscr{L}^*u = b(u,v) \tag{10.15}$$

where $b(u, v)$ is a bilinear operator called the *bilinear concomitant* of $\mathscr{L}$. Integrating (10.15), we get

$$\langle u,\mathscr{L}v \rangle - \langle v,\mathscr{L}^*u \rangle \equiv B(u,v) \tag{10.16}$$

where $B(u,v)$ is defined on $\partial\mathscr{R}$. Generally $B(0,v) = B(u,0) = 0$ or a constant. If the boundary conditions are such that $B(u,v) = 0$, $\mathscr{L}^*$ is called the *adjoint* of $\mathscr{L}$. Thus, instead of (10.7), we use

$$J(u) = \langle u,\mathscr{L}^*u \rangle + B(u,u) - 2\langle f,u \rangle \tag{10.17}$$

Clearly, for homogeneous boundary conditions, $B(u,u) = 0$, $\langle u,\mathscr{L}^*u \rangle = \langle u,\mathscr{L}u \rangle$, and (10.17) reduces to (10.7) ($\mathscr{H}$ being real).

Nonlinear operators Many of the ideas discussed in connection with linear, positive-definite operators can be generalized and extended to nonlinear operators and nonpositive linear operators. In this section, we give brief consideration to such extensions. We begin by considering a Banach space $\mathscr{V}$, the elements of which are denoted u, and a nonlinear operator $\mathscr{P}$ on $\mathscr{V}$. A few definitions are needed:

Definition 1 Let $\mathscr{P}$ be an operator on $\mathscr{V}$. If for some $u \in \mathscr{V}$ and every $h \in \mathscr{V}$

$$\lim_{\alpha \to 0} \left\| \frac{1}{\alpha} (\mathscr{P}(u + \alpha h) - \mathscr{P}(u)) - G\mathscr{P}(u,h) \right\| = 0 \tag{10.18}$$

exists, α being real, then the operator $G\mathscr{P}(u,h)$ is called the *Gateaux differential* of $\mathscr{P}$ at the point u in direction h.

It follows that every Gateaux differential is homogeneous but not necessarily linear. If $\mathscr{P}$ has a Gateaux differential at each point of some convex set $\Omega \subset \mathscr{V}$, then it can be shown to follow† that for every $u, u + h \in \Omega$, the Lipschitz condition

$$\|\mathscr{P}(u + h) - \mathscr{P}(u)\| \leq \|G\mathscr{P}(u + \alpha h, h)\|$$

holds, for $\alpha \in (0,1)$.

A stronger definition of differentials of nonlinear operators is afforded by the following:

Definition 2 If at $u \in \mathscr{V}$

$$\mathscr{P}(u + h) - \mathscr{P}(u) = \delta\mathscr{P}(u,h) + \omega(u,h) \tag{10.19}$$

wherein $\delta\mathscr{P}(u,h)$ is a linear operator in $h \in \mathscr{V}$ and

$$\lim_{\|h\|\to 0} \frac{1}{\|h\|} \|\omega(u,h)\| \to 0 \tag{10.20}$$

then $\delta\mathscr{P}(u,h)$ is called the *Frechet differential* of $\mathscr{P}$ at the point u and $\omega(u,h)$ is called the *remainder* of the differential.

Clearly, if $\delta\mathscr{P}(u,h)$ exists, so also does $G\mathscr{P}(u,h)$; however, the converse is not always true. If the Gateaux differential is linear in h, then we may write

$$G\mathscr{P}(u,h) = D_G\mathscr{P}(u)(h)$$

and $D_G\mathscr{P}(u)$ is referred to as the Gateaux derivative at u. Likewise, $\mathscr{P}'(u)$ is the Frechet derivative of $\mathscr{P}$ at u if

$$\delta\mathscr{P}(u,h) = \mathscr{P}'(u)(h)$$

If $D_G\mathscr{P}(u)$ exists in some neighborhood $\mathscr{N}(u_0,r) = \{u \mid \|u - u_0\| \leq r\}$ and is continuous at u_0 (i.e., if a sequence $\{u^n\}$ exists such that $\lim_{n\to\infty} \|D_G\mathscr{P}(u_0) - D_G\mathscr{P}(u^n)\| \to 0$), then $\delta\mathscr{P}(u_0,h)$ exists and $\delta\mathscr{P}(u_0,h) = G\mathscr{P}(u_0,h)$.

The Frechet differential is said to be locally uniform on a set $\Omega \subset \mathscr{V}$ if for every $\epsilon > 0$ there exists $\eta(\epsilon,u_0)$, $\delta(\epsilon,u_0) > 0$ such that for every $u \in \mathscr{N}(u_0,\eta)$, $\|\omega(u,h)\| < \epsilon \|h\|$ if $\|h\| < \delta$. If $\mathscr{P}'$ is bounded and locally uniform in $\mathscr{N}(u_0,r)$, then $\mathscr{P}'$ is continuous in $\mathscr{N}(u_0,r)$ [Vainberg, 1964, p. 43].

† Vainberg [1964, p. 37]; sufficient conditions for the linearity of Gateaux differentials are also given in this monograph.

Definition 3 Let $K(u)$ be a functional defined on a set $\Omega \subset \mathscr{V}$ and suppose that $K(u)$ is Frechet differentiable on Ω. Further, let $\langle u,v \rangle$ denote an inner product defined on $\mathscr{V}$ and its conjugate space $\mathscr{V}^*$. Then the operator $\mathscr{P}$ defined by the formula

$$\langle \mathscr{P}(u),h \rangle = \lim_{\alpha \to 0} \frac{1}{\alpha} [K(u + \alpha h) - K(u)]$$

is called the *gradient of the functional* $K(u)$, and we write $\mathscr{P}(u) =$ grad $K(u)$. [If $K(u)$ is Gateaux differentiable, we refer to grad $K(u)$ as the *weak gradient;* if $K(u)$ is Frechet differentiable, grad $K(u)$ is the *strong gradient* of $K(u)$.]

Definition 4 The operator $\mathscr{P}:\mathscr{V} \to \mathscr{V}^*$ is said to be a *potential operator* on some set $\Omega \subset \mathscr{V}$ if there exists a functional $K(u)$ such that grad $K(u)$ $= \mathscr{P}(u)$ for every $u \in \Omega$.

It can be shown that a necessary and sufficient condition for an operator $\mathscr{P}:\mathscr{V} \to \mathscr{V}^*$ to be potential in $\mathscr{N}(u_0,r)$ is that the functional $\langle \delta\mathscr{P}(u,h_1),h_2 \rangle$ be symmetric in the sense that $\langle \delta\mathscr{P}(u,h_1),h_2 \rangle = \langle \delta\mathscr{P}(u,h_2),h_1 \rangle$, for every h_1, $h_2 \in \mathscr{V}$ and $u \in \mathscr{N}(u_0,r)$, where $\langle \delta\mathscr{P}(u,h_1), h_2 \rangle$ is assumed to be continuous on $\mathscr{N}(u_0,r)$.

Definition 5 A point $u_0 \in \mathscr{V}$ such that $K(u) \leq K(u_0)$ or $K(u) \geq K(u_0)$ for every $u \in \mathscr{N}(u_0,r)$ is called an *extreme point* of the functional $K(u)$. If grad $K(u_0) = \theta$, where θ is the zero element of $\mathscr{V}^*$, then u_0 is called a *critical point* of the functional $K(u)$.

We now state an important theorem, the proof of which is given by Vainberg [1964, p. 58].

Theorem 10.4 *Let $\mathscr{P}$ be an operator, not necessarily linear, that is potential on $\mathscr{N}(u_0,r)$. Then there exists a functional $K(u)$, unique up to a constant, whose gradient is $\mathscr{P}$, which is given by*

$$K(u) = \int_0^1 \langle \mathscr{P}(u_0 + s(u - u_0)), u - u_0 \rangle \, ds + K_0 \tag{10.21}$$

wherein $K_0 = K(u_0)$ is a constant.

Clearly, Theorem 10.4 represents a considerable generalization of Theorem 10.2. Indeed, if a potential operator $\mathscr{P}$ is linear but not positive-definite or if $\mathscr{P}$ is nonlinear, and if solutions to (10.1) exist, then we can construct an associated functional $K(u)$ by means of (10.21), the gradient of which is $\mathscr{P}$, and view the problem of finding solutions to (10.1) as one of finding critical points of $K(u)$. If the solution u^* does not minimize $K(u)$, we refer to the idea of seeking solutions such that grad $K(u^*) = \theta$ as an extremum principle for the operator $\mathscr{P}$.

To emphasize that (10.21) is, indeed, a generalization of (10.7), set, $u_0 = 0$, $K_0 = 0$, $\mathscr{V} = \mathscr{H}$, where $\mathscr{H}$ is a real Hilbert space, and

$$\mathscr{P}(u) = \mathscr{L}u - f$$

where $\mathscr{L}$ is the symmetric positive-definite linear operator in (10.5). Then, according to Theorem 10.4, $\mathscr{P}(u)$ is the gradient of the functional

$$K(u) = \int_0^1 \langle \mathscr{L}(su) - f, u\rangle \, ds = \int_0^1 s\langle \mathscr{L}u,u\rangle \, ds - \int_0^1 \langle f,u\rangle \, ds$$

Integrating, we get

$$K(u) = \tfrac{1}{2}\langle \mathscr{L}u,u\rangle - \langle f,u\rangle = \tfrac{1}{2}J(u)$$

where $J(u)$ is the quadratic functional defined in (10.7).

While we shall not dwell on the question of the existence or uniqueness of solutions of nonlinear operator equations† or on the existence of relative minima of their corresponding functionals, it is nevertheless interesting to note that conditions similar to those imposed on the linear operator $\mathscr{L}$ exist also for a large class of nonlinear operators. In the case of the nonlinear potential operator equation $\mathscr{P}u = f$, for example, it is not difficult to prove that at least one solution exists if the Frechet differential $\delta\mathscr{P}(u,h)$ is positive-bounded below; i.e., if there exists a constant γ such that

$$\langle \delta\mathscr{P}(u,h),h\rangle \geq \gamma^2 \, \|h\|^2 \tag{10.22}$$

then $K(u)$ of (10.21) is bounded below. Moreover, there exists a point u^* at which $K(u)$ is a relative minimum, $K(u)$ being a functional on a compact set of $\mathscr{V}$, if

$$\delta^2K(u,h,h) \geq \|h\| \, \lambda(\|h\|) \tag{10.23}$$

where $0 < \lambda(s)$, $\lim_{s\to\infty} \lambda(s) = \infty$, and $\delta^2K(u,h,h)$ is the second Frechet differential of $K(u)$. Any minimizing sequence of $K(u)$ [i.e., any sequence $\{u^n\}$ such that $\lim_{n\to\infty} K(u^n) = \min K(u)$] converges in the metric of $\mathscr{H}$ to some limit called the *generalized solution* of the problem of minimizing the functional (10.21). Clearly, (10.23) generalizes the conventional requirement for a function to be a relative minimum at a critical point to Banach spaces. Indeed, it is meaningful to write

$$K(u + h) = K(u) + \delta K(u,h) + \tfrac{1}{2}\delta^2K(u,h,h) + \cdots \tag{10.24}$$

Then if grad $K(u) = \theta$, that is, if u is a critical point of $K(u)$, the positiveness

† For discussions of such questions, see Saaty [1967] or, again, Vainberg [1964].

of $K(u)$ in $\mathscr{N}(u,r)$ is determined by the quadratic form $\delta^2K(u,h,h)$ in the sense of (10.23).

In concluding this section, we point out a property of potential operators of fundamental importance. Consider the nonlinear operator equation

$$\mathscr{P}u = f \tag{10.25}$$

and let $\hat{\mathscr{P}}u = \mathscr{P}u - f$ be a potential operator from $\mathscr{V}$ into $\mathscr{V}^*$, which is the gradient of a functional $K(u)$. Suppose that a solution u^* of (10.25) exists that is a critical point of $K(u)$. Then

$$\text{grad } K(u^*) = \langle \hat{\mathscr{P}}u^*,h\rangle = 0$$

or

$$\langle \mathscr{P}u^* - f,h\rangle = 0 \tag{10.26}$$

Since h is an arbitrary element of $\mathscr{V}$, it follows that the image $\hat{\mathscr{P}}(u^*)$ of the solution of (10.25) is orthogonal to $\mathscr{V}$. This observation provides the basis of a general method for constructing finite-element approximations of solutions to nonlinear operator equations.†

10.2 FINITE-ELEMENT APPROXIMATIONS

In finite-element approximations, we are generally faced with the following problem: Let $\mathscr{H}$ denote an inner-product space, usually the Hilbert space $L_2^r(\mathscr{R})$, the elements of which are functions $F(\mathbf{x})$ with domain $\mathscr{R}$ in $\mathscr{E}^k$. [In general, we shall be interested in cases in which $F(\mathbf{x})$ and its $r+1$st derivatives are piecewise continuous and square integrable on $\mathscr{R}$.] Further, let $\overline{\mathscr{H}}$ denote another space, the elements of which are functions $\bar{F}(\mathbf{x})$, $\mathbf{x} \in \mathscr{R}$, which are of class $C^r(\mathscr{R})$, $r < p$. Construct a sequence of linear subspaces $\Phi^1, \Phi^2, \ldots$ of $\overline{\mathscr{H}}$, each of finite dimension $n_1, n_2, \ldots$, not necessarily embedded, such that for each $F(\mathbf{x}) \in \mathscr{H}$, there exists a sequence $\bar{F}^1(\mathbf{x}) \in \Phi^1$, $\bar{F}^2(\mathbf{x}) \in \Phi^2, \ldots$, such that

$$\begin{aligned}\lim_{n\to\infty} \|F(\mathbf{x}) - \bar{F}^n(\mathbf{x})\| &= \lim_{n\to\infty} \|F_{,i}(\mathbf{x}) - \bar{F}^n_{,i}(\mathbf{x})\| \\ &= \cdots = \lim_{n\to\infty} \|F_{,i_1i_2\cdots i_m}(\mathbf{x}) - \bar{F}_{,i_1i_2\cdots i_m}(\mathbf{x})\| = 0 \end{aligned} \tag{10.27}$$

Here $\|\cdot\|$ denotes the norm in $\mathscr{H}$ and $\overline{\mathscr{H}}$; $i_1, i_2, \ldots, i_m = 1, 2, \ldots, k$; $F_{,i_1i_2\cdots i_m} = \partial^m F/\partial x_{i_1}\partial x_{i_2}\cdots\partial x_{i_m}$ with $i_1 \le i_2 \le \cdots \le i_m$.

We regard $\bar{F}^n(\mathbf{x})$ as the nth member of a sequence $\{\bar{F}^n\}$ in $\overline{\mathscr{H}}$. Other members of the sequence are generated by constructing *refinements* of the model $\bar{\mathscr{R}}$. By a refinement of $\bar{\mathscr{R}}$, we shall mean an increase in the number of elements r_e and global nodes G such that every point $\mathbf{x} \in \bar{\mathscr{R}}$ is contained

† See the discussion of Galerkin's method in Sec. 10.3.

in an arbitrarily small element and the class of each element and its corresponding interpolation functions are the same for each refinement. For example, if $\bar{\mathscr{R}} \subset \mathscr{E}^2$ consists of triangular elements over which linear interpolation functions $\psi_N^{(e)}(x^1, x^2)$ are used, each refinement of $\bar{\mathscr{R}}$ will also consist of triangular elements and linear interpolation functions. If every node and interelement boundary in a model $\bar{\mathscr{R}}$ is also a node and an interelement boundary in a refinement of $\bar{\mathscr{R}}$, then the refinement is said to be *regular*; otherwise, it is *irregular*.

To cast the idea of a refinement in quantitative terms, we introduce the notion of a *diameter* of a finite element r_e. Let $\mathbf{x}_i^e$ and $\mathbf{x}_j^e$ be arbitrary points in element r_e. The diameter δ_e of r_e is the maximum distance between points in r_e; that is, $\delta_e = \max \|\mathbf{x}_i^e - \mathbf{x}_j^e\|$; $\mathbf{x}_i^e, \mathbf{x}_j^e \in r_e$. The *mesh* δ of a finite element model is the maximum diameter of all the elements: $\delta = \max\{\delta_1, \delta_2, \ldots, \delta_E\}$. Let $\bar{\mathscr{R}}$ be a finite-element model and $\bar{\mathscr{R}}'$ be a refinement; $\bar{\mathscr{R}}'$ is *uniform* if $\delta' < \delta$, where δ' is the mesh of $\bar{\mathscr{R}}'$. If $\hat{\delta}$ and $\hat{\delta}'$ are the smallest diameters of $\bar{\mathscr{R}}$ and a refinement $\bar{\mathscr{R}}'$, then $\bar{\mathscr{R}}'$ is *uniformly decreasing* if $\hat{\delta}' < \hat{\delta}$. Henceforth we assume that all refinements are regular and uniform, and for a sequence of such refinements $\bar{\mathscr{R}}^1, \bar{\mathscr{R}}^2, \ldots, \bar{\mathscr{R}}^\Gamma, \ldots$, the meshes $\delta(\Gamma)$ satisfy the obvious requirement: $\lim_{\Gamma \to \infty} \delta(\Gamma) = 0$. Further, we assume that a specific sequence of refinements is given in which the local approximate functions are of the same form for every element.

Clearly, our first concern is the development of a method for constructing the sequence $\{\bar{F}^n\}$. Toward this end, we proceed, as usual, by decomposing $\mathscr{R}$ into a model $\bar{\mathscr{R}}$ consisting of a finite number E of subdomains r_e connected at common nodal points and joined continuously together at interelement boundaries. The restriction of $\bar{F}(\mathbf{x})$ to r_e is again denoted by $\bar{f}^{(e)}(\mathbf{x})$ where, for simplicity, no distinction is made between local and global frames of reference ($\mathbf{x}^{(e)} \equiv \mathbf{X}$). The local finite-element representation $\bar{f}^{(e)}(\mathbf{x})$ is assumed to be of the form

$$\bar{f}^{(e)}(\mathbf{x}) = f_{(e)}^N \psi_N^{(e)0}(\mathbf{x}) + f_{(e),i}^N \psi_N^{(e)i}(\mathbf{x}) + \cdots + f_{(e),i_1 i_2 \cdots i_r}^N \psi_N^{(e)i_1 i_2 \ldots i_r}(\mathbf{x}) \tag{10.28}$$

Here $N = 1, 2, \ldots, N_e$; $f_{(e)}^N, f_{(e),i}^N, \ldots, f_{(e),i_1 \cdots i_r}^N$ are the values of $\bar{f}^{(e)}(\mathbf{x})$ and its partial derivatives up to order r at node N, and the local interpolation functions $\psi_N^{(e)0}(\mathbf{x}), \ldots, \psi_N^{(e)i_1 i_2 \cdots i_r}(\mathbf{x})$ are either identically zero for a certain node N or they satisfy the generalized Hermite interpolation properties discussed in Sec. 8.8. The total number of independent terms appearing in (10.28) is referred to as the *degree of freedom* of the element. Henceforth, we shall also employ the following conventions and assumptions: (1) The local representations $\bar{f}^{(e)}(\mathbf{x})$ and the global representations $\bar{F}(\mathbf{x})$ are uniquely determined by their principal nodal values $f_{(e)}^N, f_{(e),i}^N, \ldots, f_{(e),i_1 \cdots i_r}^N$ appearing in (10.28) and the corresponding global values $F^\Delta, F^\Delta_{,i}, \ldots, F^\Delta_{,i_1 \cdots i_r}$ for all decompositions of $\mathscr{R}$. (2) The element identification label (e) shall be dropped for simplicity and the dependence on $\mathbf{x}$ of the functions $\psi_N^0, \psi_N^i, \ldots, \psi_N^{i_1 \ldots i_r}$ shall be understood.

It is convenient to introduce the following definitions:

Definition 1 All finite-element approximations $\bar{f}(\mathbf{x})$ which include the specification of partial derivatives of order r at least one node are said to belong to a *family* $\mathscr{F}^r$ of finite-element approximations.

For example, $\bar{f}(\mathbf{x}) = f^N \psi_N^0 + f^1_{,ijk}\psi_1^{ijk}$ defines a family $\mathscr{F}^3$ of finite-element approximations. By allowing the parametric coefficients $f^N, f^1_{,ijk}$ to take on all real (or complex) values, we generate all functions in $\mathscr{F}^3$. The representation $\bar{f}(\mathbf{x}) = f^N \psi_N^0 + f^2_{,ijk}\psi_2^{ijk}$ defines still another $\mathscr{F}^3$ family of approximations.

Definition 2 Let $f(\mathbf{x}_e)$ and $g(\mathbf{x}_e)$ define two families of finite-element approximations. Then $f(\mathbf{x}_e)$ and $g(\mathbf{x}_e)$ are *r-equivalent* at $\mathbf{x}_e \in \imath_e$ if and only if

$$\|f(\mathbf{x}_e) - g(\mathbf{x}_e)\| \leq K\delta_e^r \tag{10.29}$$

where δ_e is the diameter of $\imath_e$ and K is a finite-positive constant independent of δ_e.

Equation (10.29) establishes an equivalence relation for certain families of functions defined on $\bar{\mathscr{R}}$ that is assumed to hold for all elements in $\bar{\mathscr{R}}$. The diameter δ_e is regarded as a parameter; i.e., for a family $\{\bar{\mathscr{R}}^\Gamma\}$ of refinements for which $\delta(\Gamma)$ is the mesh of each $\bar{\mathscr{R}}^\Gamma$, we have $\delta_e \leq \delta(\Gamma)$ so that $\delta_e \to 0$ as $\delta(\Gamma) \to 0$. The notion of r-equivalence is meaningful only with respect to a given norm. Unless noted otherwise, we shall be concerned with *uniform equivalence* of functions; that is, $\|f - g\| = |f - g| \leq K\,\delta_e^r$. For example, the function $f(x_e) = 1 + x_e + x_e^2 + 3\delta \sin x_e$ is 2-equivalent to $1 + x_e + x_e^2$ while $f(x_e) = \sin x_e$ is 3-equivalent to x_e, for δ_e sufficiently small. Two functions $f(\mathbf{x}_e)$ and $g(\mathbf{x}_e)$, which are at least 1-equivalent, are, in the limit as $\delta \to 0$, equal almost everywhere in the sense of the norm $\|\cdot\|$. Likewise, if the values of f and g at $\mathbf{x}_e$ differ only by terms of *order* δ_e^r, that is,

$$f(\mathbf{x}_e) = g(\mathbf{x}_e) + O(\delta_e^r) \tag{10.30}$$

then f and g are r-equivalent on $\imath_e$. By $O(\delta_e^r)$ we mean a function of δ_e^r (and possibly $\mathbf{x}_e$) which has the property $\lim_{\delta_e \to 0} \dfrac{1}{\delta_e^{r-1}} |O(\delta_e^r)| = 0$.

Definition 3 Let $\bar{f}^{(e)}(\mathbf{x})$ define a family $\mathscr{F}^r$ of finite-element approximations on $\imath_e$. Define

$$G(\mathbf{x}) = \bar{f}^{(e)}(\mathbf{x}) - F(\mathbf{x}) \tag{10.31}$$

Then $\mathscr{F}^r$ is *r-admissible* for approximating $F(\mathbf{x})$ with respect to $\|\cdot\|$ if and only if

$$\lim_{\delta\to 0} \|G(\mathbf{x})\| = \lim_{\delta\to 0} \|G(\mathbf{x})_{,i}\| = \cdots = \lim_{\delta\to 0} \|G(\mathbf{x})_{,i_1 i_2 \cdots i_r}\| = 0 \qquad (10.32)$$

Obviously, if $\mathscr{F}^r$ is r-admissible, $\bar{f}_{,i_1\cdots i_r}$ and $F_{,i_1\cdots i_r}$ are at least 1-equivalent on r_e.

If a finite element of family $\mathscr{F}^r$ has d degrees of freedom, the value of $\bar{f}(\mathbf{x})$ at a point $\mathbf{x}$ is obtained by superimposing the values of d linearly independent functions belonging to $\mathscr{F}^r$. These linearly independent functions are said to characterize the "unit modes" of the element. While it is generally taken for granted that the value of $\bar{f}(\mathbf{x})$ at a point $\mathbf{x}$ can be assigned an arbitrary finite value by adjusting the amplitudes of the unit modes, we are content when working in normed spaces to be able to assign $\bar{f}(\mathbf{x})$ a given value only to within terms of order δ_e. This leads us to the following theorem.

Theorem 10.5 *Let $\bar{f}$ belong to a class of r-admissible finite-element approximations and let $\bar{f}$ be* 1*-equivalent to a complete polynomial containing m terms. Then the degree-of-freedom d of element r_e is $\leq m$.*

Proof Obviously, the coefficients of a complete polynomial† containing m independent terms can be equated to m values of $\bar{f}$ and certain of its derivatives at nodes in r_e. Then, by definition, $d = m$. However, if two or more terms have a common coefficient, then $d < m$. Thus, in general, $d \leq m$. If $\bar{f}$ is only 1-equivalent to a complete polynomial containing m terms, the same argument still holds, the only difference then being that the values of $\bar{f}$ and certain of its derivatives at the nodes can be prescribed only to within terms of order δ_e.

The preceding definitions and Theorem 10.5 establish some general properties of finite-element approximations that are adequate for uniformly approximating a given function $F(\mathbf{x})$. However, in obtaining finite-element solutions to boundary-value problems, we require, as emphasized in the previous section, that the basis functions be complete in energy. To resolve questions of completeness in energy we must, of course, first identify the character of operator $\mathscr{L}$ in the definition of the energy inner product. We shall be content here to discuss the question in connection with a rather broad class of linear, positive-bounded-below, differential operators of order $2p$ of the

† By a complete polynomial of degree n in k variables, we mean a polynominal containing $(n + k)!/n!k!$ independent terms; however, the coefficients need not be independent.

form†

$$\mathscr{L} = \sum_{s=0}^{p} \left[(-1)^s \sum_{i_1, i_2, \ldots, i_k = 0}^{s} \frac{\partial^s}{\partial x_{i_1}^1 \partial x_{i_2}^2 \cdots \partial x_{i_s}^k} \times \sum_{j_1, j_2, \ldots, j_k = 0}^{s} A_{j_1 j_2 \cdots j_k}^{i_1 i_2 \cdots i_k} \frac{\partial^s}{\partial x_{j_1}^1 \partial x_{j_2}^2 \cdots \partial x_{j_s}^k} \right] \tag{10.33}$$

where $i_1 + i_2 + \cdots + i_k = j_1 + j_2 \cdots + j_k = p$ and $A_{j_1 j_2 \cdots j_k}^{i_1 i_2 \cdots i_k}$ is an array of bounded continuously differentiable functions on $\mathscr{R}$ which is symmetric in the indices $i_1, \ldots, i_k, j_1, \ldots, j_k$ and positive in the sense that for every $\mathbf{x} \in \mathscr{R}$, $x^{i_1} x^{i_2} \cdots x^{i_k} A_{j_1 \cdots j_k}^{i_1 \cdots i_k} x^{j_1} \cdots x^{j_k} \geq \gamma^2 \|\mathbf{x}\|^2$, $\gamma > 0$. Assume also that $u \in \mathscr{D}_{\mathscr{L}}$ satisfies homogeneous boundary conditions on $\partial\mathscr{R}$ that make it possible, by partial integrations, to cast the inner product $\langle u, \mathscr{L}v \rangle$ into the form

$$\langle u, \mathscr{L}v \rangle = [u, v] = \int_{\mathscr{R}} \hat{\partial}^p u \, \mathbf{A} \partial^p v \, d\mathscr{R} \tag{10.34}$$

where, to simplify notation, we have introduced the obvious symbolism

$$\sum_{s=0}^{p} \sum_{\substack{i_1, i_2, \ldots, i_k = 0 \\ j_1, j_2, \ldots, j_k = 0}}^{s} \frac{\partial^s u}{\partial x_{i_1}^1 \cdots \partial x_{i_s}^k} A_{j_1 \cdots j_k}^{i_1 \cdots i_k}(\mathbf{x}) \frac{\partial^s v}{\partial x_{j_1}^1 \cdots \partial x_{j_s}^k} \equiv \hat{\partial}^p u \mathbf{A} \, \partial^p v \tag{10.35}$$

The energy norm is then

$$\|u\|_{\mathscr{L}}^2 = [u, u] = \int_{\mathscr{R}} \hat{\partial}^p u \mathbf{A} \, \partial^p u \, d\mathscr{R} \tag{10.36}$$

and if A is an upper bound of $\mathbf{A}$ [for example, $\max_{\mathbf{x} \in \mathscr{R}} |\mathbf{A}(\mathbf{x})| \leq A$], it can be shown that

$$\|u - v\|_{\mathscr{L}}^2 \leq \mathscr{R} A \max_{u, v \in \mathscr{D}} |\partial^p (u - v)|^2 \tag{10.37}$$

The distance in energy is then $d_{\mathscr{L}}(u, v) = \|u - v\|_{\mathscr{L}}$.

We can, of course, proceed directly from (10.36) to (10.7) or (10.8) to obtain a variational formulation of the partial differential equation $\mathscr{L}u = f$ that involves the functional

$$J(u) = \int_{\mathscr{R}} (\hat{\partial}^p u \mathbf{A} \, \partial^p u - 2uf) \, d\mathscr{R} \tag{10.38}$$

Quadratic functionals of the form in (10.38) (i.e., functionals involving derivatives of order p) are said to be of *class* C^{p-1}. In a C^{p-1} variational problem, boundary conditions on u involving derivatives up to and including those of order $p - 1$ are called *principal* (or "geometric") *boundary conditions*, while those involving derivatives of order $p, p + 1, \ldots, 2p - 1$ are referred

† This represents a slight generalization of an example presented by Felippa and Clough [1968].

to as *natural boundary conditions*. The derivatives of u up to and including those of order $p - 1$ are called *principal derivatives* of u. In constructing approximations to (10.36) and (10.38), we generally require that u have continuous principal derivatives on $\mathscr{R} + \partial\mathscr{R}$ (that is, $u \in C^{p-1}(\mathscr{R} + \partial\mathscr{R})$) and that the derivatives of u of order p be square integrable [that is, $\partial^p u \in L_2(\mathscr{R})$]. Then, if $\mathscr{L}$ is positive-definite, this class of functions will be dense in $\mathscr{H}_{\mathscr{L}}$.

With this introduction, we establish an important theorem†:

Theorem 10.6 *Let d be the degree of freedom of a finite element r_e over which the local finite-element representation $\bar{f}(\mathbf{x})$ is of the family $\mathscr{F}^r$. Then the local basis functions $\psi_N^0(\mathbf{x}), \psi_N^i(\mathbf{x}), \ldots, \psi_N^{i_1\cdots i_r}(\mathbf{x})$ can generate sequences of approximations which are complete in the energy norm (10.36) if the following conditions hold:*

a. *$\bar{f}(\mathbf{x})$ is at least 1-equivalent to a polynomial containing d terms*

b· *$\mathscr{F}^r$ contains a subfamily $\mathscr{G}$ of functions that, together with their derivatives up to order s, are m-equivalent to the corresponding derivatives of complete polynomials of degree $r + 1$ with $s \le r + 1$, $m \ge p + 1 - s$*

c. *$r + 1 = p$*

Proof We observe that if the theorem holds when $\bar{f}(\mathbf{x})$ is a polynomial containing d terms that includes a complete polynomial of degree $r + 1 = p$, then in the limit as the mesh $\delta \to 0$, the theorem also holds when $\bar{f}(\mathbf{x})$ is 1-equivalent to such functions. Thus, assume that $\bar{f}(\mathbf{x})$ is a polynomial containing d terms with d independent coefficients and that we can identify in this polynomial a complete polynomial $p(\mathbf{x})$ of degree $p = r + 1$ with coefficients that can take on arbitrary constant values. Thus, $\bar{f}(\mathbf{x})$ can be used to characterize elements with d degrees of freedom (see Theorem 10.5).

Now let $F(\mathbf{x})$ be an element of $\mathscr{D}_{\mathscr{L}}$, where $\mathscr{L}$ is given by (10.33). Since $\mathscr{L}$ is of order $2p$, we can represent $F(\mathbf{x})$ over r_e by

$$F(\mathbf{x}) = F^N + F^N_{,i}(x_i - x_i^N) + \cdots + \frac{1}{(p+1)!} \times F(\mathbf{x}^*)_{,i_1\cdots i_{p+1}}(x_{i_1} - x_{i_1}^N)\cdots(x_{i_{p+1}} - x_{i_{p+1}}^N) \qquad (10.39)$$

† This theorem represents a generalization of a theorem presented by Arantes e Oliveira [1968] in which only polynomial approximations were considered. It is important to realize that this theorem provides only local conditions for completeness of finite-element approximation function; i.e., functions satisfying the conditions of this theorem insure that (10.51) is satisfied locally but they may violate the fundamental global condition of possessing finite-energy norms. To overcome this difficulty, we subsequently impose the additional condition of r-conformity of members of the family $\mathscr{F}^r$ (see Definition 4 on page 117).

where F^N, $F^N_{,i}$, . . . denote values of $F(\mathbf{x})$ and its derivatives at node $\mathbf{x}^N$ and $\mathbf{x}^*$ is "between" $\mathbf{x}^N$ and $\mathbf{x}$. Assume $\mathscr{G} \subset \mathscr{F}^r$ contains the polynomial $p(\mathbf{x})$ of degree p, the coefficients of which can be equated to the corresponding coefficients of the first p terms of (10.39). Thus, if $\bar{K}$ denotes an upper bound of $|F(\mathbf{x})_{,i_1 \cdots i_{p+1}}|$,

$$|F(\mathbf{x}) - p(\mathbf{x})| \leq \frac{b\bar{K}}{(p+1)!} (\delta_e)^{p+1} \tag{10.40}$$

wherein $b = (p+k)!/[(p+1)!(k-1)!]$ is the number of terms of degree $p+1$ in k variables x_i and δ_e is the diameter of the element. Similarly, by successive differentiations, we obtain

$$|F(\mathbf{x})_{,i_1 \cdots i_s} - p(\mathbf{x})_{,i_1 \cdots i_s}| \leq \frac{bK}{(p+1-s)!} (\delta_e)^{p+1-s} \tag{10.41}$$

where K is a positive constant.

By hypothesis, $p(\mathbf{x}) \in \mathscr{F}^r$. Hence, $p(\mathbf{x})$ can be written in the form

$$p(\mathbf{x}) = p^N \psi^0_N(\mathbf{x}) + p^N_{,i} \psi^i_N(\mathbf{x}) + \cdots + p^N_{,i_1 \cdots i_r} \psi^{i_1 \cdots i_r}_N(\mathbf{x}) \tag{10.42}$$

in which $r = p - 1$. Let $\bar{g}(\mathbf{x})$ be an element of $\mathscr{F}^r$ that can coincide with the principal derivatives of $F(\mathbf{x})$ at each node $\mathbf{x}^N$ of $\imath_e$. Then

$$p(\mathbf{x}) - \bar{g}(\mathbf{x}) = (p^N - g^N)\psi^0_N(\mathbf{x}) + (p^N_{,i} - g^N_{,i})\psi^i_N(\mathbf{x}) + \cdots$$
$$+ (p^N_{,i_1 \cdots i_r} - g^N_{,i_1 \cdots i_r})\psi^{i_1 \cdots i_r}_N(\mathbf{x}) \tag{10.43}$$

Now in order that (10.28) [or (10.42)] be dimensionally homogeneous, it is necessary that the local interpolation functions $\psi^0_N, \psi^i_N, \ldots$ be representable in the following manner:

$$\psi^0_N(\mathbf{x}) = \varphi^0_N(\boldsymbol{\xi}),\ \psi^i_N(\mathbf{x}) = l\varphi^i_N(\boldsymbol{\xi}), \ldots,\ \psi^{i_1 \cdots i_r}_N(\mathbf{x}) = l^r \varphi^{i_1 \cdots i_r}_N(\boldsymbol{\xi}) \tag{10.44}$$

Here l is a characteristic length of the element, $\boldsymbol{\xi}$ is a dimensionless local coordinate (for example, $\xi_i = x_i/l$), and the functions $\varphi^0_N, \varphi^i_N, \ldots,$ $\varphi^{i_1 \cdots i_r}_N$ are dimensionless local basis functions that do not depend upon the absolute dimensions of the element. These functions are assumed to be bounded on $\imath_e$; that is, for every $x_i/l \in \imath_e$, there exist positive constants $K^0, K^1, \ldots, K^r$ such that $|\varphi^0_N(\boldsymbol{\xi})| \leq K^0$, $|\varphi^i_N(\boldsymbol{\xi})| \leq K^1, \ldots,$ $|\varphi^{i_1 \cdots i_r}_N(\boldsymbol{\xi})| \leq K^r$. Thus, if δ_e is the diameter of the element, it follows that

$$|\psi^0_N(\mathbf{x})| \leq K^0,\ |\psi^i_N(\mathbf{x})| \leq K^1\delta_e,\ |\psi^{ij}_N(\mathbf{x})| \leq K^2\delta^2_e, \ldots,\ |\psi^{i_1 \cdots i_r}_N(\mathbf{x})| \leq K^r\delta^r_e \tag{10.45}$$

Note also that

$$|\psi^{i_1 \cdots i_s}_{N,i_1 \cdots i_m}(\mathbf{x})| \leq K^s \delta^{s-m}_e \tag{10.46}$$

Consequently, returning to (10.43), we have

$$|p(\mathbf{x}) - \bar{g}(\mathbf{x})| \leq |p^N - g^N|\, K^0 + k\, |p^N_{,i} - g^N_{,i}|\, K^1\delta_e + \cdots + \frac{(r+k-1)!}{r!(k-1)!}\, |p^N_{,i_1\cdots i_r} - g^N_{,i_1\cdots i_r}|\, K^r(\delta_e)^r \qquad (10.47)$$

for any choice of indices $i_1, i_2, \ldots, i_r$.

Thus, in view of (10.41),

$$|p^N_{,i_1\cdots i_s} - g^N_{,i_1\cdots i_s}| \leq \frac{bK}{(p+1-s)!}(\delta_e)^{p+1-s} \qquad (10.48)$$

and $s = 0, 1, 2, \ldots, r$. Introducing this result into (10.43) and taking into account (10.46), we have

$$|p(\mathbf{x}) - \bar{g}(\mathbf{x})| \leq M(\delta_e)^{p+1}$$

$$\cdots\cdots\cdots\cdots\cdots\cdots$$

$$|p(\mathbf{x})_{,i_1\cdots i_p} - \bar{g}(\mathbf{x})_{,i_1\cdots i_p}| \leq M\delta_e \qquad (10.49a)$$

where

$$M = N_e bK \sum_{s=0}^{r} \frac{K^s(s+k-1)!}{(p+1-s)!s!(k-1)!} \qquad (10.49b)$$

and $r = p - 1$. Note that in deriving (10.49*a*) we have taken into account that $\psi^{i_1\cdots i_r}_{N,i_1\cdots i_{r+1}}(\mathbf{x})$ exist in $\imath_e$ but the order of representation of $\bar{g}(\mathbf{x})$ is only r.

Combining (10.49*a*) and (10.41), we observe that†

$$\begin{aligned}|F(\mathbf{x})_{,i_1\cdots i} - \bar{g}(\mathbf{x})_{,i_1\cdots i_j}| &= |F(\mathbf{x})_{,i_1\cdots i_j} - p(\mathbf{x})_{,i_1\cdots i_j} - (\bar{g}(\mathbf{x})_{,i_1\cdots i_j} \\ &\qquad - p(\mathbf{x})_{,i_1\cdots i_j})| \\ &\leq C_j(\delta_e)^{p+1-j}\end{aligned} \qquad (10.50)$$

for any $j = 0, 1, 2, \ldots, p$, where $C_j = M + bK/(p+1-j)!$. Thus, using the symbolism of (10.35),

$$|\partial^p(F(\mathbf{x}) - \bar{g}(\mathbf{x}))| \leq C\,\delta_e \qquad (10.51)$$

where C is a constant independent of δ_e. Introducing (10.51) into (10.37), we have

$$\|F - \bar{g}\|^2_{\mathscr{L}} \leq \imath_e C^2 A(\delta_e)^2 \qquad (10.52)$$

wherein the dependence of F and $\bar{g}$ on $\mathbf{x}$ is understood. If δ is the mesh of the connected finite-element model (that is, if $\delta = \max\,\{\delta_1,$

† If $\bar{f}(\mathbf{x})$ is a polynomial of degree q and $p + 1 - j \leq 2(q + 1 - p)$, then (10.50) agrees with the error estimate of Fix and Strang [1969]. For detailed discussions of such error estimates, see Fried [1971]; for error estimates for parabolic equations, consult Fix and Nassif [1971]. See also Korneev [1967], Babuška [1971], McLay [1971], and Oden [1971*b*].

$\delta_2, \ldots, \delta_E\}$), and if, globally, the norms (10.52) lead to finite energy, then $\|F - \bar{g}\|^2_{\mathscr{L}} \leq \bar{\mathscr{R}}\bar{C}^2 A\,\delta^2$, $\bar{C}$ being the maximum C for all elements. Clearly, for a sequence of global approximations generated by representations from the class $\mathscr{F}^r$ and corresponding to regular uniform refinements for which $\delta \to 0$, $\lim_{\delta \leftarrow 0} \|F - \bar{g}\|_{\mathscr{L}} = 0$. Therefore, global finite-element basis functions so constructed are complete in energy. To now show that $\tilde{p}(\mathbf{x}),_{i_1 \ldots i_s}, \tilde{p}(\mathbf{x}) \in \mathscr{G}$, need be only m-equivalent to $p(\mathbf{x}),_{i_1 \ldots i_s}, p(\mathbf{x})$ being a complete polynomial of degree p, note that according to (10.41) and (10.48),

$$|(F - \bar{g}),_{i_1 \ldots i_s}| = |[F - p + (p - \tilde{p}) + (\tilde{p} - p) + p - \bar{g}],_{i_1 \ldots i_s}|$$
$$\leq [\hat{K} + N(\delta_e)^{n+s-p-1} + M](\delta_e)^{p+1-s}$$

Thus (10.50) still holds as $\delta_e \to 0$ so long as $m \geq p + 1 - s$. This completes the proof of the theorem.

To fix ideas, it is informative to consider a simple example. Set $\mathscr{R} = \mathscr{E}^1$ and consider the linear operator

$$\mathscr{L} \equiv -\frac{d^2}{dx^2}$$

The domain $\mathscr{D}_{\mathscr{L}}$ is assumed to contain functions $u(\mathbf{x}) \in C^2[a, b]$ such that $u(a) = u(b) = 0$. Observe that

$$\langle u, \mathscr{L}v \rangle = -\int_a^b u \frac{d^2v}{dx^2}\,dx = \int_a^b \left(\frac{du}{dx}\right)\left(\frac{dv}{dx}\right) dx$$

provided $u(a) = u(b) = 0$. Thus $\mathscr{L}$ is symmetric and positive-definite and

$$\|u - v\|^2_{\mathscr{L}} = \int_a^b \left(\frac{d(u - v)}{dx}\right)^2 dx \leq (b - a) \max \left|\frac{d(u - v)}{dx}\right|^2 \tag{10.53}$$

In this case $p = 1$ and $r = 0$.

Now suppose that $[a, b]$ is partitioned into a number E of finite elements of length $\delta_1, \delta_2, \ldots, \delta_E$ and consider a class of first-order finite-element representations of the form

$$\bar{f}_{(e)}(x) = f^N_{(e)}\psi^{(e)}_N(x) = f^1_{(e)}\left(1 - \frac{h_e(x)}{\delta_e}\right) + f^2_{(e)}\frac{h_e(x)}{\delta_e} \tag{10.54}$$

where, for convenience, we also use the symbol x for the local coordinate ($x^1_{(e)} = 0$, $x^2_{(e)} = \delta_e$). The function $h_e(x)$ is assumed to have the properties

$$h_e(0) = 0 \qquad h_e(\delta_e) = \delta_e \tag{10.55}$$

and the form of $h_e(x)$ is assumed to be the same for each element. Clearly, for a given $h(x)$, $\bar{f}_{(e)}$ is uniquely defined over $\imath_e$ by its values $f^N_{(e)}$ at each node of the element.

Let $U(x)$ be an arbitrary element of $\mathscr{D}_{\mathscr{L}}$ (or its extension) and let

$U^1, U^2, \ldots, U^G$ denote the values of $U(x)$ at the global nodes. Without loss in generality, consider the subinterval joining nodes 1 and 2. By using Taylor's theorem, it is easily shown that

$$\frac{1}{\delta_e}(U^2 - U^1) = U'(x) + \frac{1}{2\delta_e}[U''(\bar{x})(\delta_e - x)^2 - U''(x^*)x^2]$$

where $U'(x) = dU/dx$, $U''(x) = d^2U/dx^2$, $x \in [x^1, x^2]$, $x \leq \bar{x} \leq \delta_e$, and $x^1 \leq x^* \leq x$.

Returning to (10.54) and temporarily dropping the label e, we have

$$\bar{f}' = \frac{d\bar{f}}{dx} = h'(x)\frac{f^2 - f^1}{\delta}$$

Thus, if $\bar{f}$ coincides with $U(x)$ at the nodes,

$$\bar{f}'(x) - U'(x) = (h'(x) - 1)U'(x) + \frac{1}{2\delta}h'(x)[U''(\bar{x})(\delta - x)^2 - U''(x^*)x^2]$$

for $x \in \imath_e$. Let $|h'(x)| \leq A$, $|U'(x)| \leq B$, and $|U''(\bar{x})| \leq C$ for all $x \in [a, b]$. Then

$$|\bar{f}' - U'| \leq B\,|h'(x) - 1| + D\delta$$

where $D \leq AC$. Consequently, if δ is now the mesh of the connected model and $\bar{F}(x)$ is the global counterpart of $\bar{f}_{(e)}(x)$,

$$\|\bar{F} - U\|^2_{\mathscr{L}} \leq (b - a)[B\,|h' - 1| + D\delta]^2 \tag{10.56}$$

Thus, $\lim_{\delta \to 0} \|\bar{F} - U\|_{\mathscr{L}} \neq 0$ *unless* $h'(x)$ *is at least* 1-*equivalent to unity*; that is, $|h' - 1| \leq K\delta$.

In the present example, $p = 1$, so that according to Theorem 10.6, it is sufficient to use class $\mathscr{F}^0$ finite-element representations (i.e., representations for which only the values of the function and not its derivatives are specified at nodal points and that contain functions at least 1-equivalent to complete polynomials of degree $p = r + 1 = 1$; that is, a first-degree polynomial, $a + bx$). Indeed, if $\psi_1(x) = 1 - x/\delta$, $h(x) = x$ and $|h' - 1| \equiv 0$. However, $h(x)$ *need be only* 1-*equivalent to* x.

To emphasize this point, we note that we can select numerous functions that satisfy (10.55). For example,

1. $h(x) = \delta_e \sin \dfrac{\pi x}{2\delta_e}$
2. $h(x) = \delta_e \dfrac{\sin x}{\sin \delta_e}$ (10.57)
3. $h(x) = x + x(x - \delta_e)^m$
4. $h(x) = \delta_e \left(\dfrac{1 - e^{-x}}{1 - e^{-\delta_e}}\right)$

etc. In example 1, $h'(x)$ is clearly not 1-equivalent to unity; in fact, $|h'(x) - 1| = |1 - \pi/2 \cos(\pi x/2\delta_e)|$. Finite-element models constructed using functions of this type are therefore not complete in energy. On the other hand, for examples 2 and 3 we have $|h' - 1| \leq 2\delta_e^2/3 + 13\delta_e^4/90 + \cdots$ and $|h' - 1| \leq (m - 1)\delta_e^m$, respectively. Thus, if δ_e is sufficiently small, h' of 2 is 2-equivalent to unity and h' of 3 is m-equivalent to unity. Example 4 is also admissible, since it can be shown that $h'(x) = 1 + O(\delta)$.

Examples (10.57) and Theorem 10.6 lead us to some interesting observations:

1. The local interpolation functions need not be polynomials.
2. However, if the local interpolation functions are polynomials, they need not be complete polynomials. Indeed, for functionals of class C^{p-1} (or energy norms involving derivatives of order p), they need only contain complete polynomials of order $\geq p$.

A general criterion for selecting local interpolation functions that are complete in energy follows from Theorem 10.6.†

Theorem 10.7 *Let conditions* **a** *to* **c** *of Theorem 10.6 hold. Then* $\bar{f}_e(\mathbf{x})$ *and all of its partial derivatives up to order* $p = r + 1$ *can be assigned values over* $\imath_e$ *that differ from arbitrary constants only by terms of at least order* δ_e.

† In their finite-element analysis of plate bending, Bazeley, Cheung, Irons, and Zienkiewicz [1966] proposed as completeness ("convergence") criteria that the displacement function be such that "self-straining due to a rigid body motion of the element" not be permitted and that the displacement function within each element be such that constant "strain" and curvature conditions be possible over the element. Similar requirements were noted earlier by Irons and Draper [1965]. Since the functional involved in their analysis was of class C^1, their criteria amount to requiring that constant values of the displacement function and its first and second partial derivatives be possible over each element. A proof for polynomial approximations was later furnished by Arantes e Oliveira [1968]. Theorem 10.7 shows that such constant values need only be obtained in the limit as $\delta \to 0$. Unfortunately, many subsequent writers interpreted criteria proposed by Bazeley et al. for plates too literally and used the "rigid motion" and "constant strain" requirements as fundamental conditions for completeness for *all* finite-element approximations of displacement fields. It is easily shown, however, that a finite-element approximation of the displacement field in an elastic body can be constructed that satisfies the completeness requirements of Theorem 10.7, and yet does not lead to either a rigid motion or a constant strain when δ_e is finite. This can be seen by regarding the x^i in Theorem 10.6 as curvilinear coordinates (e.g., cylindrical; then u_r = constant does not produce rigid motion nor does it necessarily lead to constant strain, while $u_r = a + br$ may be admissible for a C^0 problem). Such rigid motions and constant strains need be obtained only in the limit as $\delta \to 0$. On the other hand, Murray [1970] has shown by means of numerical examples that the inclusion of rigid-body motion in elements of finite dimension may significantly improve rates of convergence.

Proof This follows immediately from Theorem 10.6 since $\mathscr{F}^r$ contains functions $p(\mathbf{x})$ that are at least 1-equivalent to complete polynomials of degree p with arbitrary coefficients. For example, to assign a constant value K to, say, the $m + n$th partial derivative $\partial^{m+n}\bar{f}/\partial(x^1)^m(\partial x^2)^n, (m + n \leq p)$, we simply equate the appropriate coefficient of the monomial $(x^1)^m(x^2)^n$ of a complete polynomial of degree p to $K/m!n!$ and equate all other coefficients of the polynomial to zero.

Theorems 10.6 and 10.7 provide a convenient basis for designing finite-element approximations. It is essential, however, to add to the requirements suggested by Theorems 10.6 and 10.7 that of *conformity*.†

Definition 4 A family $\mathscr{F}^r$ of finite-element approximations is said to be *r-conformable* (or simply *conformable*) if upon assembling elements to form the connected model the global approximations $\bar{F}(\mathbf{x})$ and all of their partial derivatives up to and including those of order r are continuous across interelement boundaries.

Thus, for conformable finite-element approximations, $\bar{F}(\mathbf{x}) \in C^r(\bar{\mathscr{R}})$. The conformability requirement arises from the fact, mentioned earlier, that in a variational problem of class C^{p-1} the domain of the functional consists of functions $u \in C^{p-1}(\mathscr{R})$ but whose partial derivatives of order $p \in L_2(\mathscr{R})$; otherwise infinite integrals (energy norms) might result (see footnote on page 111). While the use of conformable finite-element approximations guarantees an appropriate degree of smoothness for a given problem, it also imposes restrictions on choices of element geometry. In Sec. 10.4 we cite examples of acceptable finite-element approximations.

10.3 FINITE-ELEMENT MODELS OF OPERATOR EQUATIONS

In this section we consider briefly several methods for constructing finite-element models of equations of the form (10.1) and (10.5) that involve linear and nonlinear operators. The methods mentioned by no means exhaust those which can be successfully used for constructing satisfactory models.

† Use of the terms "conforming" and "nonconforming" in connection with the continuity of finite-element approximations originated with the paper of Bazeley, Cheung, Irons, and Zienkiewicz [1966]. The continuity requirement was suggested by Melosh [1963*a*]; see also Key [1966]. Bazeley et al. demonstrated that the continuity requirement may be too restrictive in certain cases; however, their results indicate that for nonconformable elements convergence may depend upon the orientation of elements in the finite-element network. Arantes e Oliveira [1968] noted that discontinuities in derivatives of order $r = p - 1$ may be admitted in certain cases; e.g., problems in which jump conditions are prescribed on surfaces which are maintained as interelement boundaries in all refinements of the network; also, for uniform regular refinements of models constructed of composite elements (see Sec. 10.4.6) discontinuities in the rth derivatives may be tolerated on certain surfaces within the element. See also Irons and Draper [1965] and Clough and Tocher [1966].

The Ritz method† Of all the techniques available for formulating finite-element approximations of various boundary-value problems, the Ritz method is the most thoroughly explored. The essential features of the method have been strongly implied in our discussion of completeness in the previous section. To further demonstrate the ideas, let $\mathscr{L}$ denote a symmetric, linear, positive-bounded-below operator and consider again the equation

$$\mathscr{L}u = f \tag{10.58}$$

where f is an element of $L_2(\mathscr{R})$ and the domain of $\mathscr{L}$ is denoted $\mathscr{D}(\mathscr{D} \equiv \mathscr{D}_{\mathscr{L}})$. The solution u is assumed to satisfy homogeneous boundary conditions on $\partial\mathscr{R}$. As pointed out earlier, the problem of solving (10.58) can also be viewed as one of minimizing the functional

$$J(u) = \langle \mathscr{L}u, u\rangle - 2\langle f, u\rangle \tag{10.59}$$

provided the solution u^* belongs to $\mathscr{D}$ or is arbitrarily close to an element v of $\mathscr{D}$ in the sense of the norm $\|\cdot\|$. Here f is an element of $\mathscr{H}$ and $\mathscr{H}$ is real. The domain $\mathscr{D}$ is dense in a complete space $\mathscr{H}_{\mathscr{L}}$ obtained by the completion of the inner-product space generated by defining for the elements of $\mathscr{D}$ the energy $[u,v] = \langle \mathscr{L}u,v\rangle$. If $\lambda = \min J(u)$, a sequence $\{u^n\} \in \mathscr{H}_{\mathscr{L}}$ is referred to as a *minimizing sequence* if $\lim_{n\to\infty} J(u^n) = J(u^*) = \lambda$. The Ritz method is characterized by the following:

1. Determine an n-dimensional subspace Φ dense in $\mathscr{H}_{\mathscr{L}}$ by identifying a collection of elements $(\varphi_1, \varphi_2, \ldots, \varphi_n)$ that are complete in energy. The elements φ_i can be taken from $\mathscr{D}$, since $\mathscr{D}$ is dense in $\mathscr{H}_{\mathscr{L}}$.
2. Construct an approximation $u^n = \sum_{i=1}^{n} \alpha_i\varphi_i$ to u^* in Φ by choosing the α_i so as to minimize $J(u)$ in Φ. It can be shown‡ that the sequence $\{u^n\}$ of approximations so obtained is a minimizing sequence for $J(u)$. Moreover, $\lim_{n\to\infty} \|u^* - u^n\|_{\mathscr{L}} = 0$ if $\mathscr{L}$ is positive-bounded-below.

Suppose that we wish to construct a finite-element solution of the linear equation (10.58). We begin by constructing a finite-element model

† The well-known Ritz method was introduced by Ritz [1908, 1909] as a generalization of a technique described by Rayleigh [1877].

‡ Cf. Mikhlin [1965]. If $f_n \equiv \mathscr{L}u^n$, and $\mathscr{L}$ is unbounded, then it may happen that $\lim_{n\to\infty} \|f_n - f\| \neq 0$. However, f_n converges weakly to f in the sense that $\lim_{n\to\infty} \langle f_n - f, g\rangle = 0$ where $g \in \mathscr{H}_{\mathscr{L}}$ or if $\|f_n\|$ is bounded for all n, g may belong to $\mathscr{H}$.

$\bar{\mathscr{R}}$ of $\mathscr{R} + \partial\mathscr{R}$ and of u of the form

$$U(\mathbf{x}) = \sum_{e=1}^{E} \bar{u}^{(e)}(\mathbf{x}) = U^{\Delta}\Phi_{\Delta}(\mathbf{x}) \qquad \Phi_{\Delta}(\mathbf{x}) = \sum_{e=1}^{E} \overset{(e)}{\Omega}{}^{N}_{\Delta}\psi^{(e)}_{N}(\mathbf{x})$$
$$u^{N}_{(e)} = \overset{(e)}{\Omega}{}^{N}_{\Delta} U^{\Delta} \tag{10.60}$$

where the functions $\Phi_{\Delta}(\mathbf{x})$ are assumed to be complete in energy. The local approximation of u over element r_e is, according to (10.60),

$$\bar{u}^{(e)}(\mathbf{x}) = u^{N}_{(e)}\psi^{(e)}_{N}(\mathbf{x}) \tag{10.61}$$

Now let $l^{(e)}$ be the restriction of $\mathscr{L}$ to functions defined on r_e. Then, if the solution of (10.58) minimizes the functional (10.59), we may define the restriction of $J(u)$ to r_e as

$$j^{(e)}(u) = \langle l^{(e)}u, u\rangle - 2\langle u, f\rangle \tag{10.62}$$

and

$$J(u) = \sum_{e=1}^{E} j^{(e)}(u) \tag{10.63}$$

wherein u and f are understood to be the restrictions of u and f to r_e. Introducing (10.61) into (10.62), we obtain

$$j^{(e)}(u^{N}) = l^{(e)}_{MN}u^{M}_{(e)}u^{N}_{(e)} - 2f^{(e)}_{N}u^{N}_{(e)} \tag{10.64}$$

where $M, N = 1, 2, \ldots, N_e$ and

$$l^{(e)}_{MN} = \langle l^{(e)}\psi^{(e)}_{M}, \psi^{(e)}_{N}\rangle \qquad f^{(e)}_{N} = \langle f, \psi^{(e)}_{N}\rangle \tag{10.65}$$

Likewise, introducing (10.60) into (10.59) gives, for the global approximation,

$$J(U^{\Delta}) = L_{\Delta\Gamma}U^{\Delta}U^{\Gamma} - 2F_{\Delta}U^{\Delta} \tag{10.66}$$

where $\Delta, \Gamma = 1, 2, \ldots, G$ and

$$L_{\Delta\Gamma} = \langle \mathscr{L}\Phi_{\Delta}, \Phi_{\Gamma}\rangle \qquad F_{\Delta} = \langle f, \Phi_{\Delta}\rangle \tag{10.67}$$

In view of (10.63),

$$\begin{aligned} L_{\Delta\Gamma}U^{\Delta}U^{\Gamma} - 2F_{\Delta}U^{\Delta} &= \sum_{e=1}^{E} (l^{(e)}_{NM}u^{N}_{(e)}u^{M}_{(e)} - 2f^{(e)}_{N}u^{N}_{(e)}) \\ &= \sum_{e=1}^{E} (l^{(e)}_{NM}\overset{(e)}{\Omega}{}^{N}_{\Delta}\overset{(e)}{\Omega}{}^{M}_{\Gamma}U^{\Delta}U^{\Gamma} - 2f^{(e)}_{N}\overset{(e)}{\Omega}{}^{N}_{\Delta}U^{\Delta}) \end{aligned} \tag{10.68}$$

Locally, $j^{(e)}(u^{N})$ is a minimum if

$$\frac{\partial j^{(e)}(u^{N})}{\partial u^{N}_{(e)}} = 0 = 2l^{(e)}_{NM}u^{M}_{(e)} - 2f^{(e)}_{N} \tag{10.69}$$

or

$$l^{(e)}_{NM} u^{M}_{(e)} = f^{(e)}_{N} \tag{10.70}$$

In view of (10.60) and (10.63), (10.70) leads to the global requirement

$$L_{\Delta\Gamma} U^{\Gamma} = F_{\Delta} \tag{10.71}$$

which, in accordance with (10.66), implies that $\partial J(U^{\Delta})/\partial U^{\Delta} = 0$. Upon solving (10.71) for U^{Γ}, the function $U^{\Gamma}\Phi_{\Gamma}(\mathbf{x})$ represents the Ritz approximation of the solution u^* of (10.58) in the subspace Φ.

We shall consider briefly the question of convergence of sequences of Ritz approximations. As in the previous section, we consider a sequence of subspaces $\Phi^1, \Phi^2, \ldots,$ of $\mathscr{H}$, each spanned by respective sets of functions $\Phi^1_{\Delta}(\mathbf{x}), \Phi^2_{\Delta}(\mathbf{x}), \ldots,$ which are built up of local finite-element approximations in the usual manner. The sequence is constructed by using regular, uniform refinements of $\bar{\mathscr{R}}$, with meshes $\delta \to 0$; and the basis functions for each subspace are designed so as to be r-conformable and complete in energy in the sense of Theorem 10.6.

Let $u^* = u^*(\mathbf{x})$ denote the function in $\mathscr{H}$ that actually minimizes the functional $J(u)$; let $\bar{U} = \bar{U}(\mathbf{x})$ denote the global finite-element approximation of u^* that minimizes $J(u)$ in an n-dimensional subspace Φ^n defined by an acceptable finite-element model; and let $\bar{G} = \bar{G}(\mathbf{x})$ denote an element of Φ^n that coincides with u^* and its principal derivatives at the nodal points. The restriction $\bar{g}_e(\mathbf{x})$ of $\bar{G}(\mathbf{x})$ to $\imath_e$ is the function described in (10.43). Since by hypothesis the sets of basis functions $\Phi^1_{\Delta}(\mathbf{x}), \Phi^2_{\Delta}(\mathbf{x}), \ldots$ generated by uniform, regular refinements are complete in energy (with respect to a class of functions containing u^*), we can, in view of (10.52), find an integer N such that for any $\epsilon > 0$ the distance in energy from u^* to $\bar{G}$ will be less than $\epsilon^{\frac{1}{2}}$ for $n > N$, n being the dimension of subspace Φ^n; i.e.,

$$d_{\mathscr{L}}(u^*,\bar{G}) < \epsilon^{\frac{1}{2}} \qquad \text{for } n > N \tag{10.72}$$

Also, since $J(\bar{G}) = d^2_{\mathscr{L}}(u^*,\bar{G}) - \|u^*\|^2_{\mathscr{L}} = d^2_{\mathscr{L}}(u^*,\bar{G}) + J(u^*)$ [see (10.8)],

$$J(\bar{G}) - J(u^*) < \epsilon \qquad \text{for } n > N \tag{10.73}$$

or $\lim_{n\to\infty} J(\bar{G}^{(n)}) = J(u^*)$. [Recall that $d_{\mathscr{L}}(u^*,\bar{G}) \geq 0$.] Clearly, $J(u^*) \leq J(\bar{U}) \leq J(\bar{G})$, so that

$$J(\bar{U}) - J(u^*) \leq J(\bar{G}) - J(u^*) \tag{10.74}$$

Thus, for $n > N$, $J(\bar{U}) - J(u^*) < \epsilon$. Consequently, $\lim_{n\to\infty} J(\bar{U}^n) = J(u^*) = \min J(u)$; that is, the sequence of finite-element approximations of the functional $J(u)$ converges to the minimum $J(u^*)$ if the basis functions $\Phi_{\Delta}(\mathbf{x})$ are complete in energy. Note also that if $\mathscr{H}$ is real, and $\mathscr{L}$ is self-adjoint,

$$\begin{aligned} d^2_{\mathscr{L}}(\bar{U},u^*) &= [\bar{U} - u^*, \bar{U} - u^*] = [\bar{U},\bar{U}] - 2[\bar{U},u^*] + [u^*,u^*] \\ &= J(\bar{U}) - J(u^*) \end{aligned} \tag{10.75}$$

Then $d^2_{\mathscr{L}}(\bar{U}^{(n)},u^*) < \epsilon$ for $n > N$ and the distance in energy from $\bar{U}^{(n)}$ to u^* approaches zero as n tends to infinity. It may happen, of course, that $\mathscr{L}\bar{U}^{(n)} = f^{(n)}$ does not converge to $f = \mathscr{L}u^*$ [see Mikhlin, 1965].

We remark that if the refinements $\bar{\mathscr{R}}^{\Gamma}$ are made in such a way that $\Phi^1 \subset \Phi^2 \subset \cdots \Phi^n \subset \cdots$, then the corresponding Ritz approximations $\bar{U}^1, \bar{U}^2, \ldots, \bar{U}^n, \ldots$ are such that

$$J(\bar{U}^1) \geq J(\bar{U}^2) \geq \cdots J(\bar{U}^n) \geq \cdots \geq J(u^*) \tag{10.76}$$

We then have monotonic convergence.† However, monotonic convergence does not imply convergence to $J(u^*)$. We may state, however, that if $\Phi^{i+1} \supset \Phi^i$, $i = 1, 2, \ldots, n, \ldots$, and if the basis functions for approximations are complete in energy with respect to the class of functions containing u^*, then $J(\bar{U}^n)$ converges monotonically to $J(u^*)$.

Weighted residuals‡ A general method for constructing finite-element models of nonlinear equations involves the idea of weighted residuals. Consider, for example, the equation

$$\mathscr{P}(u) = f \tag{10.77}$$

where $\mathscr{P}$ is a nonlinear operator on some set $\mathscr{D}$. Let

$$\bar{u}_{(e)}(\mathbf{x}) = u^N_{(e)}\psi^{(e)}_N(\mathbf{x}) \tag{10.78}$$

be a finite-element approximation of u over a typical finite-element r_e and, for the sake of discussion, suppose that $\bar{u}_{(e)}(\mathbf{x}) \in \mathscr{D}_{\mathscr{P}(e)}$, wherein $\mathscr{P}^{(e)}$ is the restriction of $\mathscr{P}$ to a domain of functions defined on r_e. The function

$$r_{(e)}(\mathbf{x}) = \mathscr{P}^{(e)}(\bar{u}_{(e)}) - f^{(e)} \tag{10.79}$$

is called the *residual* corresponding to element r_e. Obviously, if $r_{(e)}(\mathbf{x}) = 0$, then $\bar{u}_{(e)}$ is a solution of (10.77) over r_e. The method of weighted residuals involves the identification of a set of *weight functions* $w^{(e)}_N(\mathbf{x})$, $N = 1, 2, \ldots, N_e$, so that $r_e(\mathbf{x})$ vanishes in some *weighted average sense* over r_e. Generally we require that

$$\int_{r_e} r_{(e)}(\mathbf{x}) w^{(e)}_N(\mathbf{x})\, dr_e = 0 \tag{10.80}$$

or, in view of (10.79),

$$\mathfrak{p}^{(e)}_M(u^N_{(e)}) - f^{(e)}_{M(w)} = 0 \tag{10.81}$$

† Melosh [1963*a*] proposed as a sufficient condition for monotonic convergence of conformable finite-element approximations that $\Phi^{i+1} \supset \Phi^i$. This condition was also studied by Key [1966].

‡ A detailed survey of applications of the method of weighted residuals was given by Finlayson and Scriven [1966]. For summary accounts, consult Ames [1965] or Crandall [1956]. Applications of the method to finite-element approximations of linear differential equations were discussed by Leonard and Bramlette [1970].

where

$$\mathfrak{p}_M^{(e)}(u_{(e)}^N) = \int_{v_e} \mathscr{P}^{(e)}(\psi_N^{(e)} u_{(e)}^N) w_M^{(e)}(\mathbf{x})\, dv_e \tag{10.82a}$$

$$f_{M(w)}^{(e)} = \int_{v_e} f^{(e)}(\mathbf{x}) w_M^{(e)}(\mathbf{x})\, dv_e \tag{10.82b}$$

and $\mathfrak{p}_M^{(e)}(u_{(e)}^N)$ is generally nonlinear in $u_{(e)}^N$. Globally, we have

$$W_\Delta(\mathbf{x}) = \sum_{e=1}^{E} \overset{(e)}{\Omega}{}_\Delta^N w_N^{(e)}(\mathbf{x}) \qquad u_{(e)}^N = \overset{(e)}{\Omega}{}_\Delta^N U^\Delta \tag{10.83a, b}$$

$$\mathfrak{P}_\Delta(U^\Gamma) = \sum_{e=1}^{E} \overset{(e)}{\Omega}{}_\Delta^M \mathfrak{p}_M^{(e)}(\overset{(e)}{\Omega}{}_\Gamma^N U^\Gamma) \qquad F_\Delta^{(w)} = \sum_{e=1}^{E} \overset{(e)}{\Omega}{}_\Delta^N f_{N(w)}^{(e)} \tag{10.83c, d}$$

or

$$\mathfrak{P}_\Delta(U^\Gamma) - F_\Delta^{(w)} = 0 \tag{10.84}$$

Solutions of the nonlinear equations (10.84) ensure the vanishing of the weighted average of the residual globally; i.e.,

$$\int_{\overline{\mathscr{R}}} [\mathscr{P}(U^\Gamma \Phi_\Gamma(\mathbf{x})) - f(\mathbf{x})] W_\Delta(\mathbf{x})\, d\overline{\mathscr{R}} = 0 \tag{10.85}$$

Various types of averaging methods differ in their choice of the weight functions $W_\Delta(\mathbf{x})$.

Observe that (10.80) and (10.85) are orthogonality conditions; i.e., the residual is made orthogonal to some subspace spanned by the weight functions W_Δ. The nature of the subspace, again, depends upon the choice of these weight functions. The idea of seeking elements $\bar{u}$ in linear subspaces Φ such that $u - \bar{u}$ is orthogonal to Φ is based on the classical projection theorem that is fundamental to most averaging methods.

Theorem 10.8 *Let $\mathscr{H}$ be a Hilbert space and Φ a closed subspace of $\mathscr{H}$. For every $u \in \mathscr{H}$ there exists a unique element $\bar{u}^* \in \Phi$ such that $\|u - \bar{u}^*\| < \|u - \bar{u}\|$ for all $\bar{u} \in \Phi$. Moreover, a necessary and sufficient condition that the distance $\|u - \bar{u}^*\|$ be a minimum is that $u - \bar{u}^*$ be orthogonal to Φ.*

Proof A variation of this theorem was proved in Sec. 9.3. We shall consider a more elaborate proof here. First, suppose that the element $\bar{u}^* \in \Phi$ minimizes $d^2(u,\bar{u}) = \|u - \bar{u}\|$ but that $u - \bar{u}^*$ is not orthogonal to Φ. Then $\langle u - \bar{u}^*, \bar{u}\rangle = \lambda \neq 0$. Let $\bar{u}_1 = \bar{u}^* + \lambda\bar{u}$. Then

$$\|u - \bar{u}_1\|^2 = \|u - \bar{u}^*\|^2 - \langle u - \bar{u}^*, \lambda\bar{u}\rangle - \langle \lambda\bar{u}, u - \bar{u}^*\rangle + |\lambda|^2$$
$$= \|u - \bar{u}^*\|^2 - |\lambda|^2 < \|u - \bar{u}^*\|^2$$

where, without loss in generality, we have assumed that $\|\bar{u}\| = 1$. This result implies that if $u - \bar{u}^*$ is not orthogonal to Φ, $\bar{u}^*$ does not minimize $d^2(u, \bar{u})$, which is a contradiction of the original hypothesis. Thus, if $\bar{u}^*$ minimizes $d^2(u, \bar{u})$, $u - \bar{u}^*$ is orthogonal to Φ. To show that if $u - \bar{u}^*$ is orthogonal to Φ, $\bar{u}^*$ is the unique vector in Φ which minimizes $d^2(u, \bar{u})$, note that for any $\bar{u} \in \Phi$,

$$\|u - \bar{u}\|^2 = \|u - \bar{u}^* + \bar{u}^* - \bar{u}\|^2 = \|u - \bar{u}^*\|^2 + \|\bar{u} - \bar{u}^*\|^2$$

Hence $\|u - \bar{u}\| > \|u - \bar{u}^*\|$ for $\bar{u} \neq \bar{u}^*$.

Notice that the uniqueness and orthogonality of $\bar{u}^*$ do not depend upon the completeness of $\mathscr{H}$. If $\mathscr{H}$ is complete (i.e., if $\mathscr{H}$ is a Hilbert space), we may proceed to establish the existence of $\bar{u}^*$. Let $\lambda = \inf \|u - \bar{u}\|$, $\bar{u} \in \Phi$ and let $\{\bar{u}^n\}$ be a sequence in $\mathscr{H}$ such that $\lim_{n\to\infty} \|u - \bar{u}^n\| \to \lambda$. Then

$$\|(\bar{u}^m - u) + (u - \bar{u}^n)\|^2 + \|(\bar{u}^m - u) - (u - \bar{u}^n)\|^2 = 2\|\bar{u}^m - u\|^2 + 2\|u - \bar{u}^n\|^2$$

or

$$\|\bar{u}^m - \bar{u}^n\|^2 = 2\|\bar{u}^m - u\|^2 + 2\|u - \bar{u}^n\|^2 - 4\left\|u - \frac{(\bar{u}^n + \bar{u}^m)}{2}\right\|^2$$

Since Φ is a linear subspace, $\bar{u}^m + \bar{u}^n \in \Phi$. From this relation, it follows that

$$\|\bar{u}^m - \bar{u}^n\|^2 \leq 2\,\|\bar{u}^m - u\|^2 + 2\,\|\bar{u}^n - u\|^2 - 4\lambda^2$$

and, since $\|\bar{u}^m - u\| \to \lambda$ as $m \to \infty$, it follows that $\|\bar{u}^m - \bar{u}^n\| \to 0$ as $m, n \to \infty$. Therefore $\{\bar{u}^m\}$ is a Cauchy sequence in a closed subspace Φ. Hence, it has a limit point $\bar{u}^* \in \Phi$. Since $\|u - \bar{u}^n\|$ is continuous, it follows that $\|u - \bar{u}^n\| = \lambda$. This completes the proof.

Galerkins's method Perhaps the most powerful technique for generating acceptable finite-element models of nonlinear equations is the averaging method of Galerkin.† Though the method amounts to a special case of the method of weighted residuals (and a generalization of the Ritz method), it involves a rational choice of the weight functions $w_N(\mathbf{x})$ so as to be consistent with the form of the finite-element approximation employed.

† The method was developed in the papers of B. G. Galerkin [1915], and was used in finite-element applications by Oden [1967*a*, *c*, 1969*b*], Szabo [1969], Szabo and Lee [1969*a*, *b*], and others. The papers of Szabo elaborate on the application of the method to the construction of finite-element models of problems in static plane elasticity, and then elastic plates, wherein the method is essentially the same as the Ritz method. However, the method does lead naturally to more explicit statements of conditions at finite-element boundaries. Most of the finite-element formulations presented in Chaps. III and V of this book may be interpreted as applications of Galerkin's method.

The motivation of the method can best be argued for the case in which $\hat{\mathscr{P}}(u) = \mathscr{P}(u) - f$ is a potential operator and $\mathscr{P}(u)$ is the nonlinear operator in (10.77). As pointed out in Sec. 10.1, $\hat{\mathscr{P}}(u)$ can be regarded as the gradient of a functional $K(u)$ whose critical points are solutions of $\hat{\mathscr{P}}(u) = 0$. Then, as indicated in (10.26),

$$\langle \mathscr{P}(u^*) - f, h \rangle = 0 \tag{10.86}$$

where h is an arbitrary element of the space containing u, and u^* is a critical point of $K(u)$. If the finite-element approximation $\bar{u}^{(e)}(\mathbf{x})$ of (10.78) only approximately satisfies (10.77), then the residual $r_{(e)}(\mathbf{x})$ of (10.79) will not necessarily be orthogonal to h. We can, however, select the coefficients $u^N_{(e)}$ of the approximation so that $r_{(e)}(\mathbf{x})$ is orthogonal to the finite-dimensional subspace Φ spanned by the interpolation functions $\psi^{(e)}_N(\mathbf{x})$ [or, equivalently, $\Phi_\Delta(\mathbf{x})$]. Locally, this amounts to setting

$$\langle \mathscr{P}^{(e)}(\bar{u}_{(e)}) - f^{(e)}, h \rangle = 0 \tag{10.87}$$

However, since $h = h(\mathbf{x})$ is now an arbitrary element of Φ, it can be expressed in the form $h = h^N_{(e)}\psi^{(e)}_N(\mathbf{x})$. It follows that

$$\langle \mathscr{P}^{(e)}(\bar{u}_{(e)}) - f^{(e)}, \psi^{(e)}_N \rangle h^N_{(e)} = 0 \tag{10.88}$$

But (10.88) must hold for arbitrary coefficients $h^N_{(e)}$. Therefore,

$$\langle r_{(e)}, \psi^{(e)}_N \rangle \equiv \langle \mathscr{P}^{(e)}(\bar{u}_{(e)}) - f^{(e)}, \psi^{(e)}_N \rangle = 0 \tag{10.89}$$

This result represents a system of nonlinear equations for the coefficients $u^N_{(e)}$. Clearly, for the case in which $\langle u, v \rangle = \int_{\imath_e} uv \, d\imath_e$, (10.89) becomes

$$\int_{\imath_e} r_{(e)}(\mathbf{x})\psi^{(e)}_N(\mathbf{x}) \, d\imath_e = 0 \tag{10.90}$$

Comparing this equation with (10.80), we see that Galerkin's method is essentially the method of weighted residuals for a choice of the weight functions $w^{(e)}_N(\mathbf{x}) = \psi^{(e)}_N(\mathbf{x})$. Note also that in this case $\mathfrak{p}^{(e)}_M(u^N_{(e)})$ and $f^{(e)}_{M(w)}$ of (10.82) become

$$\mathfrak{p}^{(e)}_M(u^N_{(e)}) = \int_{\imath_e} \mathscr{P}^{(e)}(\psi^{(e)}_N u^N_{(e)})\psi^{(e)}_M(\mathbf{x}) \, d\imath_e \qquad f^{(e)}_M = \int_{\imath_e} f^{(e)}(\mathbf{x})\psi^{(e)}_M(\mathbf{x}) \, d\imath_e \tag{10.91}$$

We remark that if $\mathscr{P}$ involves derivatives with respect to a parameter t (time) and $\bar{u}^{(e)} = u^N_{(e)}(t)\psi^{(e)}_N(\mathbf{x})$, then Galerkin's method will lead to systems of differential equations of the form (10.81) [with $\mathfrak{p}^{(e)}_M(u^N_{(e)}(t))$ and $f^{(e)}_M(t)$ given by (10.91)] in $u^N_{(e)}(t)$ for each finite element.

Least-squares approximations As a final example of a method for constructing finite-element models, we give brief consideration to the method of least squares. In this method, we observe that a measure of the error e of an approximation $\bar{u}^{(e)}$ is provided by the square of the norm of the residual:

$$e = \|r_{(e)}\|^2 = \langle r_{(e)}, r_{(e)} \rangle \tag{10.92}$$

The method of least squares consists of selecting the coefficients $u^N_{(e)}$ so that e is a minimum. Introducing (10.79), we have

$$e = \langle \mathscr{P}^{(e)}(\psi_N u^N_{(e)}) - f^{(e)}, \mathscr{P}^{(e)}(\psi^{(e)}_M u^M_{(e)}) - f^{(e)} \rangle \tag{10.93}$$

The condition

$$\frac{\partial e}{\partial u^N_{(e)}} = 0 = 2 \left\langle r_{(e)}, \frac{\partial r_{(e)}}{\partial u^N_{(e)}} \right\rangle \tag{10.94}$$

leads to the system of equations

$$Q^{(e)}_N(u^M_{(e)}) - f^{*(e)}_N = 0 \tag{10.95}$$

where

$$Q^{(e)}_N(u^M_{(e)}) = \left\langle \mathscr{P}^{(e)}(\bar{u}^{(e)}), \frac{\partial \mathscr{P}^{(e)}(\bar{u}^{(e)})}{\partial u^N_{(e)}} \right\rangle \qquad f^{*(e)}_N = \left\langle f^{(e)}, \frac{\partial \mathscr{P}^{(e)}(\bar{u}^{(e)})}{\partial u^N_{(e)}} \right\rangle \tag{10.96a, b}$$

The corresponding global form of (10.95) is obtained in the usual manner. Comparing (10.94) with (10.80), we see that the method of least squares amounts to an application of the method of weighted residuals for a choice of the weight functions $w^{(e)}_N(\mathbf{x}) = \partial r_{(e)}/\partial u^N_{(e)} = \partial \mathscr{P}^{(e)}(\bar{u}^{(e)})/\partial u^N_{(e)}$.

10.4 EXAMPLES OF FINITE ELEMENTS

In Sec. 10.2 we established that conformable finite-element approximations that satisfy the completeness requirements implied by Theorems 10.6 and 10.7 can be used to construct acceptable approximations of solutions to a large class of physical problems. In the present section we cite several examples of local interpolation functions $\overset{0}{\psi}_N(\mathbf{x}), \overset{i}{\psi}_N(\mathbf{x}), \ldots, \overset{i_1 \cdots i_r}{\psi_N}(\mathbf{x})$ that lead to global basis functions satisfying these criteria† [i.e., $\bar{F}(\mathbf{x}) \in C^r(\bar{\mathscr{R}})$ and $\bar{F}(\mathbf{x})$ and its derivatives up to order $r + 1$ can be assigned values over each $\imath_e$ that are at least 1-equivalent to arbitrary constants]. For convenience, we list approximations of a scalar-valued function $U(\mathbf{x})$ with local approximations $\bar{u}^{(e)}(\mathbf{x})$, but it is emphasized that the same local interpolation functions $\overset{0}{\psi}_N(\mathbf{x}), \overset{i}{\psi}_N(\mathbf{x}), \ldots$ can be used for the representation of vector- and tensor-valued functions as well (see Examples 7.4 and 7.5). Also, since the discussion is to pertain to a typical finite element, the element identification label (e) is dropped temporarily for convenience. We remark that

† Catalogs of various acceptable finite elements have been given by a number of authors. See, for example, Argyris [1966a]; Argyris, Fried, and Scharpf [1968a, 1968b, 1968c]; Argyris and Scharpf [1968]; Argyris and Fried [1968]; Argyris et al. [1969]; Zienkiewicz and Cheung [1967]; Felippa [1966]; and Felippa and Clough [1968].

polynomial expansions are the most commonly used functional forms for the local interpolation functions† due to their simplicity and convenience for use in computations (e.g., numerical integrations) and to the fact that it is generally simpler to test polynomial approximations for satisfaction of the completeness and conformability requirements.

10.4.1 Simplex models‡ The simplest of all finite elements are the so-called simplex models§ that form topological simplexes in the space in which the element is embedded. In simplex representations, the local fields $u(\mathbf{x})$ are approximated by functions linear in the coordinates x^i:

$$u(\mathbf{x}) = a_0 + a_i x^i \qquad i = 1, 2, \ldots, k \tag{10.97}$$

Here $a_0, a_1, a_2, \ldots, a_k$ are $k + 1$ constants yet to be determined. For a k-dimensional space, we proceed to identify $k + 1$ nodal points ($N_e = k + 1$) and to evaluate (10.97) at each node $\mathbf{x}^N$:

$$u(\mathbf{x}^N) \equiv u^N = a_0 + a_i x^{Ni} \qquad \begin{aligned} i &= 1, 2, \ldots, k \\ N &= 1, 2, \ldots, k + 1 \end{aligned} \tag{10.98a}$$

Equations (10.98*a*) represent a system of $k + 1$ simultaneous equations in the $k + 1$ quantities $a_0, a_1, \ldots, a_k$. It is convenient to rewrite (10.98*a*) in the following matrix form

$$\begin{bmatrix} 1 & x^{11} & x^{12} & \cdots & x^{1k} \\ 1 & x^{21} & x^{22} & \cdots & x^{2k} \\ 1 & x^{31} & x^{32} & \cdots & x^{3k} \\ \cdots & \cdots & \cdots & \cdots & \cdots \\ 1 & x^{k+1,1} & x^{k+1,2} & \cdots & x^{k+1,k} \end{bmatrix} \begin{bmatrix} a_0 \\ a_1 \\ a_2 \\ \cdots \\ a_k \end{bmatrix} = \begin{bmatrix} u^1 \\ u^2 \\ u^2 \\ \cdots \\ u^{k+1} \end{bmatrix} \tag{10.98b}$$

† It is emphasized that, in general, the interpolation functions need not be polynomials. Indeed, this fact is clearly indicated by the examples (10.57). Krahula and Polhemus [1968], for example, proposed functions formed by adding to a basic polynomial an appropriate Fourier series. General discussions of polynomial approximations for finite elements were given by Felippa [1966], Dunne [1968], and Silvester [1969].

‡ A simplex in a k-dimensional space is a convex set S determined by a collection of $k + 1$ vertices (nodes) $N_1, N_2, \ldots, N_{k+1}$ that do not lie in the $(k - 1)$-dimensional hyperplane. The set S consists of all points y such that $y = \sum_{i=1}^{k+1} c_i N_i$, for which $c_i \geq 0$ and $\sum_{i=1}^{k+1} c_i = 1$. See, for example, Graves [1956, pp. 146–150], or Greub [1963, pp. 253, 254]. In euclidean spaces $\mathscr{E}^k$, a simplex is a tetrahedron (four nodes) in three-dimensional space $\mathscr{E}^3$, a triangle (three nodes) in two-dimensional space $\mathscr{E}^2$, and a straight line (two nodes) in one-dimensional space $\mathscr{E}^1$.

§ Simplex models merely involve the use of linear interpolation functions within an element and were among the first types of finite elements to be used. See, for example, Turner, Clough, Martin, and Topp [1956]; Synge [1957]; and Gallagher, Padlog, and Bijlaard [1962]. Properties of the simplex approximation have been discussed by Wissmann [1963], Oden [1967*a*, 1967*b*, 1969*a*], Felippa [1966], and Oden and Aguirre-Ramirez [1969], among others. Simplex models in higher-dimensional spaces were discussed by Oden [1969*a*] and Fried [1969*a*].

or

$$\mathbf{Ca} = \mathbf{u} \tag{10.98c}$$

where $\mathbf{C}$ is the $(k + 1) \times (k + 1)$ coefficient matrix in (10.98*b*). Equation (10.98*c*) will possess a unique solution provided $\mathbf{C}$ is nonsingular. Thus, we require that

$$\det \mathbf{C} \neq 0 \tag{10.99}$$

Equation (10.99) may be interpreted as the requirement that the $k + 1$ nodal points $\mathbf{x}^N$ must not all lie in the same $(k - 1)$-dimensional hyperplane.

Assuming that (10.99) is satisfied, we solve (10.98*c*) and obtain

$$\mathbf{a} = \mathbf{C}^{-1}\mathbf{u} \tag{10.100}$$

In view of (10.97), (10.100) gives the coefficients $a_0, a_1, \ldots, a_k$ linearly in terms of the nodal values u^N of the local function and the specific nodal coordinates x^{Ni}. Introducing (10.100) back into (10.97) and simplifying, we finally arrive at the desired local approximation,†

$$u(\mathbf{x}) = u^N \psi_N(\mathbf{x}) \qquad N = 1, 2, \ldots, k + 1 \tag{10.101a}$$

where, in this case, the interpolation functions are linear in the local coordinates x^i:

$$\psi_N(\mathbf{x}) = a_N + b_{Ni}x^i \tag{10.101b}$$

Again, the quantities a_N and b_{Ni} are known in terms of the nodal coordinates x^{Ni} by virtue of (10.100). Specifically, if C_{ij} $(i, j = 1, 2, \ldots, k + 1)$ is an element in the ith row and j the column of $\mathbf{C}$, then

$$a_N = \frac{1}{C} \text{ cofactor } (C_{N1}) \tag{10.102a}$$

$$b_{Ni} = \frac{1}{C} \text{ cofactor } (C_{N(i+1)}) \tag{10.102b}$$

where

$$C = \det \mathbf{C} \tag{10.102c}$$

The functions $\psi_N(\mathbf{x})$ have the following properties for every $\mathbf{x} \in \imath_e$:

$$\psi_N(\mathbf{x}) \geq 0 \qquad \sum_{N=1}^{k+1} \psi_N(\mathbf{x}) = 1 \tag{10.103a, b}$$

† Such linear interpolation functions are sometimes called *barycentric coordinates* of the simplex. See Grueb [1963, p. 253].

Moreover, for any nodal point $\mathbf{x}^M \in \imath_e$,

$$\psi_N(\mathbf{x}^M) = \delta_N^M \tag{10.103c}$$

We summarize the important properties of the simplex model, as follows:

1. The number N_e of nodal points belonging to a simplex element is one greater than the dimension of the space in which the element lies ($N_e = k + 1$).
2. The local approximations $u(\mathbf{x})$ [and the interpolation functions $\psi_N(\mathbf{x})$] are linear functions of the local coordinates x^i.
3. The completeness and continuity requirements are satisfied. A linear function in $\mathscr{E}^k$ is uniquely determined by specifying its values at $k + 1$ noncoincident points. Since the local functions $u(\mathbf{x})$ are linear along each boundary of the element, fitting two simplex elements together amounts to prescribing the same nodal values of adjacent local approximations at two points on their interelement boundary. Thus, the local fields coincide at *all* points on interelement boundaries in the connected model and the global basis functions $\Phi_\Delta(\mathbf{x})$ are everywhere continuous. To verify that (10.101) is capable of representing uniform values of $u^{(e)}(\mathbf{x})$ over an element, we simply set each nodal value $u^N = \lambda$, where λ is a constant. Then, in view of (10.103*b*), $u(\mathbf{x}) = \lambda$ for all $\mathbf{x} \in \imath_e$.
4. The continuity requirement mentioned in (3) is satisfied for arbitrary orientations of the element. That is, the boundaries of the element need not be coordinate lines in order that the continuity requirements be satisfied. (As will be seen subsequently, not all conformable finite elements have this property.)
5. The interpolation functions $\psi_N(\mathbf{x})$ satisfy (10.103) (these properties, of course, are not unique to simplex models).

We now examine specific forms of (10.101*b*) for spaces of dimension $k \leq 3$.

Simplex in a three-dimensional space For the case in which the space is euclidean and three dimensional, the simplex model is a tetrahedron with four nodes, as indicated in Fig. 10.1*a*. Then

$$\mathbf{C} = \begin{bmatrix} 1 & x^{11} & x^{12} & x^{13} \\ 1 & x^{21} & x^{22} & x^{23} \\ 1 & x^{31} & x^{32} & x^{33} \\ 1 & x^{41} & x^{42} & x^{43} \end{bmatrix} \tag{10.104}$$

and

$$a_N = \frac{1}{6C} \epsilon_{NRST} \epsilon_{ijk} x^{Ri} x^{Sj} x^{Tk} \tag{10.105a}$$

$$b_{Ni} = \frac{-1}{6C} \epsilon_{NRST} \epsilon_{mjk} \beta_{(i)}^{Rm} \beta_{(i)}^{Sj} \beta_{(i)}^{Tk} \tag{10.105b}$$

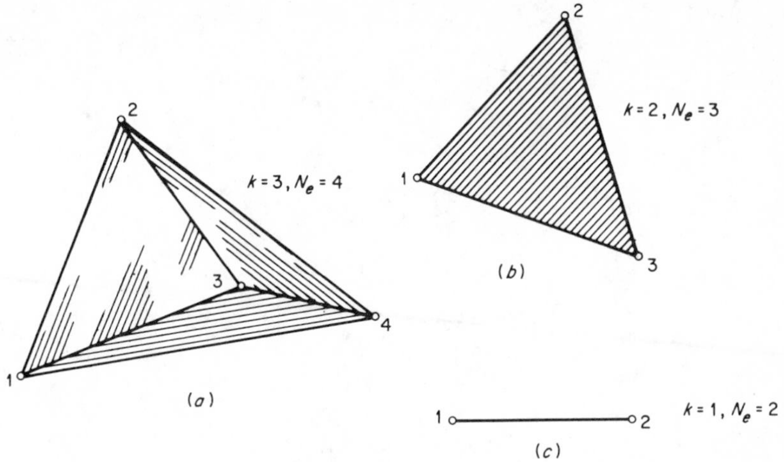

Fig. 10.1 Simplex finite elements, $N_e = k + 1$.

where ϵ_{NRST} and ϵ_{ijk} are the four-dimensional and three-dimensional permutation symbols, respectively; $N, R, S, T = 1, 2, 3, 4$; $i, j, k, m = 1, 2, 3$; and $\beta_{(i)}^{Rm} = x^{Rm}$ if $i = m$, $\beta_{(i)}^{Rm} = 1$ if $i \neq m$. Here

$$|C| = |\det \mathbf{C}| = 6v \tag{10.105c}$$

where v is the volume of the tetrahedron.

Simplex in a two-dimensional space For two-dimensional euclidean spaces, the simplex is a triangle with three nodes, as indicated in Fig. 10.1*b*. Then

$$\{a_N\} = \frac{1}{2A}\begin{bmatrix} x^{21}x^{32} - x^{31}x^{22} \\ x^{31}x^{12} - x^{11}x^{32} \\ x^{11}x^{22} - x^{21}x^{12} \end{bmatrix} \tag{10.106a}$$

$$[b_{Ni}] = \frac{1}{2A}\begin{bmatrix} x^{22} - x^{32} & x^{31} - x^{21} \\ x^{32} - x^{12} & x^{11} - x^{31} \\ x^{12} - x^{22} & x^{21} - x^{11} \end{bmatrix} \tag{10.106b}$$

$$2A = \det \mathbf{C} = \det \begin{bmatrix} 1 & x^{11} & x^{12} \\ 1 & x^{21} & x^{22} \\ 1 & x^{31} & x^{32} \end{bmatrix} \tag{10.106c}$$

If the nodes are numbered according to the right-hand rule, A in (10.106*c*) is the area of the triangle.

The interpolation functions $\psi_N(\mathbf{x})$ corresponding to a two-dimensional simplex are illustrated in Fig. 10.2. If the three functions are superimposed, they yield a uniformly constant value of unity over the triangle by virtue of

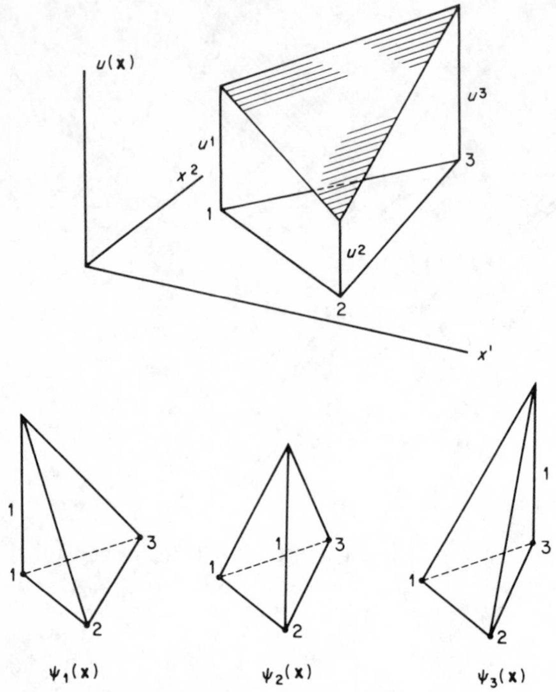

Fig. 10.2 Interpolation functions $\psi_N(\mathbf{x})$ for a two-dimensional simplex.

(10.103*b*). If, on the other hand, each is first multiplied by the corresponding nodal value u^N and the three are then superimposed, the local approximate representation $u(\mathbf{x})$ of (10.101*a*), indicated in Fig. 10.2, is obtained.

As an indication of the error involved in simplex approximations in $\mathscr{E}^2$, Synge† has shown that if the finite-element approximation $\bar{u}^{(e)}$ coincides with the given function u at the nodes,

$$|u_{,i} - \bar{u}_{,i}^{(e)}| \leq \frac{1}{2} \frac{K\delta_e}{\sin \theta} \tag{10.107}$$

where K is a bound of $|u_{,ij}|$, δ_e is the diameter of the element, and θ is the

† Synge [1957, p. 211]. Similar results were obtained by Key [1966]. Key was able to also show that for the three-dimensional simplex

$$|u_{,i} - \bar{u}_{,i}^{(e)}| \leq 18 \frac{K\delta_e}{\cos(\theta/2) \sin \varphi}$$

where θ is the largest angle formed by sides in any triangular face, and φ is the largest angle between the plane containing θ and the remaining three sides of the tetrahedron.

largest angle formed by the sides of the triangle. This result may be viewed as a word of caution in constructing uniform refinements of simplex models: avoid long, thin triangular elements with small interior angles. Apparently, the right isosceles triangle produces the least error in the sense of (10.107).

Simplex in a one-dimensional space For one-dimensional euclidean spaces, the simplex model consists of two nodes connected by a line (Fig. 10.1*c*). Then, if we set $x^1 = x$,

$$a_1 = \frac{x^2}{L} \qquad a_2 = \frac{-x^1}{L}$$
$$b_{11} = \frac{-1}{L} \qquad b_{21} = \frac{1}{L} \tag{10.108}$$

where $L = x^2 - x^1$. Also

$$\psi_1(x) = \frac{x^2 - x}{L} \qquad \psi_2(x) = \frac{x - x^1}{L} \tag{10.109}$$

which we recognize as the familiar Lagrange linear interpolation functions of (7.60).

Curvilinear simplex models In all the equations above concerning simplex models, the nature of the local coordinates x^i is completely arbitrary, and there is no reason to require that they be rectangular cartesian. Indeed, x^i may be any type of curvilinear, intrinsic coordinate system that we care to identify, and the space in which the finite element is need not be euclidean. Thus, in general, we may use curvilinear simplex models of the type indicated in Fig. 10.3. The functions $u(\mathbf{x})$ are then linear in the curvilinear coordinates x^i, and continuity is still maintained at interelement boundaries.

Natural coordinates In constructing local interpolation functions, it is often convenient to use a special system of coordinates ζ_i called *natural*, or, in the case of two-dimensional simplexes, *area coordinates*.† For a k-dimensional space, the ζ_i are $k + 1$ in number and are designed so as to have the following properties:

$$\zeta_i(\mathbf{x}^N) = \delta_i^N \qquad \zeta_1 + \zeta_2 + \cdots + \zeta_{k+1} = 1 \tag{10.110}$$

That is, ζ_i = constant is the parametric equation of a hyperplane parallel to the ith exterior face of the element and $\zeta_i = 0$ is the equation of the face itself. The natural coordinates of nodes $1, 2, \ldots, N_e = k + 1$ are then

† See, for example, Taig and Kerr [1964], Argyris [1966*a*], Zienkiewicz and Cheung [1967], Stricklin [1968], and Fried [1969*a*]. A general discussion is given in the note by Fried [1969*a*], and use of such coordinates for a variety of different types of finite elements is discussed by Felippa [1966].

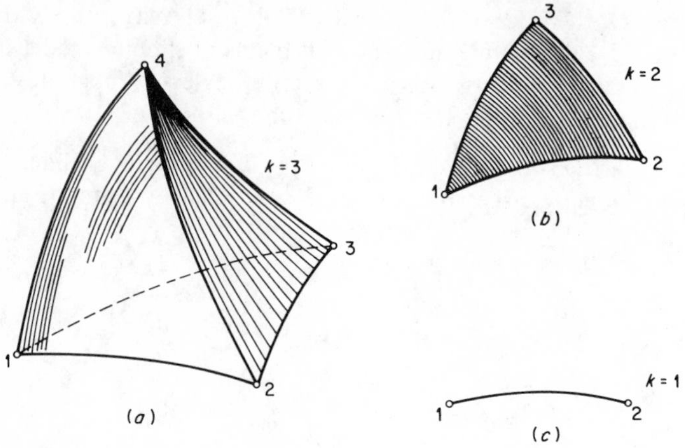

Fig. 10.3 Curvilinear simplex models.

$(1, 0, 0, \ldots, 0)$, $(0, 1, 0, \ldots, 0), \ldots, (0, 0, 0, \ldots, 1)$. In the case of triangular elements, the natural coordinates of the centroid 0 are $\zeta_1 = A_1/A$, $\zeta_2 = A_2/A$, and $\zeta_3 = A_3/A$, where A_1, A_2, A_3 are the areas formed by lines from nodes 1, 2, 3 to 0 [i.e., $A_1 =$ area $(1, 0, 3)$, $A_2 =$ area $(1, 0, 2)$, $A_3 =$ area $(2, 0, 3)$] and $A = A_1 + A_2 + A_3$ is the total area of the triangle. This property motivated the use of the term area coordinates. Clearly, such coordinates are dimensionless and intrinsic, i.e., independent of choices of external frames of reference and of the shape of the element.

As pointed out by Fried [1969*a*], it is meaningful to define a linear, affine transformation of ζ_i into x^i by the relations

$$x^i = \Upsilon^{ij}\zeta_j \qquad i = 1, 2, \ldots, k \qquad j = 1, 2, \ldots, k, k+1 \tag{10.111a}$$

and set

$$x^{k+1} = \zeta_1 + \zeta_2 + \cdots + \zeta_{k+1} \tag{10.111b}$$

We can now define a matrix C_{ij} $(i, j = 1, 2, \ldots, k + 1)$ with the properties

$$C_{1j} = \frac{\partial x^{k+1}}{\partial \zeta_j} = 1 \qquad C_{i+1,\, j} = \frac{\partial x^i}{\partial \zeta_j} = \Upsilon^{ij} \tag{10.112a, b}$$

$$\det C_{ij} = C \tag{10.112c}$$

where C is the determinant defined in (10.102*c*). Moreover, if

$$A_{ij} = \frac{1}{C} \operatorname{cofactor} (C_{i+1,j}) \tag{10.112d}$$

then it can be shown [Fried, 1969*a*] that

$$\frac{\partial}{\partial x^i} = A_{ij}\frac{\partial}{\partial \zeta_j} \qquad \begin{array}{l} i = 1, 2, \ldots, k \\ j = 1, 2, \ldots, k+1 \end{array} \tag{10.113}$$

For $k \leq 3$, $A_{ij} = A_{(j)}\alpha_{ij}$ where $A_{(j)}$ is the area (length) of the jth face of the element and α_{ij} is the direction cosine of the angle between x_i and a normal to A_j. From this fact it can be shown to follow that the directional derivative in the direction of a unit vector $\mathbf{n}$ is

$$\frac{\partial}{\partial n} = \frac{1}{C} A_j n_i \alpha_{ij} \tag{10.114}$$

Moreover, as also pointed out by Fried, a complete set of polynomials of order M can be obtained by taking all combinations of $i_1, i_2, \ldots, i_p$ of terms $\zeta_1^{i_1}\zeta_2^{i_2}\cdots\zeta_{k+1}^{i_p}$ such that $i_1 + i_2 + \cdots + i_p = M$. This fact can be used to advantage in integrating polynomials in natural coordinates over finite elements.†

10.4.2 Complex models We refer to finite elements for which $N_e > k + 1$, but for which it is possible to maintain interelement continuity without requiring that the element boundaries be coordinate lines as complex models.‡ For example, we may take the quadratic approximation

$$u(\mathbf{x}) = a_0 + a_i x^i + a_{ij} x^i x^j \tag{10.115a}$$

in which $a_{ij} = a_{ji}$ and $i, j = 1, 2, \ldots, k$. Likewise, the corresponding interpolation functions are of the form

$$\psi_N(\mathbf{x}) = a_N + b_{Ni} x^i + c_{Nij} x^i x^j \tag{10.115b}$$

where $c_{Nij} = c_{Nji}$. The coefficients a_N, b_{Ni}, c_{Nij} are obtained, as before, by solving the N_e equations generated by evaluating (10.115*a*) at each node.

For the quadratic approximation, elements of the same shape as the simplex model but with additional nodes can be used. For example, in the case of a three-dimensional complex element, a curvilinear tetrahedron of the type shown in Fig. 10.4*a* can be used. In this case $N_e = 10$, $k = 3$. The four vertices serve as four nodes, and the location of the remaining six is, to an extent, arbitrary (so long as the determinant of the coefficient matrix $\mathbf{C}$ is nonzero). For convenience and symmetry, these nodes are located at the midpoints of each side.

For the two-dimensional model, $N_e = 6$ and the triangular element with six nodes shown in Fig. 10.4*b* is obtained. Notice that along each edge $u(\mathbf{x})$ is quadratic. Since a quadratic is uniquely defined by specifying

† See Stricklin [1968].

‡ Such elements have been used by Fraeijs de Veubeke [1965], Argyris [1965*a*, 1965*b*, 1965*c*], Felippa [1966], and Zienkiewicz and Cheung [1967], among others.

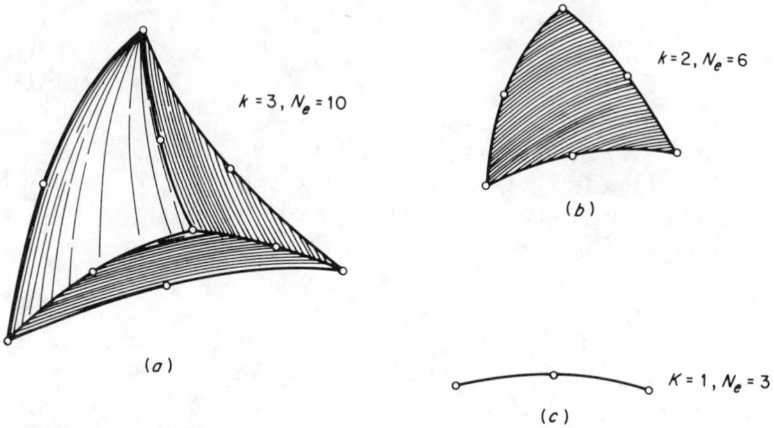

Fig. 10.4 Complex finite elements.

three independent values, full continuity is maintained when adjacent elements are fitted together. Zlamal [1968] showed that for this element

$$|u_{,i} - \bar{u}_{,i}^{(e)}| \leq \frac{2}{\sin \theta} K\delta_e^2$$

where $u_{,i}$ and $\bar{u}_{,i}^{(e)}$ coincide at the nodes, K is the bound on $|u_{,ijk}|$, δ_e is the diameter of the element, and θ is the largest angle between sides of the triangle.

In the one-dimensional case, $N_e = 3$ and the element shown in Fig. 10.4c is obtained. Note that connecting such one-dimensional elements at the end points is not sufficient to determine a quadratic variation over the element. In this case, the value of the local approximation at the interior node must be specified. The fact that one node is an interior node is a significant difference between the one-dimensional complex model and similar models for higher-dimensional spaces.

10.4.3 Multiplex models In multiplex models, element boundary must also be coordinate lines in order to achieve interelement continuity. We cite a few examples.

Bilinear approximation† For a curvilinear element with four nodes in $\mathscr{E}^2$ (Fig. 10.5a),

$$u(\mathbf{x}) = a_0 + a_\alpha x^\alpha + bx^1x^2 \tag{10.116}$$

† For rectangular coordinates, bilinear approximations have been employed by Gallagher, Rattinger, and Archer [1964]; Argyris [1964]; and Zienkiewicz and Cheung [1967], among others. Curvilinear elements have been described by Ergatoudis, Irons, and Zienkiewicz [1968a] and Oden [1969a]. Curved noneuclidean bilinear approximations were used by Wempner, Oden, and Kross [1968].

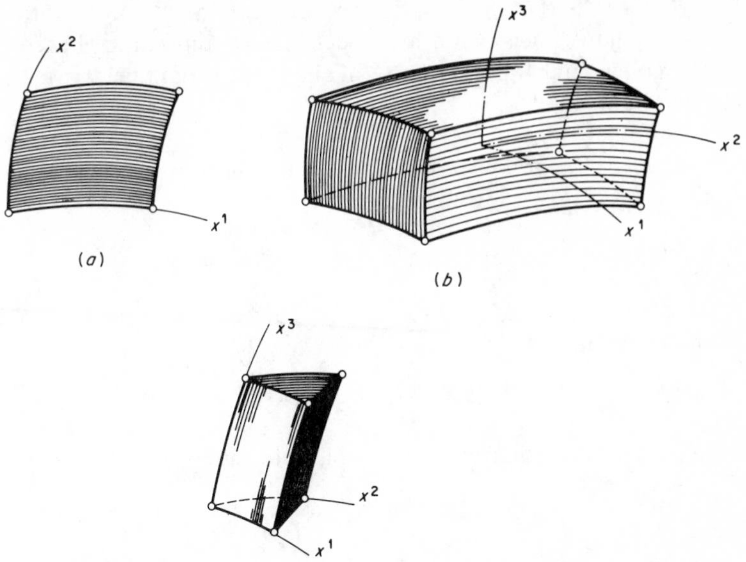

Fig. 10.5 Multiplex finite elements.

and the four coefficients a_0, a_1, a_2, b are determined from the conditions

$$u^N = a_0 + a_\alpha x^{N\alpha} + bx^{N1}x^{N2} \tag{10.117}$$

with $\alpha = 1, 2$ and $N = 1, 2, 3, 4$, no sum on N. Solving (10.117) and introducing the results into (10.116), we obtain, as usual, an equation of the form $u(\mathbf{x}) = u^N\psi_N(\mathbf{x})$, where in the present case

$$\psi_N(\mathbf{x}) = \bar{a}_N + \bar{b}_{N\alpha}x^\alpha + \bar{c}_N x^1 x^2 \tag{10.118a}$$

where

$$\bar{a}_N = \frac{1}{\bar{C}}\,\epsilon_{NMRS}x^{M1}x^{R2}y^S \tag{10.118b}$$

$$\bar{b}_{N\alpha} = \frac{1}{\bar{C}}\,\epsilon_{NMRS}\epsilon_{\alpha\beta}a^M x^{R\beta}y^S \tag{10.118c}$$

$$\bar{c}_N = \frac{1}{\bar{C}}\,\epsilon_{NMRS}a^M x^{R2}x^{S1} \tag{10.118d}$$

Here $y^S = x^{S1}x^{S2}$ (no sum), $a^M = 1$ for $M = 1, 2, 3, 4$; $\epsilon_{\alpha\beta}$ and ϵ_{NMRS} are permutation symbols, and

$$\bar{C} = \det\begin{bmatrix} 1 & x^{11} & x^{12} & y^1 \\ 1 & x^{21} & x^{22} & y^2 \\ 1 & x^{31} & x^{32} & y^3 \\ 1 & x^{41} & x^{42} & y^4 \end{bmatrix} \tag{10.118e}$$

Notice that (10.117) is linear along all boundaries of the element. Thus, a unique linear variation is defined by connecting elements together at the nodes, and full continuity of the approximate field is maintained at element boundaries.

Trilinear approximation For a curvilinear parallelopiped with eight nodes in $\mathscr{E}^k$ (Fig. 10.5*b*), we assume†

$$u(\mathbf{x}) = a_0 + a_i x^i + a_{ij} x^i x^j + b x^1 x^2 x^3 \tag{10.119}$$

where $a_{ij} = a_{ji}$ and $a_{ij} \neq 0$ for $i = j$. The eight coefficients are determined from the conditions

$$u^N = a_0 + a_i x^{Ni} + a_{ij} x^{Ni} x^{Nj} + b x^{N1} x^{N2} x^{N3} \tag{10.120}$$

(no sum on N) so that $u(\mathbf{x})$ can be put into the final form

$$u(\mathbf{x}) = u^N \psi_N(\mathbf{x}) \qquad N = 1, 2, \ldots, 8 \tag{10.121}$$

If the origin is located at the center of the elements,

$$\begin{aligned}
\psi_1(\mathbf{x}) &= \tfrac{1}{8}(1 + \xi)(1 + \eta)(1 + \zeta) \\
\psi_2(\mathbf{x}) &= \tfrac{1}{8}(1 + \xi)(1 - \eta)(1 + \zeta) \\
\psi_3(\mathbf{x}) &= \tfrac{1}{8}(1 - \xi)(1 - \eta)(1 + \zeta) \\
\psi_4(\mathbf{x}) &= \tfrac{1}{8}(1 - \xi)(1 + \eta)(1 + \zeta) \\
\psi_5(\mathbf{x}) &= \tfrac{1}{8}(1 + \xi)(1 + \eta)(1 - \zeta) \\
\psi_6(\mathbf{x}) &= \tfrac{1}{8}(1 + \xi)(1 - \eta)(1 - \zeta) \\
\psi_7(\mathbf{x}) &= \tfrac{1}{8}(1 - \xi)(1 - \eta)(1 - \zeta) \\
\psi_8(\mathbf{x}) &= \tfrac{1}{8}(1 - \xi)(1 + \eta)(1 - \zeta)
\end{aligned} \tag{10.122a}$$

where

$$\xi = \frac{x^1}{a} \qquad \eta = \frac{x^2}{b} \qquad \zeta = \frac{x^3}{c} \tag{10.122b}$$

Here the boundaries of the element are assumed to be $x^1 = \pm a$, $x^2 = \pm b$, and $x^3 = \pm c$. Notice that the function $u(\mathbf{x})$ is linear along all element boundaries so that continuity at interelement boundaries is insured when elements are connected together.

† Such approximations were first used for rectangular "brick" elements by Melosh [1963*b*]. See also Key [1966], Zienkiewicz and Cheung [1967], and Oden [1969*a*].

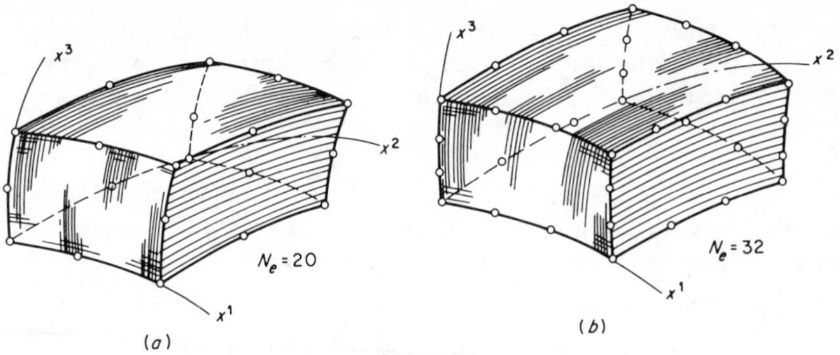

Fig. 10.6 Curvilinear parallelepiped finite elements.

Prismatic elements Local approximations of the form†

$$u(\mathbf{x}) = a_0 + a_i x^i + x^3(cx^1 + dx^2) \tag{10.123a}$$

$$\psi_N(\mathbf{x}) = \hat{a}_N + \hat{b}_{Ni} x^i + c_N x^1 x^3 + d_N x^2 x^3 \tag{10.123b}$$

correspond to prismatic elements of the type shown in Fig. 10.5*c*.

Parallelopiped elements—Curvilinear parallelopiped elements with 20 nodes, as shown in Fig. 10.6*a*, are obtained by using local approximations of the form‡

$$u(\mathbf{x}) = a_0 + a_i x^i + a_{ij} x^i x^j + (x^1)^2(b_2 x^2 + b_3 x^3) + (x^2)^2(c_1 x^1 + c_3 x^3) + (x^3)^2(d_1 x^1 + d_2 x^2) + x^1 x^2 x^3(e_0 + e_i x^i) \tag{10.124}$$

wherein $a_{ij} = a_{ji}$. Similarly, for the element with 32 nodes in Fig. 10.6*b*,§ we use all terms of (10.124) plus the following:

$$f_1(x^1)^3 + f_2(x^2)^3 + f_3(x^3)^3 + x^1 x^2 x^3(g_{ij} x^i x^j) + (x^1)^3(h_2 x^2 + h_3 x^3) + (x^2)^3(j_1 x^1 + j_3 x^3) + (x^3)^3(k_1 x^1 + k_2 x^2) \tag{10.125}$$

where $g_{ij} = 0$ for $i \neq j$.

10.4.4 Isoparametric elements The development of curvilinear finite elements described in the preceding section is based on the assumption that a set of local intrinsic, curvilinear coordinates are specified a priori, and that the local fields $u(\mathbf{x})$ can be approximated by suitable polynomials in these coordinates. In many problems, however, very irregular boundaries that follow no easily definable curvilinear coordinate lines are encountered, and element boundaries may, at best, only approximate actual curved boundaries.

† Cf. Key [1966].
‡ Cf. Ergatoudis, Irons, and Zienkiewicz [1968*a*, 1968*b*].
§ Ibid.

A better approximation of curved boundaries is provided by the notion of curved isoparametric finite elements.† The construction of such elements is based on the idea of fitting polynomial boundary curves through specified points on a boundary, in much the same way that the local function $u(\mathbf{x})$ is approximated over each element.

Suppose, for example, that x^1, x^2, x^3 denotes a system of rectangular cartesian coordinates in $\mathscr{E}^3$. Transformation of x^i into an arbitrary system of curvilinear coordinates ξ^1, ξ^2, ξ^3 is accomplished by relations of the form

$$\xi^i = \xi^i(x^1, x^2, x^3) \tag{10.126}$$

which are assumed to have inverse transformations of the form

$$x^i = x^i(\xi^1, \xi^2, \xi^3) \tag{10.127}$$

Now in applications, the coordinates x^i may be a system of local coordinates of a typical finite element, but the precise form of (10.127) required to fit a given boundary may be quite complicated and generally varies from case to case. Thus, we seek approximations to (10.127) that can be used in a wide range of boundary shapes. It is natural to consider polynomial approximations of the form

$$x^i = a^i + a^i_j \xi^j + a^i_{jk} \xi^j \xi^k + \cdots \tag{10.128}$$

Following the same procedure used in constructing the local approximations $u(\mathbf{x})$, we evaluate (10.128) at a finite number N_e nodal points, generally on the element's boundary, and obtain the system of equations

$$x^{Ni} = a^i + a^i_j \xi^{Nj} + a^i_{jk} \xi^{Nj} \xi^{(N)k} + \cdots \tag{10.129}$$

where x^{Ni} and ξ^{Nj} are the coordinates, cartesian and curvilinear, of node N. Solving (10.129) and introducing the results into (10.128), we obtain for the approximate coordinate transformation

$$x^i = \psi_N(\boldsymbol{\xi}) x^{Ni} \tag{10.130}$$

where the interpolation functions $\psi_N(\boldsymbol{\xi})$ are polynomials in ξ^i and satisfy the usual conditions

$$\sum_{N=1}^{N_e} \psi_N(\boldsymbol{\xi}) = 1 \qquad \psi_N(\boldsymbol{\xi}) \geq 0 \tag{10.131a, b}$$

The same interpolation functions $\psi_N(\boldsymbol{\xi})$ can now be used to approximate the local field $u(\boldsymbol{\xi})$ over the element $u(\boldsymbol{\xi}) = u^N \psi_N(\boldsymbol{\xi})$.

It is easily verified that the functions $\psi_N(\boldsymbol{\xi})$ automatically ensure that linear variations (and uniform first derivatives) of $u(\boldsymbol{\xi})$ can be represented

† A type of isoparametric shear element was used by Taig [1961]. The idea was generalized and thoroughly exploited by Ergatoudis [1966] and Ergatoudis, Irons, and Zienkiewicz [1968*a*, 1968*b*]. See also Ahmad, Irons, and Zienkiewicz [1969] and Felippa and Clough [1968].

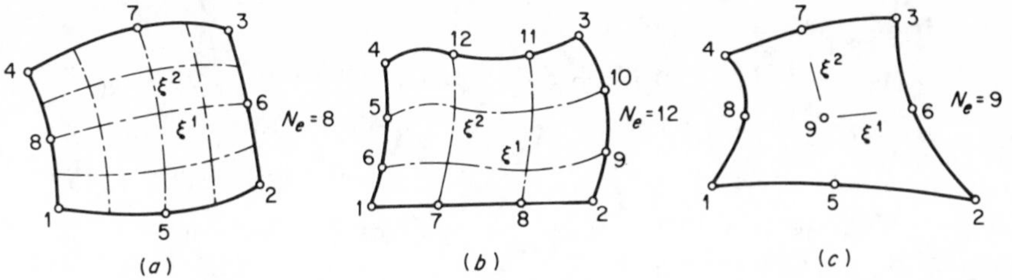

Fig. 10.7 Examples of isoparametric finite elements. (After Ergatoudis, Irons, and Zienkiewicz [1968*a*].)

over the element.† For example, consider

$$u = a_0 + a_i x^i = u^N \psi_N(\boldsymbol{\xi}) \tag{10.132}$$

Since $u^N = a_0 + a_i x^{Ni}$, we have

$$u(\mathbf{x}) = a_0 \sum_{N=1}^{N_e} \psi_N(\boldsymbol{\xi}) + a_i \psi_N(\boldsymbol{\xi}) x^{Ni} \tag{10.133}$$

The first term is a_0 by virtue of (10.131*a*) and the second term is $a_i x^i$, according to (10.130). Thus (10.132) reduces to an identity.

Ergatoudis, Irons, and Zienkiewicz [1968*a*] have used various polynomial approximations in ξ^i to develop curvilinear elements of the type indicated in Fig. 10.7. Isoparametric elements of the type indicated in Fig. 10.6 can also be developed. It is noted that the functions $\psi_N(\boldsymbol{\xi})$ corresponding to elements such as those in Fig. 10.7 in general satisfy the requirements of completeness, they ensure the inclusion of uniform derivatives, and they satisfy continuity requirements at interelement boundaries. For the element in Fig. 10.7*a*, (10.128) is of the form

$$x^k = a_0^k + a_i^k \xi^i + a_{ij}^k \xi^i \xi^j + b_1^k (\xi^1)^2 \xi^2 + b_2^k \xi^1 (\xi^2)^2 \tag{10.134}$$

where $i, j, k = 1, 2$ and $a_{ij}^k = a_{ji}^k$, and assuming that $\xi^i = \pm 1$ defines the boundaries of the element,

$$\psi_N(\boldsymbol{\xi}) = \tfrac{1}{4}(1 + \xi^1 \xi^{N1})(1 + \xi^2 \xi^{N2}) - \tfrac{1}{4}[1 - (\xi^1)^2][1 + \xi^2 \xi^{N2}] - \tfrac{1}{4}(1 + \xi^1 \xi^{N1})[1 - (\xi^2)^2] \tag{10.135}$$

with $N = 1, 2, \ldots, 8$. For the element shown in Fig. 10.7*b*,

$$\psi_N(\boldsymbol{\xi}) = \tfrac{1}{32}(1 + \xi^1 \xi^{N1})(1 + \xi^2 \xi^{N2})\{9[(\xi^1)^2 + (\xi^2)^2] - 10\} \tag{10.136a}$$

$N = 1, 2, 3, 4$ (corner nodes only), and assuming that the nodes are located at $\xi^1 = \pm 1$, $\xi^2 = \pm \tfrac{1}{3}$ along each side,

$$\psi_N(\boldsymbol{\xi}) = \tfrac{9}{32}(1 + \xi^1 \xi^{N1})[1 - (\xi^2)^2](1 + \xi^2 \xi^{N2}) \tag{10.136b}$$

† Cf. Ergatoudis, Irons, and Zienkiewicz [1968*a*].

for $\xi^2 = \pm 1$, $\xi^1 = \pm\frac{1}{3}$ and

$$\psi_N(\boldsymbol{\xi}) = \tfrac{9}{32}[1 - (\xi^1)^2](1 + \xi^1\xi^{N1})(1 + \xi^2\xi^{N2}) \tag{10.136c}$$

$$x^k = a_0^k + a_i^k\xi^i + a_{ij}^k\xi^i\xi^j + c_1^k(\xi^1)^3 + c_2^k(\xi^1)^2\xi^2 + c_3^k\xi^1(\xi^2)^2 + c_4^k(\xi^2)^3 + d_1^k\xi^1(\xi^2)^3 + d_2^k(\xi^1)^3\xi^2 \tag{10.137}$$

and $\psi_N(\boldsymbol{\xi})$ acquire a similar form. In the case of the element shown in Fig. 10.7c, we have, in matrix form,

$$\begin{bmatrix} \psi_1(\boldsymbol{\xi}) & \psi_8(\boldsymbol{\xi}) & \psi_4(\boldsymbol{\xi}) \\ \psi_5(\boldsymbol{\xi}) & \psi_9(\boldsymbol{\xi}) & \psi_7(\boldsymbol{\xi}) \\ \psi_2(\boldsymbol{\xi}) & \psi_6(\boldsymbol{\xi}) & \psi_3(\boldsymbol{\xi}) \end{bmatrix} = \boldsymbol{\beta}(\xi^1)\boldsymbol{\beta}^{\mathrm{T}}(\xi^2) \tag{10.138a}$$

where, assuming that the nodes are located at points $\xi^i = 0, \pm 1$,

$$\boldsymbol{\beta}(\xi^1) = \begin{bmatrix} \beta_1(\xi^1) \\ \beta_2(\xi^1) \\ \beta_3(\xi^1) \end{bmatrix} = \begin{bmatrix} -\frac{1}{2}\xi^1(1 - \xi^1) \\ 1 - (\xi^1)^2 \\ \frac{1}{2}\xi^1(1 + \xi^1) \end{bmatrix} \tag{10.138b}$$

10.4.5 Examples of higher-order elements In most of the applications considered in subsequent sections, we confine our attention to first-order finite-element representations. For the sake of completeness, however, we cite here a number of examples of higher-order finite element representations.†

Higher-order representations in $\mathscr{E}^1$ In one-dimensional space, we have, of course, the usual Hermite interpolation functions discussed in Art. 8 [see Eqs. (8.23) to (8.26)]:

$$u(x) = a_0 + a_1x + a_2x^2 + a_3x^3 \tag{10.139}$$

or

$$u(x) = u^N\psi_N^0(x) + u,_1^N\psi_N^1(x) \qquad N = 1, 2 \tag{10.140}$$

where

$$\begin{aligned} \psi_1^0(x) &= 1 - 3\xi^2 + 2\xi^3 \\ \psi_2^0(x) &= 3\xi^2 - 2\xi^3 \\ \psi_1^1(x) &= L(\xi - 2\xi^2 + \xi^3) \\ \psi_2^1(x) &= -L(\xi^2 - \xi^3) \end{aligned} \tag{10.141a–d}$$

and

$$\xi = \frac{x - x_1}{L} \qquad L = x_2 - x_1 \tag{10.142a, b}$$

† Felippa [1966] has cited a number of examples of higher-order elements and has classified them according to "$C - q$ compatibility." According to this classification, an element has $C - 0$ type compatibility if only the local function itself is continuous at interelement boundaries. If the first q partial derivatives are also continuous, the element is said to have "$C - q$ compatibility." Thus, according to our definition in Sec. 10.2, an element with $C - q$ compatibility is q-comformable. Felippa also classified elements according to the existence of interior nodal points, degrees of freedom, etc. See also Felippa and Clough [1968].

We obtain a third-order representation over a one-dimensional element with two nodes by using the fifth-degree polynomial.†

$$u(x) = a_0 + a_1x + a_2x^2 + a_3x^3 + a_4x^4 + a_5x^5 \tag{10.143}$$

Then $u(x)$, $du(x)/dx$, and $d^2u(x)/dx^2$ are specified at each node.

In general, for one-dimensional elements with two nodes, an nth-order representation is obtained by assuming $u(x)$ to be a complete polynomial of degree $n + 2$, and by specifying $u(x), du(x)/dx, \ldots, d^nu(x)/dx^n$ at both nodal points.

Second-order representation in $\mathscr{E}^3$ A function $u(\mathbf{x})$ and its first partial derivatives are specified at each of four vertices of a tetrahedron, and $u(\mathbf{x})$ is given by the complete cubic (Fig. 10.8*a*).‡

$$u(\mathbf{x}) = a_0 + a_ix^i + a_{ij}x^ix^j + a_{ijk}x^ix^jx^k \tag{10.144}$$

wherein $i, j, k = 1, 2, 3$; $a_{ij} = a_{ji}$, $a_{ijk} = a_{ikj} = a_{jik}$. There are 20 independent parameters $a_0, \ldots, a_{ijk}$ and $N_e = 4$. The location of the fourth node is arbitrary, but the centroid of the element is often a convenient choice.

A second-order representation in $\mathscr{E}^2$ We can develop a second-order approximation of a function $u(x^1,x^2)$ over a triangular subdomain of $\mathscr{E}^2$ by specifying $u(x^1,x^2)$ at four nodal points and its first partial derivatives $u(\mathbf{x})_{,\alpha}$ at nodes at the three vertices of the triangle§ (Fig. 10.8*b*). In this case $N_e = 4$ and $u(\mathbf{x})$ is given by the complete cubic

$$u(\mathbf{x}) = a_0 + a_ix^i + a_{11}(x^1)^2 + a_{12}x^1x^2 + a_{22}(x^2)^2 + a_{111}(x^1)^3 + a_{112}(x^1)^2x^2 + a_{122}x^1(x^2)^2 + a_{222}(x^2)^3 \tag{10.145}$$

Then, the final approximation is of the form

$$u(\mathbf{x}) = u^N\psi_N^0(\mathbf{x}) + u_{,i}^N\psi_N^i(\mathbf{x}) \tag{10.146}$$

where $N = 1, 2, 3, 4$; $i = 1, 2$; and, for arbitrary $\mathbf{x}$,‖

$$\psi_N^0(\mathbf{x}) \neq 0 \qquad \psi_4^i(\mathbf{x}) = 0 \tag{10.147}$$

The location of the fourth node is arbitrary, but the centroid of the element is a logical choice (Fig. 10.8*b*).

Zlamal [1968] proved that if the finite-element approximation $\bar{u}^{(e)}$ coincides with a continuously differentiable function u with bounded fourth

† Cf. Langhaar and Chu [1968].

‡ See, for example, Argyris [1965*a*].

§ Elements such as this were used by Best and Oden [1963]. For a detailed discussion of this and related elements, see Felippa [1966]. See also Irons [1966]; Tocher and Hartz [1967]; and Oden, Rigsby, and Cornett [1969].

‖ Alternatively, we may set $\psi_4^0(\mathbf{x}) = 0$ and, for the tenth condition, prescribe the derivative $\partial u/\partial n$ normal to the boundary at a fourth node located on the boundary. By inserting such "floating" boundary nodes at which the function or its normal derivative is specified, it is possible to develop a variety of finite-element models in which different types of elements, each having a different number of nodes, are connected together. See Felippa [1966] and Felippa and Clough [1968].

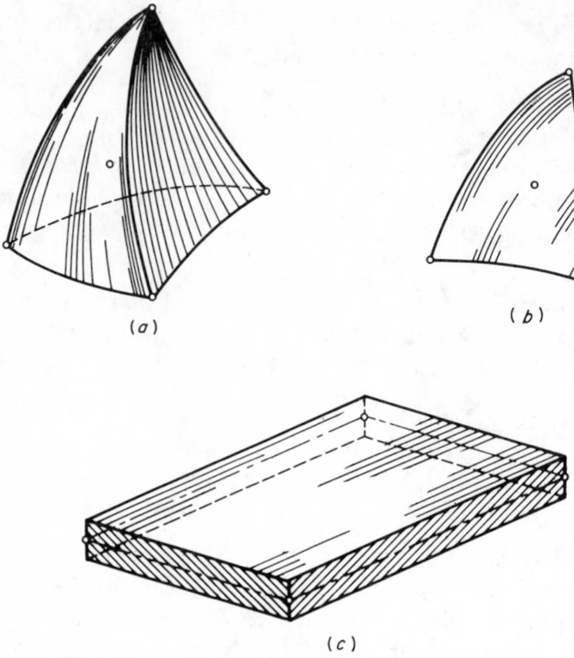

Fig. 10.8 Higher-order finite elements.

derivatives at each node (the three vertices and the centroid) and if the derivatives of $\bar{u}^{(e)}$ and u coincide at each vertex, then

$$|u_{,i} - \bar{u}_{,i}| \leq \frac{5}{\sin\theta} K\delta_e^3 \qquad \text{and} \qquad |u - \bar{u}^{(e)}| \leq \frac{3}{\sin\theta} K\delta_e^4 \tag{10.148a}$$

where θ is the largest angle between sides of the triangle, K the bound on $|u_{,ijkl}|$, and δ_e the diameter of the element.

A third-order representation in $\mathscr{E}^2$ A third-order representation over a triangular element in two-dimensional space is obtained using the complete quintic polynomial† in x^1 and x^2:

$$u(\mathbf{x}) = a_0 + a_i x^i + a_{ij} x^i x^j + a_{ijk} x^i x^j x^k + a_{ijkm} x^i x^j x^k x^m + a_{ijkmn} x^i x^j x^k x^m x^n \tag{10.148b}$$

† This element was apparently first proposed by Withum [1966]; it was employed in the analysis of elastic plates by Bosshard [1968], Bell [1969], Argyris, Fried, and Scharpf [1968*c*], and Argyris and Buck [1968]. Its convergence properties were studied by Zlamal [1968]. Higher-order representations are, of course, possible. Bramble and Zlamal [1970], for example, presented a general method for polynomial approximations over triangles in which a general polynomial in x^1 and x^2 of degree $4m + 1$ is determined by specifying all its derivatives up to order $2m$ at the vertices, its derivatives of order $2m - 1$ at the centroid, and its directional derivatives normal to each side at $r + 1$ equally spaced nodes along each side of the triangle, where $r = 1, 2, \ldots, m$.

By eliminating all dependences among the terms, we find that there are 21 independent parameters in such a fifth-degree polynomial. Twenty-one conditions are provided by specifying $u, u_{,1}, u_{,2}, u_{,11}, u_{,12}$, and $u_{,22}$ at the three vertices of the triangle and the normal derivatives $\partial u/\partial n$ at the midpoints of each side. In this case, u is of the form

$$u(\mathbf{x}) = u^N \psi_N^0(\mathbf{x}) + u_{,i}^N \psi_N^i(\mathbf{x}) + u_{,ij}^N \psi_N^{ij}(\mathbf{x}) + u_{,n}^{N'} \varphi_{N'}(\mathbf{x}) \tag{10.149}$$

where $N = 1, 2, 3$, $N' = 4, 5, 6$, $\varphi_{N'}(\mathbf{x}^N) = 0$, $\partial \varphi_{N'}(\mathbf{x}^{M'})/\partial n = \delta_{N'}^{M'}$. Here the vertices are numbered 1, 2, 3 and the midpoints 4, 5, 6. Zlamal [1968] has shown that, in this case,

$$|u_{,i} - \bar{u}_{,i}^e| \le \frac{K}{(\sin\theta)^{|i|}} \delta_e^{6-|i|} \tag{10.150}$$

where $|i| \le 4$ and K is an upper bound to the sixth derivatives of u.

Comments on other higher-order elements The most extensive use of higher-order finite element models has been in connection with applications to the problem of bending of thin plates and shells. If the Kirchhoff-Love theory is used, the deformation of a plate or shell element is described in terms of the displacement field of points on the middle surface plus the first derivatives of this field. As a result, continuity of the entire displacement field requires that not only the displacements of the middle surface be continuous at interelement boundaries but also that the first partial derivatives be continuous.† This plus the fact that the model should be capable of depicting uniform curvatures‡ leads to significant complications in constructing conformable finite elements.§ These complications represent one of numerous examples wherein assumptions originally designed to make a theory more manageable for applications (e.g., Kirchhoff-Love hypothesis,

† The problem of continuity can be avoided by abandoning the Kirchhoff-Love hypothesis and using first-order approximations of the displacements and the slopes. See, for example, Melosh [1966], Utku [1966], Utku and Melosh [1967], Wempner, Oden, and Kross [1968], and Oden and Wempner [1967].

‡ This requirement was apparently first noted by Irons and Draper [1965] and was stressed in the paper by Bazeley et al. [1966]. See footnote on p. 116.

§ In early literature on finite elements of thin plates, the requirements of continuity and completeness, including the requirement that uniform strains (curvatures and twists) be included (in the limit), were not generally recognized and a number of unacceptable elements were proposed. An interesting review of previous work has been given by Felippa [1966, pp. 215–219]. A comparison of several rectangular and triangular elements developed prior to 1965 has been given by Clough and Tocher [1966]. Bazeley, Cheung, Irons, and Zienkiewicz [1966] developed nonconformable triangular elements, and Clough and Tocher [1966] developed a fully compatible triangular plate element using a composite of three cubic polynomials. A completely compatible higher-order, rectangular plate element was devised by Bogner, Fox, and Schmit [1966]. A number of plate and shell elements have been presented wherein stresses or both stresses and displacements have been approximated; for example, consult Best [1963]; Pian [1964*b*]; Fraeijs de Veubeke [1966]; Herrmann [1966]; and Pian and Tong [1969*a*, 1969*b*].

incompressibility of solids, etc.) make the construction of suitable finite-element models considerably more difficult. Indeed, we might say that quite often the more basic the theory, the more readily we are able to construct acceptable finite-element models.

As an example of a fully compatible, two-dimensional, third-order finite-element model, corresponding to a rectangular plate element (see Fig. 10.8*c*), or a curvilinear quadrilateral shell element [Bogner, Fox, and Schmit, 1967], we cite

$$u(\mathbf{x}) = \sum_{r=0}^{3}\sum_{s=0}^{3} C_{rs}(x^1)^r(x^2)^s \tag{10.151}$$

The sixteen coefficients C_{rs} are determined so that $u(\mathbf{x})$, $\partial u(\mathbf{x})/\partial x^i$, and $\partial^2 u(\mathbf{x})/\partial x^1 \partial x^2$ take on prescribed values at four nodes, one node located at each corner. The interpolation functions are then two-dimensional Hermite interpolation functions and $u(\mathbf{x})$ can be put in the form

$$u(\mathbf{x}) = u^N \psi_N^0(\mathbf{x}) + u^N_{,i}\psi_N^i(\mathbf{x}) + u^N_{,12}\psi_N^{12}(\mathbf{x}) \tag{10.152}$$

where $N = 1, 2, 3, 4$. Precise forms of the functions ψ_N^0, ψ_N^i, ψ_N^{12} for orthonormal coordinates can be found in the paper of Bogner, Fox, and Schmit [1967].

10.4.6 Further remarks on finite-element models In concluding this section, we list several general remarks about certain properties of finite element models.

Remark 1—additional parameters We recall that the general form of the local approximation over an element is, according to (7.7),

$$u(\mathbf{x}) = \Psi(\mathbf{x}, u^N) \tag{10.153}$$

where $\Psi(\ \ )$ must satisfy

$$\Psi(\mathbf{x}_M, u^N) = u^M \tag{10.154}$$

We have used the special linear form

$$u(\mathbf{x}) = u^N \psi_N(\mathbf{x}) \tag{10.155}$$

requiring that

$$\psi_N(\mathbf{x}^M) = \delta_N^M \tag{10.156}$$

Conditions (10.154), however, are also satisfied if we take†

$$u(\mathbf{x}) = u^N \psi_N(\mathbf{x}) + \chi(\mathbf{x}) \tag{10.157}$$

where $\chi(\mathbf{x})$ is any function that vanishes at the nodes:

$$\chi(\mathbf{x}^N) = 0 \tag{10.158}$$

Pian [1964*a*] has shown that improved finite-element models can be

† An example of an approximation with this property is cited in the footnote on p. 49 of Art. 7.5.

obtained by using polynomial forms for the function $\chi(\mathbf{x})$:

$$\chi(\mathbf{x}) = b_0 + b_i x^i + b_{ij} x^i x^j + \cdots \tag{10.159}$$

For example, in the case of the bilinear approximation described in (10.116) we may use

$$u(\mathbf{x}) = u^N \psi_N(\mathbf{x}) + x^1 x^2 (a - x^1)(b - x^2)(b_0 + b_1 x^1 + b_2 x^2 + b_3 (x^1)^2 + \cdots) \tag{10.160}$$

wherein a and b denote the dimensions of the sides of the element and (0,0) is the origin of x^i. The additional parameters $b_0, b_1, b_2, \ldots$ must be determined from auxiliary conditions associated with the problem at hand (e.g., minimization of potential energy, etc.).

Krahula and Polhemus [1968] have used, instead of (10.159), Fourier series approximations; e.g., instead of (10.160),

$$u(\mathbf{x}) = u^N \psi_N(\mathbf{x}) + x^1 x^2 (a - x^1)(b - x^2) \sum_{m=1}^{\infty} \sum_{n=1}^{\infty} A_{mn} \sin \frac{\pi x^1}{a} \sin \frac{\pi x^2}{b} \tag{10.161}$$

Remark 2—non-euclidean elements We remark again that the elements described previously need not be restricted to euclidean spaces. Elements in Riemann spaces are equally easy to construct once the space itself is adequately defined. For example, two-dimensional simplex, complex, and multiplex elements embedded in a Riemann surface are obtained immediately from (10.106), (10.115), and (10.118) by simply regarding the x^i as intrinsic surface coordinates. Such models are indicated in Fig. 10.9.

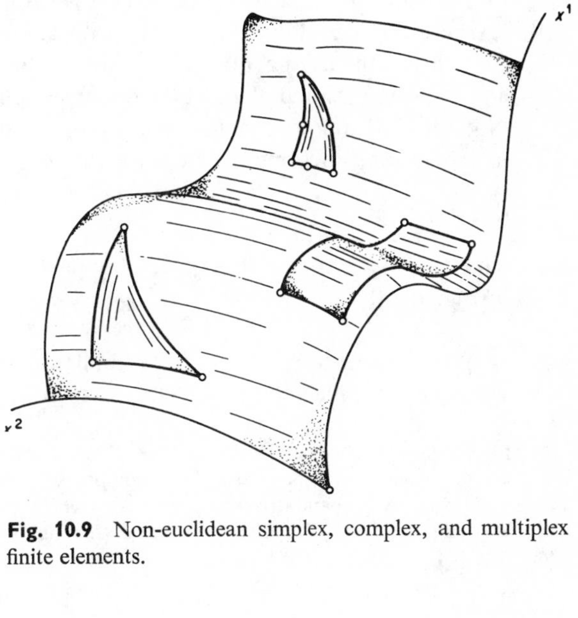

Fig. 10.9 Non-euclidean simplex, complex, and multiplex finite elements.

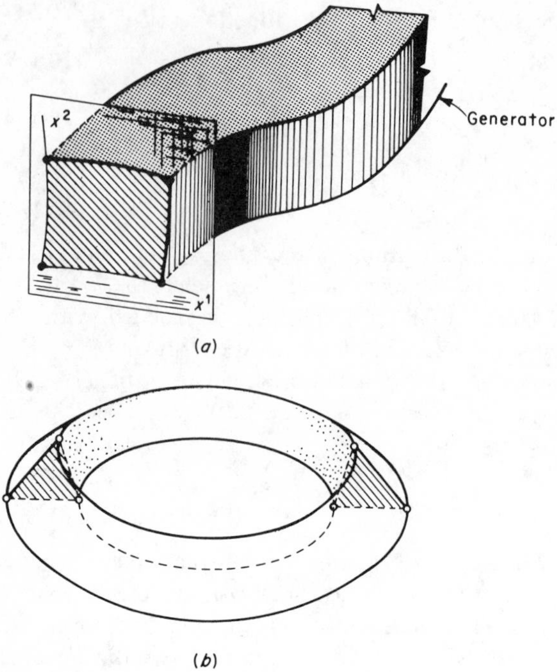

Fig. 10.10 Examples of generated elements.

Remark 3—generated elements It is often convenient to construct models in k-dimensional space (or higher) by generating a $k - 1$ dimensional region along a suitable generator in $\mathscr{E}^k$. For example, finite elements for a three-dimensional body of revolution can be generated by sweeping any two-dimensional element 2π radians about the axis of revolution.† Nodal points in $k - 1$ space then become nodal lines in k space. The interpolation functions then assume the form

$$\hat{\psi}_N(\mathbf{x},s) = \psi_N(\mathbf{x})f(s) \tag{10.162}$$

where $\mathbf{x} \in \mathscr{E}^{k-1}$ and s is the generating parameter. Examples are given in Fig. 10.10.

Remark 4—composite elements‡ As a final comment, we remark that a variety of so-called composite elements can be developed by connecting

† See, for example, Rashid [1964, 1966], Clough and Rashid [1965], and Wilson [1965].
‡ Composite elements have been used by a number of investigators. Among these are Wissmann [1962], Becker and Brisbane [1965], Argyris [1966*a*], Clough and Tocher [1966], Felippa [1966], Percy [1967], Zienkiewicz and Cheung [1967], and Hughes and Allik [1969].

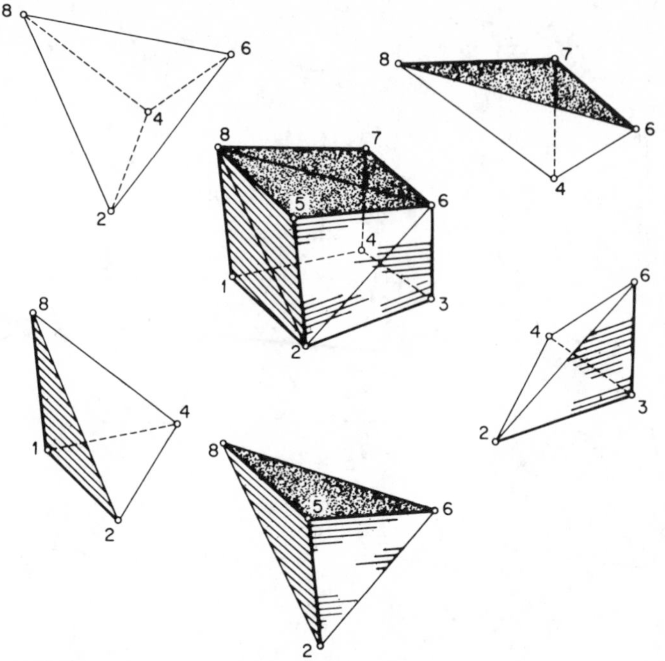

Fig. 10.11 Composite prismatic finite elements.

together two or more basic elements of the types described previously. Some examples are shown in Figs. 10.11 and 10.12: the prismatic element obtained by joining five simplex tetrahedrons (Fig. 10.11), the quadrilateral element obtained by joining four triangular simplexes (Fig. 10.12*a*), a planar quadrilateral element obtained by superimposing four triangles (Fig. 10.12*b*), and a quadrilateral element obtained by connecting four triangular complex elements together (Fig. 10.12*c*).

11 SELECTED APPLICATIONS

To apply the concepts presented previously to linear or nonlinear problems, all that is needed is some means to translate a relation that holds at a point into one that must hold over a finite region. In solving partial differential equations, this translation from point relations to regional relations can be provided by equivalent variational statements of the problem or by application of other methods, such as weighted residuals, Galerkin's method, etc.; in various problems of physics, it may also be provided by local and global forms of the balance laws of thermodynamics and electromagnetics. In this article, we consider several examples.

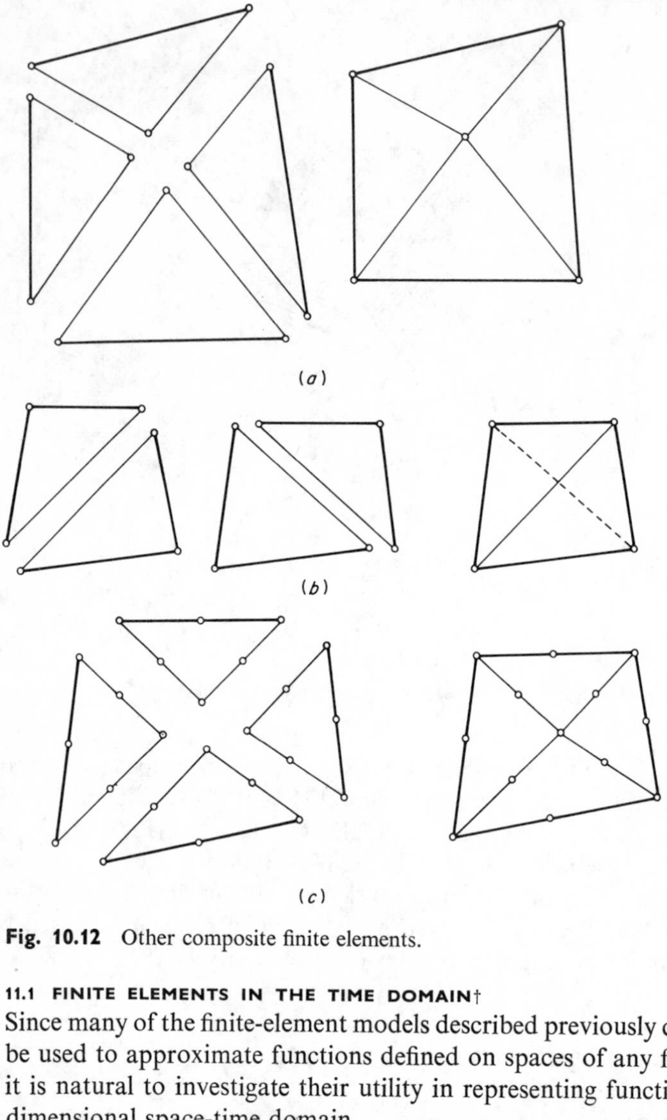

Fig. 10.12 Other composite finite elements.

11.1 FINITE ELEMENTS IN THE TIME DOMAIN†

Since many of the finite-element models described previously can, in principle, be used to approximate functions defined on spaces of any finite dimension, it is natural to investigate their utility in representing functions in the four-dimensional space-time domain.

Consider, for example, a scalar-valued function $\Phi(X^1, X^2, X^3, t)$ of position X^i and time t. Clearly, we can set $X^4 = t$ and consider $\Phi(\ )$ to be defined on a four-dimensional space, the points of which are given by X^i $(i = 1, 2, 3, 4)$. If the domain is represented by a collection of E four-dimensional subdomains, then we can concentrate on approximating the

† The use of finite-element approximations in time was suggested independently by Nickell and Sackman [1968], Oden [1969*b*], and Fried [1969*b*]; the present discussion is based on the paper by Oden. See also Argyris and Scharpf [1969].

field $\varphi^{(e)}(x^1,x^2,x^3,x^4)$ locally over a typical finite element e:

$$\varphi^{(e)} = \overset{N}{\varphi}_{(e)}\psi_N^{(e)}(\mathbf{x}) \tag{11.1}$$

where the interpolating functions $\psi_N^{(e)}(\mathbf{x})$ are the functions of x^1, x^2, x^3, x^4. The general process of assembling elements amounts to a simple application of the ideas discussed in Art. 7. However, the details of applying the model to propagation problems, hyperbolic and parabolic partial differential equations, etc., can only be appreciated through specific examples.

Two-dimensional space-time To illustrate the use of finite elements in the time domain, consider the simple example of a one-dimensional elastic bar of length L and cross-sectional area A, subjected either to a prescribed force $P(t)$ at a free end or to a prescribed initial displacement $u(x,0) = f(x)$. In conventional finite-element models, the longitudinal displacement $u = u(x,t)$ is approximated by one-dimensional interpolating functions $\psi_N(x)$ ($N = 1, 2, \ldots, N_e$) multiplied by nodal displacements which are functions of time. In the present analysis, the displacements $\overset{N}{u}_{(e)}$ are values of a local field $u^{(e)}(x,t)$ defined over a region in a two-dimensional space (x,t) and are independent of x and t. The interpolating functions are functions of both the longitudinal coordinate x and time t: $\psi_N = \psi_N(x,t)$. Thus, for a typical finite element,

$$u^{(e)}(x,t) = \overset{N}{u}_{(e)}\psi_N(x,t) \tag{11.2}$$

The lagrangian potential $\mathfrak{L}$ for a linearly elastic bar of modulus E and mass density ρ is

$$\mathfrak{L} = \frac{1}{2}\int_t\int_v\left[\rho\left(\frac{\partial u}{\partial t}\right)^2 - E\left(\frac{\partial u}{\partial x}\right)^2\right] dv\, dt + \int_t \sum_\alpha S_\alpha(t)u(x_\alpha,t)\, dt \tag{11.3}$$

where v is the volume of the bar and $-\sum_\alpha S_\alpha(t)u(x,t) = -S_1(t)u(x_1,t) - S_2(t)u(x_2,t)$ is the potential of forces $S_1(t)$ and $S_2(t)$ at the ends of the segment $[x_1,x_2]$ of the bar under consideration. Thus, for a typical finite element e,

$$\mathfrak{L}_{(e)} = \tfrac{1}{2}a_{MN}^{(e)}\overset{M}{u}_{(e)}\overset{N}{u}_{(e)} + \overset{N}{u}_{(e)}p_N^{(e)} \tag{11.4}$$

where

$$a_{MN}^{(e)} = \int_t\int_v \rho\left(\frac{\partial\psi_M}{\partial t}\frac{\partial\psi_N}{\partial t} - E\frac{\partial\psi_M}{\partial x}\frac{\partial\psi_N}{\partial x}\right) dv\, dt \tag{11.5}$$

and

$$p_N^{(e)} = \int_t \sum_\alpha S_\alpha(t)\psi_N(x_\alpha,t)\, dt \tag{11.6}$$

In these equations, the integration is taken over the portion of the time domain spanned by the element.

In view of (11.4), the lagrangian $\mathfrak{L}$ has an interesting property that differs significantly from the usual case: it is not a functional of velocity. Indeed, $\mathfrak{L}$ becomes an ordinary function of nodal values of displacements, but because of the particular type of formulation, these are independent of time. Hamilton's principle, of course, still applies, so that

$$\delta\mathfrak{L}_{(e)} = \frac{\partial \mathfrak{L}_{(e)}}{\partial u^N_{(e)}}\, \delta u^N_{(e)} = 0 \tag{11.7}$$

and we obtain

$$a^{(e)}_{MN} u^M_{(e)} + p^{(e)}_N = 0 \tag{11.8}$$

The process of assembling the elements into the total model follows the usual procedure for conventional two-dimensional finite-element models.

One-dimensional waves It is important to note that the procedure by which the above finite-element equations are solved is quite different than in the case of purely elliptic-type problems. In fact, (11.8) is the finite-element analog of the hyperbolic wave equation

$$\frac{\partial^2 u}{\partial x^2} - \alpha^2 \frac{\partial^2 u}{\partial t^2} = f(t) \tag{11.9}$$

where $\alpha = \sqrt{E/\rho}$.

To illustrate the procedure, consider the simple example in which the local field is given by the linear simplex approximation

$$u^{(e)} = u^N_{(e)} \psi^{(e)}_N(x,t) = a + bx + ct \tag{11.10}$$

where a, b, and c are constants and $N = 1, 2, 3$. In this case, the finite element is a triangle in two-dimensional space-time, such as is indicated in Fig. 11.1. We find that

$$\begin{aligned}
\psi_1(x,t) &= \frac{1}{2\Delta}\,[(x_2t_3 - x_3t_2) + (t_2 - t_3)x + (x_3 - x_2)t] \\
\psi_2(x,t) &= \frac{1}{2\Delta}\,[(x_3t_1 - x_1t_3) + (t_3 - t_1)x + (x_1 - x_3)t] \\
\psi_3(x,t) &= \frac{1}{2\Delta}\,[(x_1t_2 - x_2t_1) + (t_1 - t_2)x + (x_2 - x_1)t]
\end{aligned} \tag{11.11}$$

where Δ is the area of the element in the x, t plane. For example, introducing the geometry of the shaded element in Fig. 11.1 into (11.11) and making use of (11.8), we find that for this rather crude approximation the

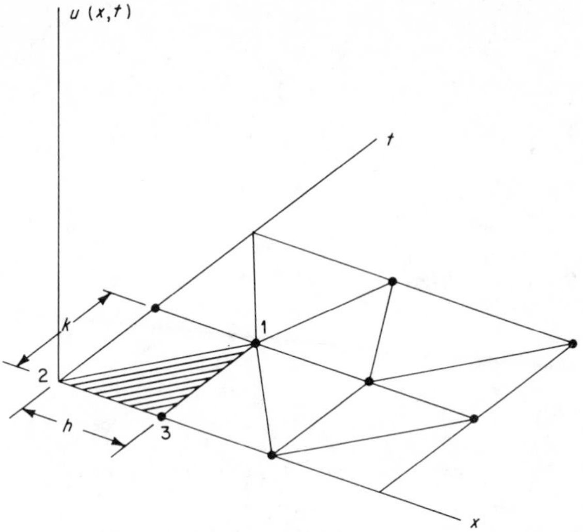

Fig. 11.1 Finite elements in the time domain.

local equations take the form

$$\begin{aligned}
\hat{p}_1^{(e)} &= u_{(e)}^1 - u_{(e)}^3 \\
\hat{p}_2^{(e)} &= \lambda^2(u_{(e)}^3 - u_{(e)}^1) \\
\hat{p}_3^{(e)} &= u_3^{(e)} - u_{(e)}^1 - \lambda^2(u_{(e)}^3 - u_{(e)}^1)
\end{aligned} \tag{11.12}$$

in which $\tilde{p}_N = -k^2 p_N / A\rho\Delta$ and $\lambda^2 = k^2\alpha^2/h^2$.

Suppose that $u(0,t) = u(L,t) = 0$, $u(x,0) = f(x)$, $\partial u(x,0)/\partial t = 0$ are the given boundary and initial conditions and that the finite-element network shown in Fig. 11.2 is used. The analysis proceeds as follows:

1. Conceptually, only one tier of elements (the first row corresponding to the interval $0 \leq t \leq k$, the second to $k \leq t \leq 2k$, etc.) need be considered to be generated at a time. Global values U^{Δ} of the displacements of boundary nodes are equated to zero in agreement with given boundary conditions: $U^1 = U^6 \ (= U^{11} = U^{16} = \cdots) = 0$, $U^5 = U^{10}(= U^{15} = U^{20} = \cdots) = 0$. Displacements at interior nodes corresponding to $t = 0$ may be given the prescribed values; that is, $U^2 = f(h)$, $U^3 = f(2h)$, $U^4 = f(3h)$, etc.
2. Since the displacements U^2, U^3, U^4 take on prescribed values, the corresponding global generalized "forces" (conjugate variables) $\hat{P}_2$, $\hat{P}_3$, $\hat{P}_4$ vanish. The only unknowns in the resulting equations

$$\hat{P}_2(U^2,U^7) = \hat{P}_3(U^3,U^8) = \hat{P}_4(U^4,U^9) = 0 \tag{11.13}$$

are the nodal values U^7, U^8, U^9, which represent the displaced profile

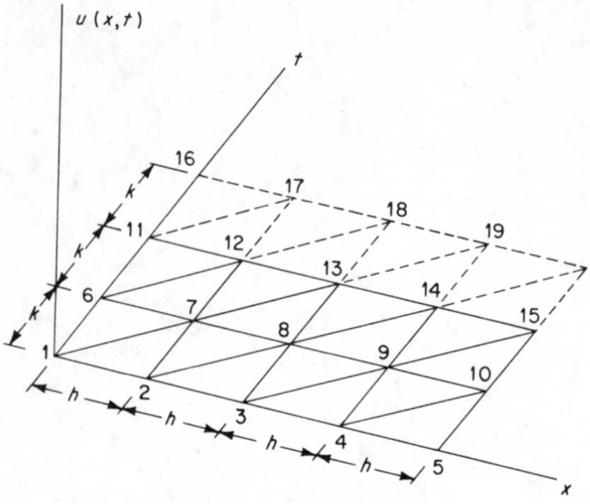

Fig. 11.2 Example of propagation of solution in time.

after k sec. Since each equation in (11.13) has only one unknown, the set can be solved immediately to give U^7, U^8, and U^9 in terms of the prescribed nodal displacements at $t = 0$.

3. Another tier of elements ($k \leq t \leq 2k$) is now considered. Displacements U^{12}, U^{13}, and U^{14} are obtained from the conditions $\hat{P}_7 = \hat{P}_8 = \hat{P}_9 = 0$. Then a third tier of elements is considered, and the process is repeated.

Thus the finite-element solution is propagated in time in the manner of conventional explicit finite difference schemes.

We remark that in the case in which a time-varying end load is applied and initial displacements $u(x,0)$ are not prescribed, the same procedure is followed except that $U^5, U^{10}, \ldots \neq 0$ and, instead of (11.13), $\hat{P}_2$, $\hat{P}_3$, $\hat{P}_4$ (and $\hat{P}_7$, $\hat{P}_8, \ldots$) take on prescribed nonzero values.

Stability The rather crude simplex model used in the above example is the most primitive finite model for the problem at hand. By using higher-order approximations or adding more degrees of freedom to the elements, greater accuracy can be expected for more difficult propagation problems. Nevertheless, it is interesting to note that for an interior node such as 8 in the mesh indicated in Fig. 11.2, we have

$$-2\hat{P}_8 = 0 = (-U^3 - 2U^8 + U^{13}) + \lambda^2(U^7 - 2U^8 + U^9) \qquad (11.14)$$

which is precisely the form of the first-order central difference approximation of (11.9). Thus, we can draw on the criteria of Courant, Friedrichs, and Lewy [1928] to obtain conclusions on the stability of the scheme outlined above. Accordingly, the solution is unstable for $\lambda > 1$ and violently unstable for increasing values of λ; for $\lambda < 1$, it is stable but the accuracy decreases with decreasing λ; for $\lambda = 1$, the solution is stable and may agree with the exact solution of (11.9).

Dynamical systems It is not difficult to apply the above ideas to more general dynamical systems. Foı example, let $\mathbf{u}(\mathbf{x},t)$ denote the displacement field and consider the local finite-element approximation

$$\mathbf{u}_e = \psi_N(\mathbf{x})\mathbf{u}^N(t) \tag{11.15}$$

Now the nodal quantities $\mathbf{u}^N$ are functions of time so that we can also construct a finite-element approximation of $\mathbf{u}^N$ over the "time element" $[t_1,t_2]$. For example, use of Hermite interpolation in time yields†

$$\mathbf{u}_e(\mathbf{x},t) = \psi_N(\mathbf{x})(\varphi_0^\alpha(t)\mathbf{u}_\alpha^N + \varphi_1^\alpha(t)\dot{\mathbf{u}}_\alpha^N) \tag{11.16}$$

where $\alpha = 1, 2$, $\mathbf{u}_\alpha^N = \mathbf{u}^N(t_\alpha)$, $\dot{\mathbf{u}}_\alpha^N = d\mathbf{u}^N(t_\alpha)/dt$, and, in view of (8.24),

$$\begin{aligned} \varphi_0^1(t) &= 2\xi^3 - 3\xi^2 + 1 \qquad & \varphi_0^2(t) &= 3\xi^2 - 2\xi^3 \\ \varphi_1^1(t) &= \xi^3 - 2\xi^2 + \xi \qquad & \varphi_1^2(t) &= \xi^3 - \xi^2 \end{aligned} \tag{11.17}$$

where $\xi = t/h$, $h = \Delta t$ is the length of the element in time, and for simplicity we used the local reference $t_1 = 0$, $t_2 = h$.

Now consider a dynamical system for which

$$\mathfrak{L}_{(e)} = \int_{t_1}^{t_2} (\kappa - U + V + D)\, dt \tag{11.18}$$

where κ is the kinetic energy, $+U - V$ is the potential energy, and D is the work of viscous forces. Since the integrand involves integrals of $\mathbf{u}$ and its gradients over the volume of the continuum, substitution of (11.16) into (11.18) leads to

$$\mathfrak{L}_{(e)} = \tfrac{1}{2}\int_{t_1}^{t_2} (m_{NM}\dot{\mathbf{u}}^N \cdot \dot{\mathbf{u}}^M - k_{NM}\mathbf{u}^N \cdot \mathbf{u}^M + 2\mathbf{p}_N \cdot \mathbf{u}^N + 2c_{NM}\mathbf{u}^N \cdot \dot{\mathbf{u}}^M)\, dt \tag{11.19}$$

wherein m_{NM}, k_{NM}, and c_{NM} and the mass, stiffness, and damping matrices for the element and $\mathbf{p}_N = \mathbf{p}_N(t)$ is the generalized force vector at node N.‡ Here

$$\mathbf{u}^N(t) = \varphi_0^\alpha(t)\mathbf{u}_\alpha^N + \varphi_1^\alpha(t)\dot{\mathbf{u}}_\alpha^N \tag{11.20}$$

If we approximate $\mathbf{p}_N(t)$ in time by

$$\mathbf{p}_N(t) = \psi_\alpha(t)\mathbf{p}_N^\alpha \tag{11.21}$$

† Cf. Fried [1969*b*].

‡ More detailed discussions of mass and stiffness matrices and generalized forces for finite elements are given in Arts. 13 and 16.

$\alpha = 1, 2$, then substitution of (11.20) and (11.21) into (11.19) gives

$$\mathfrak{L}_{(e)} = \tfrac{1}{2}(A_{NM}^{\alpha\beta}\mathbf{u}_\alpha^N \cdot \mathbf{u}_\beta^M + B_{NM}^{\alpha\beta}\dot{\mathbf{u}}_\alpha^N \cdot \mathbf{u}_\beta^M + C_{NM}^{\alpha\beta}\mathbf{u}_\alpha^N \cdot \dot{\mathbf{u}}_\beta^M + D_{NM}^{\alpha\beta}\dot{\mathbf{u}}_\alpha^N \cdot \dot{\mathbf{u}}_\beta^M + d_{\alpha 0}^{\beta}\mathbf{p}_N^\alpha \cdot \mathbf{u}_\beta^N + d_{\alpha 1}^{\beta}\mathbf{p}_N^\alpha \cdot \dot{\mathbf{u}}_\beta^N) \quad (11.22)$$

where

$$A_{NM}^{\alpha\beta} = m_{NM}a_{00}^{\alpha\beta} - k_{NM}c_{00}^{\alpha\beta} + 2c_{NM}b_{00}^{\alpha\beta} \quad (11.23a)$$

$$B_{NM}^{\alpha\beta} = m_{NM}a_{10}^{\alpha\beta} - k_{NM}c_{10}^{\alpha\beta} + 2c_{NM}b_{10}^{\alpha\beta} \quad (11.23b)$$

$$C_{NM}^{\alpha\beta} = m_{NM}a_{01}^{\alpha\beta} - k_{NM}c_{01}^{\alpha\beta} + 2c_{NM}b_{01}^{\alpha\beta} \quad (11.23c)$$

$$D_{NM}^{\alpha\beta} = m_{NM}a_{11}^{\alpha\beta} - k_{NM}c_{11}^{\alpha\beta} + 2c_{NM}b_{11}^{\alpha\beta} \quad (11.23d)$$

and

$$a_{rs}^{\alpha\beta} = \int_{t_1}^{t_2} \dot{\varphi}_r^\alpha \dot{\varphi}_s^\beta \, dt \quad (11.24a)$$

$$b_{rs}^{\alpha\beta} = \int_{t_1}^{t_2} \varphi_r^\alpha \dot{\varphi}_s^\beta \, dt \quad (11.24b)$$

$$c_{rs}^{\alpha\beta} = \int_{t_1}^{t_2} \varphi_r^\alpha \varphi_s^\beta \, dt \quad (11.24c)$$

$$d_{\alpha 0}^{\beta} = \int_{t_1}^{t_2} 2\psi_\alpha \varphi_0^\beta \, dt \quad (11.24d)$$

$$d_{\alpha 1}^{\beta} = \int_{t_1}^{t_2} 2\psi_\alpha \varphi_1^\beta \, dt \quad (11.24e)$$

with $r, s = 0, 1$.

Hamilton's principle requires that

$$\delta\mathfrak{L}_{(e)} = (\partial\mathfrak{L}_{(e)}/\partial\mathbf{u}_\alpha^N) \cdot \delta\mathbf{u}_\alpha^N + (\partial\mathfrak{L}_{(e)}/\partial\dot{\mathbf{u}}_\alpha^N) \cdot \delta\dot{\mathbf{u}}_\alpha^N = 0$$

Thus

$$\begin{aligned} A_{NM}^{\alpha\beta}\mathbf{u}_\beta^M + \tfrac{1}{2}(B_{MN}^{\alpha\beta} + C_{NM}^{\alpha\beta})\dot{\mathbf{u}}_\beta^M + d_{\beta 0}^{\alpha}\mathbf{p}_N^\beta &= \mathbf{0} \\ D_{NM}^{\alpha\beta}\dot{\mathbf{u}}_\beta^M + \tfrac{1}{2}(B_{NM}^{\alpha\beta} + C_{MN}^{\beta\alpha})\mathbf{u}_\beta^M + d_{\beta 1}^{\alpha}\mathbf{p}_N^\beta &= \mathbf{0} \end{aligned} \quad (11.25)$$

Upon connecting elements together (in space and time) and applying appropriate boundary and initial conditions, (11.25) lead to systems of linear equations in $\mathbf{u}_\alpha^N$ and $\dot{\mathbf{u}}_\alpha^N$ that must be solved simultaneously. Thus, unlike the scheme described in the wave propagation example, this particular finite-element model leads to an implicit (and hence often unconditionally stable) integration scheme.

11.2 FINITE ELEMENTS IN THE COMPLEX PLANE; SCHROEDINGER'S EQUATION

In quantum mechanics, the Schroedinger wave equation can be written in terms of wave function χ, which is complex-valued. We now examine the development of finite-element analogs of Schroedinger's equations for χ and its complex conjugate $\bar{\chi}$ for the case of a single particle of mass m acting under the influence of a potential field $V(\mathbf{x}) = V(x,y,z)$.

The wave function $\chi(\mathbf{x},t)$ can be written in the form

$$\chi(\mathbf{x},t) = u(\mathbf{x},t) + iv(\mathbf{x},t) \tag{11.26}$$

where $i = \sqrt{-1}$. The complex conjugate is $\bar{\chi}(\mathbf{x},t) = u(\mathbf{x},t) - iv(\mathbf{x},t)$, and physically $\chi(\mathbf{x},t)\bar{\chi}(\mathbf{x},t)$ represents the probability density at time t for the presence of the particle for the configuration of the system specified by the coordinates $\mathbf{x}$. Confining our attention to a typical finite element e, we approximate the real and imaginary parts of $\chi(\mathbf{x},t)$ locally by

$$u^{(e)}(\mathbf{x},t) = \psi_N(\mathbf{x})u^N_{(e)} \qquad v^{(e)}(\mathbf{x},t) = \psi_N(\mathbf{x})v^N_{(e)} \tag{11.27a, b}$$

where $u^N_{(e)}$, $v^N_{(e)}$ are the time-dependent nodal values of $u(\mathbf{x},t)$ and $v(\mathbf{x},t)$. Then

$$\chi^{(e)}(\mathbf{x},t) = \chi^N_{(e)}\psi^{(e)}_N(\mathbf{x}) \tag{11.28a}$$

$$\bar{\chi}^{(e)}(\mathbf{x},t) = \bar{\chi}^N_{(e)}\psi^{(e)}_N(\mathbf{x}) \tag{11.28b}$$

where

$$\chi^N_{(e)} = u^N_{(e)} + iv^N_{(e)} \qquad \bar{\chi}^N_{(e)} = u^N_{(e)} - iv^N_{(e)} \tag{11.29a, b}$$

The Lagrange density $L^{(e)}$ for an element is†

$$L^{(e)} = \frac{h^2}{8\pi^2 m}\operatorname{grad}\bar{\chi}\cdot\operatorname{grad}\chi - \frac{h}{4\pi i}\left(\bar{\chi}\frac{\partial\chi}{\partial t} - \frac{\partial\bar{\chi}}{\partial t}\chi\right) - \bar{\chi}V\chi \tag{11.30}$$

where h is Planck's constant and $\bar{\chi}$ and χ are to be varied independently until $\mathfrak{L} = \iiiint L^{(e)}\,d\mathscr{R}\,dt$ is a minimum.

Introducing (11.28) into (11.30) and requiring that

$$\frac{d}{dt}\left(\frac{\partial L^{(e)}}{\partial\dot{\chi}^N}\right) - \frac{\partial L^{(e)}}{\partial\chi^N} = \frac{d}{dt}\left(\frac{\partial L^{(e)}}{\partial\dot{\bar{\chi}}^N}\right) - \frac{\partial L^{(e)}}{\partial\bar{\chi}^N} = 0$$

we arrive at the pair of equations

$$\frac{h^2}{8\pi^2 m}\alpha^{(e)}_{MN}\bar{\chi}^M_{(e)} + \frac{h}{2\pi i}\beta^{(e)}_{MN}\dot{\bar{\chi}}^M_{(e)} - \gamma^{(e)}_{MN}\bar{\chi}^M_{(e)} = 0 \tag{11.31a}$$

$$\frac{h^2}{8\pi^2 m}\alpha^{(e)}_{MN}\chi^M_{(e)} - \frac{h}{2\pi i}\beta^{(e)}_{MN}\dot{\chi}^M_{(e)} - \gamma^{(e)}_{MN}\dot{\chi}^M_{(e)} = 0 \tag{11.31b}$$

† See Morse and Feshbach [1953, pp. 314–318].

where

$$\alpha_{MN}^{(e)} = \int_{\imath_e} \psi_{M,i}(\mathbf{x})\psi_{N,i}(\mathbf{x})\, d\imath_e \tag{11.32a}$$

$$\beta_{MN}^{(e)} = \int_{\imath_e} \psi_M(\mathbf{x})\psi_N(\mathbf{x})\, d\imath_e \tag{11.32b}$$

and

$$\gamma_{MN}^{(e)} = \int_{\imath_e} \psi_M(\mathbf{x})V(\mathbf{x})\psi_N(\mathbf{x})\, d\imath_e \tag{11.32c}$$

Equations (11.31) are the discrete counterparts of the Schroedinger wave equations for the finite element. The quantities $h\beta_{MN}^{(e)}\chi_{(e)}^{M}/2\pi i$ and $h\beta_{MN}^{(e)}\bar{\chi}_{(e)}^{M}/2\pi i$ are the generalized canonical momenta at node N of the element, and $(h^2\alpha_{MN}^{(e)}/8\pi^2 m) - \gamma_{MN}^{(e)}$ represents the discrete equivalent of the hamiltonian operator for the particle while in finite element e.

11.3 A NONLINEAR PARTIAL DIFFERENTIAL EQUATION

We now give brief consideration to the finite-element formulation of a nonlinear boundary-value problem which involves determining a function $u(x,y)$ over a closed region $\mathscr{R}$ of the euclidean plane which satisfies the nonlinear partial differential equation

$$2u\nabla^2 u + u_x^2 + u_y^2 = f(x,y) \tag{11.33}$$

subject to the homogeneous condition

$$u(s) = 0 \tag{11.34}$$

on the boundary curve Γ. Here $\nabla^2 = \partial^2/\partial x^2 + \partial^2/\partial y^2$, $u_x = \partial u/\partial x$, and $u_y = \partial u/\partial y$.

Following the arguments of Sec. 10.1, we introduce a nonlinear operator $\mathscr{P}(u)$ defined by

$$\mathscr{P}(u) = f - 2u\nabla^2 u - u_x^2 - u_y^2 \tag{11.35}$$

so that (11.33) can be written

$$\mathscr{P}(u) = 0 \tag{11.36}$$

It is not difficult to show that $\mathscr{P}$ is a potential operator; indeed, in accordance with (10.19),

$$\begin{aligned} \delta\mathscr{P}(u,h) &= \mathscr{P}(u+h) - \mathscr{P}(u) - \omega(u,h) \\ &= -2(\nabla^2 uh - u_x h_x - u_y h_y) \end{aligned} \tag{11.37}$$

where h is an arbitrary function satisfying (11.34). If g is another such function,

$$\langle \delta\mathscr{P}(u,h),g\rangle - \langle \delta\mathscr{P}(u,g),h\rangle$$

$$= -2\int_{\mathscr{R}} [g\nabla^2 uh - h\nabla^2 ug - g(u_x h_x + u_y h_y) + h(u_x g_x + u_y g_y)]\, d\mathscr{R}$$

$$= -2\int_{\mathscr{R}} [gu\nabla^2 h - hu\nabla^2 g + g(u_x h_x + u_y h_y) - h(u_x g_x + u_y g_y)]\, d\mathscr{R} \tag{11.38}$$

Since

$$\int_{\mathscr{R}} gu\nabla^2 h\, d\mathscr{R} = -\int_{\mathscr{R}}\left[h_x \frac{\partial}{\partial x}(gu) + h_y \frac{\partial}{\partial y}(gu)\right] d\mathscr{R} + \int_{\Gamma} gu \frac{\partial h}{\partial n}\, ds$$

etc., and the line integral is zero because g and h satisfy (11.34), the last integral in (11.38) vanishes and

$$\langle \delta\mathscr{P}(u,h),g\rangle = \langle \delta\mathscr{P}(u,g),h\rangle \tag{11.39}$$

This symmetry verifies that $\mathscr{P}$ is potential.

Consequently, according to Theorem 10.4, there exists a functional $K(u)$, given by (10.21), whose gradient is $\mathscr{P}(u)$. Observing that

$$-\langle \mathscr{P}(\tau u),u\rangle = \int_{\mathscr{R}} [\tau^2(2u^2\nabla^2 u + uu_x^2 + uu_y^2) - uf]\, d\mathscr{R}$$

where τ is a parameter, we find on successive partial integrations and use of (11.34) that

$$\langle \mathscr{P}(\tau u),u\rangle = \int_{\mathscr{R}} [3\tau^2 u(u_x^2 + u_y^2) + uf]\, d\mathscr{R} \tag{11.40}$$

Introducing (11.40) into (10.21) (with s replaced by τ and $u_0 = K_0 = 0$) and performing the indicated integration, we obtain the functional

$$K(u) = \int_{\mathscr{R}} [u(u_x^2 + u_y^2) + uf]\, d\mathscr{R} \tag{11.41}$$

It is easily verified that $\mathscr{P}(u)$ is the gradient of $K(u)$:

$$K(u + \alpha h) - K(u) = \alpha \int_{\mathscr{R}} [h(u_x^2 + u_y^2) + 2u(u_x h_x + u_y h_y) + hf]\, d\mathscr{R}$$

$$+ \alpha^2 \int_{\mathscr{R}} [u(h_x^2 + h_y^2) + 2h(u_x h_x + u_y h_y)]\, d\mathscr{R}$$

$$+ \alpha^3 \int_{\mathscr{R}} h(h_x^2 + h_y^2)\, d\mathscr{R} \tag{11.42}$$

Since $h = 0$ on Γ, the first integral on the right side of (11.42) can be written, after partial integrations, as

$$-\alpha \int_{\mathscr{R}} h(2u\nabla^2 u + u_x^2 + u_y^2 - f)\, d\mathscr{R}$$

so that

$$\lim_{\alpha\to 0} \frac{1}{\alpha} [K(u + \alpha h) - K(u)] = \int_{\mathscr{R}} \mathscr{P}(u) h\, d\mathscr{R} = \langle \mathscr{P}(u), h \rangle$$

Hence,

$$\text{grad } K(u) = \mathscr{P}(u) \tag{11.43}$$

and at the point u^* that satisfies (11.36),

$$\text{grad } K(u^*) = 0 \tag{11.44}$$

that is, u^* is a critical point of the functional $K(u)$.

Turning now to the construction of a finite-element model of (11.33), we represent $\mathscr{R}$ as a collection $\bar{\mathscr{R}}$ of E finite elements and denote by $k^{(e)}(u)$ the restriction of $K(u)$ to element e. Then, approximately,

$$K(u) = \sum_{e=1}^{E} k^{(e)}(u^{(e)}) \tag{11.45}$$

where

$$k^{(e)}(u^{(e)}) = \int_{r_e} [u^{(e)}(u_x^{(e)^2} + u_y^{(e)^2}) + u^{(e)} f]\, dr_e \tag{11.46}$$

Locally, over a typical element r_e, we have $u^{(e)}(x,y) = u_{(e)}^N \psi_N^{(e)}(x,y)$

$$k^{(e)}(u^{(e)}) = a_{NMR}^{(e)} u_{(e)}^N u_{(e)}^M u_{(e)}^R + u_{(e)}^N f_N^{(e)} \tag{11.47}$$

where

$$a_{NMR}^{(e)} = \int_{r_e} \psi_N^{(e)} \left(\frac{\partial \psi_M^{(e)}}{\partial x} \frac{\partial \psi_R^{(e)}}{\partial x} + \frac{\partial \psi_M^{(e)}}{\partial y} \frac{\partial \psi_R^{(e)}}{\partial y} \right) dr_e \tag{11.48a}$$

$$f_N^{(e)} = \int_{r_e} \psi_N^{(e)} f\, dr_e \tag{11.48b}$$

and $M, N, R = 1, 2, \ldots, N_e$. Locally, $u^{(e)}$ provides an extremum to $k^{(e)}$ in the subspace spanned by $\psi_N^{(e)}(x,y)$ if the coefficients $u_{(e)}^I$ are chosen so that

$$\frac{\partial k^{(e)}(u^{(e)})}{\partial u_{(e)}^I} = 0 \tag{11.49}$$

Introducing (11.47), we find that this leads to the system of quadratic equations

$$\hat{a}_{IMN}^{(e)} u_{(e)}^M u_{(e)}^N + f_I^{(e)} = 0 \tag{11.50}$$

where

$$\hat{a}_{IMN}^{(e)} = a_{IMN}^{(e)} + 2a_{NMI}^{(e)} \tag{11.51}$$

Since

$$u_{(e)}^{N} = \overset{(e)}{\Omega}{}_{\Delta}^{N} U^{\Delta} \tag{11.52}$$

we have, from (11.47) and (11.45),

$$K(u) = A_{\Lambda\Gamma\Delta} U^{\Lambda} U^{\Gamma} U^{\Delta} + F_{\Delta} U^{\Delta} \tag{11.53}$$

where

$$A_{\Lambda\Gamma\Delta} = \sum_{e=1}^{E} \overset{(e)}{\Omega}{}_{\Lambda}^{N} \overset{(e)}{\Omega}{}_{\Gamma}^{M} \overset{(e)}{\Omega}{}_{\Delta}^{R} a_{NMR}^{(e)} \qquad F_{\Delta} = \sum_{e=1}^{E} \overset{(e)}{\Omega}{}_{\Delta}^{N} f_{N}^{(e)} \tag{11.54a, b}$$

The final finite-element analog of (11.33) is obtained from the conditions $\partial K(u)/\partial U^{\Delta} = 0$ [or directly by introducing (11.52) into (11.50) and substituting the result into (11.54*b*)]:

$$\hat{A}_{\Lambda\Gamma\Delta} U^{\Gamma} U^{\Delta} + F_{\Lambda} = 0 \tag{11.55}$$

where

$$\hat{A}_{\Lambda\Gamma\Delta} = A_{\Lambda\Delta\Gamma} + 2A_{\Delta\Gamma\Lambda} \tag{11.56}$$

Upon applying boundary conditions (i.e., setting $U^{\Delta} = 0$ at nodes Δ on the boundary of $\bar{\mathscr{R}}$), (11.55) yields a system of quadratic equations in the unknown nodal values U^{Δ}. We discuss methods for solving such systems of nonlinear algebraic equations in Art. 17.

11.4 KINETIC THEORY OF GASES; BOLTZMANN'S EQUATION

The statistical mechanics of dilute gases (collisionless free molecular flow) involves problem areas in which finite-element models in the six-dimensional μ space may be used to advantage. Here the molecular density is assumed to be sufficiently low and the temperature sufficiently high that each molecule of gas can be considered to be a classical particle with a reasonably well-defined position and momentum. The behavior of a contained gas is characterized, according to classical kinetic theory,† by a distribution function $f(\mathbf{x},\mathbf{v},t)$ which is defined so as to represent the number of molecules at time t which have positions lying in a "volume" element $d\Omega$ in a six-dimensional velocity space, such that x_1, x_2, and x_3 denote the position of the molecule and $x_4 = v_1$, $x_5 = v_2$, $x_6 = v_3$ its components of velocity. Unlike classical mechanics, which deals only with mean velocities, the quantities

† See, for example, Huang [1963].

v_1, v_2, and v_3 are independent of x_1, x_2, and x_3. We outline briefly the finite-element approximation of such distribution functions.†

Consider the unsteady Boltzmann equation for the distribution function $f(\mathbf{x},\mathbf{v},t)$ with the Bhatanger-Gross–Krook (BGK) collision model and no external forces:

$$\frac{\partial f}{\partial t} + \mathbf{v} \cdot \operatorname{grad} f = \frac{F - f}{\tau} \tag{11.57}$$

Here $\mathbf{v}$ is the microscopic velocity, $\mathbf{x}$ is the position vector, F is the maxwellian distribution, and τ is the collision time.

In principle, we could proceed directly to the construction of a finite-element model of (11.57) built up of six-dimensional finite elements in μ space since all the machinery for such constructions was laid down in Art. 7. For example, we could represent the domain by a discrete model $\bar{\mathscr{R}}$ consisting of a collection of six-dimensional elements (say simplexes) and write for the restriction of $f(\mathbf{x},\mathbf{v},t)$ over a typical element e,

$$f^{(e)}(\mathbf{x},\mathbf{v},t) \approx f^{N}_{(e)}(t)\psi^{(e)}_{N}(x_1,x_2,\ldots,x_6) \tag{11.58}$$

$N = 1, 2, \ldots, N_e$ $(N_e > 7)$. Then, using the idea of weighted residuals, we obtain for the analog of (11.57) for a typical element

$$r^{(e)}_{NM}\dot{f}^{M}_{(e)} + s^{(e)}_{NM}f^{M}_{(e)} = y^{(e)}_{N} \tag{11.59}$$

where $\dot{f}^{M}_{(e)} = df^{M}_{(e)}/dt$ and, denoting $\mathbf{x}^* = (x_1,x_2,\ldots,x_6)$,

$$r^{(e)}_{NM} = \int_{\Omega_{(e)}} \psi^{(e)}_{N}(\mathbf{x}^*)\psi^{(e)}_{M}(\mathbf{x}^*)\,d\Omega \tag{11.60a}$$

$$s^{(e)}_{NM} = \int_{\Omega_{(e)}} \left[\psi^{(e)}_{N}(\mathbf{x}^*)\sum_{m=1}^{3} v_m \frac{\partial\psi^{(e)}_{M}(\mathbf{x}^*)}{\partial x_m} + \frac{1}{\tau}\psi^{(e)}_{N}(\mathbf{x}^*)\psi^{(e)}_{M}(\mathbf{x}^*)\right] d\Omega \tag{11.60b}$$

$$y^{(e)}_{N} = \int_{\Omega_{(e)}} \frac{1}{\tau} F\psi^{(e)}_{N}(\mathbf{x}^*)\,d\Omega \tag{11.60c}$$

The remainder of the formulation follows essentially the same procedures as those described earlier.

Unfortunately, the above equations may be quite difficult to apply in specific cases, and it is convenient to seek an alternate finite-element formulation. To illustrate one such alternate formulation, we combine the essential ideas of finite elements and Kantorovich's [1933] method and

† Such finite-element approximations were considered by Oden [1969b], Bramlette and Mallett [1969], and Aguirre-Ramirez, Oden, and Wu [1970]. The present discussion is based on the last paper.

assume that throughout Ω, $f(\mathbf{x},\mathbf{v},t)$ is approximately given by

$$f(\mathbf{x},\mathbf{v},t) = \sum_{\alpha=1}^{n} c_\alpha(\mathbf{x},t)\Omega_\alpha(\mathbf{v}) \tag{11.61}$$

where c_α and Ω_α are undetermined functions of the indicated arguments and n is some integer. Since (11.61) is only an approximation, we can compute the residual

$$R = \sum_{\alpha=1}^{n} \left(\frac{\partial c_\alpha(\mathbf{x},t)}{\partial t} + \mathbf{v} \cdot \operatorname{grad} c_\alpha + \frac{c_\alpha}{\tau} \right) \Omega_\alpha - \frac{F}{\tau} \tag{11.62}$$

using (11.57). Then, for a suitable choice of weight functions $W_\beta(\mathbf{v})$, we obtain from the requirement

$$\iiint_{-\infty}^{\infty} W_\beta(\mathbf{v}) R \, d\mathbf{v} = 0 \tag{11.63}$$

the system of partial differential equations

$$a_{\beta\alpha} \frac{\partial c_\alpha}{\partial t} + \mathbf{b}_{\beta\alpha} \cdot \operatorname{grad} c_\alpha + a^*_{\beta\alpha} c_\alpha - F^*_\beta = 0 \tag{11.64}$$

where the repeated index α is summed from 1 to n and

$$a_{\beta\alpha} = \iiint_{-\infty}^{\infty} W_\beta(\mathbf{v})\Omega_\alpha(\mathbf{v})\, d\mathbf{v} \qquad \mathbf{b}_{\beta\alpha} = \iiint_{-\infty}^{\infty} \mathbf{v} W_\beta(\mathbf{v})\Omega_\alpha(\mathbf{v})\, d\mathbf{v}$$
$$a^*_{\beta\alpha} = \iiint_{-\infty}^{\infty} \frac{1}{\tau} W_\beta(\mathbf{v})\Omega_\alpha(\mathbf{v})\, d\mathbf{v} \qquad F^*_\beta = \iiint_{-\infty}^{\infty} \frac{1}{\tau} W_\beta(\mathbf{v}) F(\mathbf{v})\, d\mathbf{v} \tag{11.65}$$

We can now proceed to construct a finite-element model of (11.64). Representing $\mathscr{R}$ [the spatial domain of $c_\alpha(\mathbf{x},t)$] by a collection of elements, we write for a typical element e,

$$c_\alpha^{(e)}(\mathbf{x},t) \approx c^N_{\alpha(e)}(t)\psi_N^{(e)}(\mathbf{x}) \tag{11.66}$$

The local residual r_e for each element is then

$$r_e = a_{\beta\alpha}\psi_N^{(e)}(\mathbf{x})\dot{c}^N_{\alpha(e)} + [\mathbf{b}_{\beta\alpha} \cdot \operatorname{grad} \psi_N^{(e)}(\mathbf{x}) + a^*_{\beta\alpha}\psi_N^{(e)}(\mathbf{x})]c^N_{\alpha(e)} - F^*_\beta \tag{11.67}$$

where $\dot{c}^N_{\alpha(e)} = dc^N_{\alpha(e)}/dt$. From the requirement

$$\int_{\mathscr{r}_e} \psi_N^{(e)}(\mathbf{x}) r_e \, d\mathscr{r}_e = 0 \tag{11.68}$$

we obtain the finite-element equation

$$a_{\beta\alpha} A_{MN} \dot{c}^N_{\alpha(e)} + (\mathbf{b}_{\beta\alpha} \cdot \mathbf{B}_{MN} + a^*_{\beta\alpha} A_{MN}) c^N_{\alpha(e)} - F^*_{\beta M} = 0 \tag{11.69}$$

where, omitting the element identification label e for simplicity,

$$A_{MN} = A_{NM} = \int_{v_{(e)}} \psi_M(\mathbf{x})\psi_N(\mathbf{x})\, dv_e$$

$$\mathbf{B}_{MN} = \int_{v_{(e)}} \psi_M(\mathbf{x}) \operatorname{grad} \psi_N(\mathbf{x})\, dv_e \tag{11.70}$$

$$F^*_{\beta M} = F^*_\beta \int_{v_{(e)}} \psi_M(\mathbf{x})\, dv_e$$

Using the relations

$$\overset{N}{c}_{\alpha(e)} = \overset{(e)}{\Omega}{}^N_\Delta C^\Delta_\alpha \qquad F^*_{\beta\Gamma} = \sum_{e=1}^{E} \overset{(e)}{\Omega}{}^M_\Gamma F^*_{\beta M} \tag{11.71}$$

we arrive at the system of discrete equations for the assembled system,

$$a_{\beta\alpha} A_{\Gamma\Delta} \dot{C}^\Delta_\alpha + (\mathbf{b}_{\beta\alpha} \cdot \mathbf{B}^*_{\Gamma\Delta} + a^*_{\beta\alpha} A_{\Gamma\Delta}) C^\Delta_\alpha - F^*_{\beta\Gamma} = 0 \tag{11.72}$$

where

$$A_{\Gamma\Delta} = A_{\Delta\Gamma} = \sum_{e=1}^{E} \overset{(e)}{\Omega}{}^M_\Gamma A^{(e)}_{MN} \overset{(e)}{\Omega}{}^N_\Delta$$

$$\mathbf{B}^*_{\Gamma\Delta} = \sum_{e=1}^{E} \overset{(e)}{\Omega}{}^M_\Gamma \mathbf{B}^{(e)}_{MN} \overset{(e)}{\Omega}{}^N_\Delta \tag{11.73}$$

Equation (11.72) represents a system of differential equations for the nodal quantities C^Δ_α. Note that if we know $C^\Delta_\alpha(t)$, then we can construct the discrete model for the distribution function from (11.61):

$$f(\mathbf{x},\mathbf{v},t) = \sum_{\alpha=1}^{n} C^\Delta_\alpha(t)\Phi_\Delta(\mathbf{x})\Omega_\alpha(\mathbf{v}) \tag{11.74}$$

where

$$\Phi_\Delta(\mathbf{x}) = \sum_{e=1}^{E} \overset{(e)}{\Omega}{}^N_\Delta \psi^{(e)}_N(\mathbf{x})$$

Transient couette flow As an example, we consider briefly the development of the description of flow between two parallel plates, separated by a distance d, which are accelerated in opposite directions from rest to some constant velocity $U/2$ (Fig. 11.3). It is convenient to introduce the constant $\beta = (m/2kT)^{\frac{1}{2}}$, where m is the molecular mass, k the Boltzmann constant, and T the temperature. Letting x be the axis perpendicular to the flow, we introduce a dimensionless coordinate ξ and a dimensionless velocity $\mathbf{c}$ by

$$\xi = \frac{x}{d} \qquad \mathbf{c} = \beta\mathbf{v} \tag{11.75}$$

We further assume the distribution function f to be of the form

$$f(\xi,\mathbf{c},t) = n_0\beta^3\pi^{-\frac{3}{2}} \exp(-\mathbf{c}^2)[1 + \varphi^*(\xi,\mathbf{c},t)] \tag{11.76}$$

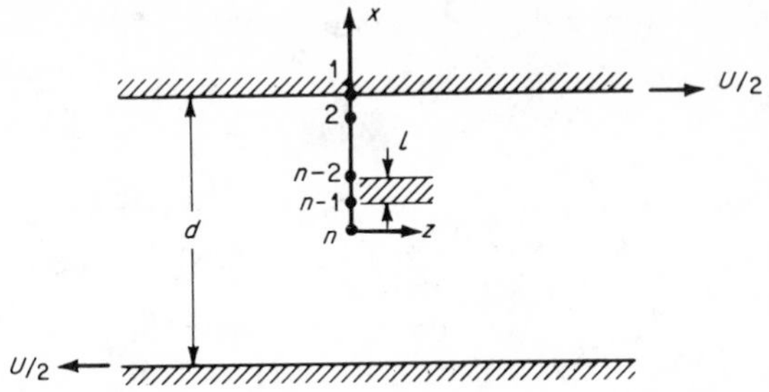

Fig. 11.3 Model of Couette flow between parallel plates.

where n_0 is the number density and φ^* is the perturbation function. For the linearized flow problem with no density and temperature variations, we may assume the form

$$\varphi^*(\xi,\mathbf{c},t) = Uc_z\varphi(\xi,c_x,t) \tag{11.77}$$

for the perturbation function, for which the linearized one-dimensional Boltzmann equation with BGK collision model becomes

$$\beta\frac{\partial\varphi}{\partial t} + c_x\frac{\partial\varphi}{\partial\xi} + \alpha\varphi = \frac{\alpha}{\sqrt{\pi}}\int_{-\infty}^{\infty}\varphi(\xi,\zeta,t)\exp\,(1-\zeta^2)\,d\zeta \tag{11.78}$$

α being the inverse Knudsen number. The quantities of interest are the macroscopic flow velocity $q_z(\xi,t)$ and viscous shearing stress $\tau_{xz}(\xi,t)$ given by

$$\begin{aligned}\frac{q_z(\xi,t)}{U/2} &= \frac{1}{\pi}\int_{-\infty}^{\infty}\exp\,(-\zeta^2)\varphi(\xi,\zeta,t)\,d\zeta\\ \frac{\tau_{xz}(\xi,t)}{\tau_{xz_{fm}}} &= \int_{-\infty}^{\infty}\zeta\exp\,(-\zeta^2)\varphi(\xi,\zeta,t)\,d\zeta\end{aligned} \tag{11.79}$$

where $\tau_{xz_{fm}} = -U/2\sqrt{\pi}$ is the free molecular value of the shear stress.

In order to obtain an approximate solution of (11.78) by finite-element methods, we consider the λth Hermite polynomial, defined by

$$H_\lambda(\zeta) = (-1)^\lambda\exp\,(\zeta^2)\frac{d^\lambda}{d\zeta^\lambda}\exp\,(-\zeta^2) \qquad \lambda = 0, 1, \ldots \tag{11.80}$$

The set of functions Ω_λ of (11.61) are now given by

$$\Omega_\lambda(\zeta) = \exp\,(-\tfrac{1}{2}\zeta^2)H_\lambda(\zeta) \tag{11.81}$$

The $H_\lambda(\zeta)$ are orthogonal on $(-\infty,\infty)$ with respect to $\exp(-\frac{1}{2}\zeta^2)$. Also

$$\int_{-\infty}^{\infty}[\Omega_\lambda(\zeta)]^2\,d\zeta = 2^\lambda\lambda!\sqrt{\pi} \tag{11.82}$$

We assume the approximation of φ to be of the form

$$\varphi(\xi,c_x,t) = \sum_{\lambda=1}^{n} c_\lambda(\xi,t)\Omega_\lambda(c_x) \tag{11.83}$$

with Ω_λ given by (11.81), and we construct the residual

$$R = \sum_{\lambda=1}^{n}\left(\beta\frac{\partial c_\lambda}{\partial t}(\xi,t) + c_x\frac{\partial c_\lambda}{\partial \xi}(\xi,t) + \alpha c_\lambda(\xi,t)\right)\Omega_\lambda(c_x) - \frac{\alpha}{\sqrt{\pi}}\sum_{\lambda=1}^{n} G_\lambda c_\lambda(\xi,t) \tag{11.84}$$

where

$$G_\lambda = \int_{-\infty}^{\infty} \exp(-\zeta^2)\Omega_\lambda(\zeta)\,d\zeta \tag{11.85}$$

Equation (11.64) now becomes

$$a_{\psi\lambda}\frac{\partial c_\lambda}{\partial t} + b_{\psi\lambda}\frac{\partial c_\lambda}{\partial \xi} + a^*_{\psi\lambda}c_\lambda = 0 \tag{11.86}$$

where

$$\begin{aligned} a_{\psi\lambda} &= \beta\int_{-\infty}^{\infty} W_\psi(\zeta)\Omega_\lambda(\zeta)\,d\zeta \\ b_{\psi\lambda} &= \alpha\int_{-\infty}^{\infty} \zeta W_\psi(\zeta)\Omega_\lambda(\zeta)\,d\zeta \\ a^*_{\psi\lambda} &= \frac{\alpha}{\beta}a_{\psi\lambda} - \frac{\alpha}{\sqrt{\pi}}\int_{-\infty}^{\infty} W_\psi(\zeta)\,d\zeta G_\lambda \end{aligned} \tag{11.87}$$

In view of the orthogonality of the functions Ω_λ, we find it convenient to choose $W_\beta(\zeta) = \Omega_\beta(\zeta)$ (that is, Galerkin's method). Also choosing, for illustration purposes, $n = 2$ in (11.83), we find that

$$\frac{\partial c_\lambda}{\partial t} + E_{\lambda\psi}\frac{\partial c_\psi}{\partial \zeta} + D_{\lambda\psi}c_\psi = 0 \tag{11.88}$$

where

$$[E_{\lambda\psi}] = \frac{\alpha}{2\beta}\begin{bmatrix} 0 & 4 \\ 1 & 0 \end{bmatrix} \qquad [D_{\lambda\psi}] = \frac{\alpha}{4\beta}\begin{bmatrix} 4 & 0 \\ 0 & b \end{bmatrix} \tag{11.89}$$

with $b = 4 + \sqrt{8/27}$. Equations (11.79) now become

$$\begin{aligned} \frac{q_z(\xi,t)}{U/2} &= -\frac{2}{3\sqrt{3}}c_2(\xi,t) \\ \frac{\tau_{xz}(\xi,t)}{\tau_{xz_{fm}}} &= +\frac{2\sqrt{2\pi}}{3}c_1(\xi,t) \end{aligned} \tag{11.90}$$

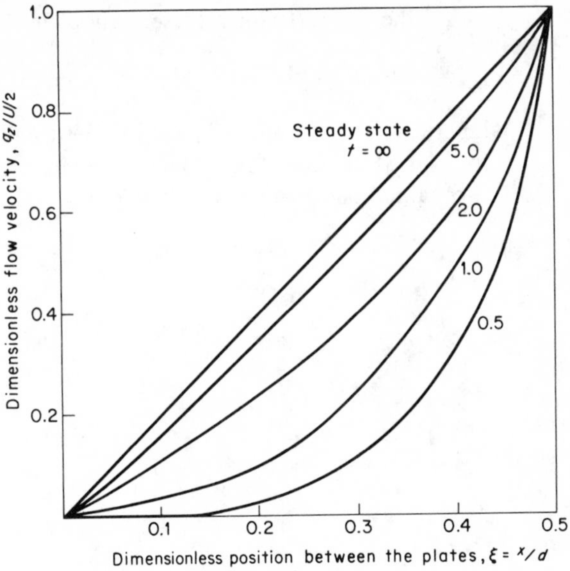

Fig. 11.4 Velocity profiles for various times—transient Couette flow.

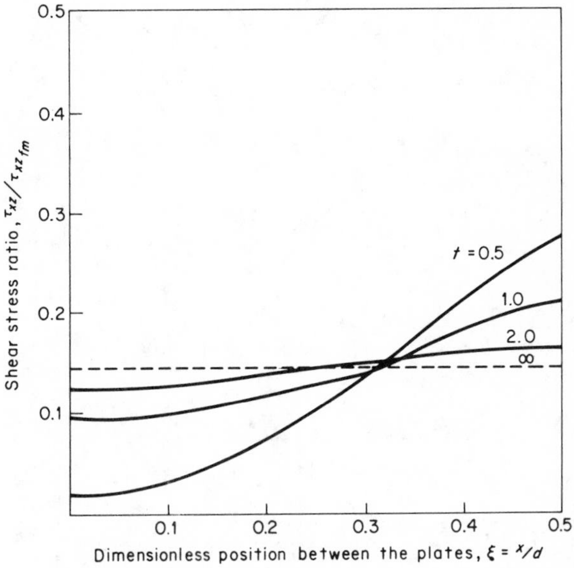

Fig. 11.5 Shear-stress profiles for various times—transient Couette flow.

Finally, the equation for a typical finite element becomes

$$A_{MN}\dot{c}^{N}_{\lambda(e)} + (E_{\lambda\psi}B_{MN} + D_{\lambda\psi}A_{MN})c^{N}_{\psi(e)} = 0 \qquad (11.91)$$

Numerical results obtained using (11.90) and (11.91) with $n = 2$ are shown in Figs. 11.4 and 11.5. These are in very good agreement with solutions obtained using other methods.†

† See, for example, Huang and Giddens [1967].

III

Thermomechanical Behavior of Finite Elements of Continuous Media

12 THERMODYNAMIC PRELIMINARIES

Whereas the general procedures for constructing discrete models discussed previously can, in principle, be applied to any continuous field, we shall primarily be concerned with thermomechanical phenomena, for it is here that most important problems of nonlinear solid mechanics await attention. In the thermomechanical principles, we find kinematic and kinetic variables bound naturally with other quantities that together depict thermodynamic states of bodies. Global energy balances may be used to augment the local balances of linear and angular momentum. These can then be used to develop finite-element equations which ensure that the basic physical laws (e.g., Cauchy's laws of motion) are satisfied, at least in an average sense, for finite volumes of the body.

To set the stage for the development of general energy balances and equations of motion for finite elements of nonlinear continua, we now review briefly certain thermodynamic concepts and principles.†

12.1 BASIC DEFINITIONS AND CONCEPTS

In the language of thermodynamics, a specific portion of the physical universe or a specific quantity of matter that we identify for study is called a *system*. If a system does not exchange matter with its surroundings, we say that it is *closed;* if it does not interact with its surroundings, we say that it is *isolated.* Conceptually, then, a certain quantity of matter which we isolate for study, independent of other quantities of matter, and which neither interacts with nor exchanges matter with its surroundings is a closed, isolated system.

The totality of information which completely characterizes a system for whatever purpose we may have in mind is said to describe the *state* of the system. In general, the information which characterizes a particular state of a system at a given time is available to us in the form of values of various functions called *state parameters* or *state variables*, each of which may describe a different property of the system.

For virtually every thermomechanical system, a natural and fundamental characteristic of its state is how hot or cold it is at a given time. The familiar concept of *temperature* abstracts this degree of hotness or coldness. The temperature θ is a real number indicated on a thermometer.‡ If a body gets hotter, its temperature rises; if it gets colder, its temperature decreases—but always the scale with which we measure the temperature is arbitrary.

Experimental evidence suggests that, regardless of the scale used for a thermometer, there is a temperature below which no system can be cooled.

† Discussions of the foundations of the subject as it pertains to continuous media have been given by Coleman [1964], Truesdell [1966*a*, 1966*b*], and Eringen [1967]. See, in particular, the monograph of Truesdell [1969]. Summary accounts of applications to deformable bodies can be found in several books and papers. Among these are Biot [1955, 1956], Boley and Weiner [1960], and Fung [1965]. For discussions of classical thermodynamics, see Hatsopoulas and Keenan [1965]. Discussions of applied thermodynamics of irreversible processes have been given by Meixner [1943], Eckart [1948], Prigogine [1947, 1961], and DeGroot and Mazur [1962].

‡ In some works on classical thermostatics, the concept of temperature is introduced in connection with the idea of equilibrium states of a system. For example, if the state variables of a system are not explicitly dependent on time and the system does not tend to change of its own accord (i.e., if further change must be brought about by external means) then the system is said to be in a state of equilibrium. Two systems which are each "in equilibrium with" a third system are assumed to be in equilibrium with each other, a postulate usually referred to as the *zeroth law of thermodynamics*. It can then be shown to follow that equilibrium of three systems implies that the three have in common a certain state variable called the temperature. Any one of the three (or more) systems then plays the role of the thermometer, which reads the temperature on a suitable but arbitrary scale. See Whaples [1952].

The temperature, in other words, is bounded below. If the greatest lower bound is assigned the value 0, then the temperature is said to be *absolute* and for any scale whatsoever

$$\theta > 0 \tag{12.1}$$

We define work in the usual way; e.g., the work done by a force $\mathbf{F}$ in moving through a displacement $d\mathbf{u}$ is $\mathbf{F} \cdot d\mathbf{u}$. In general systems, of course, work can be manifested in many ways.

In studying the mechanical behavior of continuous bodies, we encounter the *mechanical power* Ω developed by the external forces acting on the body. Power is a measure of the rate at which these forces perform work ($\Omega = \mathbf{F} \cdot d\mathbf{u}/dt$). If $\mathbf{F}$ and $\mathbf{S}$ denote the body force per unit mass and the surface traction per unit surface area acting on a continuous body of volume v, surface area A, and mass density ρ, then we assert that the mechanical power Ω of the body is given by

$$\Omega = \int_v \rho \mathbf{F} \cdot \mathbf{v} \, dv + \int_A \mathbf{S} \cdot \mathbf{v} \, dA \tag{12.2}$$

where $\mathbf{v} = \mathbf{v}(\mathbf{x},t)$ is the velocity field.

Closely related to work is the familiar concept of energy. Energy, in simple terms, is the capacity to do work. It is an additive scalar measure assigned to quantities of matter which is indicative of the capacity of all forces that may be associated with the quantity of matter (or with any of its parts) to perform work. Work is actually performed through a change in energy, and various classifications of energy reflect the nature of the forces which may be freed to perform work when there occurs a change of energy.

The energy possessed by a mechanical system by virtue of its motion is called *kinetic energy*, or energy of motion. In continuum mechanics, the kinetic energy κ of a continuous body of volume v and mass density ρ is defined by

$$\kappa = \frac{1}{2} \int_v \rho \mathbf{v} \cdot \mathbf{v} \, dv \tag{12.3}$$

In addition to kinetic energy, a body may possess *internal energy* U due to the work capacity of whatever forces may be present within it. In the case of continuous bodies, we assume that there exists an *internal-energy density* ϵ, which represents the internal energy per unit mass of the body. Then

$$U = \int_v \rho \epsilon \, dv \tag{12.4}$$

Elementary treatments of mechanics often discuss purely mechanical behavior of solid bodies which involves only changes in the kinetic and internal energy and the mechanical power developed by external forces.

In nature, however, the behavior of solid bodies seldom involves mechanical effects alone. When work is done upon a body, it generally becomes hotter. Conversely, when two noninsulated bodies at different temperatures are brought in contact with one another, experience shows that energy is transferred from one to the other, which may cause work to be done. This transition of energy due to a temperature difference is called *heat* or *heat working* and is denoted Q.

The total heat supplied to a body may arise from sources within the body plus the efflux of heat into the body from its surroundings. To measure the former, we introduce a *heat density* r, which represents the heat per unit mass introduced through heat sources within the body. To measure the latter, we consider an arbitrary element of surface area dA of the body with outer unit normal $\mathbf{n}$. The heat input across dA is denoted $\mathbf{q} \cdot \mathbf{n}\, dA$, where $\mathbf{q}$ is called the *heat-flux vector* or simply the *heat flux*. The total heat is then given by

$$Q = \int_v \rho r \, dv + \int_A \mathbf{q} \cdot \mathbf{n} \, dA \tag{12.5}$$

12.2 CONSERVATION OF ENERGY; THE FIRST LAW

Experience indicates that all the various energies of a system are additive and that whatever changes in energy occur and whatever the rate of working and heating, the internal energy is such that it balances what is left over. This law of conservation of energy is postulated as a fundamental axiom of mechanics:†

The principle of conservation of energy. The time rate of change of the kinetic energy plus the internal energy is equal to the rate of work done on the system plus the changes of all other energies of the system per unit time:

$$\dot{\kappa} + \dot{U} = \Omega + \sum_{\alpha} E_{\alpha} \tag{12.6}$$

where, as before, the superposed dot (.) indicates a time rate and where E_α ($\alpha = 1, 2, \ldots$) represents the mechanical equivalent of the αth kind of energy per unit time‡ (e.g., heat energy, electrical energy, chemical energy).

In studying thermomechanical phenomena, we are principally concerned with mechanical energies and their interconvertibility with heat. Then, of all the energies per unit time E_α, we consider only the heat Q, so that (12.6) becomes

$$\dot{\kappa} + \dot{U} = \Omega + Q \tag{12.7}$$

† Recall the three other fundamental axioms of mechanics mentioned in Art. 5.1.
‡ See Eringen [1962].

This result is known as the *first law of thermodynamics:* it is fundamental to our study of the thermomechanical behavior of continuous bodies.

Equation (12.7) is sometimes referred to as the *global form* of the first law because it pertains to a finite volume of material. Under sufficient conditions of smoothness, the Green-Gauss theorem can be used to obtain a *local form* of the first law that represents the energy balance at a point in a continuum. To obtain this local form, we follow the notation set down in Chap. I and consider a solid body in a current configuration C. A system of intrinsic coordinates x^i, originally rectangular cartesian in configuration C_0, are identified which have as natural base vectors the reciprocal vectors $\mathbf{G}_i$ and $\mathbf{G}^i$ defined previously. In the reference configuration, a set of orthonormal vectors $\mathbf{i}_i$ provides a basis, and the rectangular (spatial) coordinates of a point in C which were x^i in C_0 are, as before, denoted z_i. The velocity field $\mathbf{v}$, acceleration field $\mathbf{a}$, and heat-flux field $\mathbf{q}$ are given by

$$\mathbf{v} = v^m\mathbf{G}_m = v_m\mathbf{G}^m = \dot{u}_m\mathbf{i}_m \tag{12.8a}$$

$$\mathbf{a} = a^m\mathbf{G}_m = a_m\mathbf{G}^m = \ddot{u}_m\mathbf{i}_m \tag{12.8b}$$

$$\mathbf{q} = Q^m\mathbf{G}_m = Q_m\mathbf{G}^m = q_m\mathbf{i}_m \tag{12.8c}$$

where u_m and q_m are the rectangular cartesian components of displacement and heat flux with respect to x^i in the initial configuration C_0.

Recalling from (5.12) and (5.13) that

$$\mathbf{S} = S^j\mathbf{G}_j = \boldsymbol{\sigma} = \sigma^{ij}n_i\mathbf{G}_j \tag{12.9}$$

we see that

$$\begin{aligned}\Omega &= \int_v \rho F^j v_j\,dv + \int_A \sigma^{ij}v_j n_i\,dA \\ &= \int_v (\rho F^j v_j + \sigma^{ij}_{;i}v_j + \sigma^{ij}v_{j;i})\,dv\end{aligned} \tag{12.10}$$

where the semicolon indicates covariant differentiation with respect to the convected coordinate lines x^i. Similarly,

$$\begin{aligned}Q &= \int_v \rho r\,dv + \int_A Q^j n_j\,dA \\ &= \int_v (\rho r + \nabla\cdot\mathbf{q})\,dv\end{aligned} \tag{12.11}$$

where $\nabla\cdot\mathbf{q} = Q^k_{;k}$, and

$$\dot{\kappa} = \frac{1}{2}\frac{d}{dt}\int_v \rho v^j v_j\,dv = \int_v [\rho a^j v_j\,dv + \tfrac{1}{2}v^j v_j\overset{\cdot}{\overline{(\rho\,dv)}}]$$

which, due to the principle of conservation of mass, leads to

$$\dot{\kappa} = \int_v \rho a^j v_j \, dv \tag{12.12}$$

Introducing (12.4), (12.10), (12.11), and (12.12) into (12.7) and simplifying and collecting terms, we obtain

$$\int_v [(\sigma^{ij}_{;i} + \rho F^j - \rho a^j)v_j - (\rho\dot{\epsilon} - \sigma^{ij}v_{j;i} - Q^j_{;j} - \rho r)] \, dv = 0 \tag{12.13}$$

We recognize the terms in the first set of parentheses as those in (5.21), Cauchy's first law of motion. Thus, these terms vanish if the principle of balance of linear momentum holds. The remaining integral must still vanish for arbitrary volumes. Since the integrand is continuous by hypothesis, the terms in the second set of parentheses must also vanish, and we arrive at the local form of the first law of thermodynamics:

$$\rho\dot{\epsilon} = \sigma^{ij}v_{j;i} + Q^j_{;j} + \rho r \tag{12.14}$$

The components of vectors and tensors in (12.14) are referred to the coordinates x^i in the deformed body. Several alternate forms of (12.14) can be derived. For example, we note that the components v^i, v_i of (12.8*a*) are related to the rectangular components of velocity referred to the cartesian coordinates of the reference configuration as follows:

$$v_i = z_{m,i}\dot{u}_m \qquad v^i = \frac{\partial x^i}{\partial z_m}\dot{u}_m \tag{12.15a,b}$$

where the comma denotes ordinary partial differentiation with respect to x_i. Moreover,

$$v_{j;i} = (z_{m,j}\dot{u}_m)_{;i} = z_{m,j}\dot{u}_{m,i} \tag{12.16}$$

and in view of (4.44), (4.45), and the symmetry of σ^{ij},

$$\begin{aligned} \sigma^{ij}v_{j;i} &= \sigma^{ij}z_{m,j}\dot{u}_{m,i} \\ &= \sigma^{ij}(\delta_{mj} + u_{m,j})\dot{u}_{m,i} \\ &= \sigma^{ij}\dot{\gamma}_{ij} \end{aligned} \tag{12.17}$$

where γ_{ij} is the strain tensor of (4.16) and (4.17). The quantity $\sigma^{ij}v_{j;i} = \sigma^{ij}\dot{\gamma}_{ij}$ is sometimes referred to as the *stress power*. Further, if we define $\mathbf{q}$ so as to represent the heat-flux vector per unit area of the undeformed body, then

$$q_i = \sqrt{G}\,\delta_{im}Q^m \tag{12.18}$$

Introducing (12.17) and (12.18) into (12.14), we get

$$\rho\dot{\epsilon} = \sigma^{ij}\dot{\gamma}_{ij} + \frac{1}{\sqrt{G}}q_{i,i} + \rho r \tag{12.19}$$

In view of (12.4) and (12.5), it is clear from (12.19) that

$$\dot{U} = \int_v \sigma^{ij}\dot{\gamma}_{ij}\,dv + Q \tag{12.20}$$

Thus (12.7) can also be written in the form

$$\dot{\kappa} + \int_v \sigma^{ij}\dot{\gamma}_{ij}\,dv = \Omega \tag{12.21}$$

Another form of the local energy balance is also of interest. Let e and h denote the internal energy and the rate of heat supply per unit of undeformed volume v_0, respectively. Then

$$e = \rho_0\epsilon \qquad \text{and} \qquad h = \rho_0 r \tag{12.22a,b}$$

Recalling from (5.14) that the stress per unit area of the undeformed body referred to the coordinates x^i in the deformed body is $t^{ij} = \sqrt{G}\sigma^{ij}$, and introducing (12.22) into (12.19), we arrive at the alternate form of (12.14):

$$\dot{e} = t^{ij}\dot{\gamma}_{ij} + q_{i,i} + h \tag{12.23}$$

12.3 ENTROPY; THE CLAUSIUS-DUHEM INEQUALITY

Together with the absolute temperature θ, we place the entropy H as a fundamental property of all thermodynamic systems.† Like energy, it is important to us only through its changes, which traditionally provide a bound on Q/θ. We assume that the entropy is an additive function which is inherent in all quantities of matter. The total entropy of a body is then the sum of the entropies of its parts. For continuous bodies, we assume further that H is absolutely continuous with respect to the mass and that there exists a function η, called the *specific entropy* or *entropy per unit mass*, such that

$$H = \int_v \rho\eta\,dv \tag{12.24}$$

For a continuous body, we assume that the total change in entropy per unit time can be brought about not only by changes in H but also by the influx of entropy through the boundary surfaces of the body, plus, possibly, the supply of entropy from sources within the body. We define the *total*

† In classical thermostatics, the concept of entropy is usually introduced in an attempt to make the variation $Q\,dt$ in heat energy an exact differential. Traditionally, this is accomplished by dividing $Q\,dt$ by the absolute temperature; the result, $Q\,dt/\theta$, is referred to as dH, and H is called the entropy. Unfortunately, this definition applies only to a very restricted class of thermodynamic phenomena and is meaningless for irreversible processes. We prefer to treat entropy as a primitive concept, a fundamental property of matter, and to identify its role in the thermodynamics of deformation through the Clausius-Duhem inequality and its consequences. Similar developments have been given by Coleman and Noll [1963] and Coleman and Mizel [1964], Truesdell [1966*a*, 1969], and Eringen [1967].

entropy production of the body to be

$$\Gamma \equiv \dot{H} - \int_A \mathbf{e} \cdot \mathbf{n} \, dA - \int_v \rho s \, dv \tag{12.25}$$

where $\mathbf{e}$ is the *entropy-flow vector* and s is the *entropy supply per unit mass per unit time due to internal sources.* The postulate that Γ is nonnegative is a fundamental axiom of thermodynamics:

The Clausius-Duhem inequality. The total entropy production Γ *is always greater than or equal to zero:*

$$\Gamma \geq 0 \tag{12.26}$$

Within the present framework, (12.26) is a statement of the *second law of thermodynamics.* According to (12.25), the Clausius-Duhem inequality implies that

$$\dot{H} \geq \int_A \mathbf{e} \cdot \mathbf{n} \, dA + \int_v \rho s \, dA \tag{12.27}$$

that is, the time rate of change of the total entropy H is never less than the sum of influx of entropy through the surfaces of the body and the production of entropy by internal sources. Also, under suitable smoothness assumptions, we see that (12.24) and (12.27) yield

$$\rho \dot{\eta} \geq \nabla \cdot \mathbf{e} + \rho s \tag{12.28}$$

The quantities $\mathbf{e}$ and s can always be expressed in the form†

$$\mathbf{e} = \frac{\mathbf{q}}{\theta} + \bar{\mathbf{e}} \qquad s = \frac{r}{\theta} + \bar{s} \tag{12.29a,b}$$

where $\mathbf{q}/\theta$ is the influx of entropy due to the influx of heat, r/θ is the entropy production of internal sources due to the supply of heat from such sources, and $\bar{\mathbf{e}}$ and $\bar{s}$ represent the entropy flux and source productions from all other effects. Since we are primarily interested in thermomechanical behavior, it will suit our purposes to set $\bar{\mathbf{e}} = \mathbf{0}$ and $\bar{s} = 0$ and to take

$$\mathbf{e} = \frac{\mathbf{q}}{\theta} \qquad s = \frac{r}{\theta} \tag{12.30a,b}$$

† In most of the literature, the quantities $\mathbf{e}$ and s are taken to be $\mathbf{q}/\theta$ and h/θ *ab initio*. See, for example, Coleman and Noll [1963], Coleman and Mizel [1964], or Truesdell [1966*a*]. Eringen [1967, pp. 128–129] introduces the idea of $\bar{\mathbf{e}}$ and $\bar{s}$ and refers to a situation in which $\bar{\mathbf{e}} = \mathbf{0}$ and $\bar{s} = 0$ as a "simple thermomechanical process." Müller [1967] investigates the case in which $\bar{\mathbf{e}}, \bar{s} \neq 0$ in some detail and treats $\mathbf{e}$ as a fundamental dependent variable alongside the heat flux $\mathbf{q}$.

Introducing (12.30) into (12.28), we arrive at the inequality

$$\rho\theta\dot{\eta} \geq \nabla \cdot \mathbf{q} - \frac{1}{\theta}\mathbf{q} \cdot \nabla\theta + \rho r \tag{12.31}$$

which may be interpreted as a local form of the Clausius-Duhem inequality for cases in which Eqs. (12.30) hold.

12.4 FREE ENERGY AND INTERNAL DISSIPATION

In the theory of heat-conducting materials, it is convenient to introduce the *specific free energy* ψ, defined by

$$\psi \equiv \epsilon - \eta\theta \tag{12.32}$$

and the internal dissipation δ, defined by

$$\delta \equiv \sigma^{ij}\dot{\gamma}_{ij} - \rho(\dot{\psi} + \eta\dot{\theta}) \tag{12.33}$$

From definitions (12.32) and (12.33) follow the pair of relations

$$\rho\dot{\epsilon} = \sigma^{ij}\dot{\gamma}_{ij} + \rho\theta\dot{\eta} - \delta \tag{12.34}$$

and

$$\rho\dot{\psi} = \sigma^{ij}\dot{\gamma}_{ij} - \rho\eta\dot{\theta} - \delta \tag{12.35}$$

Substituting (12.34) into the local energy balance in (12.14), we obtain the important result

$$\rho\theta\dot{\eta} = \nabla \cdot \mathbf{q} + \rho r + \delta \tag{12.36}$$

which represents a local form of the conservation of energy in terms of the specified entropy η, internal dissipation δ, and heat variables $\mathbf{q}$ and r. Introducing (12.36) into (12.31), we arrive at the *general dissipation inequality*†

$$\frac{1}{\theta}\mathbf{q} \cdot \nabla\theta + \delta \geq 0 \tag{12.37}$$

Alternatively, it is customary to also introduce the *Clausius-Planck inequality*‡

$$\delta = \rho\theta\dot{\eta} - \nabla \cdot \mathbf{q} - \rho r \geq 0 \tag{12.38}$$

† See Coleman [1964].

‡ This terminology is used by Truesdell [1966*a*, p. 245]. The relation (12.38) does not, in general, follow from (12.37) except in certain cases, e.g., quasi-elastic response. In classical thermostatics, the term δ does not appear in (12.37) and the resulting inequality asserts that heat never flows against a temperature gradient. Truesdell [1966*a*] comments that the presence of δ in (12.37) implies that heat can indeed flow against a temperature gradient if a sufficient amount of local energy is supplied. We shall use (12.37) as the fundamental dissipation inequality.

We also remark that alternate forms of the above relations can be obtained which involve quantities referred to the undeformed (reference) configuration. Specifically, let ξ, φ, and σ denote the entropy, free energy, and dissipation per unit volume in the reference configuration:

$$\xi = \rho_0 \eta \tag{12.39}$$

$$\varphi = \rho_0 \psi \tag{12.40}$$

$$\sigma = \sqrt{G}\delta \tag{12.41}$$

Then

$$\varphi = e - \xi\theta \tag{12.42}$$

$$\sigma = t^{ij}\dot{\gamma}_{ij} - (\varphi + \xi\dot{\theta}) \tag{12.43}$$

and, instead of (12.34) to (12.36), we have

$$\dot{e} = t^{ij}\dot{\gamma}_{ij} + \theta\dot{\xi} - \sigma \tag{12.44}$$

$$\dot{\varphi} = t^{ij}\dot{\gamma}_{ij} - \xi\dot{\theta} - \sigma \tag{12.45}$$

$$\dot{\xi} = q_{i,i} + h + \sigma \tag{12.46}$$

12.5 THERMODYNAMIC PROCESSES

We define a *thermodynamic process* as a set of functions, including the motion $z_i(\mathbf{x},t)$ and the fields $\theta(\mathbf{x},t)$, $\sigma^{ij}(\mathbf{x},t)$, $\mathbf{q}(\mathbf{x},t)$, and $\eta(\mathbf{x},t)$ defined at $\mathbf{x}$ at time t, provided these satisfy the principles of linear and angular momentum and conservation of energy for every part of the body. For a given time t, these functions define the thermodynamic state of the body. We adopt the usual terminology for the classification of thermodynamic processes: a process is called *isothermal*, *adiabatic*, *isentropic*, or *isoenergetic*, respectively, if $\dot{\theta}$, Q, $\dot{H}$, or $\dot{U}$ is equal to zero. If the state functions, including the deformation gradients, are independent of $\mathbf{x}$, the process is said to be *homogeneous*.

13 THERMOMECHANICS OF A FINITE ELEMENT

With the basic mechanical and thermodynamical principles and the general concept of finite-element models of continuous fields behind us, we now bring all these ideas together to obtain general finite-element formulations describing the behavior of a finite-element model of a continuous medium.

13.1 KINEMATICS OF A FINITE ELEMENT

Consider a continuous body which, in a reference configuration C_0, occupies a region $\mathscr{R}$ in three-dimensional euclidean space. The geometry and the external constraints on the body are arbitrary, and the body may be subjected to an arbitrary system of external forces and heat supply. To describe the motion of the body, we establish, as before, a system of intrinsic *global* coordinates, denoted now X^i, which, for simplicity, we temporarily take to be rectangular cartesian in the reference configuration C_0. The rectangular cartesian coordinates of a particle $\mathbf{x}$ in an arbitrary configuration C at some time $t \geq 0$, which was initially located at the point $\mathbf{X} = \mathbf{X}(X^i)$ in C_0, are denoted Z_i, and the functions

$$Z_i = Z_i(\mathbf{X},t) \tag{13.1}$$

are said to define the motion of the body.

Following the procedures described in Chap. II, we now construct a discrete model of the body which consists of a finite number E of finite elements connected appropriately at their nodes. Within each finite element $\imath_e$ of the body, we establish a system of intrinsic *local* coordinates $x^i_{(e)}$ which, for simplicity, we temporarily take to be rectangular cartesian in the local reference configuration $C_{0(e)}$ of element e. The rectangular cartesian coordinates of a particle $\mathbf{x}$ in an arbitrary configuration $C_{(e)}$ of element $\imath_e$ at some time $t \geq 0$, which was initially at the point $\mathbf{x} = \mathbf{x}(x^i)$ in $C_{0(e)}$, are denoted $z_{i(e)}$, and the functions

$$z_{i(e)} = z_{i(e)}(\mathbf{x}_{(e)},t) \tag{13.2}$$

are said to define the *motion of element e*.

For the present, the precise shape of each element, the number of nodes, and the mode of connection are unimportant, but we take for granted that the basic compatibility requirements are satisfied and that, therefore, the collection of elements fits properly together to form a connected whole. This being the case, if $X^{\Delta i}$ denote the global coordinates of a specific node in the global system and $x^{Ni}_{(e)}$ denote the local coordinates of the *same* node in an element $\imath_e$, then the connectivity of the model is established through the mapping

$$\hat{x}^{Ni}_{(e)} = \overset{(e)}{\Omega}{}^N_\Delta X^{\Delta i} \tag{13.3}$$

where

$$\hat{x}^{Ni}_{(e)} = \overset{(e)}{\alpha}{}^i_j x^{Nj}_{(e)} \tag{13.4}$$

$\overset{(e)}{\Omega}{}^N_\Delta$ is the usual incidence array defined in (7.6), $i = 1, 2, 3$; $N = 1, 2, \ldots, N_e$; $\Delta = 1, 2, \ldots, G$; $e = 1, 2, \ldots, E$. In (13.4), $\overset{(e)}{\alpha}{}^i_j$ are the direction cosines of the angles between X^i and $\hat{x}^j_{(e)}$ and $\overset{(e)}{\alpha}{}^i_m \overset{(e)}{\alpha}{}^j_m = \delta^{ij}$, $\det \overset{(e)}{\alpha}{}^i_j = 1$. Similarly, if compatibility requirements are met throughout the motion,

$$\hat{z}^{Ni}_{(e)} = \overset{(e)}{\Omega}{}^N_\Delta Z^{\Delta i} \tag{13.5}$$

and

$$\hat{z}^{Ni}_{(e)} = \overset{(e)}{\alpha}{}^{i}_{j} z^{Nj}_{(e)} \tag{13.6}$$

where $Z^{\Delta i} = Z^{\Delta}_{i}$ are the rectangular coordinates in C of a node which was located by $X^{\Delta i}$ in C_0 and $z^{Nj}_{(e)}$ are the corresponding local coordinates of the same node in $C_{(e)}$ which was initially given by $x^{Ni}_{(e)}$ in $C_{0(e)}$.

The *global displacement* field $U_i = U_i(\mathbf{X},t)$ is defined by the displacement components

$$U_i = Z_i - X_i \tag{13.7}$$

with $X_i = X^i$, and the *local displacement field corresponding to element* $\imath_e$ is

$$u_{i(e)} = z_{i(e)} - x_{i(e)} \tag{13.8}$$

Thus, the displacement components of a node point Δ in the global system are

$$U^{\Delta}_{i} = U_i(\mathbf{X}^{\Delta}) = Z^{\Delta}_{i} - X^{\Delta}_{i} \tag{13.9}$$

and the displacement components of a node point N of element $\imath_e$ are

$$u^{N}_{i(e)} = u^{(e)}_{i}(\mathbf{x}^{N}_{(e)}) = z^{N}_{i(e)} - x^{N}_{i(e)} \tag{13.10}$$

It follows from (13.4), (13.6), and (13.10) that if $\hat{u}^{N}_{i(e)} = \overset{(e)}{\alpha}{}^{i}_{j} u^{N}_{j(e)}$, then

$$\hat{u}^{N}_{i(e)} = \overset{(e)}{\Omega}{}^{N}_{\Delta} U^{\Delta}_{i} \tag{13.11}$$

To complete the construction of the finite model of the displacement field $\mathbf{U}$, it is necessary that we now introduce appropriate approximating functions for each finite element. For simplicity, we consider only first-order representations, the development of higher-order representations being a more involved but straightforward application of procedures outlined in Art. 8. Thus, for element $\imath_e$, we take

$$u^{(e)}_{i} \approx \bar{u}^{(e)}_{i} = u^{N}_{i(e)} \psi^{(e)}_{N}(\mathbf{x}) \tag{13.12}$$

where $\psi^{(e)}_{N}(\mathbf{x})$ are the local interpolation functions for the element. Accordingly,

$$U_i \approx \bar{U}_i = \sum_e \hat{u}_{i(e)} = \sum_e \overset{(e)}{\alpha}{}^{i}_{j} u^{N}_{j(e)} \psi^{(e)}_{N}(\mathbf{x}) \tag{13.13}$$

or, in view of (13.11),

$$\bar{U}_i = \sum_e \overset{(e)}{\Omega}{}^{N}_{\Delta} \psi^{(e)}_{N}(\mathbf{x}) U^{\Delta}_{i} = \Phi_{\Delta}(\mathbf{x}) U^{\Delta}_{i} \tag{13.14}$$

Equations (13.12) and (13.14) describe the discrete model of the displacement field $\mathbf{U}$. It is now a simple matter to calculate the velocity field $\mathbf{V}$

and the acceleration field **A**:

$$V_i \equiv \dot{U}_i = \sum_e \psi_N^{(e)}(\mathbf{x}) \overset{(e)}{\Omega}{}_\Delta^N \dot{U}_i^\Delta \tag{13.15}$$

$$A_i \equiv \dot{V}_i = \ddot{U}_i = \sum_e \psi_N^{(e)}(\mathbf{x}) \overset{(e)}{\Omega}{}_\Delta^N \ddot{U}_i^\Delta \tag{13.16}$$

where $\dot{U}_i^\Delta$ and $\ddot{U}_i^\Delta$ are the components of velocity and acceleration at node Δ of the global connected system, that is,

$$V_i^\Delta = \dot{U}_i^\Delta \qquad A_i^\Delta = \ddot{U}_i^\Delta \tag{13.17a,b}$$

We also observe that

$$v_{i(e)} = v_{i(e)}^N \psi_N^{(e)}(\mathbf{x}) \qquad a_{i(e)} = a_{i(e)}^N \psi_N^{(e)}(\mathbf{x}) \tag{13.18a,b}$$

where

$$v_{i(e)}^N = \overset{(e)}{\Omega}{}_\Delta^N V_i^\Delta = \dot{u}_{i(e)}^N \tag{13.19a}$$

$$a_{i(e)}^N = \overset{(e)}{\Omega}{}_\Delta^N A_i^\Delta = \ddot{u}_{i(e)}^N \tag{13.19b}$$

In view of (13.15) and (13.16), the finite-element representation effectively separates the displacement, velocity, and acceleration fields into products of functions of spatial variables only, with the nodal values which are functions of time only. Clearly, higher-order time derivatives of **U** are obtained by successive differentiations of (13.14):

$$\overset{n}{U}_i \approx \sum_e \psi_N^{(e)}(\mathbf{x}) \overset{(e)}{\Omega}{}_\Delta^N \overset{n}{U}{}_i^\Delta \tag{13.20}$$

With the finite-element model of the displacement field now available, it is a simple matter to calculate any other quantities which characterize the deformation of the finite-element model of the body. For example, for a typical finite element $\imath_e$, the local deformation gradients are

$$z_{i,j}^{(e)} = u_{i(e)}^N \psi_N^{(e)}(\mathbf{x})_{,j} \tag{13.21}$$

The strain components, according to (4.17), are

$$\gamma_{ij}^{(e)} = \tfrac{1}{2}(u_{j(e)}^N \psi_{N,i}^{(e)} + u_{i(e)}^N \psi_{N,j}^{(e)} + u_{k(e)}^N u_{k(e)}^M \psi_{N,i}^{(e)} \psi_{M,j}^{(e)}) \tag{13.22}$$

and the components of strain rate are

$$\dot{\gamma}_{ij}^{(e)} = \tfrac{1}{2}[\dot{u}_{j(e)}^N \psi_{N,i}^{(e)} + \dot{u}_{i(e)}^N \psi_{N,j}^{(e)} + \psi_{N,i}^{(e)} \psi_{M,j}^{(e)} (\dot{u}_{k(e)}^N u_{k(e)}^M + u_{k(e)}^N \dot{u}_{k(e)}^M)] \tag{13.23}$$

Thus

$$\dot{\overline{ds_{(e)}^2}} = \psi_{N,i}^{(e)}(\delta_{jk} + u_{k(e)}^M \psi_{M,j}^{(e)}) \dot{u}_{k(e)}^N \, dx_{(e)}^i \, dx_{(e)}^j \tag{13.24}$$

The deformation tensor G_{ij} for the element is, of course,

$$G_{ij}^{(e)} = \delta_{ij} + 2\gamma_{ij}^{(e)} \tag{13.25}$$

so that the principal invariants of (4.25) can be obtained by simply introducing (13.22) into (4.26).

It is interesting to note that in the case of simplex models,

$$\psi_N^{(e)}(\mathbf{x}) = a_N^{(e)} + b_{Ni}^{(e)} x_{(e)}^i \tag{13.26}$$

where $a_N^{(e)}$ and $b_{Ni}^{(e)}$ are defined in (10.102) [and in (10.105), (10.106), and (10.108)]. Then

$$\psi_{N,j} = b_{Nj} \tag{13.27}$$

where we have omitted the element identification label (e) for simplicity. Thus, for simplex models,

$$z_{i,j}^{(e)} = b_{Nj} u_i^N \tag{13.28}$$

$$\gamma_{ij}^{(e)} = \tfrac{1}{2}(u_j^N b_{Ni} + u_i^N b_{Nj} + b_{Ni} b_{Mj} u_k^N u_k^M) \tag{13.29}$$

$$\dot{\gamma}_{ij}^{(e)} = \tfrac{1}{2}[\dot{u}_j^N b_{Ni} + \dot{u}_i^N b_{Nj} + b_{Ni} b_{Mj}(\dot{u}_k^N u_k^M + u_k^N \dot{u}_k^M)] \tag{13.30}$$

$$\overline{ds_{(e)}^2}^{\,\boldsymbol{\cdot}} = b_{Ni}(\delta_{jk} + b_{Mj} u_k^M)\dot{u}_k^N \, dx^i \, dx^j \tag{13.31}$$

$$G_{ij}^{(e)} = \delta_{ij} + b_{Ni} u_j^N + b_{Nj} u_i^N + b_{Ni} b_{Mj} u_k^N u_k^M \tag{13.32}$$

We see that in the case of simplex models, using rectangular cartesian components in the reference configuration, each finite element is in a state of *homogeneous strain*.† The deformation gradients and strains are uniform over each element.

13.2 ENERGY OF FINITE ELEMENTS

We now take advantage of the fundamental property of finite-element models: the elements $\imath_e$ can be considered disjoint and disconnected for the purpose of describing local approximations over each element. Thus, we can temporarily focus our attention on a typical finite element $\imath_e$ of an assemblage representing a continuous medium. We consider $\imath_e$ to be

† A deformation is called a *homogeneous strain* when every straight line is deformed into a straight line; see Truesdell and Toupin [1960, p. 285]. The deformation gradients in (13.28) are uniform over the element; planes are deformed into planes, and straight lines into straight lines. Simplex models of the displacement field, however, will not always lead to states of homogeneous strain. If, for example, the initial coordinates are not rectangular cartesian, the deformation corresponding to simplex models of the displacement will not, in general, be a homogeneous strain. For a cylindrical coordinate system (r, z, θ), for example, a displacement $u_r = k$ will create a circumferential strain which varies with r. There are several cases of homogeneous strain which are of interest (e.g., simple shear, uniform extension). In these cases, the finite-element model will yield an exact description of the kinematics of the deformation. See Oden [1968*b*, 1970*a*] and Oden and Aguirre-Ramirez [1969].

isolated from the rest of the system, and we again drop the element identification label (e) for simplicity and convenience, except in those instances where its omission may lead to confusion.

Consider, then, a typical finite element $\imath_e$ of mass density and volume $\rho_{(e)} = \rho_{(e)}(\mathbf{x})$ and $v_{(e)}$ in an arbitrary configuration $C_{(e)}$. In a reference configuration $C_{0(e)}$, these quantities are denoted $\rho_{0(e)} = \rho_{0(e)}(\mathbf{x})$ and $v_{0(e)}$, respectively. According to (13.12) and (13.18), the displacement, velocity, and acceleration fields for this element are of the form

$$u_i = u_i^N \psi_N(\mathbf{x}) \tag{13.33a}$$

$$\dot{u}_i = \dot{u}_i^N \psi_N(\mathbf{x}) \tag{13.33b}$$

$$\ddot{u}_i = \ddot{u}_i^N \psi_N(\mathbf{x}) \tag{13.33c}$$

the dependence of u_i^N on time being understood. Then all the kinematic relations for the element can be computed in terms of the nodal displacement components u_i^N and their time derivatives $\dot{u}_i^N$, $\ddot{u}_i^N$, as indicated in (13.22) to (13.25).

Kinetic energy It is now possible to calculate energies associated with the thermomechanical behavior of the element. Turning first to (12.3), we have for the kinetic energy $\kappa_{(e)}$ of the element,

$$\kappa_{(e)} = \frac{1}{2}\int_{v_{(e)}} \rho \mathbf{v} \cdot \mathbf{v}\, dv = \frac{1}{2}\int_{v_{0(e)}} \rho_0 \dot{\mathbf{u}} \cdot \dot{\mathbf{u}}\, dv_0 \tag{13.34}$$

In the present discussion, we prefer to refer the motion to the reference configuration $C_{0(e)}$ of the element. Thus, introducing (13.33*b*) into the second integral in (13.34), we find

$$\kappa_{(e)} = \frac{1}{2}\int_{v_{0(e)}} \rho_0 \dot{u}_i^N \psi_N \mathbf{i}_i \cdot \dot{u}_j^M \psi_M \mathbf{i}_j\, dv_0 \tag{13.35}$$

or

$$\kappa_{(e)} = \tfrac{1}{2} m_{NM} \dot{u}_i^N \dot{u}_i^M \tag{13.36}$$

where m_{NM} is a symmetric $N_e \times N_e$ matrix called the *consistent mass matrix*† for the element, defined by

$$m_{NM} = \int_{v_{0(e)}} \rho_0(\mathbf{x}) \psi_N(\mathbf{x}) \psi_M(\mathbf{x})\, dv_0 \tag{13.37}$$

† This terminology was introduced by Archer [1963], who used the term "consistent" to emphasize that the mass of the element is distributed to each node in a manner consistent with the assumed velocity field $\psi_N(\mathbf{x})\dot{\mathbf{u}}^N$, and not arbitrarily "lumped" at nodal points, as is often done. Later Archer [1965] extended these ideas to include *consistent forces*. Generalizations of these notions were examined by Oden [1969*a*], Oden and Aguirre-Ramirez [1969], and Brauchli and Oden [1971] and form the basis of the theory of conjugate approximations discussed in Art. 9.

Clearly, the matrix m_{NM} distributes the total mass of the element in the form of discrete nodal masses to each node in a manner that makes the total kinetic energy due to the motion of the N_e discrete nodal points equal to that developed by virtue of the continuous fields $\rho_0(\mathbf{x})$ and $\dot{\mathbf{u}}(\mathbf{x})$ over the element.

The time rate of change of kinetic energy for the element is

$$\dot{\kappa}_{(e)} = m_{NM}\ddot{u}_i^M\dot{u}_i^N \tag{13.38}$$

in which we used the property $m_{NM} = m_{MN}$. Notice that $\dot{\kappa}_{(e)}$ can be interpreted as the power developed by "inertia" forces $m_{NM}\ddot{u}_i^M$ at node N. Accepting this interpretation of $m_{NM}\ddot{u}_i^M$, we see that if the accelerations of all nodes are zero except (say) that of node K, then the inertia force at node N is $m_{NK}\ddot{u}_i^K$. If, in addition, $\ddot{u}_i^K = 1$, then the force at N is simply m_{NK}. Thus, *an element m_{NK} of the consistent mass matrix can be interpreted as the inertia force developed at node N due to a unit acceleration of node K.* Moreover, the symmetry property, $m_{MN} = m_{NM}$, can be interpreted as a dynamic form of Maxwell's reciprocity law: the force at N due to a unit acceleration at M is equal to the force at M due to a unit acceleration at N.

It is interesting to examine the form of (13.36) for the case in which simplex models are used. Then $\psi_N(\mathbf{x}) = a_N + b_{Ni}x^i$ and

$$m_{NM} = m_0 a_N a_M + (a_N b_{Mi} + b_{Ni}a_M)s^i + b_{Ni}b_{Mj}I^{ij} \tag{13.39}$$

where m_0 is the total mass of the element, s^i are the first moments of mass, and I^{ij} is the mass moment of inertia tensor with respect to the origin of the local coordinates x^i:

$$m_0 = \int_{v_{0(e)}} \rho_0(\mathbf{x})\,dv_0 \qquad s^i = \int_{v_{0(e)}} \rho_0(\mathbf{x})x^i\,dv_0 \tag{13.40a,b}$$

$$I^{ij} = \int_{v_{0(e)}} \rho_0(\mathbf{x})x^i x^j\,dv_0 \tag{13.40c}$$

The array $a_N a_M$ distributes the total mass appropriately to each node, whereas the array $b_{Ni}b_{Mj}$ distributes the rotary inertia of the element into discrete nodal equivalents.

We notice also that the quantity $\dot{\kappa}_{(e)}$ can be expressed as the scalar product

$$\dot{\kappa}_{(e)} = \int_{v_{0(e)}} \rho_0\mathbf{a}\cdot\dot{\mathbf{u}}^N\psi_N\,dv_0 \tag{13.41}$$

Thus, according to (9.181), the inertia force field $\rho_0\mathbf{a}$ is conjugate to the velocity field $\dot{\mathbf{u}}$ with respect to the time rate of change of kinetic energy.

Internal energy The total internal energy of the element is

$$U_{(e)} = \int_{v_{0(e)}} e\,dv_0 = \int_{v_{(e)}} \rho\epsilon\,dv \tag{13.42}$$

where e and ϵ are the specific internal energies described previously. The time rate of change of internal energy is

$$\dot{U}_{(e)} = \int_{v_{0(e)}} \dot{e}\, dv_0 \tag{13.43}$$

We recall from (12.23) that $\dot{e}$ can be expressed in terms of the stress power $t^{ij}\dot{\gamma}_{ij}$ and the heat flux and heat supply. Thus, we may also express $\dot{U}_{(e)}$ in the forms

$$\dot{U}_{(e)} = \int_{v_{0(e)}} t^{ij}\dot{\gamma}_{ij}\, dv_0 + Q_{(e)} \tag{13.44}$$

or

$$\dot{U}_{(e)} = \int_{v_{(e)}} \sigma^{ij}\dot{\gamma}_{ij}\, dv + Q_{(e)} \tag{13.45}$$

where $Q_{(e)}$ is the total heat (or heat working) of element e.

In view of (13.23),

$$t^{ij}\dot{\gamma}_{ij} = t^{ij}\psi_{N,i}(\delta_{jk} + u_k^M \psi_{M,j})\dot{u}_k^N \tag{13.46}$$

where the dependence of $\psi_{N,i}$ on $\mathbf{x}$ is understood. Thus

$$\dot{U}_{(e)} = \int_{v_{0(e)}} t^{ij}\psi_{N,i}(\delta_{jk} + u_k^M \psi_{M,j})\, dv_0 \dot{u}_k^N + Q_{(e)} \tag{13.47}$$

However, little else can be said of the specific form of $\dot{U}_{(e)}$ until the material of which the element is composed is characterized.

Mechanical power Returning to (12.2), we see that the external forces $\mathbf{F}$ and $\mathbf{S}$ are conjugate to the velocity field $\dot{\mathbf{u}}$ over the element, with respect to the mechanical power Ω. Let $\hat{\mathbf{F}}(\mathbf{x})$ denote the body force per unit of undeformed volume acting on the element, and let $\hat{\mathbf{S}}$ denote the surface force, referred to the convected coordinate lines x^i but measured per unit area A_0 of the reference configuration [$\hat{\mathbf{S}} = \mathbf{t} = t^{ij}\hat{n}_i\mathbf{G}_j$; see (5.15)]. Then the mechanical power developed by the external forces acting on the element is

$$\Omega_{(e)} = \int_{v_{0(e)}} \hat{\mathbf{F}}(\mathbf{x}) \cdot \psi_N(\mathbf{x})\dot{\mathbf{u}}^N\, dv_0 + \int_{A_{0(e)}} \hat{\mathbf{S}} \cdot \psi_N(\mathbf{x})\dot{\mathbf{u}}^N\, dA_0 \tag{13.48}$$

where $A_0 = A_{0(e)}$ is the surface area of the element while in the reference configuration. Following the procedure used to reduce the kinetic energy, we write

$$\Omega_{(e)} = \dot{\mathbf{u}}^N \cdot \mathbf{f}_N + \dot{\mathbf{u}}^N \cdot \mathbf{s}_N \tag{13.49}$$

where

$$\mathbf{f}_N = \int_{v_{0(e)}} \hat{\mathbf{F}}(\mathbf{x})\psi_N(\mathbf{x})\, dv_0 \tag{13.50}$$

and

$$\mathbf{s}_N = \int_{A_{0(e)}} \hat{\mathbf{S}}(\mathbf{x})\psi_N(\mathbf{x})\, dA_0 \tag{13.51}$$

The quantity $\mathbf{f}_N$ is the *generalized force at node N of the element due to the body forces* $\hat{\mathbf{F}}(\mathbf{x})$, *and* $\mathbf{s}_N$ *is the generalized force at node N due to* $\hat{\mathbf{S}}(\mathbf{x})$. The forces $\mathbf{s}_N$ are *the discrete analogs of stress for the finite-element model.* They represent the effects of contact of the element with its surroundings (e.g., an adjacent finite element). The generalized forces $\mathbf{f}_N$ and $\mathbf{s}_N$ are sometimes referred to as *consistent* generalized forces since, again, they are calculated in a manner consistent with the assumed velocity field; that is, the generalized forces $\mathbf{f}_N + \mathbf{s}_N$ are calculated so that mechanical power developed by the concentrated nodal forces moving through the nodal velocities $\dot{\mathbf{u}}^N$ is the same as that developed by the fields $\hat{\mathbf{F}}(\mathbf{x})$, $\hat{\mathbf{S}}(\mathbf{x})$, and $\dot{\mathbf{u}}(\mathbf{x})$.

It is important to note that the *actual surface forces* $\hat{\mathbf{S}}(\mathbf{x})$ *are available to us only in the deformed element.* Thus, the forces $\hat{\mathbf{S}}(\mathbf{x})$ are, in general, dependent on the deformation and, consequently, are functions of the nodal displacements $\mathbf{u}^N$. The precise form of this dependency is determined by the nature of the applied loading, the initial geometry of the element, and the form of the interpolation functions $\psi_N(\mathbf{x})$.† For our present purposes, it suffices to write $\hat{\mathbf{S}} = \hat{\mathbf{S}}(\mathbf{x})$ and to compute the surface tractions later in terms of the displacement gradients for each loading case and type of finite element considered.

Let $\mathbf{p}_N$ denote the total generalized force at node N:

$$\mathbf{p}_N = \mathbf{f}_N + \mathbf{s}_N \tag{13.52}$$

In component form,

$$p_{Ni} = f_{Ni} + s_{Ni} \tag{13.53}$$

or

$$p_{Ni} = \int_{v_{0(e)}} \hat{F}_i(\mathbf{x})\psi_N(\mathbf{x})\, dv_0 + \int_{A_{0(e)}} \hat{S}_i(\mathbf{x})\psi_N(\mathbf{x})\, dA_0 \tag{13.54}$$

† For example, in Chap. IV we show that in the case of a uniform pressure p,

$$S_i = -p\sqrt{G}\,\hat{n}_j(\delta_{im} + u_{i,m})G^{jm}$$

which, for a finite element, becomes

$$S_i = -p\sqrt{G_{(e)}}\,\hat{n}_j(\delta_{im} + \psi_{N,m}u_i^N)G_{(e)}^{jm}$$

Here $G_{(e)}$ and $G_{(e)}^{jm}$ are also functions of $u_{i,j}$. Fortunately, significant simplifications are possible for specific forms of $\psi_N(\mathbf{x})$. We investigate such problems in more detail later (see Arts. 16 to 20).

Thus (13.49) becomes

$$\Omega_{(e)} = \dot{\mathbf{u}}^N \cdot \mathbf{p}_N = \dot{u}_i^N p_{Ni} \tag{13.55}$$

with $N = 1, 2, \ldots, N_e$ and $i = 1, 2, 3$.

Heat We shall simply denote the total heat of the finite element by $Q_{(e)}$. For the present, the specific form of $Q_{(e)}$ in terms of approximate fields is unimportant, as it will not affect the form of the equations of motion of the element. The production of entropy and conduction of heat in the element are considered subsequently.†

13.3 CONSERVATION OF ENERGY IN A FINITE ELEMENT

The law of conservation of energy for a finite element, considering only thermomechanical behavior, is, according to (12.7),

$$\dot{\kappa}_{(e)} + \dot{U}_{(e)} = \Omega_{(e)} + Q_{(e)} \tag{13.56}$$

where, from (13.38), (13.47), and (13.55),

$$\dot{\kappa}_{(e)} = m_{NM}\ddot{u}_i^N \dot{u}_i^M \tag{13.57a}$$

$$\dot{U}_{(e)} = \int_{v_{0(e)}} t^{ij}\psi_{N,i}(\delta_{jk} + \psi_{M,j}u_k^M)\, dv_0 \dot{u}_k^N + Q_{(e)} \tag{13.57b}$$

$$\Omega_{(e)} = p_{Ni}\dot{u}_i^N \tag{13.57c}$$

$$Q_{(e)} = Q_{(e)} \tag{13.57d}$$

Substituting (13.57) into (13.56), we obtain the general energy balance for thermomechanical behavior of a finite element:

$$m_{NM}\ddot{u}_i^M \dot{u}_i^N + \int_{v_{0(e)}} t^{mj}\psi_{N,m}(\delta_{ji} + \psi_{M,j}u_i^M)\, dv_0\, \dot{u}_i^N = p_{Ni}\dot{u}_i^N \tag{13.58}$$

We observe that because we have replaced the rate of internal energy $\dot{e}$ by the right side of (12.23) in the expression for $\dot{U}_{(e)}$, the heat $Q_{(e)}$ cancels and does not appear in (13.58). We again remark that (13.58) is valid for any type of material, and no restrictions have been placed on the order of magnitude of the deformation gradients. To obtain alternate forms of (13.58) for specific materials, it is necessary to eliminate t^{ij} by introducing specific constitutive equations for the material.

13.4 GENERAL EQUATIONS OF MOTION OF A FINITE ELEMENT

It is now a simple matter to obtain general equations of motion for a typical finite element of a continuous medium. Returning to (13.58), we observe that each term has in common the nodal velocity components $\dot{u}_i^N$, which are

† See Sec. 13.8.

functions of time only. Thus, we can rewrite the general energy balance as follows:

$$\left[m_{NM}\ddot{u}_i^M + \int_{v_{0(e)}} t^{mj}\psi_{N,m}(\delta_{ji} + \psi_{M,j}u_i^M)\,dv_0 - p_{Ni}\right]\dot{u}_i^N = 0 \qquad (13.59)$$

As usual, the repeated indices are to be summed throughout their admissible ranges: $N, M = 1, 2, \ldots, N_e$; $i, j, m = 1, 2, 3$. Now (13.59) represents one form of the principle of conservation of energy for a finite element. Since it must be valid for arbitrary motions of the element, it must also hold for arbitrary values of the nodal velocities $\dot{u}_i^N$. By expanding the repeated indices N and i of (13.59) and examining each term, we see that if (13.59) is to hold for arbitrary $\dot{u}_i^N$, the term in brackets must vanish for all values of N and i. Thus,

$$m_{NM}\ddot{u}_i^M + \int_{v_{0(e)}} t^{mj}\psi_{N,m}(\delta_{ji} + \psi_{M,j}u_i^M)\,dv_0 = p_{Ni} \qquad (13.60)$$

Equation (13.60) represents the *general equation of motion of a finite element of a continuum.*† It represents $3N_e$ simultaneous equations in the nodal displacements u_i^N and their time derivatives. As before, specific forms of (13.60) can be obtained once the material has been characterized. Then t^{ij} can be written in terms of the nodal temperatures, u_i^N, $\dot{u}_i^N$, . . . , the histories of these quantities, etc., and thereby eliminated from the equations of motion. In its present form, (13.60) represents the discrete equivalent of Cauchy's first law of motion (5.27), and it ensures that linear momentum is balanced in an average sense over the element $\imath_e$.

13.5 GLOBAL FORMS OF THE EQUATIONS OF MOTION

Equation (13.60) describes the motion of a typical finite element. To obtain equations of motion for the entire assemblage of elements, let U_i^Δ, $\dot{U}_i^\Delta$, and $\ddot{U}_i^\Delta$ denote global values of the components of displacement, velocity, and acceleration at node Δ of the connected model. Then, according to (13.11)

† This equation, and various alternate forms, was derived by Oden [1967*c*, 1969*b*, 1970*a*]. See also Oden and Kubitza [1967], Oden and Aguirre-Ramirez [1969], and Oden [1970*b*]. Note that this same result can be obtained by application of the concept of "virtual work." That is, by considering the energy developed due to an arbitrary variation $\delta\dot{u}_i^N$ of the nodal velocities, we obtain an equation identical to (13.60) except that $\dot{u}_i^N$ outside the brackets is replaced by $\delta\dot{u}_i^N$. The argument then made is that the energy must balance for arbitrary variations $\delta\dot{u}_i^N$ of the nodal velocities (see Oden and Aguirre-Ramirez [1969]). Alternatively, we can regard the derivation of (13.60) as an application of Galerkin's method in which Cauchy's equations of motion (5.27) are multiplied by an "arbitrary" function $\dot{u}_i^N\psi_N(\mathbf{x})$ and integrated over the volume of an element. The interpretation given here, however, is physically more appealing and follows logically from basic physical laws.

to (13.20),

$$u^N_{i(e)} = \overset{(e)}{\Omega}{}^N_\Delta U^\Delta_i \tag{13.61a}$$

$$\dot{u}^N_{i(e)} = \overset{(e)}{\Omega}{}^N_\Delta \dot{U}^\Delta_i \tag{13.61b}$$

$$\ddot{u}^N_{i(e)} = \overset{(e)}{\Omega}{}^N_\Delta \ddot{U}^\Delta_i \tag{13.61c}$$

where, for simplicity, we have temporarily taken the local coordinate lines $x^i_{(e)}$ to be parallel to the global coordinates X^i (that is, $\overset{(e)}{\alpha}{}^i_j = \delta^i_j$).

Let $\mathbf{P}_\Delta$ denote the global value of the generalized force at node Δ of the connected model. Then the total mechanical power developed in the discrete model is

$$\Omega = \dot{\mathbf{U}}^\Delta \cdot \mathbf{P}_\Delta = \dot{U}^\Delta_i P_{\Delta i} \tag{13.62}$$

$\Delta = 1, 2, \ldots, G$. However, Ω can also be expressed as the sum

$$\Omega = \sum_{e=1}^{E} \Omega_{(e)} \tag{13.63}$$

where $\Omega_{(e)}$ is as given in (13.55). Thus

$$\mathbf{P}_\Delta \cdot \dot{\mathbf{U}}^\Delta = \sum_{e=1}^{E} \overset{(e)}{\mathbf{p}}_N \cdot \dot{\mathbf{u}}^N_{(e)} \tag{13.64}$$

Introducing (13.61*b*), we have

$$\left(\mathbf{P}_\Delta - \sum_e \overset{(e)}{\Omega}{}^N_\Delta \overset{(e)}{\mathbf{p}}_N\right) \cdot \dot{\mathbf{U}}^\Delta = \mathbf{0} \tag{13.65}$$

which leads to

$$\mathbf{P}_\Delta = \sum_{e=1}^{E} \overset{(e)}{\Omega}{}^N_\Delta \overset{(e)}{\mathbf{p}}_N \tag{13.66}$$

or, in component form,

$$P_{\Delta i} = \sum_{e=1}^{E} \overset{(e)}{\Omega}{}^N_\Delta \overset{(e)}{p}_{Ni} \tag{13.67}$$

This result, of course, also follows immediately from the definition of conjugate fields [see (9.189)]. However, the present derivation leads to a meaningful physical interpretation of the generalized forces $\mathbf{P}_\Delta$. The generalized forces $\mathbf{P}_\Delta$ and $\mathbf{p}_N$ are again seen to be conjugate to the velocities $\dot{\mathbf{U}}^\Delta$ and $\dot{\mathbf{u}}^N$ with respect to the mechanical power Ω. The forces $\mathbf{P}_\Delta$ are calculated in such a way that the mechanical power they develop due to the "discrete" motions $\dot{\mathbf{U}}^\Delta$ is the same as the total mechanical power developed over all finite elements in the model. Equation (13.64), then, may be interpreted as a statement of the invariance of mechanical power for the model. Equations (13.66) and (13.67) show that the global generalized force at node Δ of the connected model is the vector sum of the local generalized forces at nodes $\mathbf{p}_N^{(e)}$ of all elements which meet at Δ in the connected model [see (9.191)].

Before proceeding, it is important to observe an additional property of the generalized forces $\mathbf{P}_\Delta$. Introducing (13.52) into (13.66), we find

$$\mathbf{P}_\Delta = \mathbf{F}_\Delta + \mathbf{S}_\Delta \tag{13.68}$$

where

$$\mathbf{F}_\Delta = \sum_{e=1}^{E} \overset{(e)}{\Omega}{}_\Delta^N \mathbf{f}_N^{(e)} \tag{13.69}$$

$$\mathbf{S}_\Delta = \sum_{e=1}^{E} \overset{(e)}{\Omega}{}_\Delta^N \mathbf{s}_N^{(e)} \tag{13.70}$$

The quantity $\mathbf{F}_\Delta$, of course, represents the global generalized force at node Δ due to body forces $\hat{\mathbf{F}}(\mathbf{x})$ distributed throughout the medium. Assuming that the field $\hat{\mathbf{F}}(\mathbf{x})$ is given, local generalized forces $\mathbf{f}_N^{(e)}$ for each element are computed using (13.50). Then $\mathbf{F}_\Delta$ at each node of the connected model is obtained from (13.69).

The quantity $\mathbf{S}_\Delta$ represents the global generalized force at node Δ due to surface forces $\mathbf{S}(\mathbf{x})$. The local values $\mathbf{s}_N^{(e)}$, however, *are the discrete analogs of stress in the finite-element model.* Thus, for an internal node, $\mathbf{s}_N^{(e)}$ represents the action of one element on another (i.e., the discrete equivalent of contact forces). Thus, for *interior* nodes,†

$$\mathbf{S}_\Delta = \mathbf{0} \tag{13.71}$$

and (13.68) reduces to

$$\mathbf{P}_\Delta = \mathbf{F}_\Delta \tag{13.72}$$

However, for nodes on the boundary $\mathbf{S}_\Delta \neq \mathbf{0}$, in general, and (13.68) holds.

Returning now to the construction of the global equations, we transform nodal displacements, velocities, and accelerations in (13.60) with the aid of (13.61):

$$p_{Ni}^{(e)}(t) = m_{NM}^{(e)} \overset{(e)}{\Omega}{}_\Delta^M \ddot{U}_i^\Delta + \int_{v_{0(e)}} t_{(e)}^{mj} \psi_{N,m}^{(e)} (\delta_{ji} + \psi_{M,j}^{(e)} \overset{(e)}{\Omega}{}_\Delta^M U_i^\Delta)\, dv_0 \tag{13.73}$$

† It may happen that the actual stress field suffers a finite discontinuity along some surface in the body. Then $\mathbf{S}_\Delta$ may be used to represent the discrete equivalent of the jump at nodes on such surfaces. In these cases, $\mathbf{S}_\Delta = [\mathbf{S}_\Delta]$ instead of (13.71), where

$$[\mathbf{S}_\Delta] = \sum_e \overset{(e)}{\Omega}{}_\Delta^N \mathbf{s}_N^{(e)}$$

is the jump in $\mathbf{S}$ at node Δ of the connected model.

Finally, introducing (13.73) into (13.67) and simplifying, we obtain the global equations of motion for the entire assembly of finite elements:

$$P_{\Gamma i}(\mathbf{t}) = M_{\Gamma\Delta}\ddot{U}_i^\Delta + G_{\Gamma i}(\mathbf{t}) + H_{\Gamma\Delta}(\mathbf{t})U_i^\Delta \tag{13.74}$$

where

$$M_{\Gamma\Delta} = \sum_{e=1}^{E} \overset{(e)}{\Omega_\Gamma^N} \overset{(e)}{m_{NM}} \overset{(e)}{\Omega_\Delta^M} \tag{13.75}$$

$$G_{\Gamma i}(\mathbf{t}) = \sum_{e=1}^{E} \overset{(e)}{\Omega_\Gamma^N} \delta_{ji} \int_{v_{0(e)}} t_{(e)}^{mj} \overset{(e)}{\psi_{N,m}}\, dv_0 \tag{13.76}$$

$$H_{\Gamma\Delta}(\mathbf{t}) = \sum_{e=1}^{E} \overset{(e)}{\Omega_\Gamma^N} \overset{(e)}{\Omega_\Delta^M} \int_{v_{0(e)}} t_{(e)}^{mj} \overset{(e)}{\psi_{N,m}} \overset{(e)}{\psi_{M,j}}\, dv_0 \tag{13.77}$$

where the notation $G_{\Gamma i}(\mathbf{t})$, etc., indicates a dependency on the stress tensor $\mathbf{t}$. The array $M_{\Gamma\Delta}$ represents the global consistent mass matrix for the entire assembly of finite elements. The terms $G_{\Gamma i}(\mathbf{t})$ and $H_{\Gamma\Delta}(\mathbf{t})$ must be recast as functions or functionals of U_i^Δ, $\dot{U}_i^\Delta$, . . . and/or their histories after the constitutive equations for stress for the element materials have been introduced. Once these equations have been solved, local values of the nodal displacements, velocities, and accelerations are obtained for each element from (13.61). Values of these fields at arbitrary interior points are determined from (13.12) and (13.18), and if needed, the components of strain and strain rate are determined from (13.22) and (13.23). Nodal forces $p_{Ni}^{(e)}$ for each element can be calculated using (13.60). We examine more specific forms of (13.60) and (13.74) in subsequent articles.

Minor alterations of the above equations are needed in case the local rectangular coordinate systems $x_{(e)}^i$ are not parallel to the global coordinates X^i. Then, in view of (13.4) and (13.6), we use instead of (13.61) and (13.67),

$$u_{i(e)}^N = \overset{(e)}{\alpha_i^j} \overset{(e)}{\Omega_\Delta^N} U_j^\Delta \qquad \dot{u}_{i(e)}^N = \overset{(e)}{\alpha_i^j} \overset{(e)}{\Omega_\Delta^N} \dot{U}_j^\Delta \tag{13.78a,b}$$

$$\ddot{u}_{i(e)}^N = \overset{(e)}{\alpha_i^j} \overset{(e)}{\Omega_\Delta^N} \ddot{U}_j^\Delta \qquad P_{\Delta i} = \sum_{e=1}^{E} \overset{(e)}{\Omega_\Delta^N} \overset{(e)}{\alpha_i^j} p_{Nj}^{(e)} \tag{13.78c,d}$$

The remaining steps in the derivation then follow precisely as outlined above.

13.6 EQUATIONS OF MOTION IN GENERAL COORDINATES

Up to this point, all the equations derived in this section have been based on the assumption that the local coordinates $x_{(e)}^i$ are rectangular cartesian in the reference configuration of the element. General forms of the preceding equations, valid for any choice of intrinsic curvilinear coordinate system in the reference configuration, are more cumbersome but are not difficult to obtain.†

† See Oden and Aguirre-Ramirez [1969] and Oden [1970*a*].

Let $\xi^i_{(e)}$ denote a general system of curvilinear coordinates embedded in element r_e at configuration $C_{0(e)}$, with covariant and contravariant basis vectors $\mathbf{g}_{i(e)}$ and $\mathbf{g}^i_{(e)}$ [see (4.48)]. For convenience, we again drop the element identification label and write simply ξ^i, $\mathbf{g}_i$, and $\mathbf{g}^i$. The displacement field corresponding to a typical element is

$$\mathbf{u}_{(e)} = w_i(\boldsymbol{\xi},t)\mathbf{g}^i = w^i(\boldsymbol{\xi},t)\mathbf{g}_i \tag{13.79}$$

where w_i and w^i are the covariant and contravariant components of displacement. Note that the vectors $\mathbf{g}_i$ and $\mathbf{g}^i$ are also functions of ξ^i.

For the finite element, we use the representation

$$w_i = w_i^N \psi_N(\boldsymbol{\xi}) \qquad \text{or} \qquad w^i = w^{Ni}\psi_N(\boldsymbol{\xi}) \tag{13.80a,b}$$

Components of strain are then

$$\gamma_{ij}^{(e)} = \tfrac{1}{2}(\mathbf{g}_i \cdot \mathbf{u}_{,j} + \mathbf{g}_j \cdot \mathbf{u}_{,i} + \mathbf{u}_{,i} \cdot \mathbf{u}_{,j}) \tag{13.81}$$

where $\mathbf{u}_{,j} = \partial \mathbf{u}/\partial \xi^j$. Other kinematical quantities are computed in a like manner.

Noting that

$$\frac{\partial \dot{\mathbf{u}}_{(e)}}{\partial \xi^i} = \left(\frac{\partial \psi_N(\boldsymbol{\xi})}{\partial \xi^i}\mathbf{g}_m + \psi_N(\boldsymbol{\xi})\Gamma^r_{mi}\mathbf{g}_r\right)\dot{w}^{Nm} \tag{13.82}$$

and

$$\frac{\partial \dot{\mathbf{u}}_{(e)}}{\partial \xi^i} = \left(\frac{\partial \psi_N(\boldsymbol{\xi})}{\partial \xi^i}\mathbf{g}^m - \psi_N(\boldsymbol{\xi})\Gamma^m_{ri}\mathbf{g}^r\right)\dot{w}_N^m \tag{13.83}$$

where Γ^r_{mi} are the Christoffel symbols of the second kind for ξ^i in $C_{0(e)}$ [see (4.54)], we follow essentially the same procedure as that used previously and obtain, instead of (13.60), the following equation of motion for a typical finite element:

$$\begin{aligned} m_{NM}\ddot{w}^{Mi} &+ \int_{v_{0(e)}} t^{qj}(\delta^i_j\psi_{N,q} - \psi_N\Gamma^i_{jq})\,dv_0 \\ &+ \int_{v_{0(e)}} t^{qj}(\psi_{M,j}\psi_{N,q}\delta^i_m + \psi_M\psi_{N,q}\Gamma^i_{mj} - \psi_{M,j}\psi_N\Gamma^i_{mq} \\ &\qquad - \psi_M\psi_N\Gamma^r_{mj}\Gamma^i_{rq})\,dv_0 w^{Mm} = p^i_N \end{aligned} \tag{13.84}$$

where $\psi_{N,q} = \partial\psi_N(\boldsymbol{\xi})/\partial\xi^q$ and t^{qi} is the stress tensor per unit area in $C_{0(e)}$ referred to the convected coordinate lines ξ^i in $C_{(e)}$. Alternatively, we may use

$$\begin{aligned} m_{NM}\ddot{w}^M_i &+ \int_{v_{0(e)}} t^{pq}(g_{iq}\psi_{N,p} + g_{qj}\psi_N\Gamma^j_{ip})\,dv_0 \\ &+ \int_{v_{0(e)}} t^{pq}(\psi_{M,q}\psi_{N,p}\delta^m_i - \psi_M\psi_{N,p}\Gamma^m_{iq} + \psi_{M,q}\psi_N\Gamma^m_{ip} \\ &\qquad - \psi_M\psi_N\Gamma^m_{ps}\Gamma^s_{iq})\,dv_0 w^M_m = p_{Ni} \end{aligned} \tag{13.85}$$

where $g_{ij} = \mathbf{g}_i \cdot \mathbf{g}_j$ and

$$w_i^M = g_{ij} w^{Mj} \qquad p_{Ni} = g_{ij} p_N^j \tag{13.86a,b}$$

If the systems $\xi^i_{(e)}$ are parallel to global coordinate lines Ξ^i in C, then

$$w^{Ni}_{(e)} = \overset{(e)}{\Omega^N_\Delta} W^{\Delta i} \qquad w^N_{i(e)} = \overset{(e)}{\Omega^N_\Delta} W^\Delta_i \tag{13.87a,b}$$

$$P^i_\Delta = \sum_e \overset{(e)}{\Omega^N_\Delta} p^{i(e)}_N \qquad P_{\Delta i} = \sum_e \overset{(e)}{\Omega^N_\Delta} p^{(e)}_{Ni} \tag{13.88a,b}$$

If $\xi^i_{(e)}$ and Ξ^i are not parallel, then terms $\partial \xi^i_{(e)}/\partial \Xi^j$ and $\partial \Xi^i/\partial \xi^j_{(e)}$ must be inserted in (13.87) and (13.88) to refer components to a common basis.

Global equations of motion are of the form

$$M_{\Gamma\Delta} \ddot{W}^{\Delta i} + \bar{G}^i_\Gamma(\mathbf{t}) + \bar{H}^i_{\Gamma\Delta m}(\mathbf{t}) W^{\Delta m} = P^i_\Gamma \tag{13.89}$$

or

$$M_{\Gamma\Delta} \ddot{W}^\Delta_i + \bar{G}_{\Gamma i}(\mathbf{t}) + \bar{H}^m_{\Gamma\Delta i}(\mathbf{t}) W^\Delta_m = P_{\Gamma i} \tag{13.90}$$

where $M_{\Gamma\Delta}$ is as defined in (13.75) [except that $\psi_N(\boldsymbol{\xi})$ is used instead of $\psi_N(\mathbf{x})$] and

$$\bar{G}^i_\Gamma(\mathbf{t}) = \sum_e \overset{(e)}{\Omega^N_\Gamma} \int_{v_{0(e)}} t^{qj}(\delta^i_j \psi_{N,q} - \psi_N \Gamma^i_{jq})\, dv_0 \tag{13.91}$$

$$\bar{G}_{\Gamma i}(\mathbf{t}) = \sum_e \overset{(e)}{\Omega^N_\Gamma} \int_{v_{0(e)}} t^{pq}(g_{iq} \psi_{N,p} + g_{qj} \psi_N \Gamma^j_{ip})\, dv_0 \tag{13.92}$$

$$\begin{aligned}\bar{H}^i_{\Gamma\Delta m}(\mathbf{t}) = \sum_e \overset{(e)}{\Omega^N_\Gamma} \overset{(e)}{\Omega^M_\Delta} \int_{v_{0(e)}} t^{qj}(&\psi_{M,j} \psi_{N,q} \delta^i_m + \psi_M \psi_{N,q} \Gamma^i_{mj} \\ &- \psi_{M,j} \psi_N \Gamma^i_{mq} - \psi_M \psi_N \Gamma^r_{mj} \Gamma^i_{rq})\, dv_0\end{aligned} \tag{13.93}$$

We observe that in the case in which $\xi^i_{(e)}$ are rectangular cartesian coordinates, $(\xi^i_{(e)} = x^i_{(e)})$, $\Gamma^i_{jk} = 0$, $g_{ij} = g^{ij} = \delta^i_j$, $w_i^N = w^{Ni} = u_i^N$, and (13.84) and (13.85) reduce to (13.60). Likewise, $\bar{G}^i_\Gamma = \bar{G}_{\Gamma i} = G_{\Gamma i}$, and $\bar{H}^i_{\Gamma\Delta m} W^{\Delta m}$ and $\bar{H}^m_{\Gamma\Delta i} W^\Delta_m$ reduce to $H_{\Gamma\Delta} U^\Delta_i$.

We remark that a less accurate but considerably simpler form of the equations of motion in general coordinates is obtained if, instead of approximating the components w_i and w^i, we introduce a vector-valued representation $\mathbf{u}_{(e)} = \psi_N \mathbf{u}^N_{(e)}$, where $\mathbf{u}^N_{(e)} = \mathbf{u}_{(e)}(\mathbf{x}^N)$. As pointed out in Sec. 7.5 [see (7.48) to (7.52)], this is equivalent to referring $\mathbf{u}_{(e)}$ to constant-basis vectors $\overset{(e)}{\mathbf{g}_{Ni}}$, $\mathbf{g}^i_{N(e)}$, which represent an average of the values of the actual vectors $\mathbf{g}_i$ and $\mathbf{g}^i$ at the nodes. Then, instead of (13.82) and (13.83),

$$\frac{\partial \dot{\mathbf{u}}}{\partial \xi^i} = \frac{\partial \psi_N}{\partial \xi^i} w^{Nj} \mathbf{g}_{(N)j} = \frac{\partial \psi_N}{\partial \xi^i} w^N_j \mathbf{g}^j_{(N)} \tag{13.94}$$

and (13.84) and (13.85) reduce to

$$m_{NM}\ddot{w}^{Mi} + \int_{v_{0(e)}} t^{qj}\psi_{N,q}(\delta^i_j + \psi_{M,j}w^{Mi})\,dv_0 = p^i_N \tag{13.95}$$

$$m_{NM}\ddot{w}^M_i + \int_{v_{0(e)}} t^{qj}\psi_{N,q}(\delta_{ij} + \psi_{M,j}w^M_i)\,dv_0 = p_{Ni} \tag{13.96}$$

Here, again, $\psi_{N,q} \equiv \partial\psi_N(\boldsymbol{\xi})/\partial\xi^q$.

13.7 HIGHER-ORDER REPRESENTATIONS

Generation of higher-order finite-element representations follows essentially the same procedure as that used to obtain (13.60). For completeness, we consider briefly the formulation of the local equations of motion for a third-order representation.

Consider the local approximation of the displacement field given by

$$u_i = \overset{0}{\psi}_N u^N_i + \overset{j}{\psi}_N u^N_{i,j} + \overset{jk}{\psi}_N u^N_{i,jk} \tag{13.97}$$

where $\overset{0}{\psi}_N$, $\overset{j}{\psi}_N$, and $\overset{jk}{\psi}_N$ are the local interpolation functions described in Art. 8 and u^N_i, $u^N_{i,j}$, and $u^N_{i,jk}$ are the values of the displacement components, their first derivatives, and their second derivatives at node N of the element. Here $i, j, k = 1, 2, 3$, and for the $6N_e$ functions $\overset{jk}{\psi}_N$, it is understood that $j \leq k$. We recall from (8.8) that the functions $\overset{0}{\psi}_N$, $\overset{j}{\psi}_N$, and $\overset{jk}{\psi}_N$ have the properties

$$\begin{aligned}
&\overset{0}{\psi}_N(\mathbf{x}^M) = \delta^M_N && \frac{\partial\overset{0}{\psi}_N(\mathbf{x}^M)}{\partial x^i} = 0 && \frac{\partial^2\overset{0}{\psi}_N(\mathbf{x}^M)}{\partial x^i\,\partial x^r} = 0 \\
&\overset{j}{\psi}_N(\mathbf{x}^M) = 0 && \frac{\partial\overset{j}{\psi}_N(\mathbf{x}^M)}{\partial x^i} = \delta^j_i\delta^M_N && \frac{\partial^2\overset{j}{\psi}_N(\mathbf{x}^M)}{\partial x^i\,\partial x^r} = 0 \\
&\overset{jk}{\psi}_N(\mathbf{x}^M) = 0 && \frac{\partial\overset{jk}{\psi}_N(\mathbf{x}^M)}{\partial x^i} = 0 && \frac{\partial^2\overset{jk}{\psi}_N(\mathbf{x}^M)}{\partial x^i\,\partial x^r} = \delta^j_i\delta^k_r\delta^M_N
\end{aligned} \tag{13.98}$$

Note also that it is sometimes convenient to eliminate certain of the functions $\overset{0}{\psi}_N$, $\overset{j}{\psi}_N$, $\overset{jk}{\psi}_N$ in designing third-order representations of u_i.

Observing that

$$\begin{aligned}
\dot{u}_i &= \overset{0}{\psi}_N \dot{u}^N_i + \overset{j}{\psi}_N \dot{u}^N_{i,j} + \overset{jk}{\psi}_N \dot{u}^N_{i,jk} \\
\ddot{u}_i &= \overset{0}{\psi}_N \ddot{u}^N_i + \overset{j}{\psi}_N \ddot{u}^N_{i,j} + \overset{jk}{\psi}_N \ddot{u}^N_{i,jk}
\end{aligned} \tag{13.99}$$

we introduce (13.99) into (12.7) and, making use of (12.2), (12.3), (13.45), and (4.17), obtain, after some algebraic manipulations, the following statement of conservation of energy for the element:

$$P_{Ni}\dot{u}^N_i + P_{Nij}\dot{u}^N_{i,j} + P_{Nijk}\dot{u}^N_{i,jk} = 0 \tag{13.100}$$

Here P_{Ni}, P_{Nij}, and P_{Nijk} are functions of u^N_i, $u^N_{i,j}$, and $u^N_{i,jk}$ and their time derivatives [see (13.102)]. Since (13.100) must hold for arbitrary values of $\dot{u}_i^N$, $\dot{u}_{i,j}^N$, and $\dot{u}_{i,jk}^N$, we conclude that

$$P_{Ni} = P_{Nij} = P_{Nijk} = 0 \tag{13.101}$$

This leads to the three systems of equations of motion for the finite element:

$$\begin{aligned}
\bar{m}_{NM}\ddot{u}_k^M + \mathrm{I}_{NM}^r \ddot{u}_{k,r}^M + J_{NM}^{rs}\ddot{u}_{k,rs}^M + \int_{v_{0(e)}} t^{ij}\psi_{N,j}^0 z_{i,k}\, dv_0 &= \bar{p}_{Nk} \\
\mathrm{I}_{NM}^r \ddot{u}_k^M + \mathrm{II}_{NM}^{rs}\ddot{u}_{k,s}^M + \mathrm{III}_{NM}^{rst}\ddot{u}_{k,st}^M + \int_{v_{0(e)}} t^{ij}\psi_{N,j}^r z_{i,k}\, dv_0 &= d_{Nk}^r \\
J_{NM}^{rs}\ddot{u}_k^M + \mathrm{III}_{NM}^{rst}\ddot{u}_{k,t}^M + K_{NM}^{rsmn}\ddot{u}_{k,mn}^M + \int_{v_{0(e)}} t^{ij}\psi_{N,j}^{rs} z_{i,k}\, dv_0 &= \mu_{Nk}^{rs}
\end{aligned} \tag{13.102}$$

Here

$$t^{ij}z_{i,k} = t^{ij}(\delta_{ik} + \psi_{M,i}^0 u_k^M + \psi_{M,i}^r u_{k,r}^M + \psi_{M,i}^{rs} u_{k,rs}^M) \tag{13.103}$$

and

$$\begin{aligned}
\bar{m}_{NM} &= \int_{v_{0(e)}} \rho_0 \psi_N^0 \psi_M^0 \, dv_0 \qquad \mathrm{I}_{NM}^r = \int_{v_{0(e)}} \rho_0 \psi_N^0 \psi_M^r \, dv_0 \\
J_{NM}^{rs} &= \int_{v_{0(e)}} \rho_0 \psi^0{}_N \psi_M^{rs} \, dv_0 \qquad \mathrm{II}_{NM}^{rs} = \int_{v_{0(e)}} \rho_0 \psi_N^r \psi_M^s \, dv_0 \\
\mathrm{III}_{NM}^{rst} &= \int_{v_{0(e)}} \rho_0 \psi_N^r \psi_M^{st} \, dv_0 \qquad K_{NM}^{rsmn} = \int_{v_{0(e)}} \rho_0 \psi_N^{rs} \psi_M^{mn} \, dv_0 \\
\bar{p}_{Nk} &= \int_{v_{0(e)}} \rho_0 \hat{F}_k \psi_N^0 \, dv_0 + \int_{A_{0(e)}} \hat{S}_k \psi_N^0 \, dA_0 \\
d_{Nk}^r &= \int_{v_{0(e)}} \rho_0 \hat{F}_k \psi_N^r \, dv_0 + \int_{A_{0(e)}} \hat{S}_k \psi_N^r \, dA_0 \\
\mu_{Nk}^{rs} &= \int_{v_{0(e)}} \rho_0 \hat{F}_k \psi_N^{rs} \, dv_0 + \int_{A_{0(e)}} \hat{S}_k \psi_N^{rs} \, dA_0
\end{aligned} \tag{13.104}$$

The arrays $\bar{m}_{NM}$, I_{NM}^r, ..., K_{NM}^{rsmn} represent distributed mass, rotatory inertia, and "higher-order inertia" matrices, while $\bar{p}_{Nk}$, d_{Nk}^r, and μ_{Nk}^{rs} represent generalized nodal forces, doublets, and higher-order doublets corresponding to the generalized nodal displacements u_i^N and their first and second derivatives.

Global equations of motion are obtained through transformations of the form

$$\begin{aligned}
u_k^N &= \overset{(e)}{\Omega}{}_\Delta^N U_k^\Delta \qquad u_{k,i}^N = \overset{(e)}{\Omega}{}_\Delta^N U_{k,i}^\Delta \qquad u_{k,ij}^N = \overset{(e)}{\Omega}{}_\Delta^N U_{k,ij}^\Delta \\
\bar{p}_{\Delta k} &= \sum_e \overset{(e)}{\Omega}{}_\Delta^N \bar{p}_{Nk}^{(e)} \qquad D_{\Delta k}^r = \sum_e \overset{(e)}{\Omega}{}_\Delta^N d_{Nk}^{r(e)} \qquad M_{\Delta k}^{rs} = \sum_e \overset{(e)}{\Omega}{}_\Delta^N \mu_{Nk}^{rs(e)}
\end{aligned} \tag{13.105}$$

Essentially the same procedure can be used to obtain representations of any order. Since the procedure is straightforward, we need not elaborate further.

13.8 ENTROPY AND HEAT CONDUCTION IN FINITE ELEMENTS

To devise finite-element models of thermal phenomena, we must return to the concepts of entropy, heat, and temperature discussed in Art. 12. The question of which of the thermodynamic variables to use as the primary dependent variable is a natural one, for a number of choices are at our disposal (e.g., absolute temperature, heat flux, combinations of heat flux and temperature gradients, entropy flow). In all subsequent developments, we resolve this question by adopting the following convention. *In the construction of finite-element models of physical phenomena, select as primary dependent variables those quantities most naturally observable or experienced in the phenomena.* For example, in purely mechanical phenomena such as the deformation of a solid body, the most obvious and primitive characteristic is the displacement of particles from one point to another. All other mechanical quantities, e.g., strain, velocity, deformation gradients, can be calculated once the displacement field and the constitution of the material are known. In the case of thermal phenomena, the most obvious characteristic is a change in temperature. The temperature of a body is the most natural, measurable quantity available to us to describe purely thermal behavior of a system, and therefore, we shall take it to be a primary dependent variable.†

In constructing finite-element representations of temperature over a typical finite element, we may approximate the absolute temperature, locally, by

$$\theta_{(e)}(\mathbf{x},t) = \theta^N(t)\varphi_N(\mathbf{x}) \tag{13.106}$$

where $\varphi_N(\mathbf{x})$ are the local interpolation functions [we use the symbolism $\varphi_N(\mathbf{x}) = \varphi_N^{(e)}(\mathbf{x})$ here instead of the usual $\psi_N(\mathbf{x})$ to emphasize that the type of local approximation of the temperature need not be the same as that used in local approximations $\psi_N(\mathbf{x})\mathbf{u}^N$ of the displacement fields]. However, in most cases it is convenient to represent the absolute temperature as the sum of a known reference temperature $T_0(\mathbf{x})$ and a change in temperature $T(\mathbf{x},t)$:

$$\theta(\mathbf{x},t) = T_0(\mathbf{x}) + T(\mathbf{x},t) \tag{13.107}$$

† In selecting $\mathbf{u}(\mathbf{x},t)$ and $\theta(\mathbf{x},t)$ as unknowns, we have effectively employed the so-called "axiom of causality"; according to Eringen [1967, p. 145] "the motion of material points of a body and their temperatures are considered to be self-evident observable effects in every thermomechanical behavior of the body. The remaining quantities that enter the equations of balance of mass, momentum, production of energy, and conservation of energy (e.g., stress, entropy, heat flux, etc.) are the 'causes' . . . ." It is assumed, of course, that external effects such as body forces, heat sources, and certain surface tractions or displacements are given a priori. See Art. 14.

In classical thermostatics, $T_0(\mathbf{x})$ is generally taken to be uniform and it is assumed that $|T(\mathbf{x},t)| \ll T_0$. If we use (13.107), then, for a typical finite element, we take

$$T_{(e)}(\mathbf{x},t) = T^N \varphi_N(\mathbf{x}) \tag{13.108}$$

so that the absolute temperature throughout the element is

$$\theta(\mathbf{x},t) = T_0(\mathbf{x}) + T^N \varphi_N(\mathbf{x}) \tag{13.109}$$

where the dependence of T^N on t is understood. Note that $N = 1, 2, \ldots, N_e$ and that we have again omitted the element identification label for simplicity.

According to (12.46),

$$\theta \dot{\xi} = q_{i,i} + h + \sigma \tag{13.110}$$

where ξ is the entropy per unit volume in C_0, h is the heat supply per unit volume in C_0, and σ is the specific internal dissipation. To construct a finite-element model of the processes of entropy production and heat conduction in a typical finite element, we first multiply (13.109) through by $T(\mathbf{x},t)$:

$$T\theta\dot{\xi} = Tq_{i,i} + Th + T\sigma \tag{13.111}$$

noting that

$$Tq_{i,i} = (Tq_i)_{,i} - T_{,i}q_i \tag{13.112}$$

Then we rewrite (13.110) in the form

$$T\theta\dot{\xi} + T_{,i}q_i = (Tq_i)_{,i} + Th + T\sigma \tag{13.113}$$

Introducing (13.108) and (13.109) into (13.113), integrating over the element, and making use of the Green-Gauss theorem to transform the integral of the first term on the right side of (13.113) into a surface integral, we get

$$\left(\int_{v_{0(e)}} \{\varphi_N(\mathbf{x})[T_0(\mathbf{x}) + \varphi_M(\mathbf{x})T^M]\dot{\xi}_{(e)} + \varphi_{N,i}(\mathbf{x})q_i\}\, dv_0 - q_N - \sigma_N\right) T^N = 0 \tag{13.114}$$

where

$$q_N = \int_{v_{0(e)}} \varphi_N(\mathbf{x}) h\, dv_0 + \int_{A_{0(e)}} q_i \hat{n}_i \varphi_N(\mathbf{x})\, dA_0 \tag{13.115}$$

$$\sigma_N = \int_{v_{0(e)}} \sigma(\mathbf{x},t)\varphi_N(\mathbf{x})\, dv_0 \tag{13.116}$$

Here q_N is the *generalized normal component of heat flux* at node N of element e and σ_N is the *generalized dissipation* at node N.

Since (13.114) must hold for arbitrary T^N,

$$\int_{v_{0(e)}} \{\varphi_N(\mathbf{x})[T_0(\mathbf{x}) + \varphi_M(\mathbf{x})T^M]\dot{\xi}_{(e)} + \varphi_{N,i}(\mathbf{x})q_i\}\,dv_0 = q_N + \sigma_N \quad (13.117)$$

This result is the *general equation of heat conduction for a finite element.*† The global form of the heat-conduction relations for the entire assembly of elements is

$$\sum_e \left(\overset{(e)}{\Omega}{}^N_\Delta \left\{ \int_{v_{0(e)}} \overset{(e)}{\varphi}_N(\mathbf{x}) \left[T_0(\mathbf{x}) + \overset{(e)}{\varphi}_M(\mathbf{x}) \overset{(e)}{\Omega}{}^M_\Gamma \mathscr{T}^\Gamma \right] \dot{\xi}_{(e)} + \overset{(e)}{\varphi}_{N,i}(\mathbf{x}) q_i \right\} dv_0 \right) = Q_\Delta + \Sigma_\Delta \quad (13.118)$$

where $\mathscr{T}^\Gamma$ is the global value of the temperature change at node Γ of the assembled system and Q_Δ and Σ_Δ are global values of q_N and σ_N at node Δ:

$$Q_\Delta = \sum_{e=1}^{E} \overset{(e)}{\Omega}{}^N_\Delta \overset{(e)}{q}_N \quad (13.119)$$

$$\Sigma_\Delta = \sum_{e=1}^{E} \overset{(e)}{\Omega}{}^N_\Delta \overset{(e)}{\sigma}_N \quad (13.120)$$

We remark that the normal heat-flux components $q_i\hat{n}_i$ in (13.115) are, like the contact forces $\mathbf{S}(\mathbf{x})$, available to us only on material surfaces in the deformed body. Hence, q_N will, in general, depend upon the deformation of the element. A number of special forms of q_N can be identified for different boundary conditions.‡ Note also that the heat flux q_i in the volume integral in (13.117) is, in general, a function or a functional of the temperature and possibly the displacement gradients and/or their histories and that specific forms of such functionals depend upon the material of which the element is constructed. Likewise, the form of the entropy production $\dot{\xi}_{(e)}$ depends on the constitution of the material.

We present the above derivations of heat-conduction relations for finite elements only as an introduction to the subject, and we postpone a detailed examination of such relations, together with interpretations of the various quantities appearing therein, until Chap. V. As for the forms of the

† Finite-element formulations such as these were first presented by Oden and Aguirre-Ramirez [1969]. See also Oden and Kross [1969], Oden [1969*d*, 1970*b*], and Oden and Poe [1970]. Finite-element formulations of the classical heat-conduction problem were first presented by Wilson and Nickell [1966] and Becker and Parr [1967], who used a variational principle. Finite-element analyses of nonlinear heat-conduction problems were presented by Aguirre-Ramirez and Oden [1969]. In Chap. V, we examine such problems in more detail.

‡ See Oden and Kross [1969].

equations describing the conduction of heat in finite-element models, little else can be said until the element material has been characterized and constitutive equations for $\xi_{(e)}$ and $q_{i(e)}$ (or $\varphi_{(e)}$, $t^{ij}_{(e)}$, and $q_{i(e)}$) have been determined.

14 CONSTITUTIVE EQUATIONS

We have repeatedly noted that before the general equations of motion and heat conduction of a finite element [e.g., (13.60) and (13.117)] can be applied to specific problems, appropriate constitutive equations for the stress, the heat flux, the entropy, and possibly the internal energy or free energy must be introduced. In this article, we give a brief account of certain axioms from constitutive theory that provide some basis for developing constitutive equations for ideal materials.†

14.1 INTRODUCTORY COMMENTS

We recall that in the previous article we established that the dependent variables to be used in constitutive equations are the motion $\mathbf{z} = \mathbf{z}(\mathbf{x},t)$ [or, equivalently, the displacement field $\mathbf{u} = \mathbf{u}(\mathbf{x},t) = \mathbf{z}(\mathbf{x},t) - \mathbf{x}$] and the temperature $\theta = \theta(\mathbf{x},t)$ [or, equivalently, the change in temperature $T = T(\mathbf{x},t) = \theta(\mathbf{x},t) - T_0(\mathbf{x})$]. For the dependent constitutive variables, we shall use the stress tensor σ^{ij}, the heat-flux vector $\mathbf{q}$, the entropy density η, and the free-energy density ψ (the internal-energy density ϵ can then be determined as $\psi + \eta\theta$).

We are aware that two pieces of material of the same size, weight, color, and shape may respond quite differently to the same set of external forces or the same supply of heat. Hence, the response of a given specimen depends upon the physical properties of the material, the approximate character of which is generally assumed to be determined by laboratory experiments. The experimental characterization of materials, however, is in itself quite a complicated subject. Assuming that we could identify a series of experiments that would demonstrate how a certain stress component varied with temperature or with a certain component of strain or rate of strain, we could seldom hope to establish material properties on the basis of a limited number of empirical relations that would hold for a reasonable class of deformations and temperatures. In other words, an empirical formula might describe the behavior of a material under a very special prescribed deformation reproducible in the laboratory, but the same formula might give completely meaningless results for another deformation. To avoid

† For detailed treatments of constitutive theory, see Truesdell and Noll [1965] and Eringen [1962, 1967]. Good introductory accounts of the subject have been presented by Truesdell [1966*a*] and by Eringen [1967]. Summary treatments can be found in most textbooks on continuum mechanics, e.g., Jaunzemis [1967], Leigh [1968], Malvern [1970].

such difficulties, we must establish certain rules which the constitutive equations for a given material must satisfy in order that they give meaningful results for all the classes of deformations and temperatures to which we suspect it to be subjected. Secondly, we must have a precise form of the constitutive equations, consistent with our set of rules, in order to know what to measure in the laboratory to characterize the specific piece of material we have at hand. We recognize, of course, that constitutive equations characterize ideal materials; actual materials only approximately obey various classifications that we may establish from the forms of the constitutive equations. Also, we recognize that it is feasible to construct constitutive equations approximating a given material only for specified ranges of temperatures and deformations.

14.2 PHYSICAL ADMISSIBILITY, DETERMINISM, AND EQUIPRESENCE

In casting about for reasonable rules to guide us in constructing constitutive equations, three rather obvious ones immediately present themselves. First, we have devoted a good deal of attention thus far to the establishment of local forms of the five fundamental physical laws of conservation of mass, balance of linear and angular momentum, conservation of energy, and the Clausius-Duhem inequality:†

$$\begin{aligned}
&\rho_0 = \sqrt{G}\rho \\
&\sigma^{ij}_{;i} + \rho F^j = \rho a^j \\
&\sigma^{ij} = \sigma^{ji} \\
&\rho\dot{\epsilon} = \sigma^{ij}\dot{\gamma}_{ij} + Q^i_{;i} + \rho r \\
&\rho\theta\dot{\eta} - Q^i_{;i} - \rho r + \frac{1}{\theta}\mathbf{q}\cdot\nabla\theta \geq 0
\end{aligned} \tag{14.1}$$

It would be rather foolish to demand that (14.1) be always satisfied and to then select a set of constitutive equations which, for some reason, violated one of these laws. Thus, we shall establish as a general rule that all constitutive equations be consistent with (14.1). This rule is sometimes referred to as the *principle of physical admissibility:*

Rule 1 *Physical admissibility All constitutive equations must be consistent with the basic physical laws of conservation of mass, balance of momenta, conservation of energy, and the Clausius-Duhem inequality.*

The second rule which suggests itself is referred to as the *principle of determinism.* It follows from the observation that of all the values that the time parameter τ can assume, from $-\infty$ to the present time t to $+\infty$, we

† See Eqs. (5.3), (5.21), (5.31), (12.14), and (12.31).

can certainly eliminate those $>t$, for to do otherwise would mean that the behavior of the body would somehow depend upon the motion and temperature at some future time. Thus, we can exclude any dependence of the material behavior on any future events. However, the past behavior may certainly influence the current behavior of the material; in other words, the motion and temperature of the body up to and including the present time *determine* the constitutive variables $(\sigma^{ij},\mathbf{q},\eta,\psi)$:

Rule 2 *Determinism The values of the constitutive variables* $(\sigma^{ij},\mathbf{q},\eta,\psi)$ *at a material point* $\mathbf{x}$ *of a body at time t are determined by the histories of the motion and temperature of all points of the body.*

Suppose that $\mathbf{T}(\mathbf{x},t) = \sigma^{ij}\mathbf{G}_i \otimes \mathbf{G}_j$ is the stress tensor at the particle $\mathbf{x}$ at time t and that, for the moment, the constitutive equations are of the form

$$\begin{aligned} \mathbf{T}(\mathbf{x},t) &= \mathfrak{T}[\mathbf{z}(\mathbf{x}',s),\theta(\mathbf{x}',s),\mathbf{x},t] \\ \mathbf{q}(\mathbf{x},t) &= \mathbf{Q}[\mathbf{z}(\mathbf{x}',s),\theta(\mathbf{x}',s),\mathbf{x},t] \\ \eta(\mathbf{x},t) &= \mathscr{N}[\mathbf{z}(\mathbf{x}',s),\theta(\mathbf{x}',s),\mathbf{x},t] \\ \psi(\mathbf{x},t) &= \Psi[\mathbf{z}(\mathbf{x}',s),\theta(\mathbf{x}',s),\mathbf{x},t] \end{aligned} \tag{14.2}$$

where $\mathfrak{T}$ and $\mathbf{Q}$ are tensor- and vector-valued functionals and $\mathscr{N}$ and Ψ are scalar-valued functionals of the histories $\mathbf{z}(\mathbf{x}',s)$, $\theta(\mathbf{x}',s)$, with $\mathbf{x}' \in \mathscr{R}$ and $s \leq t$, and ordinary functions of $\mathbf{x}$ and t. In (14.2) we have used the same collection of independent variables $[\mathbf{z}(\mathbf{x}',s),\theta(\mathbf{x}',s),\mathbf{x},t]$ in each of the four constitutive equations so as to avoid any prejudices we may have acquired in dealing with the classical theories. For example, Fourier's law of heat conduction asserts that† $\mathbf{q} = K\nabla\theta$ while Hooke's law states that the stress components are linear combinations of the strain components. In more general theories involving dissipative phenomena, there may be cases in which both the stress and the heat flux depend upon the strain, θ, and $\nabla\theta$, etc. Generally, rule 1 concerning physical admissibility can be used to eliminate the dependence of $\mathbf{T}$, $\mathbf{q}$, η, or ψ on some of these variables, but as a precautionary measure we usually begin by assuming that all the constitutive functionals depend upon the same set of independent variables. This is called the *rule of equipresence:*

Rule 3 *Equipresence At the outset, a quantity appearing as an independent variable in one constitutive equation should appear in all constitutive equations.*

† Gradients of the motion and temperature are admitted in the arguments of the constitutive functionals by virtue of the local-action principle discussed in the following section.

14.3 LOCAL ACTION

A fourth rule, sometimes called the *principle of local action* or the *principle of neighborhood*, imposes certain restrictions on the smoothness of the constitutive functionals in the neighborhood of a material point $\mathbf{x}$.

Rule 4 *Local action* *The dependent constitutive variables at* $\mathbf{x}$ *are not appreciably affected by the values of the independent variables at material points distant from* $\mathbf{x}$.

This rule has important implications in finite-element applications. Suppose that the functions $\mathbf{u}(\mathbf{x},t) = \mathbf{z}(\mathbf{x},t) - \mathbf{x}$ and $T(\mathbf{x},t) = \theta(\mathbf{x},t) - T_0(\mathbf{x})$ are analytic in the neighborhood $\mathcal{N}(\mathbf{x}_0,r)$ of a particle $\mathbf{x}_0 \in \mathscr{R}$. Then $\mathbf{u}(\mathbf{x},t)$ and $T(\mathbf{x},t)$ admit Taylor expansions about $\mathbf{x}_0$ of the form

$$\mathbf{u}(\mathbf{x},t) = \mathbf{u}(\mathbf{x}_0,t) + \mathbf{u}_{,i}(\mathbf{x}_0,t)(x^i - x_0^i) + \tfrac{1}{2}\mathbf{u}_{,ij}(\mathbf{x}_0,t)(x^i - x_0^i)(x^j - x_0^j) + \cdots \tag{14.3}$$

$$T(\mathbf{x},t) = T(\mathbf{x}_0,t) + T_{,i}(\mathbf{x}_0,t)(x^i - x_0^i) + \tfrac{1}{2}T_{,ij}(\mathbf{x}_0,t)(x^i - x_0^i)(x^j - x_0^j) + \cdots \tag{14.4}$$

where $\mathbf{x} \in \mathcal{N}(\mathbf{x}_0,r)$. Now suppose that we identify a finite number K of points in $\mathcal{N}(\mathbf{x}_0,r)$, with $K \geq 4$, and label these points $\mathbf{x}^1, \mathbf{x}^2, \ldots, \mathbf{x}^K$. Then the set $\{\mathbf{x}^N\}$ of such points defines a set of position vectors $\mathbf{d}^N = \mathbf{x}^N - \mathbf{x}_0$ emanating from $\mathbf{x}_0$. Denoting

$$\mathbf{u}^N(t) = \mathbf{u}(\mathbf{x}^N,t) \tag{14.5}$$

$$T^N(t) = T(\mathbf{x}^N,t) \tag{14.6}$$

we see that (14.3) and (14.4) can be evaluated at each $\mathbf{x}^N \in \mathcal{N}(\mathbf{x}_0,r)$ to give the $2K$ sets of equations

$$\mathbf{u}^N(t) = \mathbf{u}(\mathbf{x}_0,t) + \mathbf{u}_{,i}(\mathbf{x}_0,t)d^{Ni} + \tfrac{1}{2}\mathbf{u}_{,ij}(\mathbf{x}_0,t)d^{Ni}d^{(N)j} + \cdots \tag{14.7}$$

$$T^N(t) = T(\mathbf{x}_0,t) + T_{,i}(\mathbf{x}_0,t)d^{Ni} + \tfrac{1}{2}T_{,ij}(\mathbf{x}_0,t)d^{Ni}d^{(N)j} + \cdots \tag{14.8}$$

In the spirit of finite-element approximations, it is clear that the values $\mathbf{u}^N(t)$ and $T^N(t)$ are merely the nodal displacements and temperatures at node $\mathbf{x}^N$ of a finite element $\mathcal{N}(\mathbf{x}_0,r)$. Thus, the axiom of local action permits us to assert that if r is taken sufficiently small, the fields $\mathbf{u}(\mathbf{x},t)$ and $T(\mathbf{x},t)$ in $\mathcal{N}(\mathbf{x}_0,r)$ can be represented with sufficient accuracy by only a finite number of terms of the series (14.3) and (14.4). It then follows that if $\mathbf{x} \in \mathcal{N}(\mathbf{x}_0,r)$, $\mathbf{u}(\mathbf{x},t)$ and $T(\mathbf{x},t)$ can be uniquely defined in terms of their values at a finite number of points in $\mathcal{N}(\mathbf{x}_0,r)$; and it is this latter observation that is the basis of finite-element approximations in continuum mechanics. Indeed, simply note that (14.3) can be rewritten

$$\mathbf{u}(\mathbf{x},t) = \mathbf{a}(t) + \mathbf{b}_i(t)x^i + \mathbf{c}_{ij}(t)x^ix^j + \cdots \tag{14.9}$$

where

$$\begin{aligned} \mathbf{a}(t) &= \mathbf{u}(\mathbf{x}_0,t) - \mathbf{u}_{,i}(\mathbf{x}_0,t)x_0^i + \tfrac{1}{2}\mathbf{u}_{,ij}(\mathbf{x}_0,t)x_0^ix_0^j + \cdots \\ \mathbf{b}_i(t) &= \mathbf{u}_{,i}(\mathbf{x}_0,t) - \mathbf{u}_{,ij}(\mathbf{x}_0,t)x_0^j + \tfrac{1}{2}\mathbf{u}_{,ijk}(\mathbf{x}_0,t)x_0^jx_0^k + \cdots \\ \mathbf{c}_{ij}(t) &= \tfrac{1}{2}\mathbf{u}_{,ij}(\mathbf{x}_0,t) - \tfrac{1}{6}\mathbf{u}_{,ijk}(\mathbf{x}_0,t)x_0^k + \tfrac{1}{24}\mathbf{u}_{,ijkm}(\mathbf{x}_0,t)x_0^kx_0^m + \cdots \end{aligned} \tag{14.10}$$

A similar equation can be written for $T(\mathbf{x},t)$ of (14.4). If, for example, only two terms of (14.9) are used, then $\mathbf{a}$ and $\mathbf{b}_i$ can be determined from just four conditions,

$$\mathbf{u}^N(t) = \mathbf{u}(\mathbf{x}^N,t) = \mathbf{a}(t) + \mathbf{b}_i(t)x^{Ni} \qquad N = 1, 2, 3, 4 \tag{14.11}$$

which we recognize as the usual simplex approximation for a tetrahedral neighborhood (finite element).

Returning to (14.3) and (14.4), we see that because of the local-action rule, the constitutive functionals can involve not only the histories of the motion $\mathbf{z}(\mathbf{x},t)$ and temperature $\theta(\mathbf{x},t)$ but also various gradients of these functions. Thus, if the constitutive functionals are sufficiently smooth, their arguments can contain, in addition to the motion $\mathbf{z}(\mathbf{x},s)$ and temperature $\theta(\mathbf{x},s)$, the gradients

$$\begin{aligned} &\mathbf{z}_{,i}(\mathbf{x},s),\ \mathbf{z}_{,ij}(\mathbf{x},s),\ \ldots,\ \mathbf{z}_{,i_1i_2\ldots i_r}(\mathbf{x},s),\ \ldots \\ &\theta_{,i}(\mathbf{x},s),\ \theta_{,ij}(\mathbf{x},s),\ \ldots,\ \theta_{,i_1i_2\ldots i_r}(\mathbf{x},s),\ \ldots \end{aligned} \tag{14.12}$$

Materials which are characterized by functionals of histories of various higher-order gradients are sometimes referred to as *nonsimple materials of the material-gradient type*.†

The subclass of many materials which involve gradients up to and including the first order are of special interest. We recall that a *homogeneous motion* relative to a reference configuration C_0 is of the form

$$\mathbf{z}(\mathbf{x},t) = \mathbf{z}_{,i}(t)(x^i - x_0^i) + \mathbf{z}_0(t) \tag{14.13a}$$

where $\mathbf{x}_0$ is a fixed particle in C_0 and $\mathbf{z}_0(t)$ is an arbitrary rigid motion independent of $\mathbf{x}$. Likewise, a homogeneous temperature is of the form

$$\theta(\mathbf{x},t) = \theta_{,i}(t)(x^i - x_0^i) + \theta_0(t) \tag{14.13b}$$

In these equations, $\mathbf{z}_{,i}$ and $\theta_{,i}$ are independent of $\mathbf{x}$ (that is, $z_{,i}$ and $\theta_{,i}$ depend on $\mathbf{x}_0$ and t). Let $\mathbf{z}_{,i}(\mathbf{x},t)$ and $\theta(\mathbf{x},t)$ be the motion and temperature at $\mathbf{x}$, and let $\bar{\mathbf{z}}(\mathbf{x},t)$ and $\bar{\theta}(\mathbf{x},t)$ be homogeneous motions and temperatures. Now,

† Truesdell and Noll [1965, p. 111] refer to materials dependent on gradients up to order N as materials of the differential type of grade N.

except possibly at certain singular points, lines, or surfaces, for every $\epsilon > 0$ there exists a $\delta > 0$ such that $\|\mathbf{z} - \bar{\mathbf{z}}\| < \epsilon$ (and $|\theta - \bar{\theta}| < \epsilon$) whenever $\|\mathbf{x} - \mathbf{x}_0\| < \delta$; that is, we can always select a neighborhood of $\mathbf{x}_0$ sufficiently small so as to make the motion and temperature at points in that neighborhood arbitrarily close to a homogeneous motion and temperature. Indeed, this is the basis of most simplex finite-element approximations which physically treat the motion and temperature as being homogeneous within a *finite* neighborhood of a material point. Now, if the constitutive functionals, for example, $\mathfrak{T}[\ \ ]$, are sufficiently smooth so that $\|\mathfrak{T}[\mathbf{z}(\mathbf{x}',s), \theta(\mathbf{x}',s),\mathbf{x},t] - \mathfrak{T}[\bar{\mathbf{z}}(\mathbf{x}',s),\bar{\theta}(\mathbf{x}',s),\mathbf{x},t]\| < \epsilon'$ whenever $\|\mathbf{z} - \bar{\mathbf{z}}\| < \epsilon$, $\|\ \ \|$ being an appropriately defined norm and $\mathbf{x}', \mathbf{x} \in \mathscr{N}(\mathbf{x}_0,\delta)$, etc., then, according to (14.13), each constitutive functional can be written as a functional of the histories† of $\mathbf{z}(\mathbf{x},t)$, $\mathbf{z}_{,i}(\mathbf{x},t)$, $\theta(\mathbf{x},t)$, and $\theta_{,i}(\mathbf{x},t)$ and as functions of $\mathbf{x}$ and t. Materials of this type are called *simple materials.* Introducing a rather standard notation‡ for histories of the motion and temperature,

$$\begin{aligned} \mathbf{z}^t(\mathbf{x},s) &= \mathbf{z}(\mathbf{x}, t - s) \\ \theta^t(\mathbf{x},s) &= \theta(\mathbf{x}, t - s) \end{aligned} \qquad s \geq 0 \tag{14.14}$$

we see that for simple materials, (14.2) becomes

$$\begin{aligned} \mathbf{T}(\mathbf{x},t) &= \mathfrak{T}[\mathbf{z}^t(s),\mathbf{z}^t_{,i}(s),\theta^t(s),\theta^t_{,i}(s),\mathbf{x},t] \\ \mathbf{q}(\mathbf{x},t) &= \mathbf{Q}[\mathbf{z}^t(s),\mathbf{z}^t_{,i}(s),\theta^t(s),\theta^t_{,i}(s),\mathbf{x},t] \\ \eta(\mathbf{x},t) &= \mathscr{N}[\mathbf{z}^t(s),\mathbf{z}^t_{,i}(s),\theta^t(s),\theta^t_{,i}(s),\mathbf{x},t] \\ \psi(\mathbf{x},t) &= \Psi[\mathbf{z}^t(s),\mathbf{z}^t_{,i}(s),\theta^t(s),\theta^t_{,i}(s),\mathbf{x},t] \end{aligned} \tag{14.15}$$

the dependence of $\mathbf{z}^t(s)$, $\mathbf{z}^t_{,i}(s)$, . . . on $\mathbf{x}$ being understood; i.e., the histories involve $\mathbf{x}$ and not $\mathbf{x}'$ due to the local-action hypothesis. Further reductions of the forms of the constitutive functionals for simple materials are to be established below.

14.4 MATERIAL FRAME INDIFFERENCE AND SYMMETRY

It is an elementary observation that the response of a material is independent of whatever spatial reference frame we as observers may establish to describe it. For example, if the stretch of a piece of material happens to be proportional to the amount of "pull," this same law of proportionality should be witnessed by different observers, with different motions, who have established different frames of reference to study the response.

To translate this observation into mathematical terms, let $\hat{\mathbf{z}}(\mathbf{x},\hat{t})$ and $\mathbf{z}(\mathbf{x},t)$ denote two spatial reference frames that differ by an orthogonal

† We eliminate the dependence of the constitutive functionals on some of these variables in the following section; see Eq. (14.21).

‡ See, for example, Coleman [1964], Truesdell and Noll [1965], or Truesdell [1966*a*].

transformation $\boldsymbol{\alpha}(t)$, a time shift $\hat{t} = t - a$, and a translation $\mathbf{c}(t)$:

$$\hat{z}_i(\mathbf{x},\hat{t}) = \alpha_{ij}(t)z_j(\mathbf{x},t) + c_i(t) \tag{14.16}$$

Here $\alpha_{ij} = \alpha^i_j$, $\alpha_{ij}\alpha_{mj} = \delta_{im}$, and $\det(\alpha_{ij}) = \pm 1$. Orthogonal transformations of the type in (14.16) are called *observer transformations* since they relate the motion of the same material point $\mathbf{x}$ viewed by two different observers. The fact that constitutive functionals should be invariant under such transformations is called the *axiom of material frame-indifference* or *material objectivity;* it provides an important means for establishing further restrictions on the forms of the constitutive functionals:

Rule 5 Material frame indifference The constitutive equations are invariant under observer transformations.

According to this rule, a constitutive functional such as $\mathfrak{T}[\ \ ]$ in (14.2) must be such that

$$\mathfrak{T}[\mathbf{z}(\mathbf{x}',s),\theta(\mathbf{x}',s),\mathbf{x},t] = \mathfrak{T}[\hat{\mathbf{z}}(\mathbf{x}',\hat{s}),\theta(\mathbf{x}',\hat{s}),\mathbf{x},\hat{t}] \tag{14.17}$$

where $\hat{\mathbf{z}}(\mathbf{x}',\hat{s})$ is given by (14.16), with $\hat{t}$ replaced by a "past-time" $\hat{s}$.

Two important consequences of the principle of material frame indifference follow immediately from two special cases of the general observer transformation in (14.16). First, consider the case in which the time shift is such that $a = t$ and $\alpha_{ij} = \delta_{ij}$, $c_i(s) = 0$. Then $\hat{\mathbf{z}}(\mathbf{x}',\hat{s}) = \mathbf{z}(\mathbf{x}', \hat{s} + t)$, $\hat{t} = t - a = 0$, and for example,

$$\mathbf{T} = \mathfrak{T}[\mathbf{z}(\mathbf{x}', \hat{s} + t),\theta(\mathbf{x}', \hat{s} + t),\mathbf{x},0] \tag{14.18}$$

However, we may introduce a new time parameter $\bar{s} = t - s$, $0 \le \bar{s} \le \infty$, so that $\hat{s} + t = t - \bar{s}$. Thus,

$$\mathbf{T} = \mathfrak{T}[\mathbf{z}(\mathbf{x}', t - \bar{s}),\theta(\mathbf{x}', t - \bar{s}),\mathbf{x},0] \tag{14.19}$$

It follows that *the constitutive functionals are independent of time* and are functionals of the histories $\mathbf{z}(\mathbf{x}', t - \bar{s}) = \mathbf{z}^t(\mathbf{x}',\bar{s})$ and $\theta(\mathbf{x}', t - \bar{s}) = \theta^t(\mathbf{x}',\bar{s})$. The second consequence follows from the special case in which (14.16) represents only a rigid translation of the reference frames: $\alpha_{ij} = \delta_{ij}$, $a = 0$, $\mathbf{c}(s) = -\mathbf{z}(\mathbf{x},s)$. Then $\hat{\mathbf{z}}(\mathbf{x}',\hat{s}) = \mathbf{z}(\mathbf{x}',s) - \mathbf{z}(\mathbf{x},s)$, and taking into account the fact that the constitutive functionals do not include time explicitly and setting $\bar{s} = s$,

$$\mathbf{T} = \mathfrak{T}[\mathbf{z}(\mathbf{x}', t - s) - \mathbf{z}(\mathbf{x}, t - s), \theta(\mathbf{x}', t - s),\mathbf{x}] \tag{14.20}$$

However, according to the principle of local action (rule 4),

$$\mathbf{z}(\mathbf{x}', t - s) - \mathbf{z}(\mathbf{x}, t - s) = \mathbf{z}_{,i}(\mathbf{x}, t - s)(x'^i - x^i) + \tfrac{1}{2}\mathbf{z}_{,ij}(\mathbf{x}, t - s)(x'^i - x^i)(x'^j - x^j) + \cdots \tag{14.20a}$$

$$\theta(\mathbf{x}', t - s) = \theta(\mathbf{x}, t - s) + \theta_{,i}(\mathbf{x}, t - s)(x'^i - x^i) + \cdots \tag{14.20b}$$

Thus, *the constitutive functionals depend upon the histories of the gradients of the motion rather than the history of the motion itself.*

Returning to (14.15), we see that as a consequence of the principle of material frame indifference, the constitutive equations for simple materials are of the form

$$\begin{aligned}\mathbf{T}(\mathbf{x},t) &= \mathfrak{T}[\mathbf{z}^t_{,i}(s),\theta^t(s),\theta^t_{,i}(s),\mathbf{x}]\\ \mathbf{q}(\mathbf{x},t) &= \mathbf{Q}[\mathbf{z}^t_{,i}(s),\theta^t(s),\theta^t_{,i}(s),\mathbf{x}]\\ \eta(\mathbf{x},t) &= \mathcal{N}[\mathbf{z}^t_{,i}(s),\theta^t(s),\theta(s^t_{,i}),\mathbf{x}]\\ \psi(\mathbf{x},t) &= \Psi[\mathbf{z}^t_{,i}(s),\theta^t(s),\theta^t_{,i}(s),\mathbf{x}]\end{aligned} \tag{14.21}$$

where, again, the dependence of the histories on $\mathbf{x}$ is understood.

A third application of rule 5, involving the special case of (14.16) in which $\mathbf{c}(s) = 0$, $a = 0$, and $\boldsymbol{\alpha}^t(s) = [\alpha_{ij}(t - s)]$ is arbitrary (i.e., an arbitrary continuous history of rigid rotations of the spatial frame of reference), may lead to important reductions in the forms of the constitutive functionals. In this case, for simple materials,

$$\mathfrak{T}[\boldsymbol{\alpha}^t(s)\mathbf{z}^t_{,i}(s),\theta^t(s),\boldsymbol{\alpha}^t(s)\theta^t_{,i}(s),\mathbf{x}] = \boldsymbol{\alpha}(t)\mathfrak{T}[\mathbf{z}^t_{,i}(s),\theta^t(s),\theta^t_{,i}(s),\mathbf{x}]\boldsymbol{\alpha}^T(t) \tag{14.22}$$

Any constitutive functional must be form-invariant under transformations of the type in (14.22).

Transformations of the spatial frame of reference as defined in (14.16) suggest that we also consider the possibility of invariance under transformations of the *material* reference frame $\mathbf{x}$. Such transformations have meaning when we consider the fact that materials may possess certain properties of symmetry with respect to directions in a reference configuration C_0. To fix ideas, let C_0 and C_0^* denote two reference configurations of a body which are characterized by material reference frames $\mathbf{x}$ and $\mathbf{x}^*$, respectively. The motion of a body is then $\mathbf{z}(\mathbf{x},t)$ or $\mathbf{z}(\mathbf{x}^*,t)$, depending upon which configuration is used as a reference. If we hold the current configuration fixed, it is clear that

$$dz_i = \frac{\partial z_i}{\partial x^j}\,dx^j = \frac{\partial z_i}{\partial x^{*j}}\,dx^{*j} \tag{14.23}$$

Now it is possible to find a nonsingular, linear, unimodular transformation $\mathbf{S}$ that maps C_0 into C_0^* in the sense that $dx^{*i} = S^i_{.j}\,dx^j$ (or $d\mathbf{x}^* = \mathbf{S}\,d\mathbf{x}$). Thus (14.23) shows that the deformation gradients with respect to C_0 and C_0^* are related by

$$\frac{\partial z_i(\mathbf{x}^*,t)}{\partial x^{*j}} = \frac{\partial z_i(\mathbf{x},t)}{\partial x^m}\,(S^m_{.j})^{-1} \tag{14.24}$$

If, for example, $\mathbf{T} = \mathfrak{T}[\mathbf{z}^t_{,i}(\mathbf{x},s),\mathbf{x}]$ (the temperature histories being temporarily omitted for simplicity), then also $\mathbf{T} = \mathfrak{T}^*[\mathbf{z}^t_{,i*}(\mathbf{x}^*,s),\mathbf{x}^*]$, which means that the functionals $\mathfrak{T}[\ \ ]$ and $\mathfrak{T}^*[\ \ ]$ must satisfy

$$\mathfrak{T}[\mathbf{z}^t_{,i}(\mathbf{x},s),\mathbf{x}] = \mathfrak{T}^*[\mathbf{z}^t_{,i}(\mathbf{x},s)\mathbf{S}^{-1}, \mathbf{S}\mathbf{x} + \mathbf{b}] \tag{14.25}$$

where $\mathbf{b}$ indicates a possible translation of $\mathbf{x}$ with respect to $\mathbf{x}^*$. Now suppose that the material properties in two distinct configurations C_0 and C_0^* are the same (the densities ρ_0 and ρ_0^* being equal). Then, omitting translations $\mathbf{b}$,

$$\mathfrak{T}[\mathbf{z}^t_{,i}(\mathbf{x},s),\mathbf{x}] = \mathfrak{T}[\mathbf{z}^t_{,i}(\mathbf{x},s)\mathbf{S}^{-1},\mathbf{S}\mathbf{x}] \tag{14.26}$$

where $\det \mathbf{S} = \pm 1$ (since $\rho_0^* = |\det \mathbf{S}|\ \rho_0$).

The particular collection of such unimodular mappings $\mathbf{S}^{-1}$, which leave the form of the functional invariant, are said to constitute the *isotropy group* of the constitutive functional. The group property† of such mappings follows from the fact that the collection of nonsingular unimodular transformations $\mathbf{S}$ constitutes a multiplicative group $\mathscr{G}$ with respect to the operation of composition of linear transformations (i.e., matrix multiplication). That is, if $\mathbf{S}_1$, $\mathbf{S}_2$, and $\mathbf{S}_3 \in \mathscr{G}$, $\mathbf{S}_1\mathbf{S}_2 \in \mathscr{G}$ (for $\det \mathbf{S}_1\mathbf{S}_2 = \det \mathbf{S}_1 \det \mathbf{S}_2 = \pm 1$), $\mathbf{S}_1^{-1}\mathbf{S}_1 = \mathbf{I}$ and $\mathbf{S}_1^{-1}$, $\mathbf{I} \in \mathscr{G}$, and $\mathbf{S}_1(\mathbf{S}_2\mathbf{S}_3) = (\mathbf{S}_1\mathbf{S}_2)\mathbf{S}_3$. Suppose that two arbitrary reference configurations C_0 and C_0^* are related by a mapping $S: C_0 \to C_0^*$ such that the response functionals $\mathfrak{T}$ in C_0 and C_0^* are $\mathfrak{T}[\ \ ]$ and $\mathfrak{T}^*[\ \ ]$. Suppose also that these functionals have isotropy groups $\mathscr{G}$ and $\mathscr{G}^*$. Then, if $\mathbf{H} \in \mathscr{G}$, $\mathfrak{T}[\mathbf{z}^t_{,i}(\mathbf{x},s)] = \mathfrak{T}^*[\mathbf{z}^t_{,i}(\mathbf{x},s)\mathbf{S}\mathbf{H}\mathbf{S}^{-1}]$, which shows that $\mathbf{S}\mathbf{H}\mathbf{S}^{-1} \in \mathscr{G}^*$. Now two groups $\mathscr{G}$ and $\mathscr{G}^*$ related by mappings of the type $\mathbf{S}\mathbf{H}\mathbf{S}^{-1} = \mathbf{H}^*$, where $\mathbf{H}^* \in \mathscr{G}^*$ and $\mathbf{S} \in \mathscr{G}$, are said to be conjugate groups. Since the mappings are one to one and the group properties are preserved under these mappings, $\mathscr{G}$ and $\mathscr{G}^*$ are isomorphic. Thus, the isotropy groups of a functional corresponding to any two reference configurations related by a unimodular transformation are isomorphic. This property is sometimes referred to as *material isomorphism*.

The group $\mathscr{G}$ for a material is a subgroup of the full group $\mathscr{U}$ of all unimodular transformations, and it may (or may not) contain the group $\mathscr{O}$ of orthogonal transformations. The response of the material can be characterized according to the nature of the isotropy group appropriate for its corresponding constitutive functionals. For example, if a material occupies a reference configuration C_0 such that the isotropy group $\mathscr{G}$ equals or contains the orthogonal group $\mathscr{O}$, then the material is said to be *isotropic* and C_0 is referred to as the *natural* or *undistorted state*. That is, for isotropic materials, we take $\mathbf{S} = \boldsymbol{\alpha}$ and find that $\boldsymbol{\alpha}\mathfrak{T}[\mathbf{z}^t_{,i}(\mathbf{x},s),\mathbf{x}]\boldsymbol{\alpha}^T = \mathfrak{T}[\boldsymbol{\alpha}\mathbf{z}^t_{,i}(\mathbf{x},s)\boldsymbol{\alpha}^T,\boldsymbol{\alpha}\mathbf{x}]$. If $\mathscr{G} \subset \mathscr{O}$, on the other hand, the material is said to be *anisotropic* in C_0.

† See Sec. 7.1.

Further, if $\mathcal{O} \supseteq \mathcal{G}$, the material is a *simple solid*, while if $\mathcal{G}$ is the full unimodular group, it is a *simple fluid*. Finally, if we abandon such *local* tests as $d\mathbf{x}^* = \mathbf{S}\,d\mathbf{x}$ and include, as in (14.25), the translation $\mathbf{x}^* = \mathbf{S}\mathbf{x} + \mathbf{b}$, then if $\mathfrak{T}[\ \ ]$ changes due to the presence of such translations $\mathbf{b}$, it is said to be *inhomogeneous*. If $\mathfrak{T}[\ \ ]$ does not depend on $\mathbf{b}$, the material is said to be *homogeneous*.

Now the constitutive functionals for all ideal materials possess some type of isotropy group. Thus, we can establish as a guiding principle for reduction of the forms of such functionals a sixth rule, called the *principle of material symmetry:*

Rule 6 Material symmetry The constitutive equations must be form-invariant with respect to a group $\mathcal{G}$ of unimodular transformations of the material frame of reference.

Observe that for homogeneous materials if we take $\mathbf{S} = \mathbf{I}$, then, for arbitrary $\mathbf{b}$, $\mathfrak{T}[\mathbf{z}^t_{,i}(\mathbf{x},s),\mathbf{x}] = \mathfrak{T}[\mathbf{z}^t_{,i}(\mathbf{x},s), \mathbf{x} + \mathbf{b}]$, which follows only if $\mathfrak{T}[\ \ ]$ *does not depend on* $\mathbf{x}$ *explicitly*. Thus, for homogeneous bodies we can omit $\mathbf{x}$ from the arguments of the constitutive functionals.

Other rules In addition to the six axioms established thus far, it is possible to lay down a number of other rules to be followed in developing constitutive equations. Since rules 1 to 6 are often considered to be the most basic, we shall only mention two such guidelines here. First, it is obvious that the constitutive functionals should be dimensionally consistent; i.e., the dimensions of terms appearing on either side of the constitutive equations should be the same. Second, an "axiom of memory" is sometimes postulated,† which asserts that the current values of the constitutive variables are not appreciably affected by their values at times distant in the past. This assertion is the counterpart of the local action principle in the time domain and effectively imposes smoothness requirements on $\mathbf{z}(\mathbf{x},t)$ in the variable t. If, for example, $\mathbf{z}(\mathbf{x},t)$ is sufficiently smooth to admit expansions of the form

$$\mathbf{z}(\mathbf{x},\hat{t}) = \mathbf{z}(\mathbf{x},t) + \dot{\mathbf{z}}(\mathbf{x},t)(\hat{t} - t) + \cdots + \frac{1}{n!}\overset{n}{\mathbf{z}}(\mathbf{x},t)(\hat{t} - t)^n \tag{14.27}$$

$t_0 \le \hat{t} \le t$, and if t_0 is sufficiently large that the behavior for $\hat{t} < t_0$ does not appreciably affect the response at t, then it may be possible to reduce the constitutive functionals to functions of the current rates $\mathbf{z}(\mathbf{x},t)$, $\dot{\mathbf{z}}(\mathbf{x},t)$, . . . , $\overset{n}{\mathbf{z}}(\mathbf{x},t)$, [and, of course, $\theta(\mathbf{x},t)$, $\dot{\theta}(\mathbf{x},t)$, . . . , $\overset{n}{\theta}(\mathbf{x},t)$]. Such materials are

† A recent formulation was given by Coleman and Noll [1961]. See also Truesdell and Noll [1965] and Coleman [1964]. According to Eringen [1967, p. 153], a concept of fading heredity was introduced formally by Straneo [1925]. We discuss the idea further in Sec. 19.1.

referred to as *materials of the rate type*. Materials of the rate type may demand uncommonly high differentiability of $\mathbf{z}(\mathbf{x},t)$ and $\theta(\mathbf{x},t)$, however, so that the axiom of memory may be more generally put to use by assuming functional forms of the constitutive equations which require only integrability of $\mathbf{z}^t_{,j}(\mathbf{x},s)$, $\theta^t(\mathbf{x},s)$, and so forth.

We do not place an importance on the so-called axiom of memory equal to that of the six rules covered previously because there are many cases in which it obviously does not apply (e.g., elastoplastic materials). However, this rule may be particularly useful in the study of certain problems in thermoviscoelasticity; we shall elaborate on it further in this context in Chap. V.

14.5 CONSTITUTIVE FUNCTIONALS FOR FINITE ELEMENTS

Given a set of constitutive functionals that satisfy all the basic principles of constitutive theory, we can obtain alternate forms of these functionals appropriate for finite-element investigations. Consider, for example, the case of a homogeneous, simple, thermomechanical material characterized by constitutive equations of the form

$$\begin{aligned}
\mathbf{T}(\mathbf{x},t) &= \mathfrak{T}[\mathbf{z}^t_{,i}(\mathbf{x},s),\theta^t(\mathbf{x},s),\theta^t_{,i}(\mathbf{x},s)] \\
\mathbf{q}(\mathbf{x},t) &= \mathbf{Q}[\mathbf{z}^t_{,i}(\mathbf{x},s),\theta^t(\mathbf{x},s),\theta^t_{,i}(\mathbf{x},s)] \\
\eta(\mathbf{x},t) &= \mathscr{N}[\mathbf{z}^t_{,i}(\mathbf{x},s),\theta^t(\mathbf{x},s),\theta^t_{,i}(\mathbf{x},s)] \\
\psi(\mathbf{x},t) &= \Psi[\mathbf{z}^t_{,i}(\mathbf{x},s),\theta^t(\mathbf{x},s),\theta^t_{,i}(\mathbf{x},s)]
\end{aligned} \tag{14.28}$$

For a typical finite element e, we have

$$\begin{aligned}
z_i(\mathbf{x},t) &= x_i + u_i^N(t)\psi_N(\mathbf{x}) \\
z_{i,j}(\mathbf{x},t) &= \delta_{ij} + u_i^N(t)\psi_{N,j}(\mathbf{x}) \\
\theta(\mathbf{x},t) &= T_0 + T^N(t)\varphi_N(\mathbf{x}) \\
\theta_{,i}(\mathbf{x},t) &= T^N(t)\varphi_{N,i}(\mathbf{x})
\end{aligned} \tag{14.29}$$

where the element identification label has been dropped for simplicity and $T_0(\mathbf{x}) = T_0$ is a uniform reference temperature. Denoting the histories of the nodal temperatures and displacements by

$$\begin{aligned}
\mathbf{u}^{N(t)}(s) &= \mathbf{u}^N(t-s) \\
T^{N(t)}(s) &= T^N(t-s)
\end{aligned} \tag{14.30}$$

we see that (14.28) can be written

$$\begin{aligned}
\mathbf{T}(\mathbf{x},t) &= \hat{\mathfrak{T}}[\mathbf{u}^{N(t)}(s),T^{N(t)}(s)] \\
\mathbf{q}(\mathbf{x},t) &= \hat{\mathbf{Q}}[\mathbf{u}^{N(t)}(s),T^{N(t)}(s)] \\
\eta(\mathbf{x},t) &= \hat{\mathscr{N}}[\mathbf{u}^{N(t)}(s),T^{N(t)}(s)] \\
\psi(\mathbf{x},t) &= \hat{\Psi}[\mathbf{u}^{N(t)}(s),T^{N(t)}(s)]
\end{aligned} \tag{14.31}$$

where

$$\hat{\mathfrak{T}}[\mathbf{u}^{N(t)}(s), T^{N(t)}(s)] = \mathfrak{T}[\delta_{ij} + \mathfrak{T}\psi_{N,j}(\mathbf{x})u_i^{N(t)}(s),\ T_0 + \varphi_N(\mathbf{x})T^{N(t)}(s),$$
$$\varphi_{N,i}(\mathbf{x})T^{N(t)}(s)] \quad (14.32)$$

$\cdots$. Thus, for finite elements, the dependent variables **T**, **q**, η, and ψ at a point **x** within an element are functionals of the total histories of the nodal displacements and changes in temperature. Similar conclusions are reached for more general materials of the material-gradient type. We investigate specific examples in Chaps. IV and V.

IV
Finite Elasticity

15 FOUNDATIONS OF THE THEORY OF ELASTICITY

In the present chapter we investigate the application of the finite-element method to problems in finite elasticity,† i.e., problems involving very large deformations of elastic bodies in which no restrictions are placed on the orders of magnitude of the displacements, displacement gradients, or strain components. In this way we obtain as special cases various discrete models of problems in classical infinitesimal elasticity. Before discussing properties of the discrete model, however, it is necessary that we establish the mechanical properties of materials which are properly classified as elastic.

† Accounts of finite-elasticity theory and solutions to certain problems can be found in the books of Green and Zerna [1968] and Green and Adkins [1960].

15.1 INTRODUCTORY REMARKS

Among the oldest theories concerning deformable bodies are those dealing with elastic materials. In 1678, for example, Hooke [1678]† gave the following description of a class of elastic bodies: "It is very evident that the Rule or Law of Nature in every springing body is that the force or power thereof to restore itself to its natural position is always proportionate to the distance or space it is removed therefrom . . . ." While we now recognize that Hooke's law of proportionality need not hold for elastic bodies (and, in fact, seldom holds at all for finite elastic deformations), such materials nevertheless do possess "natural positions" which are restored upon reversing the external forces prescribed in various processes. Thus, in setting out to describe these properties in a more precise manner, two characteristics of elastic bodies immediately suggest themselves: First, thermodynamic processes performed on elastic bodies should be reversible in the sense that there should be no net dissipation at any point in the body. Second, the response of an elastic body should be independent of the history of its deformation; e.g., the stress at each particle in an elastic body is determined by only the current deformation—otherwise, restoration of the body to a reference state by different histories of deformation might result in different stresses. We shall see that the second characteristic imposes weaker restrictions on the definition of elastic bodies than the first.

15.2 ELASTIC MATERIALS

The basic definition of elastic materials follows directly from the observation that the stress in such materials depends only on the current deformation. Indeed, *a material is said to be elastic if the stress at any time t depends only upon the local deformation at time t and not upon the history of the deformation.* It follows that constitutive equations for elastic materials are of the form

$$\sigma^{ij} = \Sigma^{ij}(G_{rs}) \tag{15.1}$$

where $\Sigma^{ij}(G_{rs})$ is a function of the current deformation, as represented here by the current value of the deformation tensor relative to some fixed reference configuration ($G_{ij} = \delta_{ij} + 2\gamma_{ij}$). Obviously, any equivalent measure of deformation such as γ_{ij}, $\mathbf{z}_{,i}$ could be used instead of G_{rs} in (15.1). It is understood, of course, that the particle $\mathbf{x}$ should also appear in the argument of the response function $\Sigma^{ij}(\;)$ in order to describe nonhomogeneous bodies, but for convenience, we shall not list this dependency here.

Since all thermal variables (for example, θ, η, q^i) are ignored in the definition of elastic materials, the theory of elasticity describes purely

† Hooke arrived at his law of proportionality in 1660 and published it as an anagram in 1676. For a historical account of the development of the theory of elasticity, see Truesdell [1952]. A reprint of this article was published in 1966. A much earlier history was written by Todhunter and Pearson [1893].

mechanical behavior of materials, and (15.1) may be referred to as a mechanical constitutive equation. Moreover, on comparing (15.1) with (14.21), we observe that elastic materials constitute a special class of simple materials: (1) which are independent of temperature and temperature histories, and (2) for which, of all deformation histories, the stress depends only upon the "rest" history, i.e., the history during which the deformation has been equal to the current deformation at all times in the past.† Thus, it is permissible to say that elastic materials are a simple class of simple materials.

In evaluating (15.1) in the light of the constitutive axioms discussed in Art. 14, it is evident that the principles of determinism and local action are satisfied and that, for the moment, equipresence has little significance as we have only one constitutive equation. We shall identify physically admissible processes for certain elastic bodies in the following section. Of the remaining requirements of material frame indifference and material symmetry, we may say that the response function Σ^{ij} must be invariant under all observer transformations of the form (14.16) and under transformation of the material frame belonging to the isotropy group of the material. In particular, if an orthogonal transformation $\boldsymbol{\alpha} = [\alpha^i_{.j}(t)]$ belongs to the isotropy group of an elastic material and if

$$\Sigma^{ij}(\boldsymbol{\alpha}\mathbf{G}\boldsymbol{\alpha}^T) = \alpha^i_m \Sigma^{mn}(\mathbf{G})\alpha^j_n \tag{15.2}$$

then the material is isotropic. Otherwise, for a transformation of material coordinates $\bar{\mathbf{x}} = \mathbf{Sx}$, we must have $\bar{\sigma}^{ij} = \bar{\Sigma}^{ij}(\bar{\mathbf{G}})$, where $\bar{\mathbf{G}} = \mathbf{SGS}^{-1}$ and $\bar{\Sigma}^{ij}(\bar{\mathbf{G}}) = \mathbf{S}\Sigma^{ij}(\mathbf{G})\mathbf{S}^{-1}$. We also observe that while G_{ij} is dimensionless, $\Sigma^{ij}(G_{rs})$ must have the dimensions of stress.

Of particular interest is the case of isotropic materials (15.2). Then σ^{ij} is an isotropic function of the symmetric tensor G_{ij}, and a relatively simple form of $\boldsymbol{\Sigma}(\mathbf{G})$ can be obtained. Assuming that $\boldsymbol{\Sigma}(\mathbf{G})$ is analytic in $\mathbf{G}$, we can expand the response function as a power series in $\mathbf{G}$:

$$\boldsymbol{\Sigma}(\mathbf{G}) = a_0\mathbf{I} + a_1\mathbf{G} + a_2\mathbf{G}^2 + \cdots + a_n\mathbf{G}^n + \cdots \tag{15.3}$$

Here $a_0, a_1, \ldots, a_n$ are scalar functions of the components G_{ij} of $\mathbf{G}$. According to the Cayley-Hamilton theorem for 3×3 matrices, $\mathbf{G}$ satisfies its own characteristic equation,

$$\mathbf{G}^3 = I_1\mathbf{G}^2 - I_2\mathbf{G} + I_3\mathbf{1} \tag{15.4}$$

where I_1, I_2, I_3 are the principal invariants‡ of $\mathbf{G}$. Thus $\mathbf{G}^3$ and all higher powers of $\mathbf{G}$ can be expressed in terms of $\mathbf{G}^2$, $\mathbf{G}$, and $\mathbf{1}$ and the principal

† Truesdell and Noll [1965, p. 123] point out that the theory of elasticity applies to *all* simple materials in the case of statics, i.e., in cases of simple materials at rest which have been at rest at all times in the past.

‡ See Eqs. (4.25).

invariants of $\mathbf{G}$. Thus, $\mathbf{\Sigma}(\mathbf{G})$ is an analytic isotropic function of $\mathbf{G}$ if and only if it can be expressed in the form

$$\mathbf{\Sigma}(\mathbf{G}) = \kappa_0 \mathbf{1} + \kappa_1 \mathbf{G} + \kappa_2 \mathbf{G}^2 \tag{15.5}$$

where κ_0, κ_1, and κ_2 are scalar-valued polynomials in the invariants I_1, I_2, I_3. Observing that for the convected coordinates x^i, the unit tensor $\mathbf{1}$ has components G^{ij}, G_{ij} or δ^i_j, it follows from (15.5) that the constitutive equation for an isotropic elastic solid can be written in the forms

$$\begin{aligned} \sigma^{ij} &= \kappa_0 G^{ij} + \kappa_1 \delta^{ir}\delta^{js} G_{rs} + \kappa_2 \delta^{ij} \\ \sigma^i_j &= \kappa_0 \delta^i_j + \kappa_1 \delta^{im}\delta^{rs} G_{jr} G_{ms} + \kappa_2 \delta^{ir} G_{rj} \end{aligned} \tag{15.6}$$

etc. Alternatively, if we select as material coordinates the curvilinear coordinates ξ^i in the reference configuration C_0, then

$$\begin{aligned} \sigma^{ij} &= \kappa_0 G^{ij} + \kappa_1 g^{ir} g^{js} G_{rs} + \kappa_2 g^{ij} \\ \sigma^i_j &= \kappa_0 \delta^i_j + \kappa_1 g^{im} g^{rs} G_{jr} G_{ms} + \kappa_2 g^{ir} G_{rj} \end{aligned} \tag{15.7}$$

where g^{ij} is the contravariant metric tensor in C_0.

15.3 HYPERELASTICITY, THERMODYNAMIC BASIS

An alternate definition of elastic materials arises from the notion of reversibility. Consider the classical example of a force $\mathbf{F}(\mathbf{x})$ acting on a particle as it moves from one point to another. We recall that the independence of the work W on the path of the particle implies the existence of a differential function of current position $W(\mathbf{x})$, called a potential, whose gradient equals the force: $\mathbf{F}(\mathbf{x}) = \nabla W(\mathbf{x})$. The work done about a closed path by such "conservative" forces is zero, and we say that $\mathbf{F}$ can be "derived" from a potential function W. By the same token, we may think of the "path" taken by a continuous body as a process of deformation described, for example, by values of the deformation tensor G_{ij} or the strain tensor γ_{ij}. The "forces" performing work are, of course, represented by the stress tensor σ^{ij}. Then the reversibility of elastic deformations and the independence of path (deformation histories) of elastic bodies lead us to assume the existence of a differentiable function of deformation, say $E(\gamma_{ij})$, from which the stress can be derived. Equivalently, we assume that there exists a potential function $E(\gamma_{ij})$, called the *strain-energy density*, such that

$$\rho \dot{E} = \sigma^{ij} \dot{\gamma}_{ij} \tag{15.8}$$

Then

$$\rho \dot{E}(\gamma_{ij}) = \rho \frac{\partial E}{\partial \gamma_{ij}} \dot{\gamma}_{ij} = \sigma^{ij} \dot{\gamma}_{ji} \tag{15.9}$$

and†,‡

$$\sigma^{ij} = \rho \frac{\partial E}{\partial \gamma_{ij}} \tag{15.10}$$

To distinguish between elastic materials, as described in the previous section, and the class of elastic materials satisfying (15.10), we refer to the latter as *hyperelastic;* i.e., *if the stresses in a body are derivable from a strain-energy function, as in* (15.10), *the material is called hyperelastic.*

To justify the existence of a potential function such as $E(\gamma_{ij})$ within the framework of the thermodynamic principles laid down in Art. 12, recall from (12.14), (12.35), and (12.36) that the local law of conservation of energy can be written in the forms

$$\rho\dot{\epsilon} = \sigma^{ij}\dot{\gamma}_{ij} + Q^i_{;i} + \rho r \tag{15.11}$$

$$\rho\dot{\psi} = \sigma^{ij}\dot{\gamma}_{ij} - \rho\eta\dot{\theta} - \delta \tag{15.12}$$

$$\rho\theta\dot{\eta} = Q^i_{;i} + \rho r + \delta \tag{15.13}$$

where the principles of conservation of mass and balance of linear momentum are locally satisfied. Since the reversibility of deformations of elastic bodies suggests that the internal dissipation δ is zero, we can rewrite (15.11) and (15.12) as

$$\rho\dot{\epsilon} = \sigma^{ij}\dot{\gamma}_{ij} + \rho\theta\dot{\eta} \tag{15.14}$$

$$\rho\dot{\psi} = \sigma^{ij}\dot{\gamma}_{ij} - \rho\eta\dot{\theta} \tag{15.15}$$

Now, from the previous discussion, it is clear that for elastic bodies the constitutive laws should depend upon only the current values of the independent constitutive variables and not upon their histories. Thus, consider a class of materials satisfying (15.14) and (15.15) for which the free energy is a differentiable function of the current values of the deformation (as represented by the strain tensor), the temperature, and the temperature gradients: §

$$\psi = \Psi(\gamma_{ij}, \theta, \theta_{,k}) \tag{15.16a}$$

† It is clear that no loss in generality is suffered by taking the strain-energy density as a function of γ_{ij} rather than G_{ij} since $\gamma_{ij} = (G_{ij} - \delta_{ij})/2$. For example, suppose $E(\gamma_{ij}) = (\frac{1}{2})\bar{E}(G_{ij})$; then $\rho\dot{E} = (\frac{1}{2})\rho\dot{\bar{E}} = (\frac{1}{2})\rho(\partial\bar{E}/\partial G_{rs})(\partial G_{rs}/\partial\gamma_{ij})\dot{\gamma}_{ij}$ and, instead of (15.10), $\sigma^{ij} = \rho(\partial\bar{E}/\partial G_{ij})$.

‡ To avoid error, it should be understood that in differentiating E with respect to γ_{ij}, all other components are held constant, including γ_{ji}, $i \neq j$. Equation (15.10), therefore, should actually be written $\sigma^{ij} = (\frac{1}{2})\rho[(\partial E/\partial\gamma_{ij}) + (\partial E/\partial\gamma_{ji})]$. We avoid this cumbersome notation with the understanding that, in all subsequent formulas, by $\partial/\partial A_{ij}$ we mean $(\frac{1}{2})[(\partial/\partial A_{ij}) + (\partial/\partial A_{ji})]$ for any tensor A_{ij}.

§ Again, for nonhomogeneous and anisotropic bodies we should also include the particle $\mathbf{x}$ and possibly certain preferred material directors $\mathbf{d}_k$ in the argument of Ψ, but for simplicity in notation, we shall list only γ_{ij}, θ, and $\theta_{,k}$, the dependence on $\mathbf{x}$ and $\mathbf{d}_k$ being understood.

According to the equipresence principle,† all other constitutive functions (σ^{ij},η,Q^i) must also be functions of this same collection of variables:

$$\sigma^{ij} = \hat{\Sigma}^{ij}(\gamma_{ij},\theta,\theta_{,k}) \tag{15.16b}$$

$$\eta = \mathcal{N}(\gamma_{ij},\theta,\theta_{,k}) \tag{15.16c}$$

$$Q^i = \mathfrak{Q}^i(\gamma_{ij},\theta,\theta_{,k}) \tag{15.16d}$$

Alternatively, since $\psi = \epsilon - \eta\theta$, we could use instead of (15.16*a*) the internal energy

$$\epsilon = \Xi(\gamma_{ij},\eta,\theta_{,k}) \tag{15.17a}$$

and

$$\sigma^{ij} = \tilde{\Sigma}^{ij}(\gamma_{ij},\eta,\theta_{,k}) \tag{15.17b}$$

$$\theta = \Theta(\gamma_{ij},\eta,\theta_{,k}) \tag{15.17c}$$

$$Q^i = \tilde{\mathfrak{Q}}^i(\gamma_{ij},\eta,\theta_{,k}) \tag{15.17d}$$

Clearly,

$$\dot{\psi} = \frac{\partial \Psi}{\partial \gamma_{ij}}\dot{\gamma}_{ij} + \frac{\partial \Psi}{\partial \theta}\dot{\theta} + \frac{\partial \Psi}{\partial \theta_{,k}}\dot{\theta}_{,k} \tag{15.18}$$

and

$$\dot{\epsilon} = \frac{\partial \Xi}{\partial \gamma_{ij}}\dot{\gamma}_{ij} + \frac{\partial \Xi}{\partial \eta}\dot{\eta} + \frac{\partial \Xi}{\partial \theta_{,k}}\dot{\theta}_{,k} \tag{15.19}$$

Introducing (15.18) and (15.19) into the Clausius-Duhem inequality (12.31), making use of (12.33) and (12.36), and rearranging terms, we find

$$\left(\sigma^{ij} - \rho\frac{\partial \Psi}{\partial \gamma_{ij}}\right)\dot{\gamma}_{ij} - \rho\left(\eta + \frac{\partial \Psi}{\partial \theta}\right)\dot{\theta} - \rho\frac{\partial \Psi}{\partial \theta_{,k}}\dot{\theta}_{,k} + \frac{1}{\theta}\mathbf{q}\cdot\nabla\theta \geq 0 \tag{15.20a}$$

$$\left(\sigma^{ij} - \rho\frac{\partial \Xi}{\partial \gamma_{ij}}\right)\dot{\gamma}_{ij} + \rho\left(\theta - \frac{\partial \Xi}{\partial \eta}\right)\dot{\eta} - \rho\frac{\partial \Xi}{\partial \theta_{,k}}\dot{\theta}_{,k} + \frac{1}{\theta}\mathbf{q}\cdot\nabla\theta \geq 0 \tag{15.20b}$$

Since these equations must hold for arbitrary $\dot{\gamma}_{ij}$, $\dot{\eta}$, $\dot{\theta}$, and $\dot{\theta}_{,k}$, we conclude that

$$\sigma^{ij} = \rho\frac{\partial \Psi}{\partial \gamma_{ij}} = \rho\frac{\partial \Xi}{\partial \gamma_{ij}} \tag{15.21a}$$

$$\eta = -\frac{\partial \Psi}{\partial \theta} \qquad \theta = \frac{\partial \Xi}{\partial \eta} \tag{15.21b,c}$$

$$\frac{\partial \Psi}{\partial \theta_{,k}} = \frac{\partial \Xi}{\partial \theta_{,k}} = 0 \tag{15.21d}$$

† See rule 3, Sec. 14.2.

Equation (15.21*d*) shows that if $\delta = 0$, the free energy of materials of the type in (15.16) and the internal energy of the materials (15.17) are independent of the temperature gradients $\theta_{,k}$. In view of (15.21*a*) and (15.21*b*), this implies that σ^{ij}, Ξ, and η are then also independent of $\theta_{,k}$; so that instead of (15.16) we may write

$$\begin{aligned} \sigma^{ij} &= \hat{\Sigma}^{ij}(\gamma_{ij},\theta) \\ \eta &= \mathcal{N}(\gamma_{ij},\theta) \\ Q^i &= \mathfrak{Q}^i(\gamma_{ij},\theta,\theta_{,k}) \\ \psi &= \Psi(\gamma_{ij},\theta) \end{aligned} \tag{15.22}$$

where

$$\hat{\Sigma}^{ij}(\gamma_{ij},\theta) = \rho \frac{\partial \Psi}{\partial \gamma_{ij}} \qquad \text{and} \qquad \mathcal{N}(\gamma_{ij},\theta) = -\frac{\partial \Psi}{\partial \theta} \tag{15.23}$$

while, instead of (15.17),

$$\begin{aligned} \sigma^{ij} &= \tilde{\boldsymbol{\Sigma}}^{ij}(\gamma_{ij},\eta) \\ \theta &= \Theta(\gamma_{ij},\eta) \\ Q^i &= \tilde{\mathfrak{Q}}^i(\gamma_{ij},\eta,\theta_{,k}) \\ \epsilon &= \Xi(\gamma_{ij},\eta) \end{aligned} \tag{15.24}$$

where

$$\tilde{\Sigma}^{ij}(\gamma_{ij},\eta) = \rho \frac{\partial \Xi}{\partial \gamma_{ij}} \qquad \text{and} \qquad \Theta(\gamma_{ij},\eta) = \frac{\partial \Xi}{\partial \eta} \tag{15.25}$$

Comparing (15.22) with (14.15), we see that (15.22) describes a special class of simple materials the response of which depends only upon the current values of γ_{ij} and θ and not upon their histories.† Materials defined by (15.24) are called *perfect*‡ materials.

Now we are seeking thermodynamic processes and constitutive equations that admit the existence of a potential function of the type in (15.10). Toward this end, consider first isothermal processes performed on materials of the types characterized by (15.22). In such processes, θ is constant, $\dot{\theta} = 0$, and we may write $\psi = \Psi(\gamma_{ij})$. Then (15.15) and (15.21) yield

$$\rho\dot{\psi} = \sigma^{ij}\dot{\gamma}_{ij} \qquad \sigma^{ij} = \rho \frac{\partial \Psi}{\partial \gamma_{ij}} \tag{15.26}$$

Secondly, consider isentropic processes performed on materials of the type

† Alternatively, (15.22) can be interpreted as the constitutive equations of any simple material in "thermal equilibrium" (Truesdell and Noll [1965, p. 300]). See the first footnote on page 168.

‡ This terminology is used by Truesdell and Noll [1965, p. 296].

characterized by (15.24). In such processes, $\dot{\eta} = 0$ and we may write $\epsilon = \Xi(\gamma_{ij})$. Then (15.14) and (15.21) yield

$$\rho\dot{\epsilon} = \sigma^{ij}\dot{\gamma}_{ij} \qquad \sigma^{ij} = \rho\,\frac{\partial \Xi}{\partial \gamma_{ij}} \tag{15.27}$$

Comparing (15.26) and (15.27) with (15.8) and (15.10), we see that a theory of hyperelasticity is appropriate for isothermal processes performed on the class of simple materials defined by (15.22) and for isentropic processes performed on perfect materials; in the former case the strain energy is associated with the free energy, while in the latter case it is associated with the internal energy. More generally, if we assume that the second member of (15.24) can be inverted to obtain η as a function of γ_{ij} and θ, then η can be eliminated from the remaining constitutive equations (15.24) for perfect materials and these can be recast as functions of γ_{ij} and θ. Then a theory of hyperelasticity is appropriate for both isothermal and isentropic processes† performed on perfect materials and simple materials of the class (15.22) (or, for that matter, isothermal and isentropic processes on any simple material in thermal equilibrium).

Of course, in view of (15.21*a*), there is nothing to prevent us from deriving σ^{ij} from the potential $\Psi(\gamma_{ij},\theta)$ for *any* thermodynamic process performed on materials characterized by (15.22). However, we would then obtain a theory of thermoelasticity, as the stress would depend upon θ as well as γ_{ij}. We prefer, instead, to postpone discussions of thermoelasticity until the next chapter and to restrict our attention here to purely mechanical behavior. Thus, in the remainder of this chapter we ignore all thermal effects and concentrate on materials defined by only one constitutive function—the strain energy $E(\gamma_{ij})$ or the function $\Sigma^{ij}(G_{rs})$ of (15.1).

15.4 TOTAL POTENTIALS

We remark that if the external forces **F** and **S** are also derivable from a potential function V, for example,

$$V = -\int_v \rho\mathbf{F}\cdot\mathbf{u}\,dv - \int_A \mathbf{S}\cdot\mathbf{u}\,dA \tag{15.28}$$

we can define the total potential energy Π of the body as the sum

$$\Pi = U + V \tag{15.29}$$

where U is the total strain energy:

$$U = \int_v \rho E(\gamma_{ij})\,dv \tag{15.30}$$

† Insofar as manipulating terms in the energy equation is concerned, we could also concoct an additional case: Consider isoenergetic process ($\dot{\epsilon} = 0$) on materials satisfying (15.14) for which $\eta = \mathscr{N}(\gamma_{ij}, \epsilon)$. Then the strain energy is associated with the specific entropy.

Following a standard procedure of computing variations in Π due to admissible variations δu_i in the displacement field (i.e., those which satisfy all kinematic constraints imposed on the deformation), we find, on integrating by parts, that the first variation in Π is

$$\delta\Pi = \int_v (\sigma^{ij}_{;i} + \rho F^j)\, \delta u_j \, dv \tag{15.31}$$

where use of (15.10) has been made. Recognizing that the quantity in parentheses in (15.31) is zero if the body is in equilibrium, we arrive at the well-known *principle of stationary potential energy: Of all the admissible displacement fields an elastic body can assume, those corresponding to states of static equilibrium make the total potential energy assume a stationary value* ($\delta\Pi = 0$). Further, it can also be shown to follow that if $\delta\Pi = 0$ corresponds to a state of stable equilibrium, Π is a minimum.

In the more general case of dynamics of conservative systems, we can expand the total potential to include the kinetic energy. The lagrangian potential L is defined by

$$L = \kappa - \Pi \tag{15.32}$$

where κ is the kinetic energy defined in (12.3). *Lagrange's principle* (*or Hamilton's principle for conservative systems*) states that *of all the motions that will carry a conservative system from one configuration to another during a time interval* (t_0,t_1), *that which occurs provides a stationary value* ($\delta A = 0$) *to the action integral* A, *where*

$$A = \int_{t_0}^{t_1} L \, dt \tag{15.33}$$

15.5 THE STRAIN-ENERGY FUNCTION

The geometry of an elastic body is generally assumed to be known when the body occupies a reference configuration C_0 which, ideally, is associated with an undistorted state of the material. For this reason, it is more convenient to use instead of $E(\gamma_{ij})$ an elastic potential $W(\gamma_{ij})$ measured per unit volume v_0 of the undistorted body. Since mass is conserved,

$$W = \rho_0 E \tag{15.34}$$

where ρ_0 is the mass density in C_0 ($\rho_0 = \sqrt{G}\,\rho$). Hence, instead of (15.10), we have

$$\sigma^{ij} = \frac{1}{\sqrt{G}} \frac{\partial W}{\partial \gamma_{ij}} \tag{15.35}$$

Alternatively, we can calculate the stress tensor t^{ij} of (5.14) by

$$t^{ij} = \frac{\partial W}{\partial \gamma_{ij}} \tag{15.36}$$

In the following we list various properties and special forms of the strain-energy function W.

Anisotropic materials The isotropy groups for anisotropic elastic materials are generally divided into 12 distinct types, each defining a specific class of anisotropy. The first 11 of these correspond to the 32 crystal classes associated with the classification of crystalline solids in mineralogy† and belong to six basic types of crystalline systems: triclinic, monoclinic, rhombic, tetragonal, cubic, and hexagonal. Briefly,‡ to describe the 11 crystalline types we need only list those right-handed rotations $\mathbf{S}$ which generate the isotropy groups $\mathscr{G}$ for each type of material, i.e., the set of elements of each group which, together with their inverses, may be multiplied together to give all elements of the group. Let $\mathbf{S}^{\mathbf{n},\beta}$ denote the right-handed rotation through β, $0 < \beta < 2\pi$, about an axis in the direction of the unit vector $\mathbf{n}$. Then the 11 classes are:§

Triclinic:

1. $\mathbf{I}, -\mathbf{I}$

Monoclinic:

2. $\mathbf{S}^{\mathbf{i}_3,\pi}$

Rhombic:

3. $\mathbf{S}^{\mathbf{i}_1,\pi}, \mathbf{S}^{\mathbf{i}_2,\pi}$

Tetragonal:

4. $\mathbf{S}^{\mathbf{i}_3,\pi/2}$
5. $\mathbf{S}^{\mathbf{i}_3,\pi/2}, \mathbf{S}^{\mathbf{i}_1,\pi}$

Cubic:

6. $\mathbf{S}^{\mathbf{i}_1,\pi}, \mathbf{S}^{\mathbf{i}_2,\pi}, \mathbf{S}^{\mathbf{n},2\pi/3}$
7. $\mathbf{S}^{\mathbf{i}_1,\pi/2}, \mathbf{S}^{\mathbf{i}_2,\pi/2}, \mathbf{S}^{\mathbf{i}_3,\pi/2}$

Hexagonal:

8. $\mathbf{S}^{\mathbf{i}_3,2\pi/3}$
9. $\mathbf{S}^{\mathbf{i}_1,\pi}, \mathbf{S}^{\mathbf{i}_3,2\pi/3}$
10. $\mathbf{S}^{\mathbf{i}_3,\pi/3}$
11. $\mathbf{S}^{\mathbf{i}_1,\pi}, \mathbf{S}^{\mathbf{i}_3,\pi/3}$

Here $\mathbf{i}_i$ are orthormal basis vectors and $\mathbf{n} = (\mathbf{i}_1 + \mathbf{i}_2 + \mathbf{i}_3)/\sqrt{3}$. To these we may add the twelfth type of anisotropy, called *transverse isotropy*, the isotropy group of which is generated by the identity transformation $\mathbf{I}$ and $\mathbf{S}^{\mathbf{i}_3,\beta}$, where β is now any angle between 0 and 2π. So-called *orthotropic* materials are contained in groups 3, 5, 6, and 7; these contain reflections on three mutually perpendicular planes.

To demonstrate only one of numerous cases in which the isotropy groups can be used to reduce the form of W, consider an anisotropic material with monoclinic symmetry for which the strain-energy function is a polynomial in the strain components γ_{11}, γ_{22}, γ_{33}, γ_{12}, γ_{13}, and γ_{23}. Since

† See Dana and Hurlbut [1959].

‡ For further details, consult Smith and Rivlin [1958], Green and Adkins [1960], and Truesdell and Noll [1965].

§ See Truesdell and Noll [1965, p. 83].

$W(\boldsymbol{\gamma}) = W(\mathbf{S}\boldsymbol{\gamma}\mathbf{S}^{-1})$ for every $\mathbf{S}$ belonging to the isotropy group, we have for the group generated by $\mathbf{S}^{\mathbf{i}_3,\pi}$

$$W(\gamma_{11},\gamma_{22},\gamma_{33},\gamma_{12},\gamma_{13},\gamma_{23}) = W(\gamma_{11},\gamma_{22},\gamma_{33},\gamma_{12},-\gamma_{13},-\gamma_{23}) \tag{15.37}$$

Employing an often-used theorem from group theory† that states that a polynomial basis for polynomials symmetric in variables $(a_1, a_2, \ldots, a_n)$ and $(b_1, b_2, \ldots, b_n)$ is formed by the quantities $(a_i + b_i)/2$ and $(a_ib_j + a_jb_i)/2$ $(i, j = 1, 2, \ldots, n)$, we set $(a_1,a_2) = -(b_1,b_2) = (\gamma_{13},\gamma_{23})$ and conclude that in the present case W can be expressed as a polynomial in seven quantities—γ_{11}, γ_{22}, γ_{33}, γ_{12}, γ_{13}^2, γ_{23}^2, and $\gamma_{13}\gamma_{23}$. Similar procedures can be used for other classes of anisotropic materials.‡

Polynomial representation If we assume that $W(\gamma_{ij})$ is analytic in its arguments γ_{ij}, we can express the strain-energy function in a series of the form

$$W(\gamma_{ij}) = \sum_{i=1}^{\infty} W_i(\gamma_{ij}) \tag{15.38}$$

where $W_i(\gamma_{ij})$ is a homogeneous polynomial of degree i in γ_{ij}. Specifically, (15.38) can be rewritten

$$W = W_0 + E^{ij}\gamma_{ij} + \tfrac{1}{2}E^{ijmn}\gamma_{ij}\gamma_{mn} + \tfrac{1}{3}E^{ijmnrs}\gamma_{ij}\gamma_{mn}\gamma_{rs} + \cdots \tag{15.39}$$

where W_0 is an arbitrary constant and the coefficients E^{ij}, E^{ijmn}, E^{ijmnrs}, ... are called *elasticities* of order $0, 1, 2, \ldots$. For homogeneous bodies, the elasticities are arrays of material constants. Since both σ^{ij} (or t^{ij}) and γ_{ij} are symmetric, elasticities of all orders are symmetric in each successive pair of indices:

$$\begin{aligned} E^{ij} &= E^{ji} \\ E^{ijmn} &= E^{jimn} = E^{ijnm} = E^{mnij} \\ E^{ijmnrs} &= E^{jimnrs} = E^{ijnmrs} = E^{ijmnsr} = E^{ijsrmn} \\ &= E^{mnrsij} = E^{rsijmn} \cdots \end{aligned} \tag{15.40}$$

From (15.36) it follows that for this class of materials,

$$t^{ij} = E^{ij} + E^{ijmn}\gamma_{mn} + E^{ijmnrs}\gamma_{mn}\gamma_{rs} + \cdots \tag{15.41}$$

Clearly, $t^{ij} = E^{ij}$ if $\gamma_{mn} = 0$; hence E^{ij} represents an "initial stress" in the reference configuration C_0. If we select C_0 so that it corresponds to a natural unstressed state, we can set $E^{ij} = 0$. Then, dropping the irrelevant constant W_0, (15.39) becomes

$$W = \tfrac{1}{2}E^{ijmn}\gamma_{ij}\gamma_{mn} + \tfrac{1}{3}E^{ijmnrs}\gamma_{ij}\gamma_{mn}\gamma_{rs} + \cdots \tag{15.42}$$

† See Weyl [1946, pp. 36, 53, 276].
‡ A complete catalog of invariants for polynomial expansions of W for the various isotropy groups has been given by Smith and Rivlin [1958].

Various approximate theories of hyperelasticity can be obtained by retaining only a finite number of terms of the expansion (15.42). For example, on the basis of the assumption that the strains γ_{ij} are "small," we can obtain the strain-energy function for classical infinitesimal elasticity by retaining only the first term in (15.42):†

$$W = \tfrac{1}{2}E^{ijmn}\gamma_{ij}\gamma_{mn} \tag{15.43a}$$

Then

$$t^{ij} = E^{ijmn}\gamma_{mn} \tag{15.43b}$$

Although the components γ_{ij} in (15.43) are assumed to be infinitesimal, the equations do not necessarily lead to a linear theory since γ_{ij} can contain nonlinear terms in the displacement gradients $u_{i,j}$. The strain-energy function for classical, linear elasticity is obtained from (15.43) by assuming that, in addition to γ_{ij}, the rotations ω_{ij} (and $u_{i,j}$) are infinitesimal. Then $\gamma_{ij} = e_{ij}$, where e_{ij} is the infinitesimal strain tensor of (4.18*a*), and

$$W = \tfrac{1}{2}E^{ijmn}e_{ij}e_{mn} \qquad t^{ij} = E^{ijmn}e_{mn} \tag{15.44a,b}$$

Materials obeying (15.43*b*) or (15.44*b*) are called *hookean*.

We obtain from (15.42) a *second-order* theory of hyperelasticity by retaining terms up to and including those involving second-order elasticities:

$$W = \tfrac{1}{2}E^{ijmn}\gamma_{ij}\gamma_{mn} + \tfrac{1}{3}E^{ijmnrs}\gamma_{ij}\gamma_{mn}\gamma_{rs} \tag{15.45a}$$

$$t^{ij} = E^{ijmn}\gamma_{mn} + E^{ijmnrs}\gamma_{mn}\gamma_{rs} \tag{15.45b}$$

Similarly, equations for various higher-order theories can be obtained by simply retaining additional terms in (15.42).

In view of (15.41), (15.43*b*), and (15.45*b*), it is clear that approximate theories of elasticity can be obtained by simply expanding the response function $\Sigma^{ij}(G_{rs}) = \Sigma^{ij}(\delta_{rs} + 2\gamma_{rs})$ in powers of G_{rs} or γ_{rs}. Such expansions are identical in form to the above equations for t^{ij}. However, the corresponding elasticities may not always coincide. We elaborate further on this latter point near the end of this section.

Isotropic materials Since the isotropy group for isotropic hyperelastic materials is the full orthogonal group, $W(\gamma_{ij}) = W(\alpha_i^m\gamma_{mn}\alpha_j^n)$, and it can be shown to follow that W is then a function of the principal invariants of γ_{ij} or, more conveniently, the principal invariants of G_{ij}:

$$W = W(I_1,I_2,I_3) \tag{15.46}$$

† It is understood that the functional form of (15.43*a*) et seq. is not the same as that of (15.42), but it is convenient to use W in both cases. Unless confusion is likely, we continue to use W to express various forms of the strain energy throughout the remainder of this chapter.

where I_1, I_2, I_3 are defined in (4.25). For materials defined by strain-energy functions of this type,

$$t^{ij} = \frac{\partial W}{\partial I_1}\frac{\partial I_1}{\partial \gamma_{ij}} + \frac{\partial W}{\partial I_2}\frac{\partial I_2}{\partial \gamma_{ij}} + \frac{\partial W}{\partial I_3}\frac{\partial I_3}{\partial \gamma_{ij}} \tag{15.47}$$

or, in view of (4.28),

$$t^{ij} = 2\frac{\partial W}{\partial I_1}\delta^{ij} + 4\frac{\partial W}{\partial I_2}[\delta^{ij}(1 + \gamma_{rr}) - \delta^{ir}\delta^{js}\gamma_{sr}]$$
$$+ 2\frac{\partial W}{\partial I_3}[\delta^{ij}(1 + 2\gamma_{rr}) - 2\delta^{ir}\delta^{js}\gamma_{sr} + 2\epsilon^{imn}\epsilon^{jrs}\gamma_{mr}\gamma_{ns}] \tag{15.48}$$

To obtain various approximate forms of W, it is again natural to consider expansions in powers of the invariants:

$$W = \sum_{r=0}^{\infty}\sum_{s=0}^{\infty}\sum_{t=0}^{\infty} C_{rst}(I_1 - 3)^r(I_2 - 3)^s(I_3 - 1)^t \qquad C_{000} = 0 \tag{15.49}$$

By retaining only linear terms in the invariants, for example, we arrive at the approximation

$$W = C_{100}(I_1 - 3) + C_{010}(I_2 - 3) + C_{001}(I_3 - 1) \tag{15.50}$$

An alternate polynomial approximation was proposed by Kavanagh [1969],† who assumed that

$$W = A_1K_1^2 + A_2K_2 + A_3K_1^3 + A_4K_1K_2 + A_5K_3 + A_6K_1^4$$
$$+ A_7K_1^2K_2 + A_8K_1K_3 + A_9K_2^2 \tag{15.51}$$

where $A_1, \ldots, A_9$ are material constants and K_1, K_2, and K_3 are the strain invariants

$$K_1 = \gamma_{rr} \qquad K_2 = \gamma_{rs}\gamma_{rs} \qquad K_3 = \gamma_{ij}\gamma_{ir}\gamma_{jr} \tag{15.52}$$

Thus (15.51) contains terms up to and including those of fourth order in the strain components. Alternatively, we can express I_1, I_2, and I_3 in terms of the strain components with the aid of (4.26) and obtain the coefficients C_{rst} in terms of the elasticities in (15.42). In the case of isotropic materials, we find that the number of independent components of the elasticities E^{ijmn}, $E^{ijmnrs}, \ldots$ is a minimum. For example, the first-order elasticities are then defined in terms of only the two Lamé coefficients λ and μ by the formula

$$E^{ijmn} = \lambda\delta^{ij}\delta^{mn} + \mu(\delta^{im}\delta^{jn} + \delta^{in}\delta^{jm}) \tag{15.53}$$

† This thesis also contains specific values of $A_1, \ldots, A_9$, which are obtained from experiments on polyurethane foam.

while the second-order elasticities for isotropic elastic materials are given by

$$E^{ijmnrs} = E_1\delta^{ij}\delta^{mn}\delta^{rs} + E_2(\delta^{ij}\delta^{mn}\delta^{sr} - \delta^{ij}\delta^{ms}\delta^{nr}) + E_3\delta^{im}\delta^{jn}\delta^{sr} + E_4\delta^{im}\delta^{sn}\delta^{rj} \tag{15.54}$$

Here E_1, E_2, E_3, and E_4 are *second-order elasticities*† for isotropic elastic materials. For isotropic hyperelastic bodies, it can be shown that $E_2 + E_3 = 2(\lambda - \mu)$; then there are only three independent second-order elasticities.

Incompressible solids Many important materials that are regarded as elastic for finite deformations are known to be capable of sustaining severe deformations without an appreciable change in volume. These materials fall into the category of incompressible elastic solids, and practically all existing solutions to specific problems in finite elasticity deal with materials of this type. Aside from the fact that all motions of incompressible materials are volume-preserving, the most significant characteristic of such materials is that *the stress tensor is not completely determined by the deformation.* Indeed, it is clear that we may add to the stresses in a deformed incompressible material any multiple of those stresses usually associated with a change in volume, i.e., an arbitrary hydrostatic pressure, without changing in any way the deformation of the body. In other words, the addition of a hydrostatic pressure to an incompressible elastic body obviously alters the stress, but it cannot affect the strain or, in the case of hyperelastic materials, the strain energy. Recalling from (4.33) that a unit value of the third principal invariant I_3 corresponds to isochoric motions, it follows that for incompressible materials the constitutive equation for stress is of the form

$$\sigma^{ij} = \bar{\Sigma}^{ij}(G_{rs}) + hG^{ij} \tag{15.55}$$

h being the hydrostatic pressure, in which G_{rs} must be such that the *incompressibility condition*

$$I_3 = 1 \tag{15.56}$$

is satisfied.

In the case of hyperelastic materials, we may regard (15.56) as a condition of constraint. Then, defining

$$\overline{W} = W(\gamma_{ij}) + \lambda(I_3 - 1) \tag{15.57}$$

† Values of such higher-order coefficients have been determined experimentally for materials such as iron and copper by Seeger and Buck [1960]. See also Seeger [1964]. Third-order theories have been investigated by John [1958], among others. A brief survey is given by Truesdell and Noll [1965, p. 230]. See also Sakadi [1949] and Sheng [1955].

where λ is a Lagrange multiplier, we have, instead of (15.36),

$$t^{ij} = \frac{\partial W}{\partial \gamma_{ij}} + \lambda \frac{\partial I_3}{\partial \gamma_{ij}} = \frac{\partial W}{\partial \gamma_{ij}} + hG^{ij} \tag{15.58}$$

where we have equated 2λ to the hydrostatic pressure h and γ_{ij} satisfies (15.56).

In view of (15.56), the strain-energy function for isotropic incompressible hyperelastic solids is a function of only the first two principal invariants:

$$W = W(I_1,I_2) \tag{15.59}$$

Then (15.47) and (15.48) become

$$\begin{aligned} t^{ij} &= \frac{\partial W}{\partial I_1}\frac{\partial I_1}{\partial \gamma_{ij}} + \frac{\partial W}{\partial I_2}\frac{\partial I_2}{\partial \gamma_{ij}} + h\frac{\partial I_3}{\partial \gamma_{ij}} \\ &= 2\frac{\partial W}{\partial I_1}\delta^{ij} + 4\frac{\partial W}{\partial I_2}[\delta^{ij}(1+\gamma_{rr}) - \delta^{ir}\delta^{js}\gamma_{sr}] \\ &\qquad + 2h[\delta^{ij}(1+2\gamma_{rr}) - 2\delta^{ir}\delta^{js}\gamma_{sr} + 2\epsilon^{imn}\epsilon^{jrs}\gamma_{mr}\gamma_{ns}] \end{aligned} \tag{15.60}$$

Various approximations of $W(I_1,I_2)$ for specific materials have been proposed. For polynomial approximations, we may take as the general form

$$W = \sum_{r=0}^{\infty}\sum_{s=0}^{\infty} C_{rs}(I_1-3)^r(I_2-3)^s \qquad C_{00} = 0 \tag{15.61}$$

The most widely used special case of (15.61) is that of the Mooney [1940] material,† which follows from (15.61) by retaining only linear terms in I_1 and I_2:

$$W = C_1(I_1-3) + C_2(I_2-3) \tag{15.62}$$

Here $C_1 = C_{10}$, $C_2 = C_{01}$. The Mooney form has been found to be appropriate for a fairly large range of deformations of certain natural and vulcanized rubbers. For strains larger than 450 to 500 percent, however, experiments on natural rubbers indicate a departure from the Mooney law.

Using a gaussian kinetic theory for rubbery materials, Treloar [1958] arrived at the so-called neo-hookean form

$$W = C(I_1-3) \tag{15.63}$$

Except for certain vulcanized rubbers highly swollen with organic solvents, however, (15.63) leads to results which only roughly agree with experiments.

† Rivlin [1949*a*, 1949*b*] solved a number of special problems using the Mooney law; for this reason, (15.62) is sometimes referred to as the strain-energy function of a Mooney-Rivlin material. See also Rivlin [1948*a*, 1948*b*].

Based on more extensive experiments with rubbers, Rivlin and Saunders [1951] suggest as a more general form of strain-energy function

$$W = C_1(I_1 - 3) + F(I_2 - 3) \tag{15.64}$$

where the form of $F(I_2 - 3)$ may vary from one type of material to another. Various polynomial approximations of $F(I_2 - 3)$ have been proposed. For example, Klosner and Segal [1969]† proposed that $F(I_2 - 3)$ be a cubic in $(I_2 - 3)$:

$$W = C_1(I_1 - 3) + C_2(I_2 - 3) + C_3(I_2 - 3)^2 + C_4(I_2 - 3)^3 \tag{15.65}$$

Not all polynomial forms of W that have been proposed, however, conform to (15.64). For example, we may also consider the quadratic in $(I_1 - 3)$,‡

$$W = C_1(I_1 - 3) + B_1(I_1 - 3)^2 + C_2(I_2 - 3) \tag{15.66}$$

Alternatively, Biderman [1958]§ suggests for a sulfur-filled rubber

$$W = C_1(I_1 - 3) + B_1(I_1 - 3)^2 + B_2(I_1 - 3)^3 + C_2(I_2 - 3) \tag{15.67}$$

Not all proposed forms of W for rubberlike materials have regarded the function $F(I_2 - 3)$ in (15.64) as a polynomial in $(I_2 - 3)$. Using a non-gaussian molecular theory‖ as a guide, Gent and Thomas [1958] assumed that $\partial F(I_2 - 3)/\partial I_2 = C/I_2$, where C is a material constant. Hart-Smith [1966]¶ elaborated on this theory and proposed the exponential-hyperbolic law

$$W = C\left(\int e^{k_1(I_1-3)^2}\, dI_1 + k_2 \ln \frac{I_2}{3}\right) \tag{15.68}$$

Similarly, Alexander [1968] proposed the forms

$$W = C_1(I_1 - 3) + C_2(I_2 - 3) + C_3 \ln \frac{I_2 - 3 + k}{k} \tag{15.69}$$

and

$$W = C_1 \int e^{k(I_1-3)^2}\, dI_1 + C_2(I_2 - 3) + C_3 \ln \frac{I_2 - 3 + k_1}{k_1} \tag{15.70}$$

† This form apparently yields good agreement with experiments on natural rubbers for $I_1, I_2 < 8$.

‡ This form of W was proposed by Isihara, Hashitsume, and Tatibana [1951] who arrived at it by using a nongaussian molecular theory for rubbers. Alexander [1968] points out that it gives poor results for certain uniaxial tension experiments. Another "three-constant" theory was proposed by Signorini [1955].

§ While (15.67) reportedly gives excellent results in uniaxial experiments, Alexander [1968] points out that it is not in good agreement with biaxial experiments of Treloar [1944].

‖ Thomas [1955].

¶ See also Hart-Smith and Crisp [1967].

wherein C_1, C_2, C_3, k, and k_1 are material constants. Equation (15.69) conforms to the Rivlin-Saunders form (15.64), whereas (15.70) combines the characteristics of (15.64) and (15.58) and apparently gives good agreement with experiments performed on neoprene film.

As a final example, we mention that Hutchinson, Becker, and Landel [1965]† proposed the following strain-energy function for a filled dimethyl siloxane rubber:

$$W = C_1(I_1 - 3) + B_1(I_1 - 3)^2 + B_2(1 - e^{k_1(I_2-3)}) + B_3(1 - e^{k_2(I_2-3)}) \tag{15.71}$$

This form of W yielded results in good agreement with experimental data obtained from both uniaxial and biaxial tests.

Integrability conditions Technically, not every elastic material need be hyperelastic. Indeed, for a material to qualify as an elastic material, all that is needed is a constitutive equation, consistent with the constitutive principles of Art. 14, that gives the stress tensor as a function of the current values of G_{ij} (or γ_{ij}). For an elastic material to be also hyperelastic, it is necessary that

$$\sqrt{G}\,\Sigma^{ij}(G_{rs}) = \frac{\partial W}{\partial \gamma_{ij}} \tag{15.72}$$

where $\Sigma^{ij}(G_{rs})$ is the response function defined in (15.1). Equations (15.72) may, therefore, be regarded as a system of partial differential equations for the strain-energy function W. If these equations are integrable, then the elastic material defined by $\Sigma^{ij}(G_{rs})$ is also hyperelastic. Consider, for example, an isotropic elastic material described by (15.6). In this case, we require that

$$\hat{\kappa}_0 G^{ij} + \hat{\kappa}_1 \delta^{in}\delta^{jm} G_{nm} + \hat{\kappa}_2 \delta^{ij} = \frac{\partial W}{\partial I_1}\frac{\partial I_1}{\partial \gamma_{ij}} + \frac{\partial W}{\partial I_2}\frac{\partial I_2}{\partial \gamma_{ij}} + \frac{\partial W}{\partial I_3}\frac{\partial I_3}{\partial \gamma_{ij}} \tag{15.73}$$

where $\hat{\kappa}_\alpha = \sqrt{G}\,\kappa_\alpha$, $\alpha = 0, 1, 2$, and $\hat{\kappa}_\alpha$ are functions of the principal invariants I_1, I_2, I_3. In view of (4.16) and (4.28),

$$\hat{\kappa}_0 = 2I_3\frac{\partial W}{\partial I_3} \qquad \hat{\kappa}_1 = -2\frac{\partial W}{\partial I_2} \qquad \hat{\kappa}_2 = 2\frac{\partial W}{\partial I_1} + 2I_1\frac{\partial W}{\partial I_2} \tag{15.74}$$

These equations can be regarded as a system of partial differential equations

† See also San Miguel [1965].

for the response coefficients $\hat{\kappa}_0$, $\hat{\kappa}_1$, $\hat{\kappa}_2$. The integrability conditions are

$$\frac{\partial \hat{\kappa}_0}{\partial I_2} + I_3 \frac{\partial \hat{\kappa}_1}{\partial I_3} = 0$$

$$I_3 \frac{\partial \hat{\kappa}_2}{\partial I_3} - \frac{\partial \hat{\kappa}_0}{\partial I_1} - I_1 \frac{\partial \hat{\kappa}_0}{\partial I_2} = 0 \tag{15.75}$$

$$\frac{\partial \hat{\kappa}_2}{\partial I_2} + \frac{\partial \hat{\kappa}_1}{\partial I_1} + I_1 \frac{\partial \hat{\kappa}_1}{\partial I_2} = 0$$

Equations (15.75) constitute a set of necessary and sufficient conditions for an isotropic elastic material to be also hyperelastic. Similar yet more involved procedures can be used to obtain integrability conditions for anisotropic elastic materials.† Since to assume otherwise might admit to the dissipation of energy in processes performed on elastic bodies, we assume in most of the developments to follow that such integrability conditions are satisfied.

16 FINITE ELEMENTS OF ELASTIC BODIES

Due to the highly nonlinear character of the governing equations of finite elasticity, quantitative solutions to almost every problem of great practical importance must be obtained numerically. The finite-element concept, with its inherent simplicity and generality, provides the most convenient apparatus for formulating nonlinear elasticity problems for numerical solutions.‡ In the present article, we develop general equations of motion and equilibrium for typical elements of elastic bodies.

16.1 NONLINEAR STIFFNESS RELATIONS

We recall from (13.60) that the general equation of motion of a typical finite element of a continuum is given by

$$m_{NM}\ddot{u}_i^M + \int_{v_{0(e)}} t^{kj}\psi_{N,j}(\delta_{ki} + \psi_{M,k}u_i^M)\, dv_0 = p_{Ni} \tag{16.1}$$

† See Truesdell and Noll [1965, p. 302].

‡ Finite-element formulations of problems in finite elasticity were presented by Oden [1967*b*]. Applications to plane stress problems were considered by Oden [1966*a*], Becker [1966], Peterson, Campbell, and Herrmann [1966], Oden and Sato [1967*a*, 1967*b*], and Oden and Kubitza [1967]. Problems of finite plane strain were solved using finite elements by Oden [1968*b*], and applications involving three-dimensional bodies and bodies of revolution were discussed by Oden and Key [1970, 1971*a*]. See also Nemat-Nasser and Shatoff [1970] and Hofmeister, Greenbaum, and Evensen [1970]. Kavanagh [1969] used the method in connection with his experimental work on finite elasticity. A survey has been given by Oden [1969*c*], which includes additional references to related work. Although some subsequent articles are based on the works cited here, the present development is slightly more general than those presented earlier.

where the element identification label e has been dropped for convenience. Here m_{NM} is the consistent mass matrix of (13.37), p_{Ni} are the generalized nodal forces of (13.54), and $\psi_{N,j} = \partial\psi_N(\mathbf{x})/\partial x^j$ are gradients of the local interpolation functions $\psi_N(\mathbf{x})$. Since, for elastic materials, t^{kj} is given by (15.1) (recall that $t^{kj} = \sqrt{G}\,\sigma^{kj}$), it follows that the equations of motion of an elastic finite element are

$$m_{NM}\ddot{u}_i^M + \int_{v_{0(e)}} \sqrt{G}\,\bar{\bar{\Sigma}}^{kj}(u_s^R)\psi_{N,j}(\delta_{ki} + \psi_{M,k}u_i^M)\,dv_0 = p_{Ni} \tag{16.2}$$

where, in accordance with (15.1), $\Sigma^{kj}[G_{mn}^{(e)}(u_s^R)] = \bar{\bar{\Sigma}}^{kj}(u_s^R)$. Alternatively, for hyperelastic finite elements we use (15.36) and write

$$m_{NM}\ddot{u}_i^M + \int_{v_{0(e)}} \frac{\partial W}{\partial \gamma_{kj}}\psi_{N,j}(\delta_{ki} + \psi_{M,k}u_i^M)\,dv_0 = p_{Ni} \tag{16.3}$$

In these equations, $N, M, R = 1, 2, \ldots, N_e$, where N_e is the total number of nodal points of element e, and $i, j, k, m, n, s = 1, 2, 3$.

In the case of static (or quasi-static) behavior of elastic bodies, we can drop the inertia terms $m_{NM}\ddot{u}_i^M$ in (16.2) and (16.3) and obtain the equilibrium equations for a typical finite element:

$$\int_{v_{0(e)}} \frac{\partial W}{\partial \gamma_{kj}}\psi_{N,j}(\delta_{ki} + \psi_{M,k}u_i^M)\,dv_0 = p_{Ni} \tag{16.4}$$

The coefficients of the nodal displacements in (16.4) are called *stiffnesses*, and (16.4) is referred to as a nonlinear stiffness relation. If the element is elastic but not hyperelastic, we must simply replace $\partial W/\partial\gamma_{ij}$ in (16.4) with $\sqrt{G}\,\bar{\bar{\Sigma}}^{kj}(u_s^R)$.

Special forms of (16.4) appropriate for any type of hyperelastic finite element can be obtained by simply introducing an appropriate form for W, such as, for example, any of those mentioned in Sec. 15.5. We list a few examples:

1. *$W(\gamma_{ij})$ analytic in γ_{ij}.* The strain-energy function is given by (15.38), and (16.4) becomes

$$\sum_{r=1}^{\infty}\int_{v_{0(e)}} \frac{\partial W_r}{\partial \gamma_{kj}}\psi_{N,j}(\delta_{ki} + \psi_{M,k}u_i^M)\,dv_0 = p_{Ni} \tag{16.5}$$

2. *Compressible anisotropic solids.* W is assumed to be given by (15.42):

$$\int_{v_{0(e)}} (E^{kjmn}\gamma_{mn} + E^{kjmnrs}\gamma_{mn}\gamma_{rs} + \cdots)\psi_{N,j}(\delta_{ki} + \psi_{M,k}u_i^M)\,dv_0 = p_{Ni} \tag{16.6}$$

Here E^{kjmn}, E^{kjmnrs}, . . . obey the symmetry properties (15.40), and for the finite element,

$$2\gamma_{mn} = \psi_{N,m}u_n^N + \psi_{N,n}u_m^N + \psi_{N,m}\psi_{M,n}u_k^N u_k^M \tag{16.7}$$

in accordance with (13.22).

3. *Second-order elasticity.* $W(\gamma_{ij})$ is given by (15.45*a*). Thus only second-order elasticities are retained in (16.6).

4. *Large rotations, infinitesimal strains.* Here $W(\gamma_{ij})$ is given by (15.43*a*) and γ_{ij} by (16.7), but the magnitudes of γ_{ij} are assumed to be "small." Thus, we arrive at nonlinear stiffness relations for hookean materials:

$$\int_{v_{0(e)}} E^{kjmn}\psi_{R,m}u_s^R(\delta_{sn} + \tfrac{1}{2}\psi_{S,n}u_s^S)\psi_{N,j}(\delta_{ki} + \psi_{M,k}u_i^M)\, dv_0 = p_{Ni} \tag{16.8}$$

wherein $M, N, R, S = 1, 2, \ldots, N_e$ and $i, j, k, m, n, s = 1, 2, 3$. Setting

$$\bar{a}_{sn} = (\delta_{sn} + \tfrac{1}{2}\psi_{S,n}u_s^S) \qquad \bar{b}_{ki} = z_{i,k} = (\delta_{ki} + \psi_{M,k}u_i^M) \tag{16.9}$$

(16.8) simplifies to

$$\int_{v_{0(e)}} E^{kjmn}\psi_{R,m}\psi_{N,j}\bar{a}_{sn}\bar{b}_{ki}\, dv_0\, u_s^R = p_{Ni} \tag{16.10}$$

5. *Classical infinitesimal elasticity.* In this case W is given by (15.44*a*), and for the finite element,

$$2e_{ij} = \psi_{N,i}u_j^N + \psi_{N,j}u_i^N \tag{16.11}$$

Moreover, the displacement gradients $\psi_{N,i}u_j^N$ are assumed to be infinitesimals. Then $\bar{a}_{ij} = \bar{b}_{ij} = \delta_{ij}$, and (16.10) reduces to the linear stiffness relation

$$\int_{v_{0(e)}} E^{ijmn}\psi_{M,m}\psi_{N,j}\, dv_0\, u_n^M = p_{Ni} \tag{16.12}$$

Introducing the notation

$$k_{NM}^{in} = \int_{v_{0(e)}} E^{ijmn}\psi_{M,m}\psi_{N,j}\, dv_0 \tag{16.13}$$

(16.12) becomes

$$k_{NM}^{in}u_n^M = p_{Ni} \tag{16.14}$$

where the components p_{Ni} of $\mathbf{p}_N^{(e)}$ are now independent of u_i^N. The array k_{NM}^{in} is the well-known stiffness matrix for linearly elastic materials. By comparing (16.13) with (16.8), we arrive at an interesting interpretation of the functions $\bar{a}_{sn}$ and $\bar{b}_{ki}$ of (16.9). In view of (16.12), the term E^{kjmn} can be regarded as the "stiffness matrix" of a unit cube of elastic material; the remaining terms represent linear transformations of this matrix into one associated with the entire finite element. In the linear theory, these pointwise transformations are brought about by the terms $\psi_{M,m}$ and $\psi_{N,j}$ in (16.12), while in the case of large deflections, they are altered due to the effects of rotations of the element by the functions $\bar{a}_{sn}$ and $\bar{b}_{ki}$. In (16.8), the local stiffnesses do not transform congruently as in the linear case, and the transformation terms are functions of the displacement gradients. This point becomes more apparent when specific forms of $\psi_N(\mathbf{x})$ are used.

6. *Isotropic elastic bodies.* The strain-energy function is now given by (15.46), so that (16.4) becomes

$$\int_{v_{0(e)}} \left(\frac{\partial W}{\partial I_1} \frac{\partial I_1}{\partial \gamma_{kj}} + \frac{\partial W}{\partial I_2} \frac{\partial I_2}{\partial \gamma_{kj}} + \frac{\partial W}{\partial I_3} \frac{\partial I_3}{\partial \gamma_{kj}} \right) \psi_{N,j}(\delta_{ki} + \psi_{M,k} u_i^M)\, dv_0 = p_{Ni} \tag{16.15}$$

or, in accordance with (15.48),

$$\int_{v_{0(e)}} \left\{ 2 \frac{\partial W}{\partial I_1} \delta^{kj} + 4 \frac{\partial W}{\partial I_2} [\delta^{kj}(1 + \gamma_{rr}) - \delta^{kr}\delta^{js}\gamma_{sr}] \right.$$
$$\left. + 2 \frac{\partial W}{\partial I_3} [\delta^{kj}(1 + 2\gamma_{rr}) - 2\delta^{kr}\delta^{js}\gamma_{sr} + 2\epsilon^{kmn}\epsilon^{jrs}\gamma_{mr}\gamma_{ns}] \right\}$$
$$\times \psi_{N,j}(\delta_{ki} + \psi_{M,k} u_i^M)\, dv_0 = p_{Ni} \tag{16.16}$$

Here γ_{rs} is given by (16.7), and

$$\gamma_{rr} = \psi_{N,r}(\delta_{mr} + \tfrac{1}{2}\psi_{M,r} u_m^M) u_m^N \tag{16.17}$$

These examples clearly demonstrate the general procedure for obtaining equations of equilibrium for any type of compressible elastic element; to obtain corresponding equations of motion, we simply add the term $m_{NM}\ddot{u}_i^M$ and regard u_i^N, p_{Ni}, etc., as functions of time. Equations governing incompressible elastic finite elements deserve special attention and are considered separately in the next section.

Simplex models† The stiffness relations assume particularly simple forms if we take for the local interpolation functions the simplex approximations

$$\psi_N(\mathbf{x}) = a_N + b_{Ni} x^i \tag{16.18}$$

where a_N and b_{Ni} are defined in (10.105), (10.106), and (10.108). Here $N_e = k + 1$, where k is the dimension of the space, and $i = 1, 2, \ldots, k$.

† See Oden [1967*b*].

Thus (16.18) represents the local basis functions for a line element with two nodes in one-dimensional problems, a triangular element with three nodes in two-dimensional problems, and a tetrahedral element with four nodes in three-dimensional problems. For the moment we consider the three-dimensional case ($N_e = 4$; $k = 3$).

Observing that

$$\psi_{N,i} = b_{Ni} \tag{16.19}$$

we recall from (13.29) and (13.39) that, in this case,

$$m_{NM} = \rho_0 v_0 a_N a_M + (a_N b_{Mj} + a_M b_{Nj}) s^j + b_{Ni} b_{Mj} I^{ij} \tag{16.20}$$

$$2\gamma_{ij} = b_{Ni} u_j^N + b_{Nj} u_i^N + b_{Ni} b_{Mj} u_k^N u_k^M \tag{16.21}$$

$$\gamma_{rr} = b_{Nr}(\delta_{rk} + \tfrac{1}{2} b_{Mr} u_k^M) u_k^N \tag{16.22}$$

where ρ_0 and v_0 are the mass density and volume of the element while in its reference configuration, s^j are the first moments of mass with respect to the local coordinate frame $x_{(e)}^i$, and I^{ij} is the mass moment of inertia tensor of the element with respect to $x_{(e)}^i$.

In (16.20) it has been assumed that the element is homogeneous, in that $\rho_0(\mathbf{x}) = \rho_0 =$ constant. While this is not in general true, it is nevertheless a good and convenient approximation if the dimensions of the element are taken sufficiently small. Indeed, in the case of simplex models of nonhomogeneous bodies, it is common practice (and one that we shall follow here) to treat each finite element as homogeneous and to account for variations in material properties by assigning different uniform properties to each finite element. Thus, for simplex elements the strain-energy function and its derivatives with respect to γ_{ij} may be regarded as being independent of x^i and factored outside the volume integral in (16.3). It follows that the nonlinear stiffness relations for simplex elements are of the form

$$v_0 \frac{\partial W(\gamma_{ij})}{\partial \gamma_{kj}} b_{Nj}(\delta_{ki} + b_{Mk} u_i^M) = p_{Ni} \tag{16.23}$$

For the various special cases considered previously, (16.23) assumes the following forms: (1) W analytic in γ_{ij} (16.5),

$$v_0 b_{Nj}(\delta_{ki} + b_{Mk} u_i^M) \sum_{r=1}^{\infty} \frac{\partial W_r}{\partial \gamma_{kj}} = p_{Ni} \tag{16.24}$$

(2) compressible anisotropic solids (16.6),

$$\begin{aligned} v_0 b_{Nj} E^{kjmn} b_{Rm}(\delta_{sn} + \tfrac{1}{2} b_{Sn} u_s^S)(\delta_{ki} + b_{Mk} u_i^M) u_s^R \\ + v_0 b_{Nj} E^{kjmnrs} b_{Rm} b_{Kr}(\delta_{pn} + \tfrac{1}{2} b_{Sn} u_p^S)(\delta_{qs} + \tfrac{1}{2} b_{Ts} u_q^T) \\ \times (\delta_{ki} + b_{Mk} u_i^M) u_p^R u_q^K + \cdots = p_{Ni} \end{aligned} \tag{16.25}$$

where $M, N, K, R, S, T = 1, 2, 3, 4$ and $i, j, k, m, n, p, q, r, s = 1, 2, 3$; (3) the equations for second-order elasticity obtained from (16.25) by retaining no elasticity coefficients of higher order than the second; and (4) the simplex stiffness relations for the case of infinitesimal strains but large rotations [that is, (16.8)] are obtained from (16.25) by retaining only the first term on the left of this equation:†

$$v_0 b_{Nj}(\delta_{sn} + \tfrac{1}{2} b_{Sn} u_s^S) E^{kjmn} b_{Rm}(\delta_{ki} + b_{Mk} u_i^M) u_s^R = p_{Ni} \tag{16.26}$$

Simplex representations for example 5, classical infinitesimal elasticity, are obtained by introducing (16.19) into (16.12).‡

$$v_0 b_{Mm} E^{ijmn} b_{Nj} u_n^M = p_{Ni} \tag{16.27}$$

In this case the stiffness matrix of (16.13) becomes

$$k_{NM}^{in} = v_0 b_{Mm} E^{ijmn} b_{Nj} \tag{16.28}$$

The observations made previously concerning the congruence of the transformations of E^{ijmn} are now more easily identified: in the linear theory, E^{ijmn} is transformed into an element stiffness relation by the matrices $\sqrt{v_0}\, b_{Mm}$ and $\sqrt{v_0}\, b_{Nj}$, whereas in the kinematically nonlinear case represented by (16.26), the transformations are brought about by different matrices $\sqrt{v_0}\, b_{Nj}(\delta_{sn} + \frac{1}{2} b_{Sn} u_s^S)$ and $\sqrt{v_0}\, b_{Rm}(\delta_{ki} + b_{Mk} u_i^M)$, which clearly depend upon the nodal displacements.

Finally, for simplex representations of isotropic elastic bodies, we have from (16.15),

$$v_0 b_{Nj}\left(\frac{\partial W}{\partial I_1}\frac{\partial I_1}{\partial \gamma_{kj}} + \frac{\partial W}{\partial I_2}\frac{\partial I_2}{\partial \gamma_{kj}} + \frac{\partial W}{\partial I_3}\frac{\partial I_3}{\partial \gamma_{kj}}\right)(\delta_{ki} + b_{Mk} u_i^M) = p_{Ni} \tag{16.29}$$

or, from (16.16),

$$\begin{aligned} v_0 b_{Nj} \Bigg(2\frac{\partial W}{\partial I_1}\delta^{kj} &+ 2\frac{\partial W}{\partial I_2}\{2\delta^{kj}[1 + b_{Rr}(\delta_{rp} + \tfrac{1}{2} b_{Sr} u_p^S) u_p^R] \\ &- \delta^{kr}\delta^{js}(b_{Rs} u_r^R + b_{Rr} u_s^R + b_{Rs} b_{Sr} u_p^R u_p^S)\} \\ &+ 2\frac{\partial W}{\partial I_3}\{\delta^{kj}[1 + 2b_{Rr}(\delta_{rp} + \tfrac{1}{2} b_{Sr} u_p^S) u_p^R] \\ &- \delta^{kr}\delta^{js}(b_{Rs} u_r^R + b_{Rr} u_s^R + b_{Rs} b_{Sr} u_p^R u_p^S) \\ &+ \tfrac{1}{2}\epsilon^{kmn}\epsilon^{jrs}(b_{Rm} u_r^R + b_{Rr} u_m^R + b_{Rm} b_{Sr} u_p^R u_p^S) \\ \times (b_{Kn} u_s^K &+ b_{Ks} u_n^K + b_{Kn} b_{Ts} u_p^K u_p^T)\}\Bigg)(\delta_{ki} + b_{Mk} u_i^M) = p_{Ni} \end{aligned} \tag{16.30}$$

where, as usual, $M, N, K, R, S, T = 1, 2, 3, 4$ and $i, j, k, m, n, p, r, s = 1, 2, 3$.

† This equation was first obtained by Wissmann [1963]. See also Wissmann [1966].

‡ This relation was presented in a different form by Gallagher, Padlog, and Bijlaard [1962]. It represents one of the first linear stiffness relations developed for three-dimensional finite elements of elastic bodies.

We note that equations (16.23) to (16.30) can be adapted to one- and two-dimensional elements by simply changing the ranges of the indices: for one-dimensional elements, set $M, N, \ldots, S, T = 1, 2$ and $i, j, \ldots, r, s = 1$; for two-dimensional elements, set $M, N, \ldots, S, T = 1, 2, 3$ and $i, j, \ldots, r, s = 1, 2$. In addition, these relations can also be used to describe the motion of one- and two-dimensional elements in spaces of higher dimension. For example, in the case of a three-dimensional collection of one-dimensional elements (e.g., a three-dimensional truss or framework), set $M, N, \ldots, S, T = 1, 2$ and $i, j, \ldots, r, s = 1, 2, 3$. For planar assemblies of one-dimensional elements, $M, N, \ldots, S, T = 1, 2$ and $i, j, \ldots, r, s = 1, 2$. For three-dimensional assemblies of two-dimensional elements (e.g., membranes), set $M, N, \ldots, S, T = 1, 2, 3$ and $i, j, \ldots, r, s = 1, 2, 3$.

Global forms Returning now to the more general stiffness relations represented by (16.4), we recognize that these relations apply to a typical finite element e by affixing to all appropriate quantities an element identification label:

$$\int_{v_{0(e)}} \frac{\partial W_{(e)}}{\partial \gamma_{kj}} \overset{(e)}{\psi}_{N,j}(\delta_{ki} + \overset{(e)}{\psi}_{M,k} u^M_{i(e)})\, dv_0 = \overset{(e)}{p}_{Ni} \tag{16.31}$$

In accordance with (13.11) and (13.67),

$$P_{\Delta i} = \sum_{e=1}^{E} \overset{(e)}{\Omega}{}^N_\Delta \overset{(e)}{p}_{Ni} \qquad \text{and} \qquad u^N_{i(e)} = \overset{(e)}{\Omega}{}^N_\Delta U^\Delta_i \tag{16.32}$$

where $\overset{(e)}{\Omega}{}^N_\Delta$ is the boolean transformation described in, for example, (6.9), E is the total number of finite elements, and $P_{\Delta i}$ and U^Δ_i are the components of generalized force and displacement at node Δ of the connected model. It follows that the global form of the nonlinear stiffness relations for assemblies of elastic finite elements is

$$\sum_{e=1}^{E} \overset{(e)}{\Omega}{}^N_\Delta \int_{v_{0(e)}} \frac{\partial W_{(e)}}{\partial \gamma_{kj}} \overset{(e)}{\psi}_{N,j}(\delta_{ki} + \overset{(e)}{\psi}_{M,k} \overset{(e)}{\Omega}{}^M_\Gamma U^\Gamma_i)\, dv_0 = P_{\Delta i} \tag{16.33}$$

in which $\Delta, \Gamma = 1, 2, \ldots, G$; $M, N = 1, 2, \ldots, N_e$; and $i, j, k = 1, 2, 3$. Dynamical equations governing assemblies of elastic elements can be obtained by simply adding to (16.33) the term $M_{\Delta\Gamma}\ddot{U}^\Gamma_i$, where $M_{\Delta\Gamma}$ is the global mass matrix defined in (13.75). Comparing (16.33) with (13.76) and (13.77), it is clear that in this case

$$G_{\Delta i} = \sum_{e=1}^{E} \overset{(e)}{\Omega}{}^N_\Delta \int_{v_{0(e)}} \frac{\partial W_{(e)}}{\partial \gamma_{kj}} \overset{(e)}{\psi}_{N,j} \delta_{ki}\, dv_0 \tag{16.34}$$

$$H_{\Delta\Gamma} = \sum_{e=1}^{E} \overset{(e)}{\Omega}{}^N_\Delta \overset{(e)}{\Omega}{}^M_\Gamma \int_{v_{0(e)}} \frac{\partial W_{(e)}}{\partial \gamma_{kj}} \overset{(e)}{\psi}_{N,k} \overset{(e)}{\psi}_{M,j}\, dv_0 \tag{16.35}$$

As a simple example of global forms of stiffness relations for specific materials, consider the two-bar framework shown in Fig. 16.1, which is

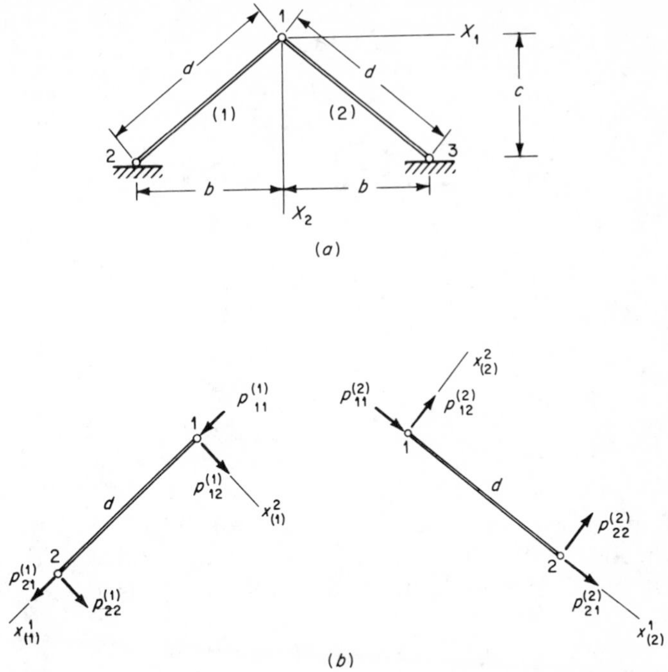

Fig. 16.1 Two-bar framework.

assumed to be constructed of an isotropic hookean material. In this case, if we take as local coordinates those indicated in Fig. 16.1*b*, we may set

$$E^{1111} = E \qquad E^{ijmn} = 0 \qquad i, j, m, n \neq 1 \tag{16.36}$$

where E is Young's modulus. Using the simplex representation (16.18), we find that

$$\psi_{N,1} = b_{N1} \qquad b_{11} = -b_{21} = -\frac{1}{d} \tag{16.37}$$

where d is the length of each bar. Noting that $v_0 = a_0 d$, where a_0 is the initial cross-sectional area of the bar, we obtain from (16.26) the generalized forces

$$\begin{aligned} p_{11}^{(e)} &= -a_0 E \gamma_{11}^{(e)} \left(1 + \frac{u_{1(e)}^{2} - u_{1(e)}^{1}}{d}\right) \\ p_{12}^{(e)} &= -a_0 E \gamma_{11}^{(e)} \frac{u_{2(e)}^{2} - u_{2(e)}^{1}}{d} \\ p_{21}^{(e)} &= a_0 E \gamma_{11}^{(e)} \left(1 + \frac{u_{1(e)}^{2} - u_{1(e)}^{1}}{d}\right) \\ p_{22}^{(e)} &= a_0 E \gamma_{11}^{(e)} \frac{u_{2(e)}^{2} - u_{2(e)}^{1}}{d} \end{aligned} \tag{16.38}$$

in which

$$\gamma_{11}^{(e)} = \frac{u_{1(e)}^{2} - u_{1(e)}^{1}}{d} + \frac{1}{2}\left[\left(\frac{u_{1(e)}^{2} - u_{1(e)}^{1}}{d}\right)^2 + \left(\frac{u_{2(e)}^{2} - u_{2(e)}^{1}}{d}\right)^2\right] \tag{16.39}$$

Note that the local forces in (16.38) comprise a self-equilibrating force system. Since the local coordinates $x_{(e)}^{i}$ are not coincident with the global coordinates X^i, we transform all local quantities to a common frame of reference in accordance with (13.11):

$$\hat{u}_{i(e)}^{N} = \overset{(e)}{\alpha}{}_{.j}^{i} u_{j(e)}^{N} \tag{16.40a}$$

$$\hat{p}_{Ni}^{(e)} = \overset{(e)}{\alpha}{}_{.j}^{i} p_{Nj}^{(e)} \tag{16.40b}$$

where

$$[\overset{(1)}{\alpha}{}_{.j}^{i}] = \frac{1}{d}\begin{bmatrix} -b & c \\ c & b \end{bmatrix} \quad \text{and} \quad [\overset{(2)}{\alpha}{}_{.j}^{i}] = \frac{1}{d}\begin{bmatrix} b & c \\ c & -b \end{bmatrix} \tag{16.41}$$

Observing that $\overset{(1)}{\Omega}{}_1^1 = \overset{(2)}{\Omega}{}_1^1 = \overset{(1)}{\Omega}{}_2^2 = \overset{(2)}{\Omega}{}_2^2 = 1$ and all other $\Omega_\Delta^N = 0$ and, from the boundary conditions, that $U_i^2 = U_i^3 = 0$, we may write the local generalized forces in terms of the global displacements U_i^1 as follows:

$$\begin{aligned}
p_{11}^{(1)} &= -a_0 E\gamma_{11}^{(1)}\left[1 + \frac{1}{d^2}(bU - cV)\right] \\
p_{12}^{(1)} &= a_0 E\gamma_{11}^{(1)} \frac{1}{d^2}(cU + bV) \\
p_{11}^{(2)} &= -a_0 E\gamma_{11}^{(2)}\left[1 - \frac{1}{d^2}(bU + cV)\right] \\
p_{12}^{(2)} &= a_0 E\gamma_{11}^{(2)} \frac{1}{d^2}(cU - bV)
\end{aligned} \tag{16.42}$$

etc., in which $U_1^1 = U$, $U_2^1 = V$, and

$$\begin{aligned}
\gamma_{11}^{(1)} &= \frac{1}{d^2}[bU - cV + \tfrac{1}{2}(U^2 + V^2)] \\
\gamma_{11}^{(2)} &= \frac{1}{d^2}[-bU - cV + \tfrac{1}{2}(U^2 + V^2)]
\end{aligned} \tag{16.43}$$

Transforming (16.42) according to (16.40b), we obtain

$$\begin{aligned}
\hat{p}_{11}^{(1)} &= \frac{a_0 E\gamma_{11}^{(1)}}{d}(U + b) \qquad & \hat{p}_{11}^{(2)} &= \frac{a_0 E\gamma_{11}^{(2)}}{d}(U - b) \\
\hat{p}_{12}^{(1)} &= \frac{a_0 E\gamma_{11}^{(1)}}{d}(V - c) \qquad & \hat{p}_{12}^{(2)} &= \frac{a_0 E\gamma_{11}^{(2)}}{d}(V - c)
\end{aligned} \tag{16.44}$$

Thus, in view of (13.67), the global forces at node 1 in the connected model are finally†

$$P_{11} = \frac{a_0 EU}{d^3} [2(b^2 - cV) + U^2 + V^2]$$
$$P_{12} = \frac{a_0 E}{d^3} (V - c)[V(V - 2c) + U^2] \tag{16.45}$$

These equations represent equilibrium equations at node 1.

In the special case in which deformations are symmetrical with respect to the x^2 axis, $U = 0$, $P_{11} = \hat{p}_{11}^{(1)} + \hat{p}_{11}^{(2)} = 0$, and we have only

$$P_{12} = \frac{a_0 E}{d^3} V(V - c)(V - 2c) \tag{16.46}$$

From (16.45) and (16.46) we see that our simplex model leads to cubic variations of the generalized forces with the nodal displacements. For the case in which (16.46) holds, P_{12} is a homogeneous cubic in V that has three zeroes: $V = 0, c, 2c$. Thus, for a given applied load P_{12}, there may be as many as three equilibrium configurations; i.e., we can no longer expect there to be a unique displacement for a given load as in the linear theory.

If V is very small, (16.46) can be linearized to give

$$P_{12} = \frac{2a_0 c^2 E}{d^3} V \tag{16.47}$$

a result which clearly depends upon the height c of the structure. It is not unreasonable to consider a perfectly flat framework, that is, $c = 0$. Then the linearized solution does not exist and the response of the structure can only be predicted by a nonlinear analysis.

In the case of nonconservative loading, the nodal forces P_{11} and P_{12} of (16.45) must generally be expressed as functions of the nodal displacements. Consider, for example, the case in which a force Q is applied at node 1 in such a way that it is always normal to element (1). Then

$$P_{11} = \frac{Q}{d^*} (c - V) \tag{16.48a}$$

$$P_{12} = \frac{Q}{d^*} (b + U) \tag{16.48b}$$

where

$$d^{*2} = d^2 + 2(bU - cV) + U^2 + V^2 \tag{16.49}$$

† For some solutions of these equations, see Oden [1971*c*].

16.2 INCOMPRESSIBLE MATERIALS†

Although in some exact solutions to problems in finite elasticity the assumption of an incompressible material leads to certain simplifications in the analysis, such is not the case in finite-element applications. We recall that the stress developed in an incompressible material is not completely determined by the strain or the strain energy; in fact, W determines the stress in incompressible hyperelastic bodies only to within an additive scalar-valued function h, called the hydrostatic pressure, which performs no work as the body deforms. In finite-element applications, the hydrostatic pressure appears as an additional unknown in the stiffness relations, and the incompressibility condition (15.56) must be maintained, at least approximately, throughout the entire discrete model. Since practically all available constitutive equations for highly elastic materials (e.g., natural and synthetic rubbers, polymers, solid propellant fuels) regard the material as incompressible, the study of finite deformations of incompressible bodies constitutes an important part of finite elasticity.

In view of (15.55), (15.58), and (16.1), the stiffness relations for incompressible finite elements are of the form

$$\int_{v_{0(e)}} [\bar{\Sigma}^{kj}(G_{rs}) + hG^{kj}]\psi_{N,j}(\delta_{ki} + \psi_{M,k}u_i^M)\, dv_0 = p_{Ni} \tag{16.50}$$

or, for hyperelastic elements,

$$\int_{v_{0(e)}} \left(\frac{\partial W}{\partial \gamma_{kj}} + hG^{kj}\right)\psi_{N,j}(\delta_{ki} + \psi_{M,k}u_i^M)\, dv_0 = p_{Ni} \tag{16.51}$$

where $h = h_{(e)}(\mathbf{x},t)$ is a local hydrostatic pressure corresponding to element e. In addition, the nodal displacements u_i^M must be such that the incompressibility condition is satisfied; for finite elements, the first approximation that suggests itself is the averaged incompressibility condition

$$\int_{v_{0(e)}} (I_3 - 1)\, dv_0 = 0 \tag{16.52}$$

† Linear finite-element analyses of incompressible bodies have been presented by a number of authors. Infinitesimal axisymmetric deformations of incompressible elastic solids of revolution were analyzed by Becker and Brisbane [1965]. Their analysis was based on a variational theorem proposed by Herrmann [1965], which can be precipitated from a more general theory given by Truesdell and Toupin [1960]. Other applications to linear incompressible materials were presented by Taylor, Pister, and Herrmann [1968], Tong [1969], Key [1969], and Hughes and Allik [1969]. Applications to finite deformations of incompressible elastic bodies were presented by Oden [1967*b*, 1968*b*] and Oden and Key [1970, 1971*a*]. Solutions to plane stress problems involving finite deformations of incompressible bodies have also been made, e.g., Oden and Sato [1967*a*] and Becker [1966], but in this class of problems the usual difficulties of maintaining the incompressibility condition and calculating hydrostatic pressures can often be avoided.

Due to our choice of a cartesian material frame in the reference configuration $C_{0(e)}$, some simplifications in the form of (16.52) are possible. Recalling that

$$I_3 = \det G_{ij} = \det z_{m,i}z_{m,j} = (\det z_{m,i})(\det z_{m,j}) = (\det z_{j,i})^2 \quad (16.53)$$

where $z_{m,i} = \delta_{mi} + \psi_{N,i}u_m^N$ are the deformation gradients, we see that (16.52) can be rewritten in the more convenient form

$$\int_{v_{0(e)}} [\det (\delta_{mi} + \psi_{N,i}u_{m}^{N})]^2 \, dv_0 - v_0 = 0 \quad (16.54)$$

We investigate other formulations of the condition of incompressibility below.

Now in neither (16.50) nor (16.51) have we indicated the manner in which $h(\mathbf{x})$ is assumed to vary over the finite element. Several possibilities suggest themselves. The simplest local approximation of $h(\mathbf{x},t)$, and the one which often leads to the least number of additional unknowns, is to assume that h is uniform over the element, i.e., the zeroth-order approximation

$$h_{(e)}(\mathbf{x}) = h_0^{(e)} \quad (16.55)$$

where $h_0^{(e)}$ is a constant. For dynamical phenomena, of course, $h_0^{(e)}$ is regarded as a function of time. In the case of a three-dimensional finite element, (16.51) together with (16.54) represents a system of $3N_e$ equations in the $3N_e + 1$ unknowns u_i^M ($M = 1, 2, \ldots, N_e$; $i = 1, 2, 3$) and $h_0^{(e)}$. The remaining equation needed to complete the system is furnished by the local incompressibility condition (16.52) or an alternate form of it, as in (16.54). Thus, for a discrete model composed of E finite elements for which the displacement components at $\bar{G}$ global nodes are unprescribed, the approximation (16.55) leads to a system of $3\bar{G} + E$ equations (e.g., equilibrium equations plus incompressibility conditions) for the $3\bar{G}$ displacement components and E element hydrostatic pressures. Notice that in the case of compressible materials, the same type of discrete model leads to only $3\bar{G}$ simultaneous equations.

Alternatively, we may use instead of (16.55) a more general local approximation of the hydrostatic pressure, such as

$$h_{(e)}(\mathbf{x}) = h^N \psi_N(\mathbf{x}) \quad (16.56)$$

where $\psi_N(\mathbf{x})$ is the same collection of local interpolation functions as that used in the approximation of the displacement fields. Then (16.51) becomes

$$\int_{v_{0(e)}} \left(\frac{\partial W}{\partial \gamma_{kj}} + \psi_R h^R G^{kj}\right) \psi_{N,j}(\delta_{ki} + \psi_{M,k}u_i^M) \, dv_0 = p_{Ni} \quad (16.57)$$

However, in this case (16.57) represents a system of $3N_e$ equations in $4N_e$ unknowns (h^R and u_i^M; M, $R = 1, 2, \ldots, N_e$; $i = 1, 2, 3$) versus the

$3N_e + 1$ obtained using the uniform approximation (16.55). Since a local incompressibility condition of the form (16.52) furnishes only one additional equation for each element, we shall not arrive at a sufficient number of equations to solve the problem if we follow the same procedure used in the case of the uniform hydrostatic pressure.

To resolve this problem, let H^Δ, $\Delta = 1, 2, \ldots, G$, denote the value of the hydrostatic pressure at node Δ of the connected model. Then

$$h^N_{(e)} = \overset{(e)}{\Omega}{}^N_\Delta H^\Delta \tag{16.58}$$

and the global form of (16.57) is

$$\sum_{e=1}^{E} \overset{(e)}{\Omega}{}^N_\Delta \int_{v_{0(e)}} \left(\frac{\partial W_{(e)}}{\partial \gamma_{kj}} + \overset{(e)}{\psi}_R \overset{(e)}{\Omega}{}^R_\Gamma H^\Gamma \right) \overset{(e)}{\psi}_{N,j} (\delta_{ki} + \overset{(e)}{\psi}_{M,k} \overset{(e)}{\Omega}{}^M_\Phi U_i^\Phi)\, dv_0 = P_{\Delta i} \tag{16.59}$$

with $\Delta, \Gamma, \Phi = 1, 2, \ldots, G$; $M, N, R = 1, 2, \ldots, N_e$; $i, j, k = 1, 2, 3$. Assuming that in the connected model m components of n nodal displacements are prescribed, (16.59) leads to a system of $3G - m$ equations in $4G - m$ unknowns. It follows that *in this case an incompressibility condition should be imposed at each node rather than for each element.* To obtain such conditions, we transform the point condition of incompressibility, $I_3 = 1$, into an equivalent local nodal statement of the condition of constraint by introducing the quantity†

$$(\mathrm{iii})^{(e)}_M \equiv \int_{v_{0(e)}} \psi^{(e)}_M (I_3^{(e)} - 1)\, dv_0 \tag{16.60}$$

If III_Δ denotes the global value of $(\mathrm{iii})^{(e)}_M$ at node Δ of the connected model, it follows from (7.16) and (9.189) that

$$(\mathrm{iii})^{(e)}_M = \overset{(e)}{\Lambda}{}^\Delta_M \mathrm{III}_\Delta \qquad \text{and} \qquad \mathrm{III}_\Delta = \sum_{e=1}^{E} \overset{(e)}{\Omega}{}^N_\Delta (\mathrm{iii})^{(e)}_N \tag{16.61a,b}$$

The fact that III_Δ rather than III^Δ appears in this equation shows that the quantities III_Δ are components of a function in the finite-dimensional subspace Φ^* which is conjugate to the space Φ spanned by the global basis functions

$$\Phi_\Delta(\mathbf{x}) = \sum_e \overset{(e)}{\Omega}{}^N_\Delta \psi^{(e)}_N(\mathbf{x})$$

Equation (16.61*b*) indicates that III_Δ represents the sum of the local contributions $(\mathrm{iii})^{(e)}_N$ from all elements sharing node Δ in the connected model. Recalling from (9.82) et seq. that such covariant quantities represent averages of values of $\mathrm{III}(\mathbf{X})$ in the neighborhood of node $\mathbf{X}^\Delta$, we see that the vanishing of III_Δ at each of the G global nodal points ensures that the incompressibility

† We interpret $(\mathrm{iii})^{(e)}_M$ in the context of variational principles in Sec. 16.4.

condition is satisfied in an average sense over the entire connected finite-element model. Thus, we set

$$\mathrm{III}_\Delta = \sum_{e=1}^{E} \overset{(e)}{\Omega}{}_\Delta^N \int_{v_{0(e)}} \psi_N^{(e)}(I_3^{(e)} - 1)\, dv_0 = 0 \tag{16.62}$$

$\Delta = 1, 2, \ldots, G$. The conditions (16.62) together with the $3G - m$ stiffness relations (16.59) represent a complete system of $4G - m$ independent equations in the $4G - m$ unknowns U_i^Δ, H^Δ.

We remark that the use of constant-element hydrostatic pressures, as in (16.55), need not lead to fewer unknowns than the more elaborate approximation (16.56). Indeed, use of (16.55) leads to $3G - m + E$ equations, while (16.56) leads to $4G - m$; and it is not difficult to construct examples for which $E > G$. On the other hand, use of (16.55) and (16.54) generally ensures that the incompressibility condition is more accurately satisfied over the connected model; in fact, in several important cases it is possible to satisfy such conditions exactly for each finite element. Computationally, the averaged incompressibility conditions of (16.62) offer some advantages over (16.54) since the connectivity of the discrete model of $h(\mathbf{x})$ is established by (16.58) in precisely the same manner as for the model of the displacement field. It may also be argued that (16.56) generally allows the hydrostatic part of the stress tensor to be determined more accurately, but the use of consistent stress approximations via the conjugate-approximation functions discussed in Art. 9 overcomes many of the disadvantages often associated with stress calculations in finite elements.

We should also note that the use of approximations of the hydrostatic pressure of higher order than that of the displacement field may lead to more unknowns without an apparent increase in accuracy.† Indeed, if conventional element stresses (i.e., stresses computed from local constitutive equations once the displacements and hydrostatic pressures are determined) can be used as a guide, the order of the local approximations of hydrostatic pressures should be *less* than that of the displacement approximations. Take, for example, the case of a homogeneous simplex model in cartesian coordinates. The displacement gradients are then constant over each finite element, as is the portion of the stress tensor derived from the strain energy. Hence, it would seem to be consistent in this case to assume a uniform hydrostatic pressure over each element. However, local stresses computed from pressure and displacement approximations of the same order can still be used to derive consistent stress distributions in the sense of the conjugate approximations discussed in Art. 9. We shall investigate such stress calculations in the following section. We also note that it is possible to construct a nonconstant approximation of $h(\mathbf{x})$, which is still of a lower order

† See the conclusions of Hughes and Allik [1969].

than the approximation of the displacements. For example, a bilinear displacement approximation over a plane quadrilateral element might be used, while the approximation for the hydrostatic pressure might be obtained by dividing the element into two (or more) triangles over which h is assumed to be constant. In view of these observations, we shall adopt as a general rule in subsequent formulations involving incompressible elements that the order of the approximation of h be less than or equal to that of the approximation of the displacement field.

Returning to the general nonlinear stiffness relations for incompressible elements (16.51), we can now write down specific forms appropriate for various types of incompressible hyperelastic materials. For isotropic materials, W is a function of the first two principal invariants I_1 and I_2. Then (16.15) becomes

$$\int_{v_{0(e)}} \left(\frac{\partial W}{\partial I_1} \frac{\partial I_1}{\partial \gamma_{kj}} + \frac{\partial W}{\partial I_2} \frac{\partial I_2}{\partial \gamma_{kj}} + h \frac{\partial I_3}{\partial \gamma_{kj}} \right) \psi_{N,j}(\delta_{ki} + \psi_{M,k} u_i^M) \, dv_0 = p_{Ni} \tag{16.63}$$

or, in accordance with (16.16),

$$\int_{v_{0(e)}} \left\{ 2 \frac{\partial W}{\partial I_1} \delta^{kj} + 4 \frac{\partial W}{\partial I_2} [\delta^{kj}(1 + \gamma_{rr}) - \delta^{kr}\delta^{js}\gamma_{sr}] + 2h[\delta^{kj}(1 + 2\gamma_{rr}) - 2\delta^{kr}\delta^{js}\gamma_{sr} + 2\epsilon^{kmn}\epsilon^{jrs}\gamma_{mr}\gamma_{ns}] \right\} \psi_{N,j}(\delta_{ki} + \psi_{M,k} u_i^M) \, dv_0 = p_{Ni} \tag{16.64}$$

in which $h = \psi_N h^N$ or $h = h_0$.

If W is expressible as a polynomial in the invariants, as in (15.61), then we must introduce in (16.64)

$$\begin{aligned} \frac{\partial W}{\partial I_1} &= \sum_{r=0}^{\infty} \sum_{s=0}^{\infty} r C_{rs} (I_1 - 3)^{r-1} (I_2 - 3)^s \\ \frac{\partial W}{\partial I_2} &= \sum_{r=0}^{\infty} \sum_{s=0}^{\infty} s C_{rs} (I_1 - 3)^{r} (I_2 - 3)^{s-1} \end{aligned} \tag{16.65}$$

The important special case of the Mooney material results when

$$\frac{\partial W}{\partial I_1} = C_1 \quad \text{and} \quad \frac{\partial W}{\partial I_2} = C_2 \tag{16.66}$$

where C_1 and C_2 are constants; while if $C_2 = 0$, we obtain the equations for neo-hookean elements. Various other forms appropriate for different types of isotropic materials can be obtained by introducing any of the special forms W given in (15.65) to (15.71).

Simplex models In the case of simplex models, more explicit forms of the stiffness relations can be written. Observing that

$$\frac{\partial I_\alpha}{\partial \gamma_{kj}} \psi_{N,j}(\delta_{ki} + \psi_{M,k} u_i^M) = \frac{\partial I_\alpha}{\partial \gamma_{kj}} \frac{\partial \gamma_{kj}}{\partial u_i^N} = \frac{\partial I_\alpha}{\partial u_i^N} \qquad \alpha = 1, 2 \tag{16.67}$$

we find that if $\gamma_{ij}^{(e)}$ is given by (16.21),

$$\frac{\partial I_1}{\partial u_i^N} = 2b_{Nk}(\delta_{ki} + b_{Mk}u_i^M) \tag{16.68a}$$

$$\frac{\partial I_2}{\partial u_i^N} = 2b_{Nm}(\delta_{ik} + b_{Mk}u_i^M)\{2\delta^{mk}[1 + b_{Ps}(\delta_{rs} + \tfrac{1}{2}b_{Rs}u_r^R)u_r^P] - b_{Im}u_k^I - b_{Ik}u_m^I - b_{Im}b_{Jk}u_j^I u_j^J\} \tag{16.68b}$$

$$\frac{\partial I_3}{\partial u_i^N} = 2b_{Nm}(\delta_{ik} + b_{Mk}u_i^M)\{\delta^{mk}[1 + 2b_{Ps}(\delta_{rs} + \tfrac{1}{2}b_{Rs}u_r^R)u_r^P] - (b_{Im}u_k^I + b_{Ik}u_m^I + b_{Im}b_{Jk}u_j^I u_j^J) + 2b_{Pj}b_{Kr}\epsilon^{mjr}\epsilon^{kst}(\delta_{sp} + \tfrac{1}{2}b_{Is}u_p^I)(\delta_{tq} + \tfrac{1}{2}b_{Jt}u_q^J)u_p^P u_q^K\} \tag{16.68c}$$

where $I, J, K, M, N, P, R = 1, 2, 3, 4$ and $i, j, k, m, n, p, q, r, s, t = 1, 2, 3$. Then the stiffness relations for homogeneous incompressible isotropic finite elements become

$$2v_0(\delta_{ki} + b_{Mk}u_i^M)\bigg(b_{Nk}\frac{\partial W}{\partial I_1} + \frac{\partial W}{\partial I_2}b_{Nm}\{2\delta^{mk}[1 + b_{Pj}(\delta_{rj} + \tfrac{1}{2}b_{Rj}u_r^R)u_r^P] - (b_{Im}u_k^I + b_{Ik}u_m^I + b_{Im}b_{Jk}u_j^I u_j^J)\} + h_0 b_{Nm}\{\delta^{mk}[1 + 2b_{Pj}(\delta_{rj} + \tfrac{1}{2}b_{Rj}u_r^R)u_r^P] - (b_{Im}u_k^I + b_{Ik}u_m^I + b_{Im}b_{Jk}u_j^I u_j^J) + 2b_{Pj}b_{Kr}\epsilon^{mjr}\epsilon^{kst}(\delta_{sp} + \tfrac{1}{2}b_{Is}u_p^I)(\delta_{tq} + \tfrac{1}{2}b_{Jt}u_q^J)u_p^P u_q^K\}\bigg) = p_{Ni} \tag{16.69}$$

in which a uniform hydrostatic pressure $h^{(e)}(\mathbf{x}) = h_0^{(e)}$ has been assumed. If, instead, we wish to assume that $h(\mathbf{x})$ varies linearly over the element [that is, $h(\mathbf{x}) = (a_N + b_{Ni}x^i)h^N$], then h_0 in (16.69) should be replaced by $m_Q h^Q$, where m_Q are the local moments defined in (9.159). For simplex representations,

$$m_Q = \int_{v_{0(e)}} \psi_Q \, dv_0 = \int_{v_{0(e)}} (a_Q + b_{Qi}x^i)\, dv_0 = v_0 a_Q + b_{Qi}s_{(v)}^i \tag{16.70}$$

where $s_{(v)}^i = \int_{v_0} x^i \, dv_0$ are the first moments of volume with respect to the local material frame $x_{(e)}^i$.

If $h^{(e)}(\mathbf{x}) = h_0^{(e)}$ is assumed, (16.69) represents a system of 12 equations in the 13 unknowns u_i^N, h_0. To these we must add the incompressibility condition (16.52) [or (16.54)]. Observing that for simplex models,

$$z_{m,i} = \delta_{mi} + b_{Ni}u_m^N \tag{16.71}$$

we have for the incompressibility condition,

$$\det(\delta_{mi} + b_{Ni}u_m^N) = 1 \tag{16.72}$$

Alternatively, a more direct approach can be used to obtain the same result. Three sides of a tetrahedral finite element in the undeformed body are formed by the three vectors

$$\mathbf{a} = (x_i^2 - x_i^1)\mathbf{i}_i \qquad \mathbf{b} = (x_i^3 - x_i^1)\mathbf{i}_i \qquad \mathbf{c} = (x_i^4 - x_i^1)\mathbf{i}_i \tag{16.73}$$

where $\mathbf{i}_i$ are orthonormal basis vectors in $C_{0(e)}$. The initial volume of the element is $\frac{1}{6}$ the scalar-triple product of these vectors:

$$v_0 = \tfrac{1}{6}\mathbf{a} \cdot (\mathbf{b} \times \mathbf{c}) = \tfrac{1}{6}\epsilon^{ijk}(x_i^2 - x_i^1)(x_j^3 - x_j^1)(x_k^4 - x_k^1) \tag{16.74}$$

The vectors forming the sides of the tetrahedron after deformation are

$$\begin{aligned} \bar{\mathbf{a}} &= (x_i^2 + u_i^2 - x_i^1 - u_i^1)\mathbf{i}_i \\ \bar{\mathbf{b}} &= (x_i^3 + u_i^3 - x_i^1 - u_i^1)\mathbf{i}_i \\ \bar{\mathbf{c}} &= (x_i^4 + u_i^4 - x_i^1 - u_i^1)\mathbf{i}_i \end{aligned} \tag{16.75}$$

and the volume of the deformed element is

$$v = \tfrac{1}{6}\bar{\mathbf{a}} \cdot (\bar{\mathbf{b}} \times \bar{\mathbf{c}}) = \tfrac{1}{6}\epsilon^{ijk}(z_i^2 - z_i^1)(z_j^3 - z_j^1)(z_k^4 - z_k^1) \tag{16.76}$$

where $z_i^N = x_i^N + u_i^N$ are the spatial coordinates of the displaced nodes. Since we require that $v = v_0$, we have, after simplifying terms,

$$\frac{1}{6v_0}\,\epsilon^{ijk}(x_i^2 + u_i^2 - x_i^1 - u_i^1)(x_j^3 + u_j^3 - x_j^1 - u_j^1) \times (x_k^4 + u_k^4 - x_k^1 - u_k^1) = 1 \tag{16.77}$$

Upon expanding the left side of this equation and comparing the result with (10.96), we can identify the coefficients b_{Ni} and verify that (16.77) and (16.72) coincide. Thus, for simplex approximations the incompressibility condition is exactly satisfied throughout the discrete model.

If the hydrostatic pressure is assumed to vary linearly over the element. incompressibility conditions must be written at each node in accordance with (16.60) and (16.62). For simplex models, we then have locally

$$(\text{iii})_M^{(e)} = m_M\{[\det(\delta_{im} + b_{Ni}u_m^N)]^2 - 1\} \tag{16.78}$$

where m_M is given by (16.70), and globally

$$\sum_{e=1}^{E}\overset{(e)}{\Omega}{}_{\Delta}^{M} m_M^{(e)}\{[\det(\delta_{im} + b_{Ni}^{(e)}\overset{(e)}{\Omega}{}_{\Gamma}^{N}U_m^{\Gamma})]^2 - 1\} = 0 \tag{16.79}$$

16.3 GENERALIZED FORCES AND STRESSES

We have frequently noted that the generalized nodal forces $p_{Ni}^{(e)}$ are the result of stresses which are developed on material surfaces in the deformed body and which are referred to the convected material coordinate lines $x_{(e)}^i$ and, of course, body forces. Thus, these forces may change in both magnitude and direction as the body deforms and, in general, must be expressed in terms of the displacements and their gradients. In the present section we address ourselves to the problem of determining forms of these forces which properly account for their dependence on the deformation.

We recall from Art. 13 that consistent forms of the generalized forces arise from the invariance of the local mechanical power in the sense that

$$\Omega_{(e)} = \int_{v_{(e)}} \rho \mathbf{F} \cdot \dot{\mathbf{u}}\, dv + \int_{A_{(e)}} \mathbf{S} \cdot \dot{\mathbf{u}}\, dA = \mathbf{p}_N \cdot \dot{\mathbf{u}}^N \tag{16.80}$$

where $\mathbf{F}$ is the body force per unit mass in $C_{(e)}$, $\mathbf{S}$ is the surface traction per unit deformed area, $\dot{\mathbf{u}}$ is the velocity vector, $\mathbf{p}_N$ is the generalized force at node N, and $\dot{\mathbf{u}}^N$ is the velocity of node N of the element. Since $\mathbf{u}(\mathbf{x},t) = \mathbf{u}^N(t)\psi_N(\mathbf{x})$ and $\dot{\mathbf{u}} = \dot{\mathbf{u}}^N(t)\psi_N(\mathbf{x})$, the generalized nodal force vectors $\mathbf{p}_N$ are given by†

$$\mathbf{p}_N = \mathbf{f}_N + \mathbf{s}_N \tag{16.81}$$

where $\mathbf{f}_N$ and $\mathbf{s}_N$ are the generalized forces due to body forces and surface forces, respectively:

$$\mathbf{f}_N = \int_{v_{(e)}} \rho(\mathbf{x})\psi_N(\mathbf{x})\mathbf{F}(\mathbf{x},t)\, dv \qquad \text{and} \qquad \mathbf{s}_N = \int_{A_{(e)}} \psi_N(\mathbf{x})\mathbf{S}(\mathbf{x},t)\, dA \tag{16.82}$$

Now the vectors $\mathbf{p}_N$ can be expressed as linear combinations of the orthonormal basis vectors $\mathbf{i}_i$ in the reference configuration or the covariant basis vectors $\mathbf{G}_i$ which are tangent to the convected coordinate lines $x^i_{(e)}$ in the deformed element; that is,

$$\mathbf{p}_N = p_{Ni}\mathbf{i}_i = t^i_N \mathbf{G}_i \tag{16.83}$$

with $N = 1, 2, \ldots, N_e$ and $i = 1, 2, 3$. Recalling that

$$\mathbf{G}_i = z_{m,i}\mathbf{i}_m = (\delta_{mi} + u_{m,i})\mathbf{i}_m = (\delta_{mi} + \psi_{M,i}u^M_m)\mathbf{i}_m \tag{16.84}$$

it follows that

$$p_{Ni} = t^m_N(\delta_{mi} + \psi_{M,m}u^M_i) \tag{16.85}$$

Similarly, we may write

$$\mathbf{F} = \hat{F}_i \mathbf{i}_i = F^j \mathbf{G}_j \tag{16.86}$$

$$\hat{F}_i = F^j(\delta_{ij} + u_{j,i}) \tag{16.87}$$

so that

$$\mathbf{f}_N = \int_{v_{0(e)}} \rho_0 \hat{F}_i \psi_N\, dv_0\, \mathbf{i}_i = \int_{v_{(e)}} \rho F^j \psi_N \mathbf{G}_j\, dv \tag{16.88}$$

† See Eqs. (13.52) and (13.54).

where the dependence of the functions ρ_0, ρ, $\hat{F}_i$, F^j, ψ_N, and $\mathbf{G}_j$ on $\mathbf{x}$ is understood.

Turning now to the contribution to $\mathbf{p}_N$ of the surface forces, we note that the tractions can be measured per unit undeformed surface area dA_0 or deformed surface area dA, so that the net surface force is

$$\int_{A_{0(e)}} \hat{\mathbf{S}}\, dA_0 = \int_{A_{(e)}} \mathbf{S}\, dA \tag{16.89}$$

Here

$$\hat{\mathbf{S}} = t^{ij}\hat{n}_i\mathbf{G}_j \qquad \text{and} \qquad \mathbf{S} = \sigma^{ij}n_i\mathbf{G}_j \tag{16.90}$$

where t^{ij} and σ^{ij} are the stress tensors referred to the convected coordinates and measured per unit undeformed and deformed area, respectively ($t^{ij} = \sqrt{G}\,\sigma^{ij}$), $\hat{n}_i$ are the components of a unit normal to dA_0, and n_i are the covariant components of a unit normal to dA. Thus

$$\mathbf{s}_N = \int_{A_{0(e)}} t^{ij}\hat{n}_i\mathbf{G}_j\psi_N\, dA_0 = \int_{A_{(e)}} \sigma^{ij}n_i\mathbf{G}_j\psi_N\, dA \tag{16.91}$$

or, in view of (16.84),

$$\begin{aligned}\mathbf{s}_N &= \int_{A_{0(e)}} t^{ij}\hat{n}_i(\delta_{jm} + \psi_{M,j}u_m^M)\psi_N\, dA_0\mathbf{i}_m \\ &= \int_{A_{(e)}} \sigma^{ij}n_i(\delta_{jm} + \psi_{M,j}u_m^M)\psi_N\, dA\mathbf{i}_m\end{aligned} \tag{16.92}$$

An important aspect of the theory of deformable solids is that the geometry of a solid body is usually assumed to be known when the body occupies certain reference configurations. Consequently, it is often both natural and convenient to refer, as we have generally done, the motion to the reference configuration and to describe the behavior of the body in terms of its initial (reference) geometry. We then have for the cartesian components of the generalized force at node N,

$$p_{Ni} = \int_{v_{0(e)}} \rho_0\hat{F}_i\psi_N\, dv_0 + \int_{A_{0(e)}} \hat{S}^j(\delta_{ji} + \psi_{M,j}u_i^M)\psi_N\, dA_0 \tag{16.93}$$

in which $\hat{S}^j = t^{ij}\hat{n}_i$.

The contribution of body forces to p_{Ni} of (16.93) presents no special problems since the components $\hat{F}_i(\mathbf{x},t)$ are assumed to be prescribed for all $\mathbf{x}$ and t. The evaluation of the surface integral in (16.93), however, deserves further consideration. Consider, for example, the case in which the material surface of a finite element is subjected to a normal traction $q = q(\mathbf{x},t)$; for example, $-q$ would represent a variable pressure. Then the net force on an elemental area dA in the deformed element is

$$q\,dA\,\mathbf{n} = \hat{S}^j \mathbf{G}_j\,dA_0 \tag{16.94}$$

where $\mathbf{n} = n_i \mathbf{G}^i$ is a unit vector normal to dA. We recall that

$$\sqrt{G^{(ii)}}\,n_i\,dA = dA_i \qquad dA_0\,\hat{n}_i = dA_{0i} \tag{16.95a,b}$$

$$dA_i = \sqrt{GG^{(ii)}}\,dA_{0i} \tag{16.95c}$$

where dA_i and dA_{0i} are the projections of dA and dA_0 on the material co-ordinate surfaces in deformed body and undeformed body, respectively. Introducing (16.95) into (16.94), we find after some algebraic manipulations that

$$\hat{S}^j = q\sqrt{G}\,G^{jk}\hat{n}_k \tag{16.96}$$

where, for the finite element,

$$\begin{aligned} G_{kj}^{(e)} &= (\delta_{ik} + \psi_{N,k}u_i^N)(\delta_{ij} + \psi_{M,j}u_i^M) \\ G^{(e)} &= \det G_{kj}^{(e)} \qquad G_{(e)}^{kj} = (G_{kj}^{(e)})^{-1} \end{aligned} \tag{16.97}$$

Thus, ignoring effects due to body forces, (16.93) gives for this case

$$p_{Ni} = \int_{A_{0(e)}} q\sqrt{G}\,G^{jk}\hat{n}_k(\delta_{ji} + \psi_{M,j}u_i^M)\psi_N\,dA_0 \tag{16.98}$$

In the case of simplex models, it is convenient to assume that $\rho_0\hat{F}_i$ and $\hat{S}^j$ are constant for each element. Then (16.93) reduces to

$$p_{Ni} = \rho_0\hat{F}_i m_N + \hat{S}^j(\delta_{ji} + b_{Mj}u_i^M)m_N^{(A_0)} \tag{16.99}$$

where m_N is the first moment defined in (16.70) and

$$m_N^{(A_0)} = \int_{A_{0(e)}} \psi_N\,dA_0 \tag{16.100}$$

If q of (16.98) is also uniform, we find that for this type of loading

$$p_{Ni} = q\sqrt{G}\,G^{jk}\hat{n}_k(\delta_{ji} + \psi_{M,j}u_i^M)m_N^{(A_0)} \tag{16.101}$$

Approximations† From the forms of (16.93) and (16.99) it is clear that even in the case of relatively simple loading and finite-element geometry, the generalized nodal forces are quite complicated functions of the nodal displacements. It is natural to seek simplified approximations of these forces. We shall describe one rather general method of approximation here which amounts to simply representing the deformed surface area of an arbitrary finite element by a collection of flat triangular or quadrilateral elements over which the loading is assumed to be uniform.

† We follow here the note of Oden [1970*d*].

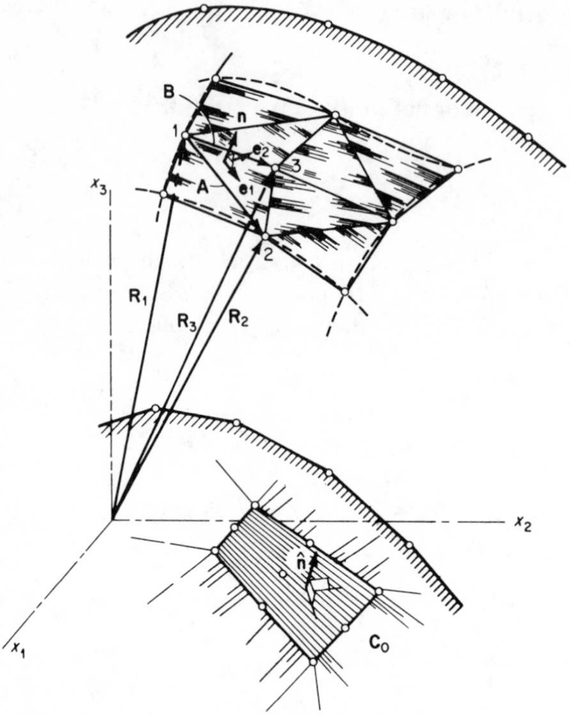

Fig. 16.2 Approximation of deformed element surface.

Consider the surface of an arbitrary type of finite element, as shown in Fig. 16.2, which undergoes large displacements and distortions due to a general system of applied surface tractions. We ignore body forces here since they are adequately accounted for by (16.88). Regardless of the actual form of the local displacement field $u_i = \psi_N u_i^N$, we represent the deformed material surface by a collection of flat elements, generally triangular in shape, the nodal points of which are coincident with those of the original material surface of the element. The applied surface tractions per unit surface area are assumed to be uniform over each flat element. Ideally, if the element is sufficiently small, the network of subelements can provide a close approximation to rather general applied loads.

Confining our attention to a typical flat triangular subelement of area a, let $\mathbf{n}$, $\mathbf{e}_1$, and $\mathbf{e}_2$ denote an orthonormal triad of vectors, $\mathbf{n}$ being normal to a in the deformed element. The total applied force on a can then be written in the form

$$\mathbf{S} = a(q_0\mathbf{n} + S_1\mathbf{e}_1 + S_2\mathbf{e}_2) \tag{16.102}$$

where q_0, S_1, and S_2 are prescribed uniform normal and tangential components of surface force per unit deformed area a. Suppose that the triangular surface on which **S** is applied passes through nodes 1, 2, and 3 of the deformed surface of the element; then, following a procedure similar to that used in deriving special incompressibility conditions in the previous section, we observe that position vectors drawn from the reference configuration to the displaced positions of nodes 1, 2, and 3 are simply

$$\mathbf{R}^N = \mathbf{r}^N + \mathbf{u}^N = (x_i^N + u_i^N)\mathbf{i}_i \qquad i, N = 1, 2, 3 \tag{16.103}$$

The sides of the triangle are formed by the two vectors

$$\mathbf{A} = \mathbf{R}^2 - \mathbf{R}^1 \qquad \text{and} \qquad \mathbf{B} = \mathbf{R}^3 - \mathbf{R}^1 \tag{16.104}$$

so that the surface area is given by

$$a = \tfrac{1}{2}\,|\mathbf{A} \times \mathbf{B}| = \tfrac{1}{2}\sqrt{A_iA_iB_jB_j - A_iB_iA_jB_j} \tag{16.105}$$

while the unit normal **n** is given by

$$\mathbf{n} = \frac{1}{2a}\,\mathbf{A} \times \mathbf{B} = \frac{1}{2a}\,\epsilon^{ijk}A_iB_j\mathbf{i}_k \tag{16.106}$$

If, for example, we set

$$\mathbf{e}_1 = \frac{\mathbf{A}}{|\mathbf{A}|} = \frac{\mathbf{A}}{\sqrt{A_iA_i}} \tag{16.107}$$

then

$$\mathbf{e}_2 = \mathbf{e}_1 \times \mathbf{n} = \frac{1}{2a}\,[(\mathbf{e}_1 \cdot \mathbf{B})\mathbf{A} - |\mathbf{A}|\mathbf{B}] \tag{16.108}$$

and (16.102) becomes

$$\mathbf{S} = \tfrac{1}{2}[q_0\mathbf{A} \times \mathbf{B} + \frac{1}{|\mathbf{A}|}\,(2aS_1 + \mathbf{A} \cdot \mathbf{B}S_2)\mathbf{A} - S_2|\mathbf{A}|\mathbf{B}] \tag{16.109}$$

The associated generalized nodal forces $\hat{\mathbf{p}}_N$ can be obtained by simply distributing **S** equally at each node; that is, $\hat{\mathbf{p}}_N = (\frac{1}{3})\mathbf{S}$, $N = 1, 2, 3$. Alternatively, we can write $\hat{\mathbf{p}}_N = \mathbf{S}m_N^{(a)}$, where $m_N^{(a)} = \int_a (a_N + b_{Ni}x^i)\,da$. Since the first alternative is generally more straightforward, we choose to write here

$$\begin{aligned}\hat{p}_{Nk} = \frac{1}{6}\Big\{&q_0\epsilon_{ijk}(x_i^2 + u_i^2 - x_i^1 - u_i^1)(x_j^3 + u_j^3 - x_j^1 - u_j^1)\\ &+ \frac{2aS_1 + S_2(x_i^2 + u_i^2 - x_i^1 - u_i^1)(x_i^3 + u_i^3 - x_i^1 - u_i^1)}{[(x_r^2 + u_r^2 - x_r^1 - u_r^1)(x_r^2 + u_r^2 - x_r^1 - u_r^1)]^{\frac{1}{2}}}\\ &\times (x_k^2 + u_k^2 - x_k^1 - u_k^1) - S_2[(x_r^2 + u_r^2 - x_r^1 - u_r^1)\\ &\times (x_r^2 + u_r^2 - x_r^1 - u_r^1)]^{\frac{1}{2}}(x_k^3 + u_k^3 - x_k^1 - u_k^1)\Big\}\end{aligned} \tag{16.110}$$

for all N, $N = 1, 2, 3$.

We emphasize that the forces $\hat{\mathbf{p}}_N$ are not to be confused with the generalized forces $\mathbf{p}_N$ appearing in the stiffness relations of the element. If R nodes of the element are on the exterior loaded surface, then the net local generalized force at a node N of the element is

$$\mathbf{p}_N = \sum_{e=1}^{E'\,(e)} \Omega_N^{N'} \hat{\mathbf{p}}_{N'} \tag{16.111}$$

where E' is the number of flat subelements used in modeling the surface of the element; $N = 1, 2, \ldots, R$; and $N' = 1, 2, 3$. We also note that it may not always be possible or convenient to use (16.107) as the definition of $\mathbf{e}_1$; depending upon the nature of the applied loading, different choices of $\mathbf{e}_1$ may be desirable for different cases. In fact, the basis vectors $\mathbf{e}_1$ and $\mathbf{e}_2$ in the plane of a need not be orthogonal, and we may simply take $\mathbf{e}_1 = \mathbf{A}/|\mathbf{A}|$ and $\mathbf{e}_2 = \mathbf{B}/|\mathbf{B}|$. Then (16.110) reduces to

$$\hat{p}_{Nk} = \frac{1}{6}\Bigg[q_0\epsilon_{ijk}(x_i^2 + u_i^2 - x_i^1 - u_i^1)(x_j^3 + u_j^3 - x_j^1 - u_j^1) + \frac{aS_1^*(x_k^2 + u_k^2 - x_k^1 - u_k^1)}{|\mathbf{A}|} + \frac{aS_2^*(x_k^3 + u_k^3 - x_k^1 - u_k^1)}{|\mathbf{B}|}\Bigg] \tag{16.112a}$$

where S_1^* and S_2^* are the components of $\mathbf{S}/a$ in the directions of $\mathbf{e}_1$ and $\mathbf{e}_2$ and

$$|\mathbf{A}| = [(x_r^2 + u_r^2 - x_r^1 - u_r^1)(x_r^2 + u_r^2 - x_r^1 - u_r^1)]^{\frac{1}{2}} \tag{16.112b}$$

$$|\mathbf{B}| = [(x_r^3 + u_r^3 - x_r^1 - u_r^1)(x_r^3 + u_r^3 - x_r^1 - u_r^1)]^{\frac{1}{2}} \tag{16.112c}$$

The important special case of a uniform normal force $\mathbf{S} = q_0 a\mathbf{n}$ leads to particularly simple expressions for the generalized forces $\hat{\mathbf{p}}_N$. From (16.110),

$$\hat{p}_{Nk} = \tfrac{1}{6}q_0\epsilon_{ijk}(x_i^2 + u_i^2 - x_i^1 - u_i^1)(x_j^3 + u_j^3 - x_j^1 - u_j^1) \tag{16.113}$$

In the case of two-dimensional problems, we may employ essentially the same procedure as that described above; i.e., a curved boundary in the deformed body is replaced by a collection of straight-line segments, as indicated in Fig. 16.3. In this case, if $\mathbf{S}$ represents a uniform applied surface traction per unit length and $\mathbf{n}$ and $\mathbf{e}$ are unit vectors normal and tangent to the deformed surface,

$$\mathbf{S} = L^*(q_0\mathbf{n} + S\mathbf{e}) \tag{16.114}$$

where q_0 and S are normal and tangential tractions per unit length, and

$$L^* = |\mathbf{R}^2 - \mathbf{R}^1| = [(x_\alpha^2 + u_\alpha^2 - x_\alpha^1 - u_\alpha^1)(x_\alpha^2 + u_\alpha^2 - x_\alpha^1 - u_\alpha^1)]^{\frac{1}{2}} \tag{16.115}$$

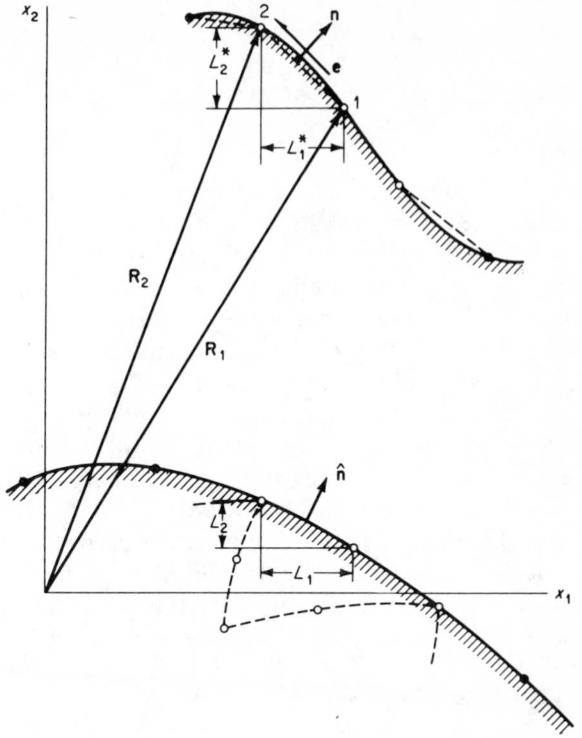

Fig. 16.3 Approximation of deformed boundary of two-dimensional body.

where $\alpha = 1, 2$. It is easily verified that

$$\mathbf{n} = \frac{1}{L^*}(L_2^*\mathbf{i}_1 + L_1^*\mathbf{i}_2) \tag{16.116a}$$

$$\mathbf{e} = \frac{-1}{L^*}(L_1^*\mathbf{i}_1 - L_2^*\mathbf{i}_2) \tag{16.116b}$$

where

$$L_1^* = L_1 + u_1^2 - u_1^1 \qquad L_2^* = L_2 + u_2^2 - u_2^1 \tag{16.116c}$$

Here $L_1 = x_1^2 - x_1^1$, $L_2 = x_2^2 - x_2^1$ are the projections of the line drawn from node 1 to 2 on the x_1 and x_2 axes in the undeformed element. Then the associated generalized forces are simply

$$\begin{aligned} \hat{p}_{N1} &= \tfrac{1}{2}[q_0(L_2 + u_2^2 - u_2^1) - S(L_1 + u_1^2 - u_1^1)] \\ \hat{p}_{N2} &= \tfrac{1}{2}[q_0(L_1 + u_1^2 - u_1^1) + S(L_2 + u_2^2 - u_2^1)] \end{aligned} \tag{16.117}$$

for all N, $N = 1, 2$, and in the special case of purely normal loading ($S = 0$), they reduce to

$$\hat{p}_{N\alpha} = \frac{q_0}{2}(-1)^{(\alpha)}\epsilon_{\beta\alpha}L^*_\beta \tag{16.118}$$

where α, $\beta = 1, 2$, and $\epsilon_{\alpha\beta}$ is the two-dimensional permutation symbol.

Stresses† Once the nodal displacements U_i^Δ and $u_{i(e)}^N$ have been determined, the local strains and strain energy can be obtained by simply introducing the local displacement fields into the strain-displacement relations and incorporating the result into the appropriate equations for the local strain-energy function for the element. Element stresses can then be computed by means of the constitutive equations (15.1) or (15.36). Stresses computed in this manner, however, will generally suffer finite discontinuities at interelement boundaries and will represent, at best, only rough averages of the actual stresses in the element. However, continuous stress distributions can be easily obtained by employing the notions of conjugate-approximation functions developed in Art. 9.

Let $t_{(e)}^{ij}$ denote the local stress tensor in element e derived from the local constitutive equations for the element (e.g., for hyperelastic elements $t_{(e)}^{ij} = \partial W^{(e)}/\partial \gamma_{ij}^{(e)}$). The discontinuous stress distribution over the connected model generated by such local stresses is, except on a set of measure zero,

$$\bar{T}^{ij} = \sum_{e=1}^{E} t_{(e)}^{ij}(\mathbf{x},t) \tag{16.119}$$

where appropriate transformations from local to global coordinates have been assumed. Being discontinuous, the global stresses $\bar{T}^{ij} = \bar{T}^{ij}(\mathbf{X},t)$ do not belong to the finite-dimensional subspace Φ or to its conjugate Φ^* spanned by the global basis functions $\Phi_\Delta(\mathbf{X})$ or $\Phi^\Delta(\mathbf{X})$. However, we can easily calculate a projection $T^{ij} = \Pi\bar{T}^{ij}$ of $\bar{T}^{ij}$ into the conjugate space Φ^* by writing

$$T^{ij} = T_\Delta^{ij}\Phi^\Delta(\mathbf{X}) \tag{16.120}$$

where

$$T_\Delta^{ij} = \langle \bar{T}^{ij},\Phi_\Delta\rangle = \int_{\bar{\mathscr{R}}} \bar{T}^{ij}\Phi_\Delta \, d\mathscr{R} \tag{16.121}$$

and the integration is taken over the entire connected domain. The quantities T_Δ^{ij}, $\Delta = 1, 2, \ldots, G$, are referred to as the *consistent components of stress at node Δ of the connected finite-element model.*

Local forms of (16.120) can be obtained by recalling that‡

$$\Phi_\Delta(\mathbf{X}) = \sum_e^{(e)} \Omega_\Delta^N \psi_N^{(e)}(\mathbf{x}) \qquad \text{and} \qquad \Phi^\Delta(\mathbf{X}) = C^{\Delta\Gamma}\Phi_\Gamma(\mathbf{X}) \tag{16.122a,b}$$

† See Oden and Brauchli [1971*a*] and Oden and Key [1971*a*].
‡ See Eqs. (9.138), (9.144), and (9.148).

where $C^{\Delta\Gamma}$ is the inverse of $C_{\Delta\Gamma}$:

$$C_{\Delta\Gamma} = \langle \Phi_\Delta, \Phi_\Gamma \rangle \tag{16.122c}$$

Then, for a typical element, we set

$$t^{ij}_{N(e)} = \langle t^{ij}_{(e)}, \psi_N \rangle = \int_{v_{0(e)}} t^{ij}_{(e)} \psi_N \, dv_0 \tag{16.123}$$

and obtain for the local approximation†

$$\bar{t}^{ij}_{(e)} = t^{ij}_{N(e)} \psi^N_{(e)}(\mathbf{x}) \tag{16.124}$$

where, in agreement with (9.154) and (9.155),

$$\psi^N_{(e)}(\mathbf{x}) = \overset{(e)}{\Omega}{}^N_\Delta \Phi^\Delta(\mathbf{X}) = \overset{(e)}{\Omega}{}^N_\Delta C^{\Delta\Gamma} \sum_{f=1}^{E} \overset{(f)}{\Omega}{}^M_\Gamma \psi^{(f)}_M(\mathbf{x}) \tag{16.125}$$

In applications, the local fields (16.124) are of only secondary interest since consistent stresses at each node can be obtained directly from (16.121).

16.4 POTENTIAL-ENERGY FORMULATIONS

In many cases it is convenient to recast the finite-element formulation of elasticity problems in terms of the total potential energy of each finite element. Returning to the nonlinear stiffness relations (16.4) for hyperelastic elements, we observe that if both sides of this equation are multiplied by arbitrary variations δu^N_i in the nodal displacements, we obtain

$$\int_{v_{0(e)}} \frac{\partial W}{\partial \gamma_{kj}} \psi_{N,j}(\delta_{ki} + \psi_{M,k} u^M_i) \delta u^N_i \, dv_0 = p_{Ni} \, \delta u^N_i \tag{16.126}$$

Noting also that

$$\begin{aligned} \delta U_{(e)} &= \int_{v_{0(e)}} \delta W_{(e)} \, dv_0 = \int_{v_{0(e)}} \frac{\partial W_{(e)}}{\partial u^N_i} \delta u^N_i \, dv_0 \\ &= \int_{v_{0(e)}} \frac{\partial W_{(e)}}{\partial \gamma_{kj}} \frac{\partial \gamma_{kj}}{\partial u^N_i} \delta u^N_i \, dv_0 \\ &= \int_{v_{0(e)}} \frac{\partial W_{(e)}}{\partial \gamma_{kj}} \psi_{N,j}(\delta_{ki} + \psi_{M,k} u^M_i) \delta u^N_i \, dv_0 \end{aligned} \tag{16.127}$$

we see that the integral in (16.126) can be interpreted as the variation in the total strain energy $U_{(e)}$ of the element due to a variation in the generalized displacements u^N_i. Likewise, if we write

$$\delta V_{(e)} = -p_{Ni} \delta u^N_i \tag{16.128}$$

then the right side of (16.126) appears to be the variation in a local potential

† See Eq. (9.153).

energy of the external forces $V_{(e)} = -p_{Ni}u_i^N$. We understand, of course, that an actual potential $V_{(e)}$ may not exist, but we can treat the term in (16.128) artificially as a potential if the correct generalized forces developed in the previous section are used and if these are not allowed to be influenced by variations in u_i^N. With this provision, we may rewrite (16.126) in the form

$$\delta\pi^{(e)} = \frac{\partial\pi^{(e)}}{\partial u_i^N}\,\delta u_i^N = 0 \tag{16.129}$$

where $\pi^{(e)}$ is the total potential energy of element e:

$$\pi^{(e)} = U_{(e)} + V_{(e)} = \int_{v_{0(e)}} W_{(e)}\,dv_0 - p_{Ni}u_i^N \tag{16.130}$$

The stiffness equations for a typical element then become

$$\frac{\partial\pi^{(e)}}{\partial u_i^N} = 0 = \int_{v_{0(e)}} \frac{\partial W_{(e)}}{\partial u_i^N}\,dv_0 - p_{Ni} \tag{16.131}$$

and the total potential energy of the entire collection of elements is

$$\Pi = \sum_{e=1}^{E} \pi^{(e)} = U + V \tag{16.132}$$

where

$$U = \sum_{e=1}^{E} U_{(e)} \qquad V = \sum_{e=1}^{E} V_{(e)} \tag{16.133}$$

Since

$$u_{i(e)}^N = \overset{(e)}{\Omega_\Delta^N} U_i^\Delta \qquad \text{and} \qquad P_{\Delta i} = \sum_{e=1}^{E}\overset{(e)}{\Omega_\Delta^N} p_{Ni}^{(e)} \tag{16.134}$$

it is clear that

$$V = -\sum_e p_{Ni}^{(e)}u_{i(e)}^N = -\sum_e \overset{(e)}{\Omega_\Delta^N} p_{Ni}^{(e)} U_i^\Delta = -P_{\Delta i}U_i^\Delta \tag{16.135}$$

Hence,

$$\Pi = \sum_{e=1}^{E} \int_{v_{0(e)}} W_{(e)}\,dv_0 - P_{\Delta i}U_i^\Delta \tag{16.136}$$

Global stiffness relations are obtained from the equilibrium conditions

$$\delta\Pi = \frac{\partial\Pi}{\partial U_i^\Delta}\,\delta U_i^\Delta = 0 \tag{16.137}$$

which implies

$$\frac{\partial\Pi}{\partial U_i^\Delta} = 0 = \sum_{e=1}^{E} \int_{v_{0(e)}} \frac{\partial W_{(e)}}{\partial u_{k(e)}^N}\frac{\partial u_{k(e)}^N}{\partial U_i^\Delta}\,dv_0 - P_{\Delta i} \tag{16.138}$$

Since

$$\frac{\partial u_{k(e)}^N}{\partial U_i^\Delta} = \overset{(e)}{\Omega_\Delta^N}\delta_{ki} \tag{16.139}$$

we see that (16.138) can be rewritten in the form

$$P_{\Delta i} = \sum_{e=1}^{E} \overset{(e)}{\Omega_{\Delta}^{N}} \int_{v_{0(e)}} \frac{\partial W_{(e)}}{\partial \gamma_{kj}} \psi_{N,j}(\delta_{ki} + \psi_{M,k} \overset{(e)}{\Omega_{\Gamma}^{M}} U_i^{\Gamma})\, dv_0 \tag{16.140}$$

which is in agreement with (16.33).

In the case of incompressible materials, we must use instead of $W_{(e)}$ the modified strain-energy function $\overline{W} = W_{(e)} + h(I_3 - 1)$. Then

$$\overline{\Pi} = \sum_{e=1}^{E} \int_{v_{0(e)}} [W_{(e)} + h(I_3 - 1)]\, dv_0 - P_{\Delta i} U_i^{\Delta} \tag{16.141}$$

and the condition $\delta\overline{\Pi} = 0$ leads to both the equilibrium conditions and the incompressibility conditions. For if $h = h^N \psi_N$, then

$$\delta\overline{\Pi} = \frac{\partial\overline{\Pi}}{\partial u_i^N}\, \delta u_i^N + \frac{\partial\overline{\Pi}}{\partial h^N}\, \delta h^N = 0 \tag{16.142}$$

which implies that both $\partial\overline{\Pi}/\partial u_i^N$ and $\partial\overline{\Pi}/\partial h^N$ vanish. The condition $\partial\overline{\Pi}/\partial u_i^N = 0$ gives (16.57) whereas we recognize

$$\frac{\partial\overline{\Pi}}{\partial h^N} = \int_{v_{0(e)}} \psi_N(I_3 - 1)\, dv_0 = (iii)_N = 0 \tag{16.143}$$

as the incompressibility condition of (16.60).

For dynamic phenomena, we construct the local lagrangian potential $l_{(e)}$ of each element by subtracting $\pi^{(e)}$ from the kinetic energy:

$$l_{(e)} = \kappa_{(e)} - \pi^{(e)} = \tfrac{1}{2} m_{NM} \dot{u}_i^N \dot{u}_i^M - \pi^{(e)} \tag{16.144}$$

Globally,

$$L = \sum_{e=1}^{E} l_{(e)} = \kappa - \Pi = \tfrac{1}{2} M_{\Delta\Gamma} \dot{U}_i^{\Delta} \dot{U}_i^{\Gamma} - \Pi \tag{16.145}$$

Minimization of the action integral A of (15.33) leads to Lagrange's equations of motion:

$$\frac{d}{dt}\frac{\partial L}{\partial \dot{U}_i^{\Delta}} - \frac{\partial L}{\partial U_i^{\Delta}} = 0 \tag{16.146}$$

which, for the assembly of finite elements, are

$$M_{\Delta\Gamma} \ddot{U}_i^{\Gamma} + \frac{\partial \Pi}{\partial U_i^{\Delta}} = 0 \tag{16.147}$$

Here $\partial\Pi/\partial U_i^{\Delta}$ is as given on the right side of (16.138). In (16.147), of course, $\partial\Pi/\partial U_i^{\Delta} \neq 0$ and $U_i^{\Delta} = U_i^{\Delta}(t)$.

16.5 INCREMENTAL FORMS†

In this section, we consider the important problem of deriving incremental forms of the equations of motion of a finite element. We temporarily remove the restriction that the element be elastic and return to the general case of motion of an element of an arbitrary continuum. The incremental equations are then specialized so as to apply to cases in which the incremental response of the element is elastic.

General equations For large deformations and general material properties, the equations of motion of a finite element are highly nonlinear. In many applications, however, it is convenient to consider linearized versions of these equations which involve small perturbations in the motion superimposed on an arbitrary motion of the element. Such incremental forms are particularly useful in problems of static and dynamic stability, plasticity, and static and quasi-static behavior of elastic bodies.

Consider a motion of a typical finite element which carries it from its reference configuration $C_{0(e)}$ to a configuration $C_t^{(e)}$ at time t. At a time $t + \Delta t$ later, the element occupies a configuration $C_{t+\Delta t}^{(e)}$, which is arbitrarily close to $C_t^{(e)}$. Let $\mathring{u}_i^N$, t_0^{ij}, S_i^0, and $\hat{F}_i^0$ and u_i^N, t^{ij}, $\bar{S}_i$, and $\hat{F}_i$ denote the nodal displacements, stress, surface tractions, and body forces corresponding to $C_t^{(e)}$ and $C_{t+\Delta t}^{(e)}$, respectively. Then

$$
\begin{aligned}
u_i^N &= \mathring{u}_i^N + \zeta_i^N & \qquad t^{ij} &= t_0^{ij} + s^{ij} \\
\bar{S}_i &= S_i^0 + R_i & \qquad \hat{F}_i &= \hat{F}_i^0 + B_i
\end{aligned}
\tag{16.148}
$$

where ζ_i^N, s^{ij}, R_i, and B_i are incremental changes in $\mathring{u}_i^N$, t_0^{ij}, S_i^0, and $\hat{F}_i^0$ during the motion from $C_t^{(e)}$ to $C_{t+\Delta t}^{(e)}$.

By $\bar{S}_i$ we mean the cartesian components of the surface traction per unit initial (undeformed) area referred to the reference configuration $C_{0(e)}$; that is, according to (16.92) and (16.93),

$$
\bar{S}_i = t^{mj}\hat{n}_m(\delta_{ji} + \psi_{M,j}u_i^M) = \hat{S}^j(\delta_{ji} + \psi_{M,j}u_i^M) \tag{16.149}
$$

† Although frequently rediscovered, the basic ideas of deriving linear incremental forms of the equations governing the behavior of deformable bodies are due to Cauchy [1829] and Saint Venant [1868]. A more recent comprehensive treatment of incremental deformations has been given by Biot [1965]. Incremental procedures have found wide use in finite-element applications. Among the first to discuss applications of the procedure in connection with finite-element analyses of geometrically nonlinear elasticity problems and problems of elastic stability were Turner [1959] and Argyris [1959]. A survey of related work up to 1965 was given by Martin [1966*b*]. Many of the incremental finite-element formulations presented prior to 1968 are incomplete in that variations in loading and rotations of the element are not properly represented. Relatively general forms of the incremental equations for finite elements have been presented by Yaghmai [1968], Hibbit, Marcal, and Rice [1970], Oden [1969*c*, 1970*b*], and Martin [1970]. For surveys, consult Oden [1969*c*], Martin [1970], and Kawai [1970]. The present article is based on a paper by Oden and Key [1971*b*].

Thus, if

$$S_i^0 = \hat{S}_0^j(\delta_{ji} + \psi_{M,j}\mathring{u}_i^M) \tag{16.150}$$

then

$$\bar{S}_i = (\hat{S}_0^j + \delta\hat{S}^j)[\delta_{ji} + \psi_{M,j}(\mathring{u}_i^M + \zeta_i^M)] \tag{16.151}$$

where the increments $\delta\hat{S}^j$ in $\hat{S}^j$ are, in general, also functions of $\mathring{u}_i^M$, ζ_i^M, and changes in loading. For example, in the case of a uniform pressure q_0, we find from (16.96) that

$$\delta\hat{S}^j = \delta q_0\sqrt{G_0}\, G_0^{kj}\hat{n}_k + q_0[(\delta\sqrt{G})G_0^{kj}\hat{n}_k + \sqrt{G_0}\,(\delta G^{kj})\hat{n}_k] + 0(\epsilon^2) \tag{16.152}$$

where G_0 and G_0^{kj} are the determinant and the contravariant components of the deformation tensor in $C_t^{(e)}$; δq_0, $\delta\sqrt{G}$, and δG^{kj} are increments in q_0, $\sqrt{G}$, and G^{kj}, δq_0 being prescribed and $\delta\sqrt{G}$ and δG^{kj} being due to increments in the deformation. The term $0(\epsilon^2)$ denotes terms of higher order in the increments δq_0, δG, and δG^{kj}. Since, in general, we can set

$$\delta\sqrt{G} = \frac{\partial\sqrt{G(\mathbf{x},\mathring{u}_i^M)}}{\partial u_i^N}\,\delta\zeta_i^N \qquad \delta G^{kj} = \frac{\partial G^{kj}(\mathbf{x},\mathring{u}_i^M)}{\partial u_i^N}\,\delta\zeta_i^N \tag{16.153}$$

the terms in brackets in (16.150) are linear functions of ζ_i^N. Consequently, up to terms of first order in the increments in the nodal displacements and applied forces, we can always write $\delta\hat{S}^j$ in the form

$$\delta\hat{S}^j = Q_i^j(\mathring{u}_i^M, T_0^i)\delta T^i + P_M^{ji}(\mathring{u}_i^M, T_0^i)\zeta_i^M \tag{16.154}$$

Here T_0^i denotes the actual prescribed applied tractions in $C_t^{(e)}$ which are independent of deformation, δT^i is a prescribed increment in T_0^i, and Q_i^j and P_M^{ji} are functions of only the state of the body in $C_t^{(e)}$ (prior to applying the increments δT^i). It follows from (16.150) and (16.151) that

$$\begin{aligned} R_i &= \bar{S}_i - S_i^0 \\ &= \hat{S}_0^j\psi_{M,j}\zeta_i^M + (\delta_{ji} + \psi_{M,j}\mathring{u}_i^M)(Q_r^j\delta T^r + P_M^{jr}\zeta_r^M) \end{aligned} \tag{16.155}$$

Hence, the generalized nodal forces in $C_{t+\Delta t}^{(e)}$ can be expressed as

$$p_{Ni} = p_{Ni}^0 + r_{Ni} \tag{16.156}$$

where

$$p_{Ni}^0 = \int_{v_{0(e)}} \rho_0\hat{F}_i^0\psi_N\,dv_0 + \int_{A_{0(e)}} \hat{S}_0^j(\delta_{ji} + \psi_{M,j}\mathring{u}_i^M)\psi_N\,dA_0 \tag{16.157}$$

and

$$\begin{aligned} r_{Ni} = \int_{v_{0(e)}} \rho_0 B_i\psi_N\,dv_0 &+ \int_{A_{0(e)}} [\hat{S}_0^j\psi_{M,j}\delta_{ri} + (\delta_{ji} + \psi_{R,j}\mathring{u}_i^R)P_M^{jr}] \\ &\times \psi_N\,dA_0\zeta_r^M + \int_{A_{0(e)}} \psi_N(\delta_{ji} + \psi_{M,j}\mathring{u}_i^M)Q_r^j\delta T^r\,dA_0 \end{aligned} \tag{16.158}$$

Turning now to the general equations of motion (13.60), we introduce (16.148) and (16.156) and obtain

$$m_{NM}(\ddot{\overset{\circ}{u}}{}_i^M + \ddot{\zeta}_i^M) + \int_{v_{0(e)}} (t_0^{mj} + s^{mj})\psi_{N,m}(\delta_{ji} + \psi_{M,j}\overset{\circ}{u}{}_i^M + \psi_{M,j}\zeta_i^M)\, dv_0 = \overset{0}{p}_{Ni} + r_{Ni} \quad (16.159)$$

which can be rewritten in the form

$$\left[m_{NM}\ddot{\overset{\circ}{u}}{}_i^M + \int_{v_{0(e)}} t_0^{mj}\psi_{N,m}(\delta_{ji} + \psi_{M,j}\overset{\circ}{u}{}_i^M)\, dv_0 - \overset{0}{p}_{Ni}\right] + m_{NM}\ddot{\zeta}_i^M$$
$$+ \int_{v_{0(e)}} t_0^{mj}\psi_{N,m}\psi_{M,j}\, dv_0\, \zeta_i^M + \int_{v_{0(e)}} s^{mj}\psi_{N,m}(\delta_{ji} + \psi_{M,j}\overset{\circ}{u}{}_i^M)\, dv_0$$
$$+ \int_{v_{0(e)}} s^{mj}\psi_{N,m}\psi_{M,j}\, dv_0\, \zeta_i^M = r_{Ni} \quad (16.160)$$

Since the quantities $\overset{\circ}{u}{}_i^M$, t_0^{ij}, and $\overset{0}{p}_{Ni}$ must satisfy the equations of motion of the element while it occupies configuration $C_t^{(e)}$, the terms in brackets in (16.160) must vanish. Moreover, the last integral on the left side of (16.160) involves the product of the incremental stress tensor s^{mj}, which is linear in ζ_i^N or its histories, and the increment nodal displacements ζ_i^N. Consequently, this term is of higher order than the first in the increments ζ_i^N and can be omitted. We are then left with the system of linear equations

$$m_{NM}\ddot{\zeta}_i^M + \int_{v_{0(e)}} t_0^{mj}\psi_{N,m}\psi_{M,j}\, dv_0\, \zeta_i^M + \int_{v_{0(e)}} s^{mj}\psi_{N,m} \times (\delta_{ji} + \psi_{M,j}\overset{\circ}{u}{}_i^M)\, dv_0 = r_{Ni} \quad (16.161)$$

Equations (16.161) are the general incremental equations of motion of a finite element.

In general, the stress tensor at any particle will be a functional of the deformation gradients, their histories, the temperature, the temperature histories, possibly gradients of the temperature, etc. For finite-element models, the response functional for stress can always be written in terms of the nodal values of displacements, temperatures, and their histories and as functions of the particle $\mathbf{x}$ and possibly certain preferred directions $\mathbf{d}_k$ in the reference configuration. Consequently, the incremental stress tensor s^{mj} can be defined so as to be a linear functional of the values and histories of the nodal displacements and temperatures, etc. For example, consider a special class of simple materials for which

$$t^{ij} = \mathop{\mathfrak{T}^{ij}}_{s=0}^{\infty}(\mathbf{u}^{N(t)}(s)) \quad (16.162)$$

where the dependence on $\mathbf{x}$ is understood. If the histories $\mathbf{u}^{N(t)}(s)$ and the current nodal displacements $\mathbf{u}^N$ take on small variations $\boldsymbol{\zeta}^{N(t)}(s)$ and $\boldsymbol{\zeta}^N$, then

t^{ij} undergoes a corresponding variation s^{ij}, where

$$s^{ij} = \underset{s=0}{\overset{\infty}{\mathfrak{T}}}{}^{ij}(\mathbf{u}^{N(t)}(s) + \boldsymbol{\zeta}^{N(t)}(s)) - t^{ij} \tag{16.163}$$

If $\underset{s=0}{\overset{\infty}{\mathfrak{T}}}{}^{ij}(\;)$ satisfies appropriate conditions of smoothness, we may write

$$s^{ij} = \underset{s=0}{\overset{\infty}{\mathfrak{T}}}{}^{ij}(\mathbf{u}^{N(t)}(s)) + \delta \underset{s=0}{\overset{\infty}{\mathfrak{T}}}{}^{ij}(\mathbf{u}^{N(t)}(s)|\boldsymbol{\zeta}^{N(t)}(s)) + \cdots - t^{ij} \tag{16.164}$$

where $\delta \underset{s=0}{\overset{\infty}{\mathfrak{T}}}{}^{ij}(\;|\;)$ is the first Frechet derivative of the functional $\underset{s=0}{\overset{\infty}{\mathfrak{T}}}{}^{ij}(\;)$ and the vertical stroke in the argument of $\delta \underset{s=0}{\overset{\infty}{\mathfrak{T}}}{}^{ij}(\;)$ indicates that it is linear in the quantity following the stroke—in this case, the histories $\boldsymbol{\zeta}^{N(t)}(s)$. Thus, introducing (16.162) into (16.163) and retaining only terms of first order in $\boldsymbol{\zeta}^N$ and $\boldsymbol{\zeta}^{N(t)}(s)$, we find that s^{ij} is a *linear functional* of the histories of the incremental nodal displacements

$$s^{ij} = \underset{s=0}{\overset{\infty}{\mathbf{S}}}{}^{ij}(\mathbf{u}^{N(t)}(s)|\boldsymbol{\zeta}^{N(t)}(s)) \tag{16.165}$$

where

$$\underset{s=0}{\overset{\infty}{\mathbf{S}}}{}^{ij}(\mathbf{u}^{N(t)}(s)|\boldsymbol{\zeta}^{N(t)}(s)) = \delta \underset{s=0}{\overset{\infty}{\mathfrak{T}}}{}^{ij}(\mathbf{u}^{N(t)}(s)|\boldsymbol{\zeta}^{N(t)}(s)) \tag{16.166}$$

For finite elements of materials of this type, the incremental equations of motion (16.161) become

$$m_{NM}\ddot{\zeta}_i^M + \int_{v_{0(e)}} \underset{s=0}{\overset{\infty}{\mathfrak{T}}}{}^{mj}(\mathring{\mathbf{u}}^{N(t)}(s))\psi_{N,m}\psi_{M,j}\,dv_0\,\zeta_i^M + \int_{v_{0(e)}} \underset{s=0}{\overset{\infty}{\mathbf{S}}}{}^{mj}(\mathring{\mathbf{u}}^{N(t)}(s)|\boldsymbol{\zeta}^{N(t)}(s))\psi_{N,m}(\delta_{ji} + \psi_{M,j}\mathring{u}_i^M)\,dv_0 = r_{Ni} \tag{16.167}$$

Elastic elements In the important special case of elastic finite elements, the stress tensor is a function of the current values of G_{ij} or γ_{ij} and the incremental stress tensor can be expressed as a linear function of the increments ζ_i^N. For example, consider the case of a hyperelastic element with a strain-energy function $W(\gamma_{ij})$ analytic in the strains. If $W(\gamma_{ij})$ is the strain energy per unit volume while the element is in configuration $C_t^{(e)}$, then the strain energy in an arbitrarily close configuration $C_{t+\Delta t}^{(e)}$ is

$$W(\gamma_{ij} + \delta\gamma_{ij}) = W(\gamma_{ij}) + \frac{\partial W}{\partial \gamma_{ij}}\,\delta\gamma_{ij} + \frac{1}{2!}\frac{\partial^2 W}{\partial\gamma_{ij}\partial\gamma_{kr}}\,\delta\gamma_{ij}\delta\gamma_{kr} + \cdots \tag{16.168}$$

where $\delta\gamma_{ij}$ is the variation in the strain tensor due to an incremental change in u_k^N:

$$\delta\gamma_{ij} = \frac{\partial\gamma_{ij}}{\partial u_k^N}\zeta_k^N = \tfrac{1}{2}(\psi_{N,i}\delta_k^j + \psi_{N,j}\delta_k^i + \psi_{N,i}\psi_{M,j}\overset{\circ}{u}{}_k^M + \psi_{M,i}\psi_{N,j}\overset{\circ}{u}{}_k^M)\zeta_k^N \tag{16.169}$$

Thus, the stress tensor in the perturbed configuration is

$$t^{ij} = \frac{\partial W(\gamma_{ij} + \delta\gamma_{ij})}{\partial\gamma_{ij}} = \frac{\partial W}{\partial\gamma_{ij}} + \frac{\partial^2 W}{\partial\gamma_{ij}\partial\gamma_{mn}}\delta\gamma_{mn} + \cdots \tag{16.170}$$

or to within first-order terms in the strain increments,

$$t^{ij} = t_0^{ij} + \frac{\partial^2 W}{\partial\gamma_{ij}\partial\gamma_{mn}}\delta\gamma_{mn} \tag{16.171}$$

where $t_0^{ij} = \partial W/\partial\gamma_{ij}$. Thus

$$s^{ij} = t^{ij} - t_0^{ij} = C^{ijmn}\delta\gamma_{mn} \tag{16.172}$$

where

$$C^{ijmn} = \frac{\partial^2 W}{\partial\gamma_{ij}\partial\gamma_{mn}} \tag{16.173}$$

Observing that C^{ijmn} has the symmetries

$$C^{ijmn} = C^{jimn} = C^{ijnm} = C^{mnij} \tag{16.174}$$

we introduce (16.169) into (16.172) and obtain

$$s^{ij} = C^{ijmn}\psi_{N,m}(\delta_{nk} + \psi_{M,n}\overset{\circ}{u}{}_k^M)\zeta_k^N \tag{16.175}$$

Thus, the incremental stress tensor for an elastic element is a linear function of the incremental nodal displacements and is linear in the strain increments $\delta\gamma_{ij}$. However, *the incremental response is not elastic in the sense of the classical theory, for the additional stress does not depend only upon an additional infinitesimal strain e_{ij} but depends also on the displacements $\overset{\circ}{u}{}_k^M$*.† Indeed, if $C^{ijmn} = E^{ijmn}$, we observe that to assume $s^{ij} = E^{ijmn}e_{mn}$ would lead to a stress of $E^{ijmn}\psi_{N,m}\zeta_n^N$. This is in error by a significant quantity, $E^{ijmn}\psi_{N,m}\psi_{M,n}\overset{\circ}{u}{}_k^M\zeta_k^N$, which represents, in part, the effects of a rotation of the element.

Incremental stiffnesses It is often convenient to identify various incremental stiffness matrices (arrays) which operate linearly on the incremental displacements ζ_i^N. Referring to (16.158), (16.161), and (16.175),

† See Truesdell and Noll [1965, p. 250]. Unfortunately, the contrary is often erroneously assumed in most incremental finite-element formulations.

we introduce the following abbreviated notation:

$$\tilde{k}^{ik}_{NM} = \int_{v_{0(e)}} C^{ijmk}\psi_{N,j}\psi_{M,m}\,dv_0 \tag{16.176}$$

$$G^{ik}_{NM} = \int_{v_{0(e)}} t_0^{mj}\psi_{N,m}\psi_{M,j}\delta^{ik}\,dv_0 \tag{16.177}$$

$$D^{ik}_{NM} = \int_{v_{0(e)}} C^{mjrs}\psi_{N,m}\psi_{M,r}[\psi_{R,j}\overset{\circ}{u}{}^R_i(\delta_{sk} + \psi_{P,s}\overset{\circ}{u}{}^P_k) + \psi_{R,s}\overset{\circ}{u}{}^R_k\delta_{ij}]\,dv_0 \tag{16.178}$$

$$R^{ik}_{NM} = \int_{A_{0(e)}} \psi_N[\hat{S}^j_0\psi_{M,j}\delta^k_i + (\delta_{ji} + \psi_{R,j}\overset{\circ}{u}{}^R_i)P^{jk}_M]\,dA_0 \tag{16.179}$$

$$\delta p_{Ni} = \int_{v_{0(e)}} \rho_0 B_i\psi_N\,dv_0 + \int_{A_{0(e)}} \psi_N(\delta_{ji} + \psi_{M,j}\overset{\circ}{u}{}^M_i)Q^j_r\delta T^r\,dA_0 \tag{16.180}$$

Here $\tilde{k}^{ik}_{NM}$ is the *incremental stiffness matrix*, G^{ik}_{NM} *is the initial-stress matrix*,† D^{ik}_{NM} *is* the *initial-rotation matrix*,‡ R^{ik}_{NM} is the *initial-load matrix*,§ and δp_{Ni} is the increment in generalized nodal forces due to prescribed increments in external forces. With this notation, the incremental equations of motion of an element become

$$m_{NM}\ddot{\zeta}^M_i + (\tilde{k}^{ik}_{NM} + G^{ik}_{NM} + D^{ik}_{NM} + R^{ik}_{NM})\zeta^M_k = \delta p_{Ni} \tag{16.181}$$

It should be noted that, contrary to implications in much of the literature, the *incremental stiffness matrix* $\tilde{k}^{ik}_{NM}$ *is not equal to the usual stiffness matrix* k^{ik}_{NM} *of classical infinitesimal elasticity given in* (16.13). Indeed, $\tilde{k}^{ik}_{NM}$ reduces to k^{ik}_{NM} only in the special case in which $C^{ijmn} = E^{ijmn}$.

† This matrix is also referred to as the "geometric" stiffness matrix or the "tangent" stiffness; see, for example, Turner, Dill, Martin, and Melosh [1960], Turner, Martin, and Weikel [1964], or Argyris, Kelsey, and Kamel [1964] and Argyris [1965*a*]. A detailed discussion of this matrix in connection with special types of elements was given by Martin [1966*b*], and a general formula was obtained by Oden [1966*b*] by using a different approach than the one followed here. Much of the earlier literature on finite-element applications to geometrically nonlinear problems used G^{ik}_{NM} to "correct" the linear stiffness matrix at the end of each successive load increment. Unfortunately, unless a new material frame of reference is established in the deformed element at the end of each increment—as done, for example, by Murray and Wilson [1969*a*, 1969*b*]—this procedure is obviously incorrect.

‡ The importance of D^{ik}_{NM} in certain nonlinear problems was discussed by Marcal [1968], who referred to it as the "initial displacement matrix." The term appeared in the incremental formulation of Oden and Kubitza [1967]. See also Oden [1969*c*, 1970*b*] and Hibbitt, Marcal, and Rice [1969].

§ The importance of including such load-correction terms in incremental formulations was pointed out by Oden [1969*e*] and Oden and Key [1970]. Specific forms of R^{ik}_{NM} were first presented by Hibbitt, Marcal, and Rice [1969]; however, their results do not coincide with (16.179).

A number of special problems can be formulated as special cases of (16.181). For example, if $\tilde{k}^{ik}_{NM} = k^{ik}_{NM}$ and the element undergoes no appreciable rotations in the unperturbed configuration, the classical static buckling load can be computed from the condition

$$\det (\tilde{k}^{ik}_{NM} + G^{ik}_{NM}) = 0 \tag{16.182}$$

However, cases in which moderately large rotations of the element occur prior to buckling are not uncommon, and it is generally necessary to add to the matrices in (16.182) D^{ik}_{NM} and R^{ik}_{NM}. If δp_{Ni} is set equal to zero and $m_{NM}\ddot{\zeta}^M_i$ is retained, the resulting equations can be used to study problems of dynamic stability.†

Incompressible elements We remark that in the case of incompressible materials we must also consider increments in the element hydrostatic pressures. If we follow essentially the same procedure as that used to derive (16.161) except that we use the equations of motion of incompressible elements (16.57) as a starting point, we obtain, instead of (16.181),

$$m_{NM}\ddot{\zeta}^M_i + (\tilde{k}^{ik}_{NM} + G^{ik}_{NM} + D^{ik}_{NM} + R^{ik}_{NM} + H^{ik}_{NM})\zeta^M_k + J_{Ni}(\delta h) = p_{Ni} \tag{16.183}$$

where H^{ik}_{NM} is the *initial-pressure matrix*

$$H^{ik}_{NM} = \int_{v_{0(e)}} \bar{h}_0 \left[\frac{\partial G^{mj}(\overset{\circ}{\mathbf{u}}{}^R)}{\partial u^M_k} \psi_{N,j}(\delta_{im} + \psi_{P,m}\overset{\circ}{u}{}^P_i) + G^{mj}_0 \psi_{N,j}\psi_{M,m}\delta^{ik} \right] dv_0 \tag{16.184}$$

and

$$J_{Ni}(\delta h) = \int_{v_{0(e)}} \delta h G^{kj}_0 \psi_{N,j}(\delta_{ik} + \psi_{M,k}\overset{\circ}{u}{}^M_i)\, dv_0 \tag{16.185}$$

Here $\bar{h}_0$ is the hydrostatic pressure in the unperturbed element (for example, $\bar{h}_0 = \psi_R \bar{h}^R_0$) and δh is the increment in hydrostatic pressure (for example, $\delta h = \psi_R \delta h^R$). The stress t^{ij}_0 in G^{ik}_{NM} is not the total stress; it is the part of the initial stress tensor determined by the deformation at time t.

In addition to (16.183), we must, of course, have incremental forms of the incompressibility conditions:

$$\left[\int_{v_{0(e)}} (\det \overset{0}{z}_{i,j})^2 \epsilon^{ijk} \epsilon^{rst} \overset{0}{z}_{j,s} \overset{0}{z}_{k,t} \psi_{M,r}\, dv_0 \right] \zeta^M_i = 0 \tag{16.186a}$$

† Special cases concerning dynamic stability of elements under a periodic initial stress were considered by Hutt [1968] and Brown, Hutt, and Salama [1968]. A large literature exists on the use of discrete models in the calculation of flutter speeds of aircraft; see, for example, Bisplinghoff, Ashley, and Halfman [1955] or the more recent work of Kariappa, Somashekar, and Shah [1970]. Barsoum [1970] presented a discussion of certain aspects of dynamic stability connected with finite-element approximations of thin-walled beams.

where

$$z^0_{i,j} = \delta_{ij} + \psi_{N,j}\overset{\circ}{u}{}^N_i \tag{16.186b}$$

Alternatively, if nodal incompressibility conditions are used, we employ an incremental form of (16.62):

$$\sum_{e=1}^{E\,(e)} \Omega^N_\Delta \delta(\text{iii})^{(e)}_N = 0 \tag{16.187}$$

where

$$\delta(\text{iii})^{(e)}_N = \int_{v_{0(e)}} \psi_N \frac{\partial G(\overset{\circ}{\mathbf{u}}{}^R)}{\partial u^M_i}\, dv_0\, \zeta^M_i \tag{16.188}$$

16.6 CURVILINEAR COORDINATES

Up to this point, all the equations presented in this article have been based on the assumption that the material reference frames $x^i_{(e)}$ are rectangular cartesian in the reference configurations $C_{0(e)}$ of each finite element. It is not difficult, however, to recast the stiffness relations in forms valid for any arbitrary system of material coordinates. Recalling from (13.79) and (13.80) that, in general,

$$\mathbf{u}_{(e)} = w^i\mathbf{g}_i = w_i\mathbf{g}^i \tag{16.189}$$

$$w^i = w^{Ni}\psi_N(\boldsymbol{\xi}) \qquad w_i = w^N_i\psi_N(\boldsymbol{\xi}) \tag{16.190}$$

where $\mathbf{g}_i$ and $\mathbf{g}^i$ are covariant and contravariant basis vectors relative to a system of local material curvilinear coordinates ζ^i, and that

$$t^{ij} = \frac{\partial W_{(e)}}{\partial \gamma_{ij}} \qquad \text{where } 2\gamma^{(e)}_{ij} = G^{(e)}_{ij} - g^{(e)}_{ij} \tag{16.191}$$

we obtain immediately from (13.84) and (13.85) the nonlinear stiffness relations for a hyperelastic element

$$\int_{v_{0(e)}} \frac{\partial W}{\partial \gamma_{qj}} [\delta^i_j\psi_{N,q} - \psi_N\Gamma^i_{jq} + (\psi_{M,j}\psi_{N,q}\delta^i_m + \psi_M\psi_{N,q}\Gamma^i_{mj} - \psi_{M,j}\psi_N\Gamma^i_{mq} - \psi_M\psi_N\Gamma^r_{mj}\Gamma^i_{rq})w^{Mm}]\, dv_0 = p^i_N \tag{16.192}$$

and

$$\int_{v_{0(e)}} \frac{\partial W}{\partial \gamma_{pq}} [g_{iq}\psi_{N,p} + g_{qj}\psi_N\Gamma^j_{ip} + (\psi_{M,q}\psi_{N,p}\delta^m_i - \psi_M\psi_{N,p}\Gamma^m_{iq} + \psi_{M,q}\psi_N\Gamma^m_{ip} - \psi_M\psi_N\Gamma^m_{qs}\Gamma^s_{ip})w^M_m]\, dv_0 = p_{Ni} \tag{16.193}$$

Here

$$\psi_{N,i} = \frac{\partial \psi_N(\boldsymbol{\xi})}{\partial \xi^i} \qquad \Gamma^i_{jk} = \frac{\partial \xi^i}{\partial x_m}\frac{\partial^2 x_m}{\partial \xi^j\,\partial \xi^k} \tag{16.194}$$

For dynamic problems, we must, of course, add the inertia terms $m_{NM}\ddot{w}^{Mi}$ and $m_{NM}\ddot{w}_i^M$ to (16.192) and (16.193). Since the process of transforming all the special forms of these equations to various curvilinear coordinate systems is a routine exercise, we shall not elaborate further on this subject.

17 NUMERICAL SOLUTIONS OF NONLINEAR EQUATIONS

It is clear from the theory presented thus far that the static behavior of finite elements subjected to finite elastic deformations is described by large systems of nonlinear algebraic and transcendental equations. Although systems of nonlinear equations have been encountered in problems of applied mechanics for centuries, no general method for obtaining exact solutions exists. Thus, we take for granted that numerical methods must be used and that, in general, it will be possible to obtain solutions only to within a preassigned degree of accuracy. In the present section, we discuss the basic ideas underlying several important numerical schemes for handling such large systems of nonlinear equations.†

17.1 INTRODUCTORY REMARKS

The problem can be stated concisely by considering a system of n simultaneous independent nonlinear equations of the form

$$\begin{aligned} f_1(X_1, X_2, \ldots, X_n) &= 0 \\ f_2(X_1, X_2, \ldots, X_n) &= 0 \\ \cdots\cdots&\cdots\cdots \\ f_n(X_1, X_2, \ldots, X_n) &= 0 \end{aligned} \tag{17.1}$$

It is required to find the values of $\mathbf{X}$ which satisfy (17.1), where $\mathbf{X}$ is the n vector of independent unknowns:

$$\mathbf{X} = \begin{Bmatrix} X_1 \\ X_2 \\ \cdot \\ \cdot \\ \cdot \\ X_n \end{Bmatrix} \tag{17.2}$$

Typically, in finite-element applications, the components of $\mathbf{X}$ will be components of nodal displacements and possibly nodal or element hydrostatic

† We give here only a relatively brief survey. For more detailed discussions consult the comprehensive text of Ortega and Rheinboldt [1970]. General properties and methods of solution of nonlinear equations are surveyed in the books of Saaty and Bram [1964] and Saaty [1967] and a number of survey and review articles are available; e.g., Brooks [1958*a*, 1958*b*], Spang [1962], Fletcher [1965], Box [1966], Zoutendijk [1966], Issacson and Keller [1966], Kowalik and Osborne [1968], and Powell [1969].

pressures, and (17.1) will then represent the global stiffness relations and incompressibility conditions.

We shall often use the notation

$$\mathbf{f}(\mathbf{X}) = \{f_i(\mathbf{X})\} \tag{17.3}$$

where $\mathbf{f}$ is the $n \times 1$ vector of nonlinear equations, so that (17.1) can be written as an n-dimensional vector equation

$$\mathbf{f}(\mathbf{X}) = \mathbf{0} \tag{17.4}$$

An alternate statement of the problem can be obtained by considering the norm

$$F(\mathbf{X}) = \|\mathbf{f}(\mathbf{X})\|^2 = \mathbf{f}^T(\mathbf{X})\mathbf{f}(\mathbf{X}) = f_1^2(\mathbf{X}) + \cdots + f_n^2(\mathbf{X}) \tag{17.5a}$$

or

$$F(\mathbf{X}) = \sum_{i=1}^{n} f_i^2(\mathbf{X}) \tag{17.5b}$$

If $\mathbf{X}$ and $f_i(\mathbf{X})$ are real, then for arbitrary $\mathbf{X}$,

$$F(\mathbf{X}) \geq 0 \tag{17.6}$$

Indeed, (17.6) becomes an equality only if each $f_i(\mathbf{X}) = 0$, and this can happen only if $\mathbf{X}$ satisfies (17.4). We see that the problem of solving the nonlinear equations (17.4) is equivalent to the problem of finding values of $\mathbf{X}$ for which the function $F(\mathbf{X})$ of (17.5) is a minimum. In applications to nonlinear elasticity problems, of course, there is no need to construct artificially an auxiliary function $F(\mathbf{X})$ to be minimized in order to obtain solutions to the stiffness (equilibrium) equations, for we already have such a function in the total potential energy Π of (16.136). Thus, instead of attempting to find solutions to sets of nonlinear equilibrium equations of the type in (17.4), we can always view the problem as one of finding an $\mathbf{X}$ such that $\Pi(\mathbf{X})$ [or $F(\mathbf{X})$] is a minimum.†

Geometrical properties It is convenient to regard $\mathbf{X}$ as a point or vector in an n-dimensional space $\mathscr{E}^n$. Then, for every pair of elements $\mathbf{X}, \mathbf{Y} \in \mathscr{E}^n$, $\mathbf{X} + \mathbf{Y} = \mathbf{Z} = (X_1 + Y_1, \ldots, X_n + Y_n) \in \mathscr{E}^n$ and $\alpha\mathbf{X} = (\alpha X_1, \ldots, \alpha X_n) \in \mathscr{E}^n$, where α is a real number. We can also introduce an inner product $\mathbf{X} \cdot \mathbf{X} \equiv \mathbf{X}^T\mathbf{X} = X_i X_i$, $i = 1, 2, \ldots, n$, the euclidean norm $\|\mathbf{X}\| = \sqrt{\mathbf{X} \cdot \mathbf{X}}$, the idea of orthogonality ($\mathbf{X} \cdot \mathbf{Y} = 0 \Rightarrow \mathbf{X} \perp \mathbf{Y}$), and angles in $\mathscr{E}^n$($\cos\theta = \mathbf{X} \cdot \mathbf{Y}/\|\mathbf{X}\|\,\|\mathbf{Y}\|$, θ being the "angle between" $\mathbf{X}$ and $\mathbf{Y}$). We remark that in some cases it is more convenient to use instead of the euclidean

† This approach toward solving problems in nonlinear elasticity, using finite-element models, was exploited by Bogner, Mallett, Minich, and Schmit [1965], who referred to it as an "energy search" approach. See also Mallett and Schmit [1967], Schmit, Bogner, and Fox [1968], Bogner [1968], Fox and Stanton [1968], and Schmit and Monforton [1969]. A similar approach, but a different minimization method, was used by Becker [1966].

norm $\sqrt{X_i X_i}$ the maximum norm $\|\mathbf{X}\|_{\max} = \max_i |X_i|, i = 1, 2, \ldots, n$; however, several concepts to be discussed hold (or are easily extended to) Banach spaces in general. $\mathbf{X}$ is a unit vector if $\|\mathbf{X}\| = 1$, and a system of n mutually orthogonal vectors $\mathbf{e}_i$ provides an orthonormal basis to $\mathscr{E}^n$ so that every $\mathbf{X} \in \mathscr{E}^n$ can be expressed in a unique manner as a linear combination of the vectors $\mathbf{e}_i$, that is, $\mathbf{X} = X_i \mathbf{e}_i$. We can also cast most of the geometric notions of 3-space into terms appropriate for $\mathscr{E}^n$. For example, if $\mathbf{X}_0$ is a prescribed point and λ is an arbitrary real number, the set of all vectors $\mathbf{X}$ satisfying

$$\mathbf{X} = \mathbf{X}_0 + \lambda \mathbf{N} \tag{17.7}$$

defines a *straight line* through $\mathbf{X}_0$ in the direction of the vector $\mathbf{N}$. Likewise, the set satisfying

$$(\mathbf{X} - \mathbf{X}_0) \cdot \mathbf{N} = 0 \tag{17.8}$$

defines an $(n - 1)$-dimensional hyperplane with normal $\mathbf{N}$.

The space $\mathscr{E}^n$, or a subspace of $\mathscr{E}^n$, is the domain $\mathscr{D}$ of the functions $f_i(\mathbf{X})$ defined in (17.1). If, for any two points $\mathbf{X}_1, \mathbf{X}_2 \in \mathscr{D}$, the line segment $\mathbf{X} = \alpha \mathbf{X}_1 + (1 - \alpha)\mathbf{X}_2 (0 \le \alpha \le 1)$ connecting them also belongs to $\mathscr{D}$, we refer to $\mathscr{D}$ as a *convex domain*. Ordinarily, a convex domain is defined so as to also be open; i.e., for every $\mathbf{X}_0 \in \mathscr{D}$ there exists a ball $\|\mathbf{X} - \mathbf{X}_0\| < r$, the elements of which also belong to $\mathscr{D}$. The set $\mathscr{N}(\mathbf{X}_0, r) = \{\mathbf{X} \mid \|\mathbf{X} - \mathbf{X}_0\| < r\}$ is referred to as the *open ball* with center $\mathbf{X}_0$ and radius r, while $\bar{\mathscr{N}}(\mathbf{X}_0, r) = \{\mathbf{X} \mid \|\mathbf{X} - \mathbf{X}_0\| \le r\}$ is the closed ball. The set $\mathcal{S}(\mathbf{X}_0, r) = \{\mathbf{X} \mid \|\mathbf{X} - \mathbf{X}_0\| = r\}$ is the *sphere* with center $\mathbf{X}_0$ and radius r. A function of the form $f(\mathbf{X}) = C$, C being a constant, describes a surface in $\mathscr{E}^n$. Thus, (17.1) describes a collection of n surfaces in $\mathscr{E}^n$, their intersection being the point $\mathbf{X}$ which is the desired solution of the system of equations.

Alternatively, we can regard the function $F(\mathbf{X})$ [or $\Pi(\mathbf{X})$] of (17.5) as a scalar-valued point function with domain $\mathscr{D}$. By setting $F(\mathbf{X}) = C =$ constant, we define an $(n - 1)$-dimensional level surface, which is called a *contour* of the function $F(\mathbf{X})$. Solutions of the nonlinear equations (17.1) correspond to points in the domain of $F(\mathbf{X})$ at which $F(\mathbf{X})$ assumes a relative minimum value. Now the rate at which $F(\mathbf{X})$ changes at a point $\mathbf{X}$ in moving from $\mathbf{X}$ in the direction of a unit vector $\mathbf{N}$ is, as usual, given by the directional derivative $\mathbf{g} \cdot \mathbf{N}$, where $\mathbf{g}$ is the gradient of $F(\mathbf{X})$:

$$\mathbf{g} = \nabla F = \left\{ \frac{\partial F}{\partial X_1}, \frac{\partial F}{\partial X_2}, \ldots, \frac{\partial F}{\partial X_n} \right\} \tag{17.9}$$

The equation for a curve in $\mathscr{E}^n$ is given by parametric equations of the form $\mathbf{X} = \mathbf{X}(\lambda)$, and the vectors $d\mathbf{X}(\lambda)/d\lambda$ are tangent to the curve. Thus, on a level surface $F(\mathbf{X}) = C$, the rate of change of $F(\mathbf{X})$ along a curve on the surface is $dF/d\lambda = 0 = \mathbf{g} \cdot d\mathbf{X}/d\lambda$, from which it follows that $\mathbf{g}$ is normal to the level surface. In addition, if $d\mathbf{X}$ is arbitrary and differs in the direction of $\mathbf{g}$ by an angle θ, then $dF = \|\mathbf{g}\| \, \|d\mathbf{X}\| \cos \theta$. Hence, the maximum rate of change of $F(\mathbf{X})$ occurs in the direction of the gradient ($\theta = 0$).

Methods of solution Most numerical schemes used to solve nonlinear equations begin with an initial estimate $\mathbf{X}_0$ of the solution, called a *test point*. In general, some process is then used to compute a sequence of such points which, hopefully, converges to the solution of the system of equations.

The numerical methods that we wish to consider can be divided into two categories: *sequential methods* and *nonsequential methods*. In sequential methods, sequences of approximate solutions are generated by a fixed set of operations. In nonsequential methods, points are picked at random. Typically, in sequential methods, the first test point is picked arbitrarily and a corrected value is obtained by recurrence formulas of the form

$$\mathbf{X}^{r+1} = \mathbf{X}^r + \lambda^r \mathbf{D}^r \tag{17.10}$$

Here $\mathbf{X}^r$ is the estimate of the solution at the rth step of the process, called the rth iterate, and $\lambda^r \mathbf{D}^r$ is the correction, $\mathbf{D}^r$ being an n-dimensional vector determining the direction of the correction and λ^r a scalar governing the magnitude of the correction. Comparing (17.10) with (17.7), it is clear that recurrence formulas of this type progress toward the solution along straight-line segments connecting successive test points. Higher-order recurrence formulas based on quadratic, cubic, or higher-order extrapolations can be obtained, of course, by adding to (17.10) terms containing the test points $\mathbf{X}^{r-1}, \mathbf{X}^{r-2}, \ldots$.

Certain iterative schemes used to determine minima of the functions $F(\mathbf{X}) = \mathbf{f}^T(\mathbf{X})\mathbf{f}(\mathbf{X})$ of (17.5) are referred to as *descent methods*; if $F(\mathbf{X}^{r+1}) \leq F(\mathbf{X}^r)$, then one "descends" from a test point $\mathbf{X}_0$ to a minimizing point $\mathbf{X}^*$ for which $F(\mathbf{X}^*) = 0$. A number of the more important descent methods are further classified as *gradient methods* because they involve some measure of the gradient of $F(\mathbf{X})$ as an indication of the direction toward the minimum. *Sequential search methods*, on the other hand, do not involve the computation of gradients; rather, these methods are structured so as to search for minimizers of $F(\mathbf{X})$ among a number of test points and to then discard points and generate new points in such a way that each new search need involve only test points that are collectively closer to the minimum than before. To these methods we may also add several general iterative methods that are based on a direct attack of the nonlinear system (17.1), or upon successive linearizations of (17.1). These include the powerful Newton-Raphson method and continuation methods such as incremental loading. Finally, we note that many iterative schemes can be regarded as some variant of the ancient method of successive approximations, which is intrinsically related to the fundamental concepts of fixed points of operators and the principle of contraction mappings. Because of their historical importance and their relation to other iterative schemes, we give a brief account of these concepts in the following section. We then devote the remainder of this article to several of the more important sequential and nonsequential methods for solving nonlinear equations.

17.2 FIXED POINTS AND CONTRACTION MAPPINGS

Consider a nonlinear operator $\mathscr{P}$ that maps $\mathscr{E}^n$ (or, more generally, a Banach space $\mathscr{V}$) into itself. Then any $\mathbf{X} \in \mathscr{E}^n$ (or $\mathscr{V}$) such that

$$\mathscr{P}(\mathbf{X}) = \mathbf{X} \tag{17.11}$$

is called a *fixed point* of the operator $\mathscr{P}$. The problem of finding solutions of (17.11) is referred to as the fixed point problem for the operator $\mathscr{P}$.

Any given equation can usually be reformulated as a fixed point problem. For example, if

$$Q(\mathbf{X}) = \mathbf{Y} \tag{17.12}$$

we may set

$$\mathscr{P}(\mathbf{X}) = \mathbf{X} + Q(\mathbf{X}) - \mathbf{Y} \tag{17.13}$$

Then the solution of (17.12) is a fixed point of $\mathscr{P}(\mathbf{X})$ of (17.13). Likewise, (17.4) can be viewed as a fixed point problem if we set $\mathscr{P}(\mathbf{X}) = \mathbf{X} - \mathbf{f}(\mathbf{X})$.

Once a fixed point problem is formulated, it may be possible to determine fixed points of $\mathscr{P}(\mathbf{X})$ by the *method of successive approximations:* Let $\mathbf{X}_0$ be a test point; i.e., a trial solution to (17.11). If $\mathscr{P}(\mathbf{X}_0)$ does not differ appreciably from $\mathbf{X}_0$, it is natural to consider $\mathbf{X}^1 = \mathscr{P}(\mathbf{X}_0)$ as an improvement to $\mathbf{X}_0$. So also, then, is $\mathbf{X}^2 = \mathscr{P}(\mathbf{X}^1)$ a possible improvement over $\mathbf{X}^1$. Continuing in this manner, we generate a sequence $\{\mathbf{X}^r\}$ of successive approximations to a fixed point $\mathbf{X}$ of $\mathscr{P}(\mathbf{X})$ by the formula

$$\mathbf{X}^{r+1} = \mathscr{P}(\mathbf{X}^r) \qquad r = 0, 1, 2, \ldots \tag{17.14}$$

Note also that if we define the compositions

$$\mathscr{P}^0 = I \qquad \mathscr{P}^{r+1} = \mathscr{P}(\mathscr{P}^r) \qquad r = 0, 1, 2, \ldots$$

then

$$\begin{aligned} &\mathbf{X}_0 = \mathscr{P}^0(\mathbf{X}_0) \qquad \mathbf{X}^1 = \mathscr{P}(\mathscr{P}^0(\mathbf{X}_0)) \\ &\mathbf{X}^2 = \mathscr{P}(\mathbf{X}^1) = \mathscr{P}(\mathscr{P}(\mathbf{X}_0)) = \mathscr{P}^2(\mathbf{X}_0) \qquad \cdots \end{aligned}$$

i.e., (17.14) can also be written

$$\mathbf{X}^{r+1} = \mathscr{P}^{r+1}(\mathbf{X}_0) \tag{17.15}$$

Definition† Let $\mathscr{P}$ be an operator from a Banach space $\mathscr{V}$ into itself. Then $\mathscr{P}$ is called a *contraction mapping* of the closed ball

$$\bar{\mathscr{N}}(\mathbf{X}_0,r) = \{\mathbf{X} \mid \|\mathbf{X} - \mathbf{X}_0\| \leq r\}$$

if a real number λ, $0 \leq \lambda < 1$, exists such that the Lipschitz condition

$$\|\mathscr{P}(\mathbf{X}) - \mathscr{P}(\mathbf{Y})\| \leq \lambda \, \|\mathbf{X} - \mathbf{Y}\| \tag{17.16}$$

is satisfied for all $\mathbf{X}$, $\mathbf{Y} \in \bar{\mathscr{N}}(\mathbf{X}_0,r)$.

The Lipschitz constant λ is called the *contraction factor*.

Theorem 17.1 (*The Contraction Mapping Principle*) *Let $\mathscr{P}(\mathbf{X})$ be a contraction mapping on $\bar{\mathscr{N}}(\mathbf{X}_0,r)$ corresponding to a contraction factor λ and let $\mathbf{X}_0$ be such that*

$$\frac{1}{1-\lambda} \|\mathscr{P}(\mathbf{X}_0) - \mathbf{X}_0\| = r_0 \leq r \tag{17.17}$$

Then

a. *The sequence $\{\mathbf{X}^r\}$ defined by $\mathbf{X}^{r+1} = \mathscr{P}(\mathbf{X}^r)$ converges to a point $\mathbf{X}^* \in \bar{\mathscr{N}}(\mathbf{X}_0,r_0)$.*
b. *$\mathbf{X}^*$ is a fixed point of the operator $\mathscr{P}$.*
c. *$\mathbf{X}^*$ is the unique fixed point of $\mathscr{P}$ in $\bar{\mathscr{N}}(\mathbf{X}_0,r)$.*

Proof To prove **a**, we will first show that if (17.17) is satisfied, all iterates $\mathbf{X}^{r+1} = \mathscr{P}(\mathbf{X}^r)$ will be in $\bar{\mathscr{N}}(\mathbf{X}_0,r_0)$. We shall use mathematical induction (i.e., prove that the postulate holds for $r = 1$ and show that its validity for $n = r$ implies that it is true for $n = r + 1$). Since $\mathbf{X}^1 = \mathscr{P}(\mathbf{X}_0)$,

$$\|\mathbf{X}^1 - \mathbf{X}_0\| = (1 - \lambda)r_0 \leq r_0$$

by (17.17). Thus $\mathbf{X}^1 \in \bar{\mathscr{N}}(\mathbf{X}_0,r_0)$. Now assume that $\|\mathbf{X}_0 - \mathbf{X}^r\| \leq r_0$

† While we are primarily interested here in nonlinear operators on $\mathscr{E}^n$, this definition and all the theorems of this section hold in general Banach spaces. Note also that the concept of contraction mappings is easily adapted to metric spaces: let $\mathscr{X}$ and $\mathscr{Y}$ be metric spaces with metrics $d(X_1,X_2)$ and $\rho(Y_1,Y_2)$ respectively and suppose that $\mathscr{P}$ is a mapping from $\mathscr{X}$ into $\mathscr{Y}$. Then $\mathscr{P}$ is a contraction mapping of $\bar{\mathscr{N}}(X_0,r) \subset \mathscr{X} = \{X \mid d(X_0,X) \leq r\}$ if a contraction factor $0 \leq \lambda < 1$ exists such that $\rho(\mathscr{P}(X_1),\mathscr{P}(X_2)) \leq \lambda \, d(X_1,X_2)$ for every X_1, $X_2 \in \bar{\mathscr{N}}(X_0,r)$.

for iterates $\mathbf{X}^1, \mathbf{X}^2, \ldots, \mathbf{X}^r$; then

$$\begin{aligned}\|\mathbf{X}^{r+1} - \mathbf{X}^r\| &= \|\mathscr{P}(\mathbf{X}^r) - \mathscr{P}(\mathbf{X}^{r-1})\| \\ &\leq \lambda\, \|\mathbf{X}^r - \mathbf{X}^{r-1}\| = \lambda\, \|\mathscr{P}(\mathbf{X}^{r-1}) - \mathscr{P}(\mathbf{X}^{r-2})\| \\ &\leq \lambda^2\, \|\mathbf{X}^{r-1} - \mathbf{X}^{r-2}\| \\ &\quad\vdots \\ &\leq \lambda^r\, \|\mathbf{X}^1 - \mathbf{X}_0\| = \lambda^r(1 - \lambda)r_0\end{aligned}$$

It follows that

$$\begin{aligned}\|\mathbf{X}^{r+1} - \mathbf{X}_0\| &= \|(\mathbf{X}^{r+1} - \mathbf{X}^r) + (\mathbf{X}^r - \mathbf{X}^{r-1}) + \cdots + (\mathbf{X}^1 - \mathbf{X}_0)\| \\ &\leq \|\mathbf{X}^{r+1} - \mathbf{X}^r\| + \|\mathbf{X}^r - \mathbf{X}^{r-1}\| + \cdots + \|\mathbf{X}^1 - \mathbf{X}_0\| \\ &\leq \lambda^r(1 - \lambda)r_0 + \lambda^{r-1}(1 - \lambda)r_0 + \cdots + (1 - \lambda)r_0\end{aligned}$$

or

$$\|\mathbf{X}^{r+1} - \mathbf{X}_0\| \leq (1 - \lambda^{r+1})r_0 \leq r_0$$

Thus $\mathbf{X}^{r+1} \in \bar{\mathscr{N}}(\mathbf{X}_0, r_0)$ and we have proved that all iterates belong to $\bar{\mathscr{N}}(\mathbf{X}_0, r_0)$.

It will now be shown to follow that the iterates $\{\mathbf{X}^r\}$ form a Cauchy sequence. Indeed, for any index m,

$$\begin{aligned}\|\mathbf{X}^m - \mathbf{X}^{m+n}\| &= \|(\mathbf{X}^m - \mathbf{X}^{m+1}) + (\mathbf{X}^{m+1} - \mathbf{X}^{m+2}) \\ &\qquad + \cdots + (\mathbf{X}^{m+n-1} - \mathbf{X}^{m+n})\| \\ &\leq \|\mathbf{X}^m - \mathbf{X}^{m+1}\| + \|\mathbf{X}^{m+1} - \mathbf{X}^{m+2}\| \\ &\qquad + \cdots + \|\mathbf{X}^{m+n-1} - \mathbf{X}^{m+n}\| \\ &\leq \lambda^m(1 - \lambda)r_0 + \lambda^{m+1}(1 - \lambda)r_0 \\ &\qquad + \cdots + \lambda^{m+n-1}(1 - \lambda)r_0 \\ &\leq \lambda^m(1 - \lambda^n)r_0 \end{aligned} \tag{17.18}$$

Since λ is fixed and $0 \leq \lambda < 1$, we can, for any $\epsilon > 0$, find an integer $N(\epsilon)$ such that $\|\mathbf{X}^m - \mathbf{X}^{m+n}\| < \epsilon$ for $n > 0$ and $m > N$ (e.g., for a given ϵ examine N such that $\lambda^N < \epsilon/r_0$; then $\|\mathbf{X}^m - \mathbf{X}^{m+n}\| < \epsilon(1 - \lambda^n) < \epsilon$ for $m > N$). Therefore $\{\mathbf{X}^m\}$ is a Cauchy sequence.

If $\mathbf{X}^*$ is the limit of this sequence, then $\mathbf{X}^* \in \bar{\mathscr{N}}(\mathbf{X}_0, r_0)$. This completes the proof of **a**.

Since

$$\|\mathscr{P}(\mathbf{X}^r) - \mathbf{X}^*\| = \|\mathbf{X}^{r+1} - \mathbf{X}^*\|$$

and, as has been proven, $\lim_{r \to \infty} \|\mathbf{X}^{r+1} - \mathbf{X}^*\| = 0$, it is clear that

$\lim_{r\to\infty} \|\mathscr{P}(\mathbf{X}^r) - \mathscr{P}(\mathbf{X}^*)\| = 0$. Moreover, $\|\mathscr{P}(\mathbf{X}^r) - \mathscr{P}(\mathbf{X}^*)\| \leq \|\mathbf{X}^r - \mathbf{X}^*\|$; hence $\lim_{r\to\infty} \mathscr{P}(\mathbf{X}^r) = \mathscr{P}(\mathbf{X}^*) = \mathbf{X}^*$; that is, $\mathbf{X}^*$ is a fixed point of $\mathscr{P}$. This completes **b**. We remark that **b** also follows from **a** due to continuity of $\mathscr{P}$ established by (17.16). In other words, (17.16) shows that $\mathscr{P}$ is Lipschitz continuous; thus, if $\lim_{r\to\infty} \mathbf{X}^r = \mathbf{X}^*$, then $\lim_{r\to\infty} \mathscr{P}(\mathbf{X}^r) = \mathscr{P}(\mathbf{X}^*) = \mathbf{X}^*$.

It remains to be shown that $\mathbf{X}^*$ is unique in $\bar{\mathscr{N}}(\mathbf{X}_0,r_0)$. Suppose $\bar{\mathbf{X}}$ is another fixed point in $\bar{\mathscr{N}}(\mathbf{X}_0,r_0)$. Then

$$\begin{aligned}\|\mathbf{X}^* - \bar{\mathbf{X}}\| &= \|\mathscr{P}(\mathbf{X}^*) - \mathscr{P}(\bar{\mathbf{X}})\| \\ &\leq \lambda \|\mathbf{X}^* - \bar{\mathbf{X}}\| \\ &< \|\mathbf{X}^* - \bar{\mathbf{X}}\|\end{aligned}$$

This, of course, is impossible, and the proof is completed.

Note that if the conditions of Theorem 17.1 are satisfied, (17.15) leads to the conclusion that

$$\mathbf{X}^* = \lim_{r\to\infty} \mathscr{P}^r(\mathbf{X}_0) \tag{17.19}$$

If (17.19) holds, $\mathbf{X}^*$ is said to be *accessible* from $\mathbf{X}_0$, and the set of all points from which $\mathbf{X}^*$ is accessible is referred to as the *region of accessibility*. It is not difficult to show that if the conditions of Theorem 17.1 are satisfied, $\mathbf{X}^*$ is accessible from *any* point in $\bar{\mathscr{N}}(\mathbf{X}_0,r_0)$. In fact, since

$$\begin{aligned}\|\mathscr{P}(\mathbf{X}) - \mathscr{P}(\mathbf{X}_0)\| &= \|\mathscr{P}(\mathbf{X}) - \mathbf{X}^1\| \leq \lambda \|\mathbf{X} - \mathbf{X}_0\| \\ &\leq \lambda r_0 = \frac{\lambda}{1-\lambda} \|\mathbf{X}^1 - \mathbf{X}_0\|\end{aligned}$$

we see that $\|\mathscr{P}(\mathbf{X}) - \mathbf{X}^1\| \leq \lambda r_0$; that is, $\mathscr{P}$ maps $\bar{\mathscr{N}}(\mathbf{X}_0,r_0)$ into the smaller ball $\bar{\mathscr{N}}(\mathbf{X}^1,\lambda r_0)$ and $\bar{\mathscr{N}}(\mathbf{X}^1,\lambda r_0) \subset \bar{\mathscr{N}}(\mathbf{X}_0,r_0)$. In fact, by allowing $n \to \infty$ in (17.18), we have

$$\|\mathbf{X}^r - \mathbf{X}^*\| \leq \lambda^r r_0 \tag{17.20}$$

and $\mathscr{P}^r$ maps $\bar{\mathscr{N}}(\mathbf{X}_0,r_0)$ into $\bar{\mathscr{N}}(\mathbf{X}^r,\lambda^r r_0)$. Note also that any sequence $\{\hat{\mathbf{X}}^r\} \in \bar{\mathscr{N}}(\mathbf{X}^r,\lambda^r r_0)$ will also converge to $\mathbf{X}^*$. This makes it possible to use approximations of $\mathscr{P}$ in generating successive approximations to $\mathbf{X}^*$.

Iterative solution of fixed point problems is illustrated in Fig. 17.1 for the one-dimensional case. Figure 17.1*a* and *b* indicate convergent schemes, the process in *a* being a monotonically increasing sequence of iterates while that in *b* is oscillatory. The operator $\mathscr{P}$ in Fig. 17.1*c* has two fixed points, X_1^* and X_2^*. However, X_2^* is inaccessible by the method of successive

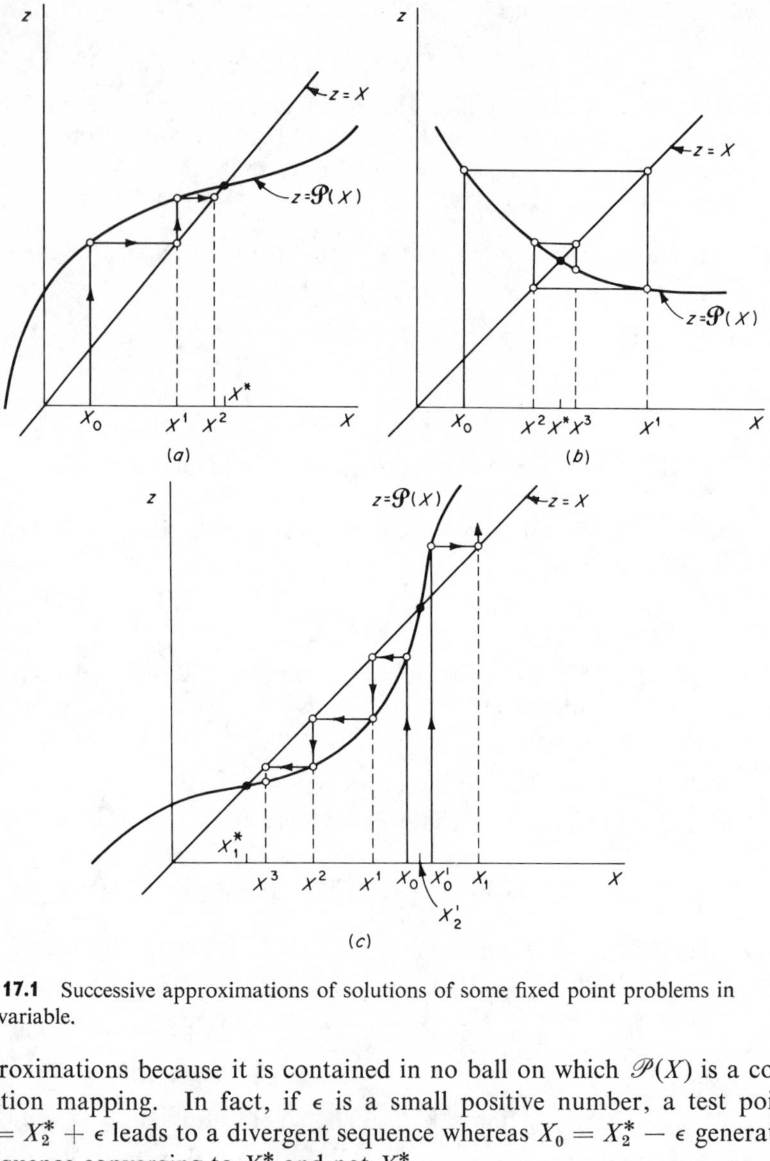

Fig. 17.1 Successive approximations of solutions of some fixed point problems in one variable.

approximations because it is contained in no ball on which $\mathscr{P}(X)$ is a contraction mapping. In fact, if ϵ is a small positive number, a test point $X_0 = X_2^* + \epsilon$ leads to a divergent sequence whereas $X_0 = X_2^* - \epsilon$ generates a sequence converging to X_1^* and not X_2^*.

An examination of the one-dimensional fixed-point problems illustrated in Fig. 17.1 suggests that the convergence of the method of successive approximations (and all the conclusions of Theorem 17.1) are valid if the slope of the curve $z = \mathscr{P}(X)$ is less than unity (i.e., $|\mathscr{P}'(X)| \leq \lambda < 1$). This is indeed true for the one-dimensional case as can be proven without much difficulty. The question arises as to whether or not similar statements can be made in the n-dimensional case. The answer to this question is also affirmative, for if $\mathscr{P} \equiv \mathbf{P}$ denotes an operator from $\mathscr{E}^n$ into itself such that $\mathbf{P}(\mathbf{X})$ has continuous derivative in $\bar{\mathscr{N}}(\mathbf{X}_0,r)$, it follows from Taylor's theorem that†

$$|p_i(\mathbf{X}) - p_i(\mathbf{Y})| \leq \sum_{j=1}^{n} \left| \frac{\partial p_i(\boldsymbol{\xi}^i)}{\partial X_j} \right| |X_j - Y_j|$$

for any two points $\mathbf{X}$ and $\mathbf{Y} \in \bar{\mathscr{N}}(\mathbf{X}_0,r)$, wherein $\boldsymbol{\xi}^i$ is a point in $\bar{\mathscr{N}}(\mathbf{X}_0,r)$ on the segment joining $\mathbf{X}$ and $\mathbf{Y}$. Thus, if we use the maximum norm $\|\mathbf{X}\| = \max_{1 \leq i \leq n} |X_i|$,

$$|\mathbf{P}(\mathbf{X}) - \mathbf{P}(\mathbf{Y})| \leq \|\mathbf{X} - \mathbf{Y}\| \sum_{j=1}^{n} \left| \frac{\partial p_i(\boldsymbol{\xi}^i)}{\partial X_j} \right|$$

Hence, if $|\partial p_i(\boldsymbol{\xi}^i)/\partial X_j| \leq \lambda/n$, $0 \leq \lambda < 1$, we have

$$\|\mathbf{P}(\mathbf{X}) - \mathbf{P}(\mathbf{Y})\| \leq \lambda \|\mathbf{X} - \mathbf{Y}\|$$

that is, $\mathbf{P}$ is a contraction mapping on $\bar{\mathscr{N}}(\mathbf{X}_0,r)$ and all the conclusions of Theorem 17.1 follow.

Returning to (17.4), we observe that it is possible to construct a number of operators $\mathbf{P}$ (or $\mathscr{P}$) such that solutions of (17.4) are fixed points of $\mathbf{P}$. For example, let $\mathbf{A}(\mathbf{X})$ be any $n \times n$ nonsingular matrix. Then if we set

$$\mathbf{P}(\mathbf{X}) = \mathbf{X} - \mathbf{A}(\mathbf{X})\mathbf{f}(\mathbf{X})$$

it is clear that $\mathbf{P}(\mathbf{X}) = \mathbf{X} \Rightarrow \mathbf{A}(\mathbf{X})\mathbf{f}(\mathbf{X}) = \mathbf{0}$. But $\mathbf{A}(\mathbf{X})\mathbf{f}(\mathbf{X}) = \mathbf{0} \Rightarrow \mathbf{f}(\mathbf{X}) = \mathbf{0}$ because $\mathbf{A}(\mathbf{X})$ is assumed to be nonsingular for all $\mathbf{X}$. The recurrence formula is then

$$\mathbf{X}^{r+1} = \mathbf{P}(\mathbf{X}^r) = \mathbf{X}^r - \mathbf{A}(\mathbf{X}^r)\mathbf{f}(\mathbf{X}^r) \tag{17.21}$$

Many different iterative schemes result from (17.21) by different choices of $\mathbf{A}(\mathbf{X})$. For example, if $\mathbf{A}(\mathbf{X}) = \mathbf{A} =$ constant matrix, (17.21) defines the so-called *chord method*. The Newton-Raphson method (see Sec. 17.4) results if $\mathbf{A}(\mathbf{X})$ is chosen to be the inverse jacobian matrix, $\mathbf{A}(\mathbf{X}) = \mathbf{J}^{-1}(\mathbf{X}) = [\partial f_i(\mathbf{X})/\partial X_j]^{-1}$. If $\mathbf{A} = \mathbf{J}^{-1}(\mathbf{X}_0)$, the modified Newton-Raphson process is obtained. A number of other examples could be cited.

17.3 DESCENT AND GRADIENT MINIMIZATION METHODS

Using (17.10) as a starting point, we shall now investigate several sequential methods designed to minimize $F(\mathbf{X}) = \mathbf{f}^T(\mathbf{X})\mathbf{f}(\mathbf{X})$. These methods differ in their choice of λ^r and $\mathbf{D}^r$; some are properly classified as descent methods $[F(\mathbf{X}^{r+1}) \leq F(\mathbf{X}^r)]$, while others involve gradients (or "conjugate" gradients) of $F(\mathbf{X})$. Since most of these techniques use some approximation of $\mathbf{g}$, they are generally regarded as gradient methods.

† See Isaacson and Keller [1966]. For additional properties of contraction mappings, consult, for example, Rall [1969] or Ortega and Rheinboldt [1970].

Univariant methods One of the simplest choices of the directional vector $\mathbf{D}^r$ of (17.10) is

$$\mathbf{D}^r = \mathbf{e}_j^r \tag{17.22}$$

where $\mathbf{e}_j$ is the n-dimensional unit vector

$$\mathbf{e}_j^r = \{0, 0, \ldots, 0, 1, 0, \ldots, 0\} \tag{17.23}$$

$$\uparrow$$

$$j\text{th row}$$

Thus, only one component of $\mathbf{X}^r$ is changed at a time. The choice of X_j may be arbitrary, or additional conditions can be introduced to govern the selection of component j for each iteration. Methods of this type are called *univariant* or *univariant relaxation methods.*

In univariant methods it is common to determine λ^r as follows: Assume that $F(\mathbf{X})$ is analytic in the neighborhood of $\mathbf{X}^r$. Then

$$F(\mathbf{X}^r + \lambda^r\mathbf{D}^r) = F(\mathbf{X}^r) + \lambda^r \sum_{i=1}^{n} \frac{\partial F(\mathbf{X}^r)}{\partial X_i} D_i^r + \tfrac{1}{2}(\lambda^r)^2 \sum_{i,j=1}^{n} \frac{\partial^2 F(\mathbf{X}^r)}{\partial X_i\, \partial X_j} D_i^r D_j^r + \cdots \tag{17.24}$$

The quantities $\partial^2 F(\mathbf{X})/\partial X_i\, \partial X_j$ are components of an $n \times n$ symmetric matrix $\mathbf{H}$ called the *hessian matrix* of $F(\mathbf{X})$:

$$H_{ij} = \frac{\partial^2 F(\mathbf{X})}{\partial X_i\, \partial X_j} \tag{17.25}$$

and

$$\sum_{i=1}^{n} \frac{\partial F(\mathbf{X}^r)}{\partial X_i} D_i^r = \mathbf{g}^r \cdot \mathbf{D}^r \tag{17.26}$$

where $\mathbf{g}^r = \nabla F(\mathbf{X}^r)$ is the gradient of $F(\mathbf{X})$ at $\mathbf{X}^r$. Thus, to within quadratic terms,

$$F(\mathbf{X}^r + \lambda^r\mathbf{D}^r) = F(\mathbf{X}^r) + \lambda^r\mathbf{g}^r \cdot \mathbf{D}^r + \tfrac{1}{2}(\lambda^r)^2\mathbf{D}^{rT}\mathbf{H}\mathbf{D}^r \tag{17.27}$$

We assume that the level surface in the neighborhood of $\mathbf{X}^r$ is strictly convex in the sense that for all $\mathbf{X}$ and $\mathbf{Y}$ in this neighborhood, $F[\alpha\mathbf{X} + (1 - \alpha)\mathbf{Y}] \leq \alpha F(\mathbf{X}) + (1 - \alpha)F(\mathbf{Y})$ for $0 \leq \alpha \leq 1$. Then if λ^r is to be such that the maximum step is taken toward the minimum for a given $\mathbf{D}^r$,

$$\frac{\partial F(\mathbf{X}^r + \lambda^r\mathbf{D}^r)}{\partial \lambda^r} = 0 = \mathbf{g}^r \cdot \mathbf{D}^r + \lambda^r\mathbf{D}^{rT}\mathbf{H}\mathbf{D}^r$$

or

$$\lambda^r = -\frac{\mathbf{g}^r \cdot \mathbf{D}^r}{\mathbf{D}^{rT}\mathbf{H}\mathbf{D}^r} = -\frac{\displaystyle\sum_{i=1}^{n} \frac{\partial F(\mathbf{X}^r)}{\partial X_i} D_i^r}{\displaystyle\sum_{i}^{n}\sum_{j}^{n} \frac{\partial^2 F(\mathbf{X}^r)}{\partial X_i\, \partial X_j} D_i^r D_i^r} \tag{17.28}$$

Since, for univariant methods, $\mathbf{D}^r$ is given by (17.22), we have

$$\mathbf{g}^r \cdot \mathbf{D}^r = g_j^r \qquad \mathbf{D}^{rT}\mathbf{H}\mathbf{D}^r = H_{(j)(j)}^r = \frac{\partial^2 F(\mathbf{X}^r)}{\partial X_{(j)}\, \partial X_{(j)}} \tag{17.29}$$

where g_j^r is the jth component of $\mathbf{g}$ at $\mathbf{X}^r$ and $H_{(j)(j)}^r$ is the element on the jth row and jth column of $\mathbf{H}$ at $\mathbf{X}^r$. Thus, in univariant methods,

$$\lambda^r = \frac{-g_j^r}{H_{(j)(j)}^r} \tag{17.30}$$

In principle, the direction j in (17.22) and (17.30) is arbitrary. An illustration of the iterative moves in a univariant process is shown in Fig. 17.2. A univariant technique in which $\mathbf{e}_j^r$ is not selected arbitrarily is the well-known Southwell relaxation method.† In this method, the row j in (17.23) which contains unity corresponds to the unknown for which the local decrease is maximum; that is, j is chosen so that $\|\partial F(\mathbf{X}^r)/\partial X_i\| = \max_{1 \leq i \leq n} |\partial F(\mathbf{X}^r)/\partial X_i|$. Numerous modifications and refinements of this method are possible.

The method of steepest descent We recall that the direction of maximum rate of increase of $F(\mathbf{X})$ is that of its gradient $\mathbf{g}(\mathbf{X})$. Consequently, $-\mathbf{g}(\mathbf{X})$ is the direction of maximum descent at $\mathbf{X}$. Setting

$$\mathbf{D}^r = -\mathbf{g}(\mathbf{X}^r) \tag{17.31}$$

we obtain Cauchy's [1847] method of steepest descent.

The value of the step λ^r corresponding to (17.31) can be obtained from (17.28). Thus, for each r, the vector $\mathbf{X}^{r+1} - \mathbf{X}^r$ is in the direction of the most rapid decrease in $F(\mathbf{X})$.

It is not difficult to show existence of solutions and convergence of the iterative process if $F(\mathbf{X})$ is a quadratic with continuous derivatives up to second order in the domain $\mathscr{D}$ and if the hessian is positive definite.

Theorem 17.2 *Let $F(\mathbf{X})$ be a quadratic function of $\mathbf{X}$ of the form*

$$2F(\mathbf{X}) = A + 2\mathbf{b}^T\mathbf{X} + \mathbf{X}^T\mathbf{H}\mathbf{X} \tag{17.32}$$

where $\mathbf{H}$ is positive definite. Further, let $\{\mathbf{X}^r\}$ denote a sequence of points such that

$$\mathbf{X}^{r+1} = \mathbf{X}^r + \lambda^r \mathbf{g}^r$$

where λ^r is given by (17.28) and $\mathbf{g}^r$ is the gradient

$$\mathbf{g}^r = \nabla F = \mathbf{H}\mathbf{X}^r + \mathbf{b}$$

† See, for example, Southwell [1940, 1946] and Allen [1954].

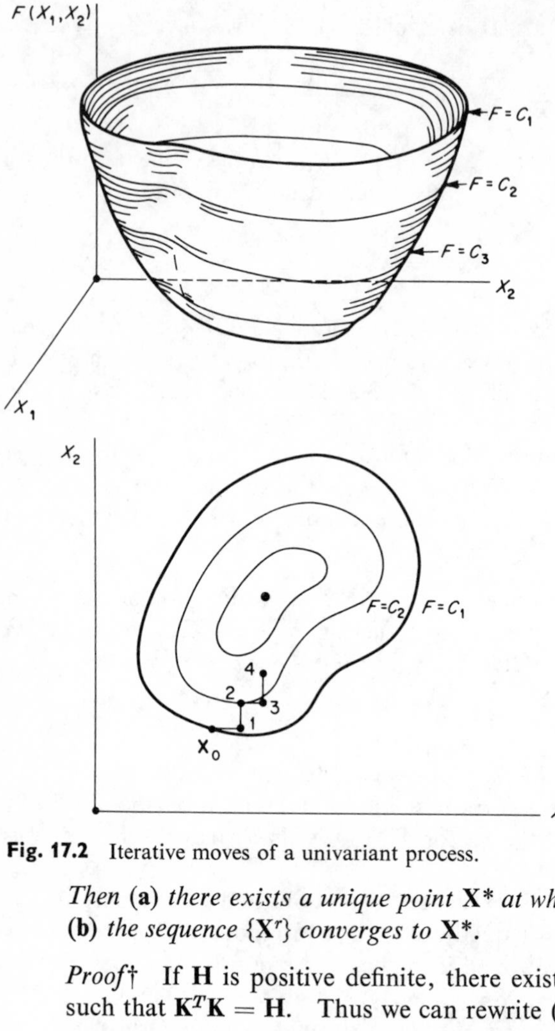

Fig. 17.2 Iterative moves of a univariant process.

Then (**a**) *there exists a unique point* $\mathbf{X}^*$ *at which* $F(\mathbf{X})$ *is a minimum and* (**b**) *the sequence* $\{\mathbf{X}^r\}$ *converges to* $\mathbf{X}^*$.

Proof† If $\mathbf{H}$ is positive definite, there exists a nonsingular matrix $\mathbf{K}$ such that $\mathbf{K}^T\mathbf{K} = \mathbf{H}$. Thus we can rewrite (17.32) in the form

$$2F(\mathbf{X}) = (\mathbf{Y}^T + \mathbf{b}^T\mathbf{K}^{-1})(\mathbf{Y} + \mathbf{K}^{-1T}\mathbf{b}) + (A - \mathbf{b}^T\mathbf{H}^{-1}\mathbf{b})$$

where $\mathbf{Y} = \mathbf{KX}$. Thus $2F(\mathbf{X}) \geq A - \mathbf{b}^T\mathbf{H}^{-1}\mathbf{b}$ for all $\mathbf{X}$, and the equality holds if and only if $\mathbf{Y} = -\mathbf{K}^{-1T}\mathbf{b}$, that is, if $\mathbf{X} + \mathbf{H}^{-1}\mathbf{b} = \mathbf{0}$. Therefore, the minimum $F(-\mathbf{H}^{-1}\mathbf{b})$ exists and is unique. This proves part **a** of the theorem.

To prove part **b**, let $\mathbf{X}^* = -\mathbf{H}^{-1}\mathbf{b}$ denote the point at which $F(\mathbf{X})$ is a minimum. If $F(\mathbf{X})$ is continuous at $\mathbf{X}^*$, then $\lim_{r\to\infty} \|\mathbf{X}^* - \mathbf{X}^r\| = 0$ implies that

$$\lim_{r\to\infty} [F(\mathbf{X}^r) - F(\mathbf{X})] = 0$$

† We follow there the proof given by Saaty and Bram [1964].

or, for monotone convergence, we have for all r

$$F(\mathbf{X}^r) - F(\mathbf{X}^*) \leq \alpha\, |F(\mathbf{X}^{r-1}) - F(\mathbf{X}^*)|$$

where $\alpha < 1$. If $\mathbf{E}^r = \mathbf{X}^r - \mathbf{X}^*$ is the error at the rth cycle,

$$F(\mathbf{X}^r) - F(\mathbf{X}^*) = F(\mathbf{X}^* + \mathbf{E}^r) - F(\mathbf{X}^*) = \tfrac{1}{2}\mathbf{E}^{rT}\mathbf{H}\mathbf{E}^r$$

If $\bar{\omega}$ is an eigenvalue of $\mathbf{H}$, then $\mathbf{HX} = \bar{\omega}\mathbf{X}$ and $\bar{\omega} = \mathbf{X}^T\mathbf{HX}/\mathbf{X}^T\mathbf{X}$. If ω is the ratio of the smallest to the largest eigenvalue of $\mathbf{H}$, then $\omega\mathbf{X}^T\mathbf{X} \leq \mathbf{X}^T\mathbf{HX} \leq \mathbf{X}^T\mathbf{X}$ for arbitrary $\mathbf{X}$. Consequently,

$$\frac{F(\mathbf{X}^r) - F(\mathbf{X}^{r+1})}{F(\mathbf{X}^r) - F(\mathbf{X}^*)} = \frac{(\mathbf{g}^{rT}\mathbf{g}^r)^2}{(\mathbf{g}^{rT}\mathbf{H}\mathbf{g}^r)(\mathbf{g}^{rT}\mathbf{H}^{-1}\mathbf{g}^r)} \geq \omega$$

where $|\omega| < 1$. In deriving this inequality, we have used the fact that $\mathbf{HE}^r = \mathbf{H}(\mathbf{X}^r - \mathbf{X}^*) = \mathbf{HX}^r + \mathbf{b} = \mathbf{g}^r$. Subtracting each side of the inequality above from unity and simplifying, we get

$$F(\mathbf{X}^{r+1}) - F(\mathbf{X}^*) \leq (1 - \omega)[F(\mathbf{X}^r) - F(\mathbf{X}^*)]$$

Hence, if $\mathbf{X}_0$ is an initial test point,

$$F(\mathbf{X}^r) - F(\mathbf{X}^*) \leq (1 - \omega)^r[F(\mathbf{X}_0) - F(\mathbf{X}^*)] \tag{17.33}$$

Since $\omega < 1$, the right side of this inequality vanishes as r becomes indefinitely large and $F(\mathbf{X}^r) - F(\mathbf{X}^*) \to 0$. This completes the proof.

If $F(\mathbf{X})$ is not a quadratic, the iterative process may still converge in the sense of (17.33) provided that $\mathbf{X}_0$ is sufficiently close to $\mathbf{X}^*$ that the quadratic terms in a Taylor series expansion of $F(\mathbf{X})$ dominate those of higher order in the neighborhood of $\mathbf{X}_0$. Convergence proofs based on this assumption are said to be based on quadratic convergence. In addition, it is possible to generalize Theorem 17.2 so as to obtain criteria for arbitrary functions $F(\mathbf{X})$ so long as $F(\mathbf{X})$ has twice continuously differentiable derivatives and the hessian is positive definite at the minimum point $\mathbf{X}^*$.

Theorem 17.3† *Let $F(\mathbf{X})$ be a real-valued, twice continuously differentiable function defined on a convex domain of $\mathscr{E}^n$. Let $\mathbf{X}_0$ be any point such that the closed set $\mathscr{S} = \{\mathbf{X}: F(\mathbf{X}) \leq F(\mathbf{X}_0)\}$ is bounded and such that for any vector $\mathbf{X} \in \mathscr{S}$, $|\mathbf{X}^T\mathbf{HX}| \leq \alpha\mathbf{X}^T\mathbf{X}$, where $\mathbf{H}$ is the hessian of $F(\mathbf{X})$ and α is a positive scalar. Further, let $\mathbf{X}^{r+1} = \mathbf{X}^r + \lambda^r\mathbf{g}^r$, where $\mathbf{g}^r$ is the gradient of $F(\mathbf{X})$ at $\mathbf{X}^r \in \mathscr{S}$ and λ^r satisfies $F(\mathbf{X}^r + \lambda^r\mathbf{g}^r) \leq F(\mathbf{X}^r + \lambda\mathbf{g}^r)$ for all $\lambda \geq 0$. Then* **(a)** *a subsequence $\{\mathbf{X}^{r,m}\}$ converges to a point $\mathbf{X}^* \in \mathscr{S}$ at which $\mathbf{g}(\mathbf{X}^*) = \mathbf{0}$,* **(b)** *$F(\mathbf{X}^{r,m})$ decrease monotonically to $F(\mathbf{X}^*) = \mathbf{0}$, and* **(c)** *if $\mathbf{X}^*$ is unique in $\mathscr{S}$, $\{\mathbf{X}^r\}$ converges to $\mathbf{X}^*$.*

† For the proof of this theorem, consult Goldstein [1962]. See also Saaty and Bram [1964].

The choice of λ^r of (17.28) ensures the optimum step size so long as $F(\mathbf{X})$ is quadratic. Then $F(\mathbf{X})$ is decreased as much as possible during a single step. If $F(\mathbf{X})$ is not quadratic or if we require that λ^r be optimum for a given number n of iterations rather than stepwise, different choices of λ^r may lead to faster convergent sequences.

Conjugate-gradient methods† The so-called conjugate-gradient methods can be shown to minimize quadratic functions in n variables in p iterations, where $p \leq n$. To demonstrate the ideas, we rewrite the quadratic function in (17.32) in the form

$$F(\mathbf{X}) = F_0 + \tfrac{1}{2}(\mathbf{X} - \mathbf{X}^*)^T\mathbf{H}(\mathbf{X} - \mathbf{X}^*) \tag{17.34}$$

where $\mathbf{X}^*$ is the point at which $F(\mathbf{X})$ is a minimum. Then

$$\mathbf{g}(\mathbf{X}) = \nabla F = \mathbf{H}(\mathbf{X} - \mathbf{X}^*) \tag{17.35}$$

As usual, $\mathbf{X}^{r+1} = \mathbf{X}^r + \lambda^r\mathbf{D}^r$. We assume that the successive direction vectors $\mathbf{D}^0, \mathbf{D}^1, \ldots, \mathbf{D}^n$ are linearly independent, and set out $\boldsymbol{\sigma}^r = \lambda^r\mathbf{D}^r$. Then

$$\begin{aligned}\mathbf{X}^r &= \mathbf{X}^{r-1} + \boldsymbol{\sigma}^{r-1}\\ &= \mathbf{X}^{r-2} + \boldsymbol{\sigma}^{r-2} + \boldsymbol{\sigma}^{r-1}\\ &= \mathbf{X}^{r-q} + \boldsymbol{\sigma}^{r-q} + \boldsymbol{\sigma}^{r-q+1} + \cdots + \boldsymbol{\sigma}^{r-1}\end{aligned}$$

or, for the nth iteration,

$$\mathbf{X}^n = \mathbf{X}^{r+1} + \sum_{q=r+1}^{n-1} \boldsymbol{\sigma}^q \tag{17.36}$$

Thus

$$\mathbf{g}^n = \mathbf{H}(\mathbf{X}^n - \mathbf{X}^*) = \mathbf{H}\mathbf{X}^{r+1} + \mathbf{H}\sum_{q=r+1}^{n-1} \boldsymbol{\sigma}^q - \mathbf{H}\mathbf{X}^*$$

In view of the fact that $\mathbf{H}\mathbf{X}^* = \mathbf{0}$ and $\mathbf{H}\mathbf{X}^{r+1} = \mathbf{g}^{r+1}$, we have

$$\mathbf{g}^n = \mathbf{g}^{r+1} + \sum_{q=r+1}^{n-1} \lambda^q\mathbf{H}\mathbf{D}^q$$

It follows that

$$\mathbf{g}^{n^T}\mathbf{D}^r = \mathbf{g}^{(r+1)^T}\mathbf{D}^r + \sum_{q=r+1}^{n-1} \lambda^q\mathbf{D}^{q^T}\mathbf{H}\mathbf{D}^r \tag{17.37}$$

We now assume that the vectors $\mathbf{D}^0, \mathbf{D}^1, \ldots, \mathbf{D}^{n-1}$ are $\mathbf{H}$-conjugate in the sense that

$$\mathbf{D}^{q^T}\mathbf{H}\mathbf{D}^r = 0 \qquad q \neq r \tag{17.38}$$

† See Hestenes and Stiefel [1952] and Fletcher and Reeves [1964]. A more general treatment has been given by Daniel [1967].

Then, from (17.37),

$$\mathbf{g}^{n^T}\mathbf{D}^r = 0 \tag{17.39}$$

for all r. Since $\mathbf{D}^0, \mathbf{D}^1, \ldots, \mathbf{D}^{n-1}$ are linearly independent, they form a basis for all vectors of the type $\mathbf{g}^n$. Consequently, (17.39) implies that

$$\mathbf{g}^n = \mathbf{0} \tag{17.40}$$

Hence

$$\mathbf{H}(\mathbf{X}^n - \mathbf{X}^*) = \mathbf{g}^n = \mathbf{0}$$

and, since $\mathbf{H}$ is positive definite,

$$\mathbf{X}^n = \mathbf{X}^* \tag{17.41}$$

Therefore, the minimum is located after n iterations or earlier if, in any particular cycle, the value of λ^n happens to be zero.

The basic requirement (17.38) of conjugate-gradient methods can be satisfied in a number of different ways. In general, we set, as usual,

$$\mathbf{D}^0 = -\mathbf{g}^0 = -\nabla F(\mathbf{X}_0) \tag{17.42a}$$

and

$$\mathbf{D}^{r+1} = -\mathbf{g}^{r+1} + \beta^r\mathbf{D}^r \tag{17.42b}$$

where β^r is selected so that (17.38) is satisfied; that is,

$$\mathbf{D}^{r+1^T}\mathbf{H}\mathbf{D}^r = -\mathbf{g}^{r+1^T}\mathbf{H}\mathbf{D}^r + \beta^r\mathbf{D}^{r^T}\mathbf{H}\mathbf{D}^r = 0$$

Thus

$$\beta^r = \frac{\mathbf{g}^{r+1^T}\mathbf{H}\mathbf{D}^r}{\mathbf{D}^{r^T}\mathbf{H}\mathbf{D}^r} \tag{17.43}$$

Direction vectors $\mathbf{D}^r$ computed in this manner are linearly independent and $\mathbf{H}$-conjugate.

Since $F(\mathbf{X})$ is generally not a quadratic function, as has been assumed in (17.34), the hessian $\mathbf{H}$ must be computed for each cycle. In such cases, our conclusions concerning convergence in n iterations are not strictly valid, but relatively rapid convergence is often experienced after n cycles for nonquadratic $F(\mathbf{X})$. Likewise, in the general case in which $F(\mathbf{X})$ is nonquadratic, steps sizes λ^r different from that in (17.28) may prove more convenient.

The variable-metric method The variable-metric method, introduced by Davidon [1959] and generalized by Fletcher and Powell [1963], is a powerful iterative-descent method for finding local minima of nonlinear functions of several variables. The method is based on the usual premise that second- and lower-order terms in the Taylor series expansions of nonquadratic functions dominate all others in the vicinity of a local minimum.

We recall that for a general quadratic function of n variables of the form (17.32), the minimum point is $\mathbf{X}^* = -\mathbf{H}^{-1}\mathbf{b}$. In the variable metric method, however, the inverse $\mathbf{H}^{-1}$ is not evaluated directly. Instead, an approximating matrix $\mathbf{G}$ is introduced which, initially, may be any positive-definite symmetric matrix. The matrix $\mathbf{G}$ is modified at the rth iteration by using information obtained by moving in the direction

$$\mathbf{D}^r = -\mathbf{G}^r\mathbf{g}^r \tag{17.44}$$

The modification is such that the step to the minimum $\boldsymbol{\sigma}^i$ along the line $\mathbf{X} = \mathbf{X}^r + \lambda\mathbf{D}^r$ is effectively an eigenvector of the matrix $\mathbf{G}^{r+1}\mathbf{H}$. This ensures that $\mathbf{G}$ approaches $\mathbf{H}^{-1}$ evaluated at the minimum. Usually $\mathbf{G}_0$ is set equal to the identity matrix $\mathbf{I}$, which causes the initial step to be taken in the direction of steepest descent.

After obtaining $\mathbf{D}^r$ from equation (17.44), λ^r is determined such that $F(\mathbf{X}^r + \lambda^r\mathbf{S}^r)$ is a minimum with respect to λ along the line $\mathbf{X}^r + \lambda\mathbf{S}^r$. The process is developed as follows: First, we set

$$\mathbf{X}^{r+1} = \mathbf{X}^r + \boldsymbol{\sigma}^r \tag{17.45}$$

where $\boldsymbol{\sigma}^r = \lambda^r\mathbf{D}^r$. We then evaluate $F(\mathbf{X}^{r+1})$, noting that $\boldsymbol{\sigma}^{r^T}\mathbf{g}^{r+1} = 0$, and define

$$\mathbf{Y}^r = \mathbf{g}^{r+1} - \mathbf{g}^r \tag{17.46}$$

Then

$$\mathbf{G}^{r+1} = \mathbf{G}^r + \mathbf{A}^r + \mathbf{B}^r \tag{17.47}$$

where

$$\mathbf{A}^r = \frac{\boldsymbol{\sigma}^r\boldsymbol{\sigma}^{r^T}}{\boldsymbol{\sigma}^{r^T}\mathbf{Y}^r} \tag{17.48a}$$

$$\mathbf{B}^r = -\frac{\mathbf{G}^r\mathbf{Y}^r\mathbf{Y}^{r^T}\mathbf{G}^r}{\mathbf{Y}^{r^T}\mathbf{G}^r\mathbf{Y}^r} \tag{17.48b}$$

The entire process is repeated, proceeding from point $\mathbf{X}^{r+1}$, with gradient $\mathbf{g}^{r+1}$ and matrix $\mathbf{G}^{r+1}$. The predicted absolute distance from the minimum at the rth iterative is

$$\hat{d}^r = (\mathbf{g}^{r^T}\mathbf{H}^{-1^T}\mathbf{H}^{-1}\mathbf{g}^r)^{\frac{1}{2}} \tag{17.49}$$

which can be approximated by

$$\hat{d}^r \approx (\mathbf{D}^{r^T}\mathbf{D}^r)^{\frac{1}{2}} \tag{17.50}$$

Equation (17.50) follows from (17.49) and the fact that $\mathbf{G}^r$ approaches $\mathbf{H}^{-1}$ as $r \to \infty$. The iterative procedure may be terminated when the distance $\hat{d}^r$ is less than some prescribed value or when every component of $\mathbf{D}^r$ is less than a prescribed constant.

It can be shown that this method converges in $m \leq n$ iterations in the case in which $F(\mathbf{X})$ is a quadratic functional.

From (17.46) it follows that

$$\mathbf{Y}^r = \mathbf{H}\mathbf{X}^{r+1} - \mathbf{H}\mathbf{X}^r = \mathbf{H}\boldsymbol{\sigma}^r$$

Thus

$$\begin{aligned}\mathbf{G}^{r+1}\mathbf{H}\boldsymbol{\sigma}^r &= \mathbf{G}^r\mathbf{Y}^r + \boldsymbol{\sigma}^r\left(\frac{\boldsymbol{\sigma}^{r^T}\mathbf{H}\boldsymbol{\sigma}^r}{\boldsymbol{\sigma}^{r^T}\mathbf{H}\boldsymbol{\sigma}^r}\right) - \frac{\mathbf{G}^r\mathbf{Y}^r(\mathbf{G}^r\mathbf{Y}^r)^T\mathbf{Y}^r}{\mathbf{Y}^{r^T}\mathbf{G}^r\mathbf{Y}^r} \\ &= \mathbf{G}^r\mathbf{Y}^r + \boldsymbol{\sigma}^r - \mathbf{G}^r\mathbf{Y}^r \\ &= \boldsymbol{\sigma}^r \end{aligned} \tag{17.51}$$

Now we assume throughout that $\mathbf{H}$ and $\mathbf{G}^r$ are well-defined symmetric positive-definite matrices. Thus, each iterate $\mathbf{X}^r \neq \mathbf{H}^{-1}\mathbf{b} = \mathbf{X}^*$ is well defined for $r = 0, 1, \ldots, n-1$. Otherwise $\mathbf{X}^r = \mathbf{X}^*$ and the process converges for $r < n$. Thus, we may assume that $\mathbf{X}^r \neq \mathbf{H}^{-1}\mathbf{b}$ for $r = 0, 1, \ldots, n-1$. We know from the previous discussion of conjugate gradients that $\mathbf{X}^n = \mathbf{H}^{-1}\mathbf{b}$ if $\mathbf{D}^0, \ldots, \mathbf{D}^r \ldots$ are $\mathbf{H}$-conjugate. Hence, we need to verify that (17.51) is true for all $r = 0, 1, 2, \ldots, n$. In order to infer that $\mathbf{G}^r$ approaches $\mathbf{H}^{-1}$, it must be shown that (17.51) is true for all r.

Toward this end, the following two equations are considered:

$$\boldsymbol{\sigma}^{i^T}\mathbf{H}\boldsymbol{\sigma}^j = 0 \qquad 0 \leq i < j < k \tag{17.52a}$$

$$\mathbf{G}^k\mathbf{H}\boldsymbol{\sigma}^i = \boldsymbol{\sigma}^i \qquad 0 \leq i < k \tag{17.52b}$$

It can be seen from (17.51) that these two equations are true if $k = 1$. It will be proved that if they hold for k, they are also true for $k + 1$.

From the original equation for the gradient $\mathbf{g}^k = \mathbf{a} + \mathbf{H}\mathbf{X}^k$ it follows that $\mathbf{g}^k = \mathbf{a} + \mathbf{H}(\mathbf{X}^{i+1} + \boldsymbol{\sigma}^{i+1} + \boldsymbol{\sigma}^{i+2} + \cdots + \boldsymbol{\sigma}^{k-1})$ and therefore,

$$\mathbf{g}^k = \mathbf{g}^{i+1} + \mathbf{H}(\boldsymbol{\sigma}^{i+1} + \boldsymbol{\sigma}^{i+2} + \boldsymbol{\sigma}^{k-1})$$

Hence, for all $i \leq k$, $i \neq k$,

$$\boldsymbol{\sigma}^{i^T}\mathbf{g}^k = \boldsymbol{\sigma}^{i^T}\mathbf{g}^{i+1} = 0$$

so that (17.52) hold for $k \leq n$. It follows that $\mathbf{X}^n = \mathbf{H}^{-1}\mathbf{b}$. Equation (17.52$a$) shows that $\mathbf{G}^n\mathbf{H}$ has n linearly independent eigenvectors with unit eigenvalues. Therefore $\mathbf{G}^n\mathbf{H} = \mathbf{I}$, which was to be proven.

17.4 THE NEWTON-RAPHSON METHOD

We now discuss the powerful and well-known Newton-Raphson method and its modified version. The method has found wide application in nonlinear problems of solid and structural mechanics.

Consider again the system of nonlinear equations (17.4) and suppose that an initial test point $\mathbf{X}_0$ differs from the solution by a small vector $\Delta\mathbf{X}$. If $\mathbf{f}(\mathbf{X})$ is sufficiently differentiable at $\mathbf{X}_0$, we may employ the Taylor expansion,

$$\mathbf{f}(\mathbf{X}) = \mathbf{f}(\mathbf{X}_0 + \Delta\mathbf{X}) = \mathbf{f}(\mathbf{X}_0) + \mathbf{J}_0(\mathbf{X} - \mathbf{X}_0) + \cdots = \mathbf{0} \tag{17.53}$$

where $\mathbf{J}_0$ is the *jacobian matrix* at $\mathbf{X}_0$:

$$\mathbf{J}_0 = \left[\frac{\partial f_i(\mathbf{X}_0)}{\partial X_j}\right] \tag{17.54}$$

If (17.53) is truncated to only linear terms, we get

$$\mathbf{0} \cong \mathbf{f}(\mathbf{X}_0) + \mathbf{J}_0(\mathbf{X} - \mathbf{X}_0)$$

so that

$$\mathbf{X} = \mathbf{X}_0 - \mathbf{J}_0^{-1}\mathbf{f}(\mathbf{X}_0) \tag{17.55}$$

This value of $\mathbf{X}$ is, of course, only approximate, but it can be used as a starting value for another step in the iteration process. In general, we have the recurrence formula

$$\mathbf{X}^{r+1} = \mathbf{X}^r - \mathbf{J}^{r^{-1}}\mathbf{f}(\mathbf{X}^r) \tag{17.56}$$

This process is continued until $|\Delta X_j| < \epsilon$ for some prescribed accuracy ϵ.

The classical case of only one unknown provides an illustration of the Newton-Raphson process. Starting with a trial solution X_0, we evaluate $f(X_0)$ and $f'(X_0)$ and compute a new value X^1 as $X_0 - f(X_0)/f'(X_0)$. As is indicated in Fig. 17.3*a*, this amounts to a descent along a tangent to $f(X)$ at X_0. If $f'(X)$ vanishes at some point in the iteration, as indicated in Fig. 17.3*b*, the Newton-Raphson process may diverge.

We notice in (17.56) that in each step of the iteration process it is necessary to invert an $n \times n$ jacobian matrix $\mathbf{J}^r$. This difficulty can be overcome by the *modified Newton-Raphson* method, in which only the jacobian matrix of the first cycle is used. Then, instead of (17.56) we have

$$\mathbf{X}^{r+1} = \mathbf{X}^r - \mathbf{J}_0^{-1}\mathbf{f}(\mathbf{X}^r) \tag{17.57}$$

We recognize the modified Newton-Raphson method as a special case of the chord method, mentioned in Sec. 17.2, for the choice of the $\mathbf{A}$ matrix of (17.21) of $\mathbf{J}_0^{-1}$. Although the number of iterations required for convergence is generally much greater when the modified process is used, the actual number of calculations and amount of computing time required may be significantly less than in the straight Newton-Raphson procedure. Conceptually, the modified version is compared with the unmodified procedure in Fig. 17.3*a* and *c*, where we see that the original slope $f'(X_0)$ is used in each cycle of the iteration process.

The Newton-Raphson method and its modified version have been used successfully to solve large systems of nonlinear equations. However, these iterative procedures are also important for their theoretical value in that theorems are available which make it possible to estimate the existence and

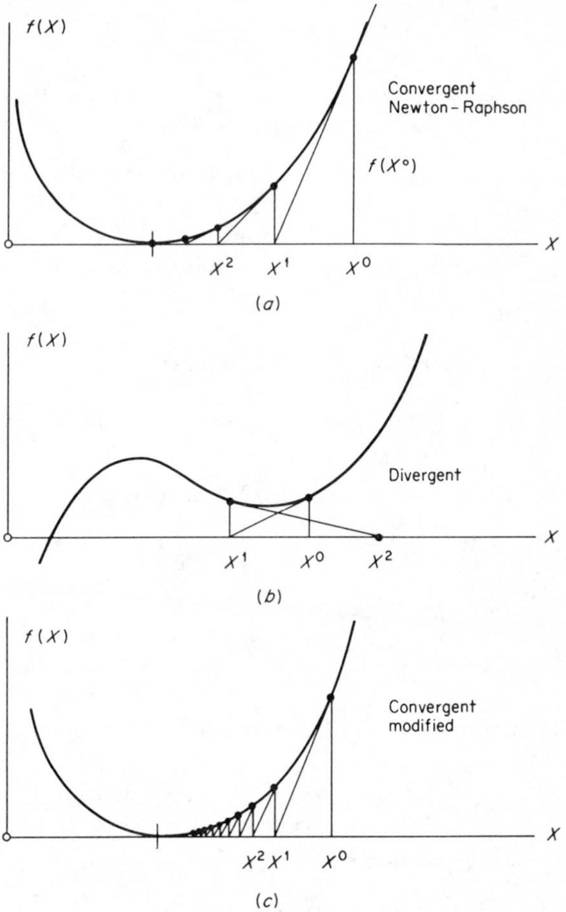

Fig. 17.3 Standard and modified Newton-Raphson methods.

uniqueness of solutions, the convergence, and the rate of convergence of the iteration process on the basis of the values of the initial test point $\mathbf{X}_0$ and the values of $|\partial f_i/\partial X_j|$ and $|\partial^2 f_i/\partial X_j\, \partial X_k|$. In addition, conclusions on existence and uniqueness of solutions and on the convergence of the Newton-Raphson schemes for nonlinear algebraic and transcendental equations can be extended to Banach spaces so as to apply to general nonlinear operators.

We list representative theorems, without proof, which are applicable to nonlinear algebraic and transcendental equations.†

† See, for example, Rall [1969].

Theorem 17.4 *The system of nonlinear equations*

$$f_i(\mathbf{X}) = 0 \qquad i = 1, 2, \ldots, n$$
$$\mathbf{X}^T = \{X_1, \ldots, X_n\}$$

has a solution which can be obtained by the Newton-Raphson method if the following conditions hold:

a. *The jacobian matrix* $\mathbf{J}_0$ *is nonsingular and*

$$\|\mathbf{J}_0^{-1}\| \leq B \tag{17.58}$$

where B is a positive number.

b. $\|\mathbf{X}^1 - \mathbf{X}_0\| \leq A$ (17.59)

where A is a positive number.

c. $\dfrac{\partial^2 f_i}{\partial X_j\, \partial X_k} \leq C \qquad \textit{for } i, j, k = 1, 2, \ldots, n$ (17.60)

Moreover, the solution $\mathbf{X}^*$ *satisfies*

$$\|\mathbf{X}^* - \mathbf{X}_0\| \leq (1 - \sqrt{1 - 2c_0})\,\frac{A}{c_0} \tag{17.61}$$

where

$$c_0 = Bn\sqrt{n}\, CA \leq \tfrac{1}{2} \tag{17.62}$$

Theorem 17.5 *If all the conditions of Theorem* 17.4 *are satisfied and if*

$$\frac{\partial^2 f_i}{\partial X_j\, \partial X_k} \leq A \qquad i, j, k = 1, 2, \ldots, n \tag{17.63}$$

in the region

$$\|\mathbf{X} - \mathbf{X}_0\| < (1 + \sqrt{1 - 2c_0})\,\frac{B}{c_0} \tag{17.64}$$

then $\mathbf{X}^*$ *is unique.*

At the rth iteration,

$$\|\mathbf{X}^{r+1} - \mathbf{X}^r\| = \|\mathbf{J}^{r^{-1}}\mathbf{f}(\mathbf{X}^r)\|$$

and we require that for a given positive ϵ there exists an integer m such that

$$\|\mathbf{X}^* - \mathbf{X}^r\| < \epsilon \qquad \text{for } r > m$$

or

$$\lim_{r \to \infty} \|\mathbf{X}^* - \mathbf{X}^r\| = 0$$

This requires that we keep a running account of $\|\mathbf{J}^{r^{-1}}\mathbf{f}(\mathbf{X}^r)\|$, starting with the initial point and carrying convenient bounds over successive iterations. Once the existence of $\|\mathbf{J}_1^{-0}\|$ is established, the bound of the inverse jacobian of the next cycle depends upon the bound of the second derivatives $|f_{i,jk}|$ and is obtained by use of the mean-value theorem. Hence, the requirement (17.60).

The Newton-Raphson method can also be regarded as another minimization technique in much the same way as the steepest-descent methods. Suppose that

$$F(\mathbf{X}) = \mathbf{f}^T(\mathbf{X})\mathbf{f}(\mathbf{X}) = f_i f_i$$

and, to within quadratic terms,

$$\begin{aligned} F(\mathbf{X}^{r+1}) &= F(\mathbf{X}^r + \boldsymbol{\sigma}^r) \\ &= F(\mathbf{X}^r) + \mathbf{g}^{rT}\boldsymbol{\sigma}^r + \tfrac{1}{2}\boldsymbol{\sigma}^{rT}\mathbf{H}^r\boldsymbol{\sigma}^r \end{aligned}$$

Select $\boldsymbol{\sigma}^r$ so that

$$\frac{\partial[F(\mathbf{X}^{r+1}) - F(\mathbf{X}^r)]}{\partial\sigma_i^r} = 0 = \mathbf{g}^r + \mathbf{H}^r\boldsymbol{\sigma}^r$$

But

$$g_i^r = \frac{\partial F(\mathbf{X}^r)}{\partial X_i} = 2f_i(\mathbf{X}^r) = 2f_i^r$$

$$H_{ij}^r = \frac{\partial^2 F(\mathbf{X}^r)}{\partial X_i\,\partial X_j} = 2J_{ij}(\mathbf{X}^r) = 2J_{ij}^r$$

Thus, the Newton-Raphson method is characterized by a correction vector

$$\boldsymbol{\sigma}^r = \lambda^r\mathbf{D}^r = -\mathbf{H}^{r^{-1}}\mathbf{g}^r = [-4J_{ij}^{r^{-1}}f_i^r] \tag{17.65}$$

and

$$\mathbf{X}^{r+1} = \mathbf{X}^r - \mathbf{H}^{r^{-1}}\mathbf{g}^r = \mathbf{X}^r - \mathbf{J}^{r^{-1}}\mathbf{f}^r \tag{17.66}$$

It should be noted that in the Newton-Raphson method (or, for that matter, in most gradient methods), the quantities $\partial f_i/\partial X_j$ are required at each cycle of the iterative process. Except in special cases, however, the calculation of a large array of derivatives is impossible or, at best, extremely time-consuming, and it is convenient to resort to finite-difference approximations of $\mathbf{J}$. Using, for example, a forward difference approximation of $\partial f_i/\partial X_j$, we prescribe a small $\Delta X_i = h$ and write

$$\frac{\partial f_i(\mathbf{X}^r)}{\partial X_j} \approx \frac{f_i(X^1, X^2, \ldots, X^j + h, \ldots, X^n) - f_i(\mathbf{X}^r)}{h} \tag{17.67}$$

Then the derivatives $\partial f_i/\partial X_j$ need not be calculated analytically. In designing programs that use the method, we need to store only the n functions $f_i(\mathbf{X})$ and not their gradients.

17.5 THE METHOD OF INCREMENTAL LOADING†

Perhaps the most appealing method for solving the systems of nonlinear equations encountered in nonlinear elasticity problems is the method of incremental loading. While it has some features similar to the Newton-Raphson method, it has certain features that make it more useful in applications to physical problems. First of all, each step of the iterative process generally admits to clear physical interpretation: The load acting on a deformable body is considered to be applied in increments $\delta\mathbf{p}$ which are sufficiently small that during each increment the response of the body is linear. At the end of each load increment, a new updated stiffness relation is obtained and another increment of load is applied. By continuing this procedure, the complete nonlinear response of the body is generated as a sequence of piecewise-linear steps. Since the body is often in a natural unstressed state prior to applying loads, the initial test point need not be selected arbitrarily. Indeed, if $\mathbf{X}$ is a vector of unknown nodal displacements, we simply set $\mathbf{X}_0 = \mathbf{0}$ to obtain an initial test point corresponding to an undeformed state of the body. Alternately, if the body is incompressible, we set the displacements equal to zero and calculate the hydrostatic pressures in the undeformed state. These, then, become components of the initial test point $\mathbf{X}_0$. Because of the physical significance generally associated with each step in the iterative process, the method yields more information concerning the overall response than just the values of displacements corresponding to a given system of loads; indeed, a complete family of solutions, each corresponding to a different fraction of the load, is obtained naturally along the way. This makes it possible to test intermediate states for stability and to modify the procedure so as to obtain bifurcation points and multivalued solutions.

To describe the method, we begin with a system of n nonlinear equations, as in (17.4), except that now we assume that each equation involves not only the unknown vector $\mathbf{X}$, but also a real parameter p; that is, we consider a system of the form

$$\mathbf{h}(\mathbf{X},p) = \mathbf{0} \tag{17.68}$$

Here $\mathbf{h}$ is an n vector of nonlinear equations in the components of the n vector

† A variety of terms are used in the literature to describe methods of incremental loading: for example, the schemes we describe here are often referred to as imbedding methods, continuation methods, or methods of variation of parameters. The imbedding methods have been used to study existence problems for nonlinear operator equations and for the solution of nonlinear functional equations. See, for example, the work of Gavurin [1958], Davidenko [1965], and Meyer [1968]. A historical summary has been presented by Ficken [1951] and a summary account of continuation methods and additional references are contained in the book of Ortega and Rheinboldt [1970].

$\mathbf{X}$; the parameter p generally corresponds to a prescribed load. If n independent loads are encountered, we may still obtain a system of the form (17.68) by setting the total load vector $\mathbf{P}$ equal to $p\boldsymbol{\Lambda}$, where $\boldsymbol{\Lambda}$ is a constant vector. If in formulating a nonstructural problem the choice of the parameter p does not present itself in a natural way, a system of equations of the form (17.68) can be obtained by subjecting the nonlinear system (17.4) to a one-parameter imbedding, generally referred to as an *operator homotopy*, such as†

$$\mathbf{h}(\mathbf{X},p) = \alpha(p)\mathbf{g}(\mathbf{X}) + \beta(p)\mathbf{f}(\mathbf{X}) \tag{17.69a}$$

where $\alpha(p)$ and $\beta(p)$ are real-valued functions and $\mathbf{g}(\mathbf{X})$ is a vector-valued function, chosen so that if $p \in [0,Q]$, Q being a real number,

$$\mathbf{h}(\mathbf{X}_0,0) = \mathbf{0}$$

and

$$\mathbf{h}(\mathbf{X}^*,Q) = \mathbf{0},\ \mathbf{X}^*$$

being the solution of (17.4). For example, we may set

$$\mathbf{h}(\mathbf{X},p) = (Q - p)(\mathbf{X} - \mathbf{X}_0) + p\mathbf{f}(\mathbf{X}) \tag{17.69b}$$

Then

$$\mathbf{h}(\mathbf{X}_0,0) = \mathbf{0}$$

and

$$\mathbf{h}(\mathbf{X}^*,Q) = Q\mathbf{f}(\mathbf{X}^*) = \mathbf{0}$$

implies that the solution $\mathbf{X}^*$ of (17.4) is reached at $p = Q$. Imbeddings such as (17.69) are rarely necessary in mechanics problems since equations of the form (17.68) are encountered quite naturally. We also remark that if a load parameter $p \in [0,Q]$, we can always normalize p by using $\hat{p} = p/Q \in [0,1]$.

For a given value of p, (17.68) defines a point in $\mathscr{E}^n$ which lies at the intersection of n surfaces, $h_i(\mathbf{X},p) = 0$. Thus (17.68) represents parametric equations of a path or curve in $\mathscr{E}^n$ with parameter p. If we could solve (17.68) for each $p \in [0,1]$, then the solutions

$$\mathbf{X} = \mathbf{X}(p)$$

are explicit parametric equations for a curve in $\mathscr{E}^n$ connecting a known initial point $\mathbf{X}_0 = \mathbf{X}(0)$ to the solution of the given system of nonlinear equations $\mathbf{X}^* = \mathbf{X}(1)$ [or $\mathbf{X}^* = \mathbf{X}(Q)$ if $p \in [0,Q]$]. This is illustrated for $n = 3$ in Fig. 17.4.

† Cf. Meyer [1968].

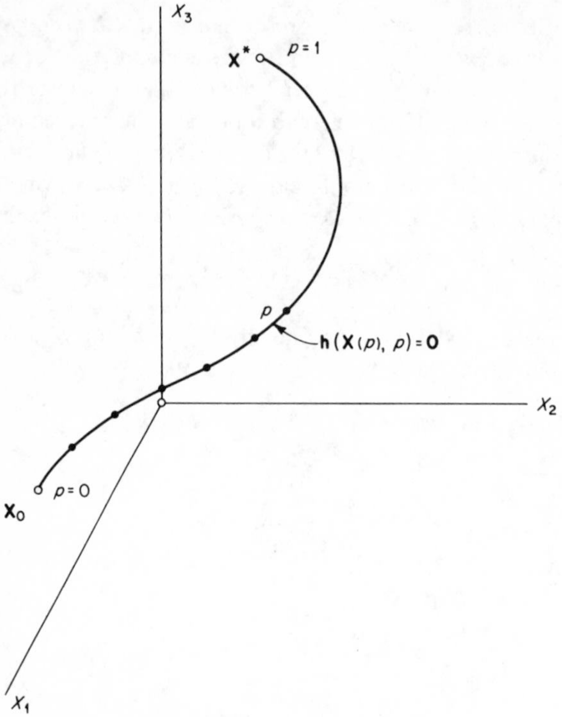

Fig. 17.4 Path in $\mathscr{E}^3$ defined by the nonlinear system $h_i(X_1,X_2,X_3;p) = 0$, $i = 1, 2, 3$, $p \in [0,1]$.

Now $\mathbf{X} = \mathbf{X}(p)$ describes a mapping from $[0,1]$ into a set $\mathscr{D} \subset \mathscr{E}^n$. We shall henceforth assume that this mapping satisfies (17.68) for all $p \in [0,1]$, that it is continuously differentiable with respect to p on $[0,1]$, and that $\mathbf{h}: \mathscr{E}^n \times [0,1] \to \mathscr{E}^n$ of (17.68) has continuous partial derivatives with respect to $\mathbf{X}$ and p. This being the case, (17.68) holds for all values of p in $[0,1]$. Consequently, if $\mathbf{X}$ is a solution corresponding to a particular choice of p, and if $\mathbf{X} + \delta\mathbf{X}$ is another solution corresponding to $p + \delta p$, δp being an incremental change in p, then

$$\mathbf{h}(\mathbf{X},p) = \mathbf{h}(\mathbf{X} + \delta\mathbf{X}, p + \delta p) = \mathbf{0} \tag{17.70}$$

From the assumed differentiability of $\mathbf{h}(\mathbf{X},p)$, we have

$$\mathbf{h}(\mathbf{X} + \delta\mathbf{X}, p + \delta p) = \mathbf{h}(\mathbf{X},p) + \mathbf{H}_x(\mathbf{X},p)\,\delta\mathbf{X} + \mathbf{h}_p(\mathbf{X},p)\,\delta p + \boldsymbol{\omega}(\mathbf{X},p,\delta\mathbf{X},\delta p) \tag{17.71}$$

where $\mathbf{H}_x$ is the $n \times n$ matrix and $\mathbf{h}_p$ and $\boldsymbol{\omega}$ are the n vectors defined by

$$\mathbf{H}_x(\mathbf{X},p) = \left[\frac{\partial h_i(\mathbf{X},p)}{\partial X_j}\right] \tag{17.72a}$$

$$\mathbf{h}_p(\mathbf{X},p) = \left[\frac{\partial h_i(\mathbf{X},p)}{\partial p}\right] \tag{17.72b}$$

and

$$\boldsymbol{\omega}(\mathbf{X},p,\delta\mathbf{X},\delta p) = \left\{\frac{1}{2!}\left[\frac{\partial^2 h_i(\boldsymbol{\xi}^1,\theta)}{\partial X_j\,\partial X_k}\,\delta X_j\,\delta X_k + 2\,\frac{\partial^2 h_i(\boldsymbol{\xi}^1,\theta)}{\partial X_j\,\partial p}\,\delta X_j\,\delta p + \frac{\partial^2 h_i(\boldsymbol{\xi}^1,\theta)}{\partial p^2}\,\delta p^2\right]\right\} \tag{17.72c}$$

Here

$$\boldsymbol{\xi}^1 \in \{\bar{\mathbf{X}} \mid \bar{\mathbf{X}} = \alpha\mathbf{X} + (1-\alpha)(\mathbf{X}+\delta\mathbf{X});\, 0 \le \alpha \le 1\},\ \theta \in [p, p+\delta p]$$

An iterative scheme for the approximate solution of (17.68) results immediately from (17.70) and (17.71) if we assume that δp and $|\delta X_i|$ are sufficiently small that only linear terms in δp and $\delta\mathbf{X}$ need be retained in (17.71). That is,

$$\mathbf{H}_x(\mathbf{X},p)\,\delta\mathbf{X} + \mathbf{h}_p(\mathbf{X},p)\,\delta p \approx \mathbf{0}$$

Hence, if $\mathbf{H}_x^{-1}(\mathbf{X},p)$ exists at $(\mathbf{X},p)$, we have

$$\delta\mathbf{X} = -\mathbf{H}_x^{-1}(\mathbf{X},p)\mathbf{h}_p(\mathbf{X},p)\,\delta p \tag{17.73}$$

This result leads directly to the method of incremental loading: Let $p \in [0,Q]$, $\mathbf{X}(0) = \mathbf{X}^0$, and $\mathbf{H}_x(\mathbf{X},p)$ be nonsingular for all p under consideration; divide the interval $[0,Q]$ into M partitions

$$0 = p_0 < p_1 < p_2 < \cdots < p_M = Q$$

so that

$$\delta p_r = p_r - p_{r-1} \qquad \text{and} \qquad Q = \sum_{r=1}^{M} \delta p_r \tag{17.74a,b}$$

Then the path $\mathbf{X} = \mathbf{X}(p)$ is generated by the process

$$\mathbf{X}^{r+1} = \mathbf{X}^r - \mathbf{H}_x^{-1}(\mathbf{X}^r,p_r)\mathbf{h}_p(\mathbf{X}^r,p_r)\,\delta p_{r+1} \tag{17.75}$$

with $r = 0, 1, \ldots, M-1$.

This process is illustrated graphically in Fig. 17.4 for the one-dimensional case $h(X,p) = 0$. Unless $h(X,p)$ is perfectly linear, some error ϵ will always result after a finite number M of load increments. If at some point $(\mathbf{X}^r,p_r)$, $\mathbf{H}_x(\mathbf{X}^r,p_r)$ is singular, the process will, of course, fail. It is then necessary to modify the equations by, for example, changing the scale of the parameter or by determining a new starting point beyond $(\mathbf{X}^r,p_r)$ by using, say, the Newton-Raphson method. Should $\det \mathbf{H}_x(\mathbf{X}_0,0) = 0$, a new initial point

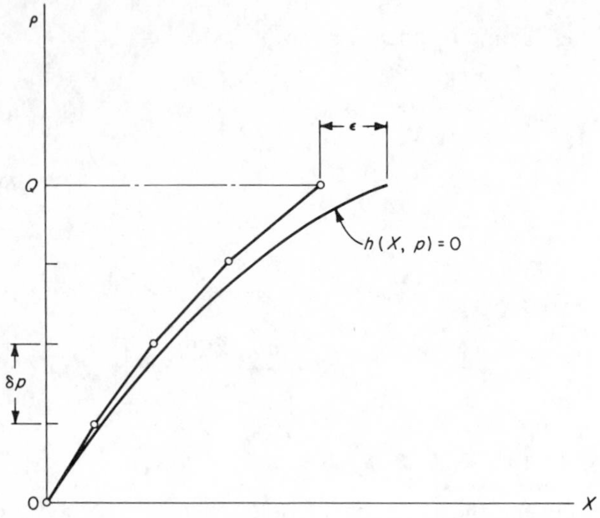

Fig. 17.5 The incremental loading process.

$\mathbf{X}_0'$ must be chosen. Again, $\mathbf{X}_0'$ may correspond to a zero value of a new parameter $p' = p'(\mathbf{X},p)$.

It is also of interest to note that all of the incremental stiffnesses discussed in Sec. 16.5 are embodied in (17.72). The instantaneous stiffness $\tilde{k}_{NM}^{ik}$ of (16.176), the initial stress matrix G_{NM}^{ik} of (16.177), and the initial rotation matrix D_{NM}^{ik} of (16.178) can be identified with the matrix $\mathbf{H}_x$ of (17.72*a*). The initial load matrix R_{NM}^{ik} of (16.179) is associated with the matrix $\mathbf{h}_p$ defined in (17.72*b*), and (16.180) can be used to generate the incremental load δp.

Modifications Several modifications and refinements of the procedure defined by (17.75) are possible. If δp_r is taken sufficiently small, it is reasonable to expect that the vectors $\mathbf{X}^r$, $r = 0, 1, \ldots$, generated by (17.75) will represent reasonable approximations to (17.68). For nonlinear systems, however, some error will inevitably enter each step of the process, as indicated in Fig. 17.5, and this error is likely to accumulate as one attempts to proceed along the path $\mathbf{X} = \mathbf{X}(p)$. One way of improving the accuracy of the incremental loading process is to use each $\mathbf{X}^r$ generated by (17.75) as an initial test point for initiating Newton-Raphson iterations. Since $\mathbf{X}^r$ is generally close to $\mathbf{X}(p_r)$, only a few cycles of iteration are usually needed to reduce the error ϵ_r to within preassigned limits. We then have for each r the process

$$\mathbf{X}^{r,m+1} = \mathbf{X}^{r,m} - \mathbf{H}_x^{-1}(\mathbf{X}^{r,m},p_r)\mathbf{h}(\mathbf{X}^{r,m},p_r) \tag{17.76}$$

with $\mathbf{X}^{1,0} = \mathbf{X}_0$, $\mathbf{X}^{r+1,0} = \mathbf{X}^{r,m_r}$, $m = 0, 1, 2, \ldots, m_r - 1$ if $r < M$; for $r = M$, $m = 0, 1, 2, \ldots$; i.e., after each increment δp_r, m_r cycles of Newton-Raphson iteration are performed, whereas after the last load increment ($r = M$), a sufficient number of iterations are used to produce a value of $\mathbf{X}$ arbitrarily close to $\mathbf{X}^* = \mathbf{X}(Q)$. Indeed, if $\mathbf{H}_x(\mathbf{X},p)$ exists, is nonsingular, and is continuous, and $\mathbf{X}(p)$ exists and is continuous for all p, it can be shown† that there exists a partition of $[0,Q]$ and integers $m_1, m_2, \ldots, m_{M-1}$ such that $\lim_{m\to\infty} \{\mathbf{X}^{M,m}\} = \mathbf{X}(Q) = \mathbf{X}^*$. With certain modifications, essentially the same process can be used to determine multiple solutions of (17.68): if $\mathbf{H}_x(\mathbf{X}^r,p_r)$ is nearly singular [that is, if $\det \mathbf{H}_x(\mathbf{X}^r,p_r) \approx 0$] for some p_r, then holding p_r constant, we introduce a point $\hat{\mathbf{X}}_0 = \mathbf{X}^r + \mathbf{A}$, where $\mathbf{A}$ is a constant vector. $\hat{\mathbf{X}}_0$ then serves as a starting point for Newton-Raphson iterations. This process, if convergent, defines a new initial point $(\mathbf{X}^*,p_r)$ which can be used to reinitiate the incremental loading process, generally with smaller increments than before and possibly with $\delta p_{r+1} < 0$. Obviously, a number of tests can be introduced to determine the sign and magnitude of δp_{r+1} and

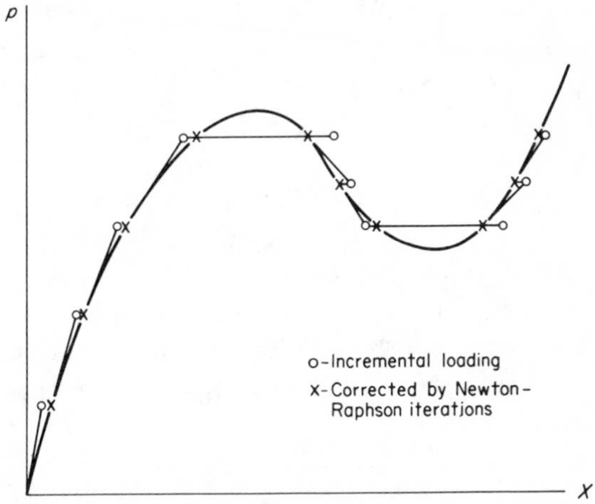

Fig. 17.6 Calculation of multiple solutions using a combination of the method of incremental loading and the Newton-Raphson method.

† For a proof of this theorem, see Ortega and Rheinboldt [1970, pp. 336, 337]. In the case of uniform increments $\delta p_r = Q/M$, Meyer [1968] developed estimates of the step size required so that the solution $\mathbf{X}^*$ would be accessible from the last iterate $\mathbf{X}^M$ via the Newton-Raphson method.

A. Figure 17.6 illustrates the continuation of the method for the determination of multivalued solutions in a one-dimensional problem.

We remark that each $\mathbf{X}^r$ generated by (17.75) can be corrected at each step by any of a number of iterative schemes other than the Newton-Raphson method. For example, if we can construct a function $\mathbf{P}(\mathbf{X},p)$ such that†

$$\mathbf{h}(\mathbf{X},p) = \mathbf{X} - \mathbf{P}(\mathbf{X},p) = \mathbf{0} \tag{17.77}$$

then each $\mathbf{X}^r$ in (17.75) can be corrected by successive approximations; i.e., for any p, $\mathbf{X}(p)$ is a fixed point of $\mathbf{P}(\mathbf{X},p)$ and

$$\mathbf{X}^{r,m+1} = \mathbf{P}(\mathbf{X}^{r,m},p_r) \tag{17.78}$$

with $\mathbf{X}^{1,0} = \mathbf{X}_0, \mathbf{X}^{r+1,0} = \mathbf{X}^{r,m_r}$, $m = 0, 1, 2, \ldots, m_r - 1$ if $r < M$, and for $r = M, m = 0, 1, 2, \ldots.$ Many other choices of corrective iterative schemes are possible.

Numerical integration Let us assume that the load parameter p is itself a known continuous function of a real variable $s \in [0,\infty)$. Then, returning to (17.73), if we divide both sides of this equation by δs and take the limit as $\delta s \to 0$, we obtain the system of nonlinear first-order differential equations

$$\dot{\mathbf{X}} = \mathbf{K}(\mathbf{X},s)\dot{p} \tag{17.79}$$

where

$$\dot{\mathbf{X}} = \frac{d\mathbf{X}}{ds} \qquad \dot{p} = \frac{dp}{ds} \tag{17.80a,b}$$

and

$$\mathbf{K}(\mathbf{X},s) = -\mathbf{H}_x^{-1}(\mathbf{X},p(s))\mathbf{h}_p(\mathbf{X},p(s)) \tag{17.80c}$$

Thus, the incremental loading process can be viewed as resulting from the numerical integration of (17.79).

Indeed, one of the simplest methods of numerical integration is Euler's method, in which $[0,S]$ is partitioned according to $0 = s_0 < s_1 < s_2 \cdots < s_M = S$ and $\dot{\mathbf{X}}$ is replaced by the forward-difference approximation,

$$\dot{\mathbf{X}}(s_r) \approx \frac{\mathbf{X}^{r+1} - \mathbf{X}^r}{\delta s_{r+1}}$$

This leads to the integration formula

$$\mathbf{X}^{r+1} = \mathbf{X}^r + \mathbf{K}(\mathbf{X}^r,s_r)\, \delta s_{r+1}\dot{p}(s_r) \tag{17.81}$$

Clearly, if $p \equiv s$ and $\dot{p} = 1$, (17.81) is identical to (17.75); that is, the incremental loading scheme (17.75) is merely Euler's method applied to the differential equation (17.79).

This observation suggests that instead of (17.75) and (17.81), the use of more sophisticated integration schemes might result in more accurate solutions of the nonlinear system (17.68). This, of course, is generally true, for Euler's method involves a local truncation error of order $h = \delta s$ whereas a number of other integration schemes can be cited which involve errors of order h^m, $m > 1$. As examples, we mention the fourth-order Runge-Kutta methods† in which $\mathbf{K}(\mathbf{X}^r,s_r)h$ in (17.81) (with $\dot{p} = 1$, $\delta s_{r+1} = h$) is replaced by

$$\Delta \mathbf{X}^r = \sum_{i=0}^{3} \mu_i \mathbf{Z}_i \tag{17.82a}$$

where

$$\mathbf{Z}_0 = \mathbf{K}(\mathbf{X}^r,s_r) \qquad \text{and} \qquad \mathbf{Z}_i = \mathbf{K}\left(\mathbf{X}^r + \sum_{j=0}^{i-1} \beta_{ij}\mathbf{Z}_j;\, s_r + \alpha_i h\right) \tag{17.82b}$$

The Runge-Kutta method with Runge's coefficients results by taking $\mu_0 = \mu_3 = \mu_1/2 = \mu_2/2 = \frac{1}{6}$, $\alpha_1 = \alpha_2 = \alpha_3/2 = \frac{1}{2}$, and $\beta_{10} = \beta_{21} = \beta_{32}/2 = \frac{1}{2}$, $\beta_{20} = \beta_{30} = 0$; Kutta's coefficients are $\mu_0 = \mu_1/3 = \mu_2/3 = \mu_3 = \frac{1}{8}$, $\alpha_1 = \alpha_2/2 = \alpha_3/3 = \frac{1}{3}$, and $\beta_{10} = -\beta_{20} = \frac{1}{3}$, $\beta_{30} = \beta_{21} = -\beta_{31} = \beta_{32} = 1$ and Gill's coefficients are $\mu_0 = \mu_3 = \frac{1}{6}$, $\mu_1 = (1 - 1/\sqrt{2})/3$, $\mu_2 = (1 + 1/\sqrt{2})/3$. $\alpha_3 = 2\alpha_2 = 2\alpha_1 = 2\beta_{10} = 1$, $\beta_{31} = -1/\sqrt{2}$, $\beta_{20} = -\beta_{10} - \beta_{31}$, $\beta_{21} = 1 + \beta_{31}$, $\beta_{32} = 1 - \beta_{31}$, and $\beta_{30} = 0$.

Alternately, we can use Euler's method or a Runge-Kutta method to determine starting points $\mathbf{X}_0, \mathbf{X}^1, \ldots, \mathbf{X}^m$ for a corrector-predictor method. Then we proceed from the observation that (17.79), and hence the method of incremental loading, can also be viewed as a problem of solving the system of nonlinear integral equations

$$\mathbf{X}(s) = \mathbf{X}_0 + \int_0^s \mathbf{K}(\mathbf{X}(\zeta),\zeta)\, d\zeta \tag{17.83a}$$

Here we have taken $\dot{p} = 1$, for simplicity. Alternately, if $\mathbf{X}^r = \mathbf{X}(s_r)$, we can write

$$\mathbf{X}^{r+1} = \mathbf{X}^{r-m} + \int_{s_{r-m}}^{s_{r+1}} \mathbf{K}(\mathbf{X}(\zeta),\zeta)\, d\zeta \tag{17.83b}$$

that is, the "history" $\mathbf{X}_0, \mathbf{X}^1, \ldots, \mathbf{X}^{r-m}$ of $\mathbf{X}$ to $\mathbf{X}^{r-m}$ is assumed to be known. The integral in (17.83b) can be replaced by quadrature formulas of the type‡

$$\int_{s_{r-m}}^{s_{r+1}} \mathbf{K}(\mathbf{X}(\zeta),\zeta)\, d\zeta \approx h \sum_{j=0}^{q} \beta_j \mathbf{K}(\mathbf{X}^{r-j+\alpha}, s_{r-j+\alpha}) \tag{17.84}$$

Here $q + 1$ stations between s_{r-m} and s_{r+1} are used, β_j are weight coefficients for the quadrature, and $\alpha = 1$ or 0. If $\alpha = 0$, (17.84) is said to be *open on the right;* if $\alpha = 1$, (17.84) is *closed on the right.* Let $r \geq \max\{m,n,q,t\}$;

† For a detailed discussion of various Runge-Kutta methods, including a historical account, see Collatz [1966]. Use of Runge-Kutta integration in incremental loading (continuation) methods was discussed by Kisner [1964] and Davidenko [1965]. See also Haisler, Stricklin, and Stebbins [1971] and Oden [1971c].

‡ We use quadrature formulas of this type in Art. 20 to solve systems of nonlinear integrodifferential equations.

then corrector-predictor methods of numerical integration are defined by the rules

$$\mathbf{X}^{r+1} = \mathbf{X}^{r-m} + h \sum_{j=1}^{q} \beta_j \mathbf{K}(\mathbf{X}^{r-j+1}, s_{r-j+1}) + h\beta_0 \mathbf{K}(\mathbf{X}_*^{r+1}, s_{r+1}) \tag{17.85a}$$

$$\mathbf{X}_*^{r+1} = \mathbf{X}^{r-n} + h \sum_{j=0}^{t} \mu_j \mathbf{K}(\mathbf{X}^{r-j}, s_{r-j}) \tag{17.85b}$$

That is, a $t + 1$ point open on the right quadrature formula is used for the *predictor* (17.85*b*) and a $q + 1$ point, closed on the right quadrature formula, is used for the *corrector* (17.85*a*). For example, if $t = n = 3$, $q = m = 0$, $24\mu = (55, -59, 37, -9)$, and $24\beta = (9, 19, -5, 1)$, then (17.85) corresponds to the Adams-Moulton method.†

It appears that the use of appropriate numerical integration schemes to carry out the incremental loading process, together with the use of the Newton-Raphson method to correct the results at the end of various steps, constitutes one of the most powerful methods for solving large systems of nonlinear equations.

17.6 SEARCH METHODS

We now return to the viewpoint of solving nonlinear equations by minimizing a functional $F(\mathbf{X})$ of the type in (17.5). We consider here representative examples of sequential and nonsequential search methods which may be used to advantage in cases in which it is inconvenient (or impossible) to compute gradients of $F(\mathbf{X})$.

Simplex search method The simplex search method utilizes the idea of a simplex.‡ A simplex in k-dimensional space is a set of $k + 1$ points (sometimes called vertices) $p_0, p_1, \ldots, p_k$ which do not lie in the $(k - 1)$-dimensional hyperplane. The simplex $\mathscr{S}$ consists of all points y,

$$y = \sum_{i=0}^{k} c_i p_i \tag{17.86a}$$

where

$$c_i \geq 0 \qquad \sum_{i=0}^{k} c_i = 1 \tag{17.86b,c}$$

When we consider the problem of finding an n-dimensional vector $\mathbf{X}^*$ which minimizes the function $F(\mathbf{X})$, the vector $\mathbf{X}$ can be considered as a point in n-dimensional space. In the simplex search method, $n + 1$ trial points $\mathbf{X}^0, \mathbf{X}^1, \ldots, \mathbf{X}^n$ are selected as vertices of a simplex in n-dimensional space. We then compute

$$y^0 = F(\mathbf{X}^0),\ y^1 = F(\mathbf{X}^1),\ \ldots,\ y^n = F(\mathbf{X}^n) \tag{17.87}$$

† Cf. Isaacson and Keller [1966, p. 388].

‡ We recall that in one-dimensional space, a simplex is a line connecting two points $[p_0, p_1]$. For $k = 2$, a simplex consists of the points within and on a triangle. For $k = 3$, a simplex consists of the points within and on a tetrahedron, etc. See Nelder and Meed [1964].

Of these, let $y^h = F(\mathbf{X}^h)$ and $y^l = F(\mathbf{X}^l)$ denote the maximum and minimum, respectively. Excluding the point $\mathbf{X}^h$ which corresponds to the highest value of y (that is, y^h), we next compute an average point $\bar{\mathbf{X}}$ as

$$\bar{\mathbf{X}} = \frac{1}{n} \sum_{\substack{i=0 \\ i \neq h}}^{n} \mathbf{X}^i \tag{17.88}$$

A line in n-dimensional space is introduced which connects $\mathbf{X}^h$ and $\bar{\mathbf{X}}$. If $\hat{\mathbf{X}}$ is an arbitrary point on this line, then

$$\hat{\mathbf{X}} = (1 + \alpha)\bar{\mathbf{X}} - \alpha \mathbf{X}^h \tag{17.89}$$

where α is a positive number called the *reflection coefficient*. Ordinarily, α is specified at the onset and $\hat{\mathbf{X}}$ is on the opposite side of $\bar{\mathbf{X}}$ from $\mathbf{X}^h$ (that is, it is closer to the minimum). We now compute

$$\hat{y} = F(\hat{\mathbf{X}}) \tag{17.90}$$

With $\hat{y}$ now known, several possibilities arise. For example:

1. $\hat{y} < y^l$—then we have determined a new minimum, and it is advisable to move further in the same direction in which we encountered $\hat{\mathbf{X}}$. We then obtain a new point

$$\hat{\hat{\mathbf{X}}} = (1 + \gamma)\hat{\mathbf{X}} - \gamma \bar{\mathbf{X}} \tag{17.91}$$

where $\gamma > \alpha > 0$. Then if $\hat{\hat{y}} = F(\hat{\hat{\mathbf{X}}}) < y^l$, we can develop a new (refined) simplex by setting $\hat{\hat{y}} = y^{h'}$, a new high vertex of a simplex with all other points unaltered, and the process is repeated.

2. $\hat{\hat{y}} > y$ (or $\hat{y} > y^l$)—the process has failed. In this case, the point $\hat{\hat{\mathbf{X}}}$ (or $\hat{\mathbf{X}}$) replaces $\mathbf{X}^h$ as the high point and the process is repeated.

3. $\hat{y} > y^i$ and $i \neq h$—then (17.91) has progressed in the wrong direction, and $\mathbf{X}^h$ must be redefined as either the original $\mathbf{X}^h$ or as $\hat{\mathbf{X}}$, whichever corresponds to the lowest y value. Then we compute a new point

$$\hat{\hat{\mathbf{X}}} = \beta \mathbf{X}^h + (1 - \beta)\bar{\mathbf{X}} \tag{17.92}$$

where $0 < \beta < 1$ is a contraction coefficient. If $\hat{y} = F(\hat{\hat{\mathbf{X}}}) \leq y^h$, then y^h is replaced by $\hat{\hat{y}}$ and the process is repeated. If $\hat{\hat{y}} > y^h$, the contraction has failed; in this case, the entire procedure can be renewed with new starting values $\tilde{\mathbf{X}}^i$, which, for example, may be taken as

$$\tilde{\mathbf{X}}^i = \tfrac{1}{2}(\mathbf{X}^i + \mathbf{X}^l) \tag{17.93}$$

The iteration continues until the minimum is reached.

Although the simplex method involves numerous tests and iteration cycles, it has proved to be effective in the solution of certain nonlinear systems in which other methods have failed. The method is most effective

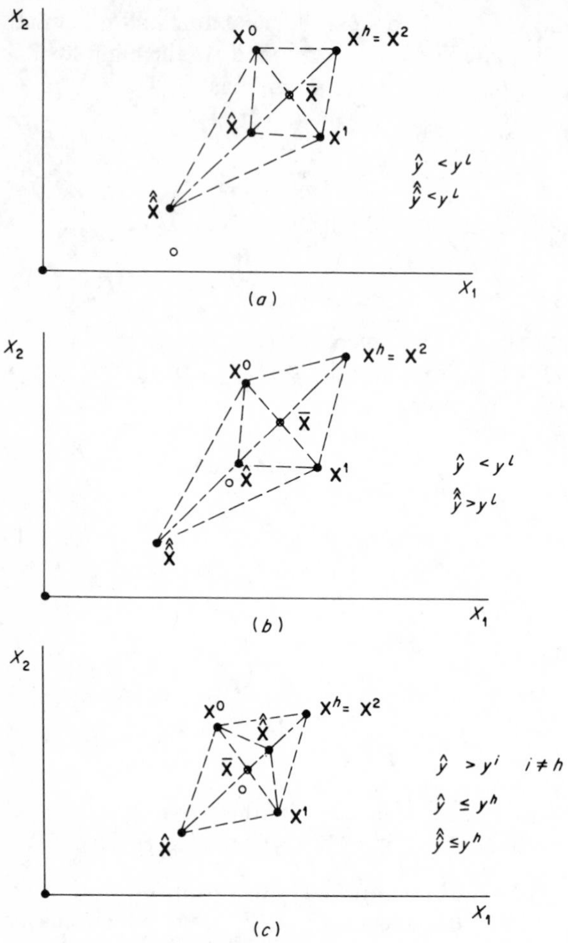

Fig. 17.7 The simplex search method.

in solving large systems of linear inequalities, however, and for this reason, it has found wider application in the field of nonlinear programming than in problems involving systems of nonlinear equations. Many modifications and refinements of the simplex search method are possible. To mention one, provisions can be made to ensure that the simplex method will not seek an incorrect minimum point, by defining a unit cube in n space which contains the desired minimum. The function is redefined to some arbitrarily high value on the outside of the cube, and any attempt to wander outside this region will cause the process to return to the interior of the cube so as to converge to the desired minimum.

Figure 17.7*a* to *e* illustrates several possible moves for a simplex search in two-dimensional space. We begin, as is indicated in Fig. 17.7*a*, by

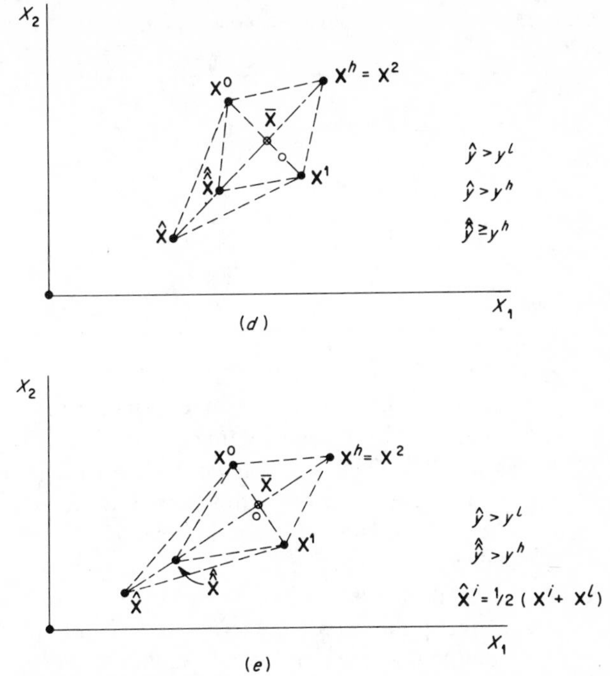

Fig. 17.7 *(Continued).*

selecting three starting test points $\mathbf{X}^0$, $\mathbf{X}^1$, $\mathbf{X}^2$ which define a simplex. Assuming $\mathbf{X}^2 = \mathbf{X}^h$ corresponds to the largest y value $[y^2 = f(\mathbf{X}^2) > y^0, y^2 > y^1]$, the average $\bar{\mathbf{X}}$ of $\mathbf{X}^0$ and $\mathbf{X}^1$ is computed and the line given by (17.89) is constructed to a new point $\hat{\mathbf{X}}$. In Fig. 17.7*a*, it is assumed that $\hat{y} > y$, so that a new point $\hat{\hat{\mathbf{X}}}$ is computed. The expansions are thus successful in this case. In Fig. 17.7*b*, $\hat{y} < y^l$, but the extension to $\hat{\hat{\mathbf{X}}}$ fails because $\hat{\hat{y}} < y^l$. Hence, in this case, $\hat{\mathbf{X}}$ is taken as the new low value. Figure 17.7*c* depicts the case in which a contraction is necessary since $\hat{y} > y^l$. A new $\mathbf{X}^h$ is set equal to $\hat{\mathbf{X}}$ because $\hat{y} < y^h$; then $\mathbf{X}^h$ is again redefined as $\hat{\hat{\mathbf{X}}}$ since it is found that $\hat{\hat{y}} \leq y^h$. The case illustrated in Fig. 17.7*d* is similar to that in 17.7*c* except that $\mathbf{X}^h$ is kept at its original value instead of $\hat{\mathbf{X}}$ because $\hat{y} > y^h$. Finally, in Fig. 17.7*e*, the contraction fails, and it is necessary to compute new starting values. Here, the original values $\mathbf{X}^0$, $\mathbf{X}^1$, $\mathbf{X}^2 = \mathbf{X}^i$ are replaced by $\bar{\mathbf{X}}^i = \frac{1}{2}(\mathbf{X}^i + \mathbf{X}^l)$, where $\mathbf{X}^l$ corresponds to the lowest initial value $y^l = F(\mathbf{X}^l)$.

Nonsequential (random) search methods Random search methods do not determine the test points exactly and, therefore, are categorized as nonsequential methods. In a random procedure, the test points are selected

Table 17.1 Effect of a and S on number of test points p†

a	S			
	0.8	0.9	0.95	0.99
0.1	16	22	29	44
0.05	32	45	59	90
0.025	64	91	119	182
0.01	161	230	299	459
0.005	322	460	598	919

† After Brooks [1958*a*].

at random according to an n-dimensional probability-density function. Usually, the shape of the density function is chosen as flat, which means that all test points in a given region have the same likelihood of being chosen.

A purely random procedure simply selects test points at random in the region where the minimum exists according to a fixed distribution. After testing a specified number of points, the location of the minimum is assumed to be where the smallest value of the function was obtained. The required number of test points can be determined in a probabilistic sense.

Assuming that the minimum is equally likely to be anywhere within an n-dimensional hypercube with all sides equal to d and that the uncertainty of each independent variable is to be reduced to δ_N, the ratio of the smaller hypercube to the original hypercube is

$$a = \left(\frac{\delta_N}{d}\right)^n \tag{17.94}$$

The probability that a test point is not in the smaller hypercube is $1 - a$. It follows from elementary statistics that the probability that p trials will fail is $(1 - a)^p$. The probability of at least one point being in the smaller hypercube is

$$S = 1 - (1 - a)^p \tag{17.95}$$

Therefore,

$$p = \frac{\log (1 - S)}{\log (1 - a)} \tag{17.96}$$

The term S is the confidence level, and p is the required number of points. Table 17.1 gives some typical values. The required number of points increases rapidly with a reduction in a.

It has been postulated† that the number of test points is independent of the number of independent variables. This means that a must be a constant.

† Brooks [1958*a*].

Hence, δ_N/d must increase as n increases. Thus, the accuracy per independent variable is decreased since the smaller hypercube is expanding. Combining (17.94) and (17.96), we get

$$p = \frac{\log (1 - S)}{\log [1 - (\delta_N/d)^n]} \tag{17.97}$$

Assuming that $(\delta_N/d)^n$ will be very small, the denominator can be approximated by the first term of a series expansion

$$\log \left[1 - \left(\frac{\delta_N}{d}\right)^n\right] = -\left(\frac{\delta_N}{d}\right)^n + \frac{1}{2}\left(\frac{\delta_N}{d}\right)^{2n} - \frac{1}{3}\left(\frac{\delta_N}{d}\right)^{3n} + \cdots$$

or

$$\log \left[1 - \left(\frac{\delta_N}{d}\right)^n\right] \approx -\left(\frac{\delta_N}{d}\right)^n$$

It follows that

$$p \approx -\left(\frac{d}{\delta_N}\right)^n \log (1 - S) \tag{17.98}$$

The number of required points increases approximately exponentially with the number of dimensions. For a confidence level S of 0.9,

$$p \approx 2.3\left(\frac{d}{\delta_N}\right)^n \tag{17.99}$$

Spang [1962] compares this number with the number required in a simple grid test. The test points are equally spaced a distance δ_N apart in a grid test. All the points are tested, and the lowest value is considered the minimum. In this case, the required number of points is

$$p = \left(\frac{d}{\delta_N}\right)^n \tag{17.100}$$

This is approximately one-half of the points required for a purely random search. This grid search is not practical for large n since p also increases exponentially.†

† A quasi-sequential method has been developed by Spang [1962] in the hope of reducing the number of necessary points. This method assumes that the minimum will be in the neighborhood of the group of lowest test points. After a specified number of tests have been made, a reduced hypercube is selected around this group. The process is continued in this fashion. Although the number of test points picked at random per iteration does not have to be as large as in the purely random procedure, it is no longer possible to estimate the confidence level since the probabilities depend on the function being tested.

18 SELECTED APPLICATIONS

In this section we present applications of the theory developed thus far to selected problems in finite elasticity. All examples considered involve isotropic bodies.

18.1 PLANE STRESS

The theory of plane stress in finite elasticity deals with thin bodies in which the stress components in one direction are prescribed and the state of stress at any particle is essentially two-dimensional.

Elastic membranes Consider a thin sheet of homogeneous elastic material which is bounded by the surfaces $x^3 = \pm d_0/2$, the initial thickness d_0 being, in general, a function of x^1 and x^2. In the initially flat, undeformed sheet, an arbitrary particle P is located by the position vector

$$\mathbf{r} = \hat{\mathbf{r}}(x^1,x^2) + x^3\hat{\mathbf{n}} \tag{18.1}$$

where $\hat{\mathbf{r}}(x^1,x^2)$ is a position vector of a particle Q in the x^1, x^2 plane and $\hat{\mathbf{n}}$ is a unit vector normal to the undeformed middle surface.

The sheet is now assumed to undergo a general motion which carries it from its reference configuration C_0 to another configuration C. The location of the particle P relative to a fixed reference frame in C_0 is given by the position vector

$$\mathbf{R} = \hat{\mathbf{R}}(x^1,x^2) + \mathbf{M}(x^1,x^2,x^3) \tag{18.2}$$

wherein $\hat{\mathbf{R}}(x^1,x^2)$ is the position vector of a particle Q on the deformed middle surface and $\mathbf{M}(x^1,x^2,x^3)$ is a vector joining particles P and Q in the deformed body. The displacement vector $\mathbf{w}$ of an arbitrary particle P is defined by

$$\mathbf{w} = \mathbf{R} - \mathbf{r} = \mathbf{u} + \mathbf{M} - x^3\hat{\mathbf{n}} \tag{18.3}$$

where now $\mathbf{u}$ denotes the displacements of particles on the middle surface:

$$\mathbf{u} = \mathbf{u}(x^1,x^2) = \hat{\mathbf{R}}(x^1,x^2) - \hat{\mathbf{r}}(x^1,x^2) \tag{18.4}$$

The geometry of this motion is illustrated in Fig. 18.1.

We can now proceed to compute the basis vectors $\mathbf{G}_i$ tangent to the deformed coordinate lines x^i:

$$\mathbf{G}_\alpha = \mathbf{A}_\alpha + \mathbf{M}_{,\alpha} \qquad \mathbf{G}_3 = \mathbf{M}_{,3} \qquad \alpha = 1, 2 \tag{18.5}$$

Here $\mathbf{A}_\alpha$ are the vectors tangent to the surface coordinates x^1, x^2 of the deformed middle surface. Here and henceforth we use both Greek and Latin indices, with the understanding that Greek indices range from 1 to 2 and Latin indices range from 1 to 3.

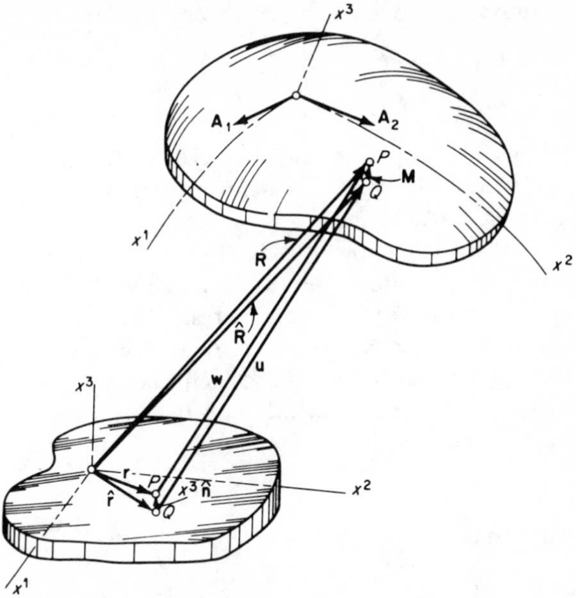

Fig. 18.1 Geometry of deformation of a thin sheet.

To obtain a theory of plane stress, we now assume that a material line normal to the undeformed middle surface of the sheet remains straight and normal to the deformed middle surface. Such material lines, however, may undergo finite extensional strains during the motion. Thus, if λ is the extension ratio of x^3 at $x^3 = 0$ (i.e., the ratio of deformed to undeformed length), we choose to write the vector $\mathbf{M}$ of (18.2) in the form

$$\mathbf{M}(x^1,x^2,x^3) = \lambda x^3 \mathbf{n} \tag{18.6}$$

where $\mathbf{n}$ is a unit vector normal to the deformed middle surface. Both λ and $\mathbf{n}$ are functions of x^1 and x^2. With this definition of $\mathbf{M}$, (18.5) becomes

$$\mathbf{G}_\alpha = \mathbf{A}_\alpha + x^3(\lambda\mathbf{n})_{,\alpha} \qquad \mathbf{G}_3 = \lambda\mathbf{n} \tag{18.7}$$

Noting that $\mathbf{n} \cdot \mathbf{A}_\alpha = 0$, we obtain for the components G_{ij} of the deformation tensor,

$$\begin{aligned} G_{\alpha\beta} &= \mathbf{G}_\alpha \cdot \mathbf{G}_\beta = A_{\alpha\beta} - 2\lambda x^3 B_{\alpha\beta} + \lambda^2 (x^3)^2 B_\alpha^\lambda B_{\beta\lambda} + (x^3)^2 \lambda_{,\alpha}\lambda_{,\beta} \\ G_{\alpha 3} &= \mathbf{G}_\alpha \cdot \mathbf{G}_3 = \lambda x^3 \lambda_{,\alpha} \\ G_{33} &= \mathbf{G}_3 \cdot \mathbf{G}_3 = \lambda^2 \qquad \alpha, \beta, \lambda = 1, 2 \end{aligned} \tag{18.8}$$

Here $A_{\alpha\beta}$ and $B_{\alpha\beta}$ are the covariant components of the first and second fundamental tensors, respectively, of the deformed middle surface and B_α^λ are the mixed components of the second fundamental tensor:

$$A_{\alpha\beta} = \mathbf{A}_\alpha \cdot \mathbf{A}_\beta \qquad B_{\alpha\beta} = -\mathbf{A}_\alpha \cdot \mathbf{n}_{,\beta}$$
$$B_\alpha^\lambda = A^{\lambda\beta} B_{\beta\alpha} \tag{18.9}$$

In this last equation, $A^{\lambda\beta}$ are the contravariant components of the first fundamental tensor of the deformed middle surface (that is, $A^{\alpha\lambda}A_{\lambda\beta} = \delta^\alpha_\beta$ or $A^{\alpha\beta} = \mathbf{A}^\alpha \cdot \mathbf{A}^\beta$, where $\mathbf{A}^\alpha \cdot \mathbf{A}_\beta = \delta^\alpha_\beta$).

In principle, the introduction of (18.8) into (4.16) leads to a form of the strain tensor appropriate for a Kirchhoff theory of finite deformations of plates. However, we are interested here in a reduced version of (18.8) which corresponds to a theory of very thin membranes. Toward this end, we restrict our attention to deformations symmetric about the middle surface of bodies sufficiently thin that G_{ij} is essentially uniform throughout the thickness. Then $\lambda = d/d_0$, and instead of (18.8), we have

$$G_{\alpha\beta} = A_{\alpha\beta} \qquad G_{\alpha 3} = 0 \qquad G_{33} = \lambda^2 \tag{18.10}$$

The strain components are then

$$\gamma_{\alpha\beta} = \tfrac{1}{2}(A_{\alpha\beta} - \delta_{\alpha\beta}) \qquad \gamma_{\alpha 3} = 0 \qquad \gamma_{33} = \tfrac{1}{2}(\lambda^2 - 1) \tag{18.11}$$

The principal invariants of the deformation tensor are given by

$$I_1 = \lambda^2 + \delta^{\alpha\beta} A_{\alpha\beta}$$
$$I_2 = \delta_{\alpha\beta}\lambda^2 \epsilon^{\alpha\lambda}\epsilon^{\beta\mu} A_{\mu\lambda} + A \tag{18.12}$$
$$I_3 = \lambda^2 A$$

where $\epsilon^{\alpha\lambda}$, $\epsilon^{\beta\mu}$ are the two-dimensional permutation symbols ($\epsilon^{12} = -\epsilon^{21} = 1$, $\epsilon^{11} = \epsilon^{22} = 0$) and

$$A = \det(A_{\alpha\beta}) = \tfrac{1}{2}\epsilon^{\alpha\beta}\epsilon^{\lambda\mu} A_{\alpha\lambda} A_{\beta\mu} \tag{18.13}$$

Note also that since

$$\mathbf{A}_\alpha = \hat{\mathbf{R}}_{,\alpha}(x^1, x^2) = \mathbf{i}_\alpha + \mathbf{u}_{,\alpha} \tag{18.14}$$

where $\mathbf{u}$ is the displacement vector of particles on the middle surface,

$$\gamma_{\alpha\beta} = \tfrac{1}{2}(u_{\alpha,\beta} + u_{\beta,\alpha} + u_{k,\alpha}u_{k,\beta})$$
$$\gamma_{33} = \tfrac{1}{2}(\lambda^2 - 1) \qquad \gamma_{\alpha 3} = 0 \tag{18.15}$$

where $\alpha, \beta = 1, 2$ and $k = 1, 2, 3$.

Suppose that the sheet is composed of a homogeneous isotropic hyperelastic material with a strain energy per unit undeformed volume W. Then locally

$$\dot{W}(I_1, I_2, I_3) = t^{ij}\dot{\gamma}_{ij} = t^{\alpha\beta}\dot{\gamma}_{\alpha\beta} + t^{33}\dot{\gamma}_{33} \tag{18.16}$$

and the stress tensor t^{ij} is determined by

$$t^{\alpha\beta} = \frac{\partial W}{\partial \gamma_{\alpha\beta}} \qquad t^{\alpha 3} = 0 \qquad t^{33} = \frac{\partial W}{\partial \gamma_{33}} \tag{18.17}$$

In view of (18.11) and (18.12), we have

$$t^{\alpha\beta} = 2\left(\frac{\partial W}{\partial I_1} + \lambda^2\right)\frac{\partial W}{\partial I_2}\delta^{\alpha\beta} + 2\left(\frac{\partial W}{\partial I_2} + \lambda^2\frac{\partial W}{\partial I_3}\right)AA^{\alpha\beta} \tag{18.18a}$$

$$t^{\alpha 3} = 0 \tag{18.18b}$$

$$t^{33} = 2\frac{\partial W}{\partial I_1} + 2\frac{\partial W}{\partial I_2}\delta_{\alpha\beta}AA^{\alpha\beta} + 2\frac{\partial W}{\partial I_3}A \tag{18.18c}$$

where

$$AA^{\alpha\beta} = \epsilon^{\alpha\lambda}\epsilon^{\beta\mu}A_{\lambda\mu} \tag{18.19}$$

The normal stress component t^{33} can be determined from boundary conditions on the deformed middle surface. For example, let q_1 and q_2 denote applied pressures per unit area of the deformed middle surface which act at $x^3 = -d/2$ and $x^3 = +d/2$, respectively. Then, approximately, $\lambda t^{33} = -q_1\sqrt{A}$ at $-d/2$ and $\lambda t^{33} = -q_2\sqrt{A}$ at $+d/2$. For thin membranes, we can also use the approximation $\lambda t^{33} = -q\sqrt{A}$, where $q = (q_1 + q_2)/2$ is the average applied pressure. However, since for very thin membranes the magnitudes of the membrane stresses $t^{\alpha\beta}$ are generally many times greater than t^{33}, it is customary to simply set $t^{33} = 0$ and to reduce the problem to one of determining only a two-dimensional state of stress. We shall adopt the approximation $t^{33} \approx 0$ in subsequent developments.

In the case of incompressible materials, $I_3 = 1$ and W is a function of only I_1 and I_2. Then, instead of (18.12), we write

$$\begin{aligned} I_1 &= \lambda^2 + \delta^{\alpha\beta}A_{\alpha\beta} \\ I_2 &= \frac{1}{\lambda^2} + \lambda^2\delta_{\alpha\beta}\epsilon^{\alpha\lambda}\epsilon^{\beta\mu}A_{\mu\lambda} \\ \lambda^2 &= (\tfrac{1}{2}\epsilon^{\alpha\lambda}\epsilon^{\beta\mu}A_{\alpha\beta}A_{\lambda\mu})^{-1} \end{aligned} \tag{18.20}$$

The constitutive equations for incompressible membranes are

$$t^{\alpha\beta} = 2\left(\frac{\partial \hat{W}}{\partial I_1} + \lambda^2\frac{\partial \hat{W}}{\partial I_2}\right)\delta^{\alpha\beta} + 2\left(h + \frac{1}{\lambda^2}\frac{\partial \hat{W}}{\partial I_2}\right)A^{\alpha\beta} \tag{18.21a}$$

$$t^{33} = 2\frac{\partial \hat{W}}{\partial I_1} + \frac{2}{\lambda^2}\frac{\partial \hat{W}}{\partial I_2}\delta_{\alpha\beta}A^{\alpha\beta} + \frac{2}{\lambda^2}h \tag{18.21b}$$

where $W = \hat{W}(I_1, I_2)$ and h is the hydrostatic pressure. The hydrostatic pressure h can be determined immediately from the condition $t^{33} \approx 0$:

$$h = -\lambda^2 \frac{\partial \hat{W}}{\partial I_1} - \frac{\partial \hat{W}}{\partial I_2} \delta_{\alpha\beta} A^{\alpha\beta} \tag{18.22}$$

Thus (18.21*a*) can be rewritten

$$t^{\alpha\beta} = 2(\delta^{\alpha\beta} - \lambda^2 A^{\alpha\beta}) \frac{\partial \hat{W}}{\partial I_1} + 2\left[\lambda^2 \delta^{\alpha\beta} + A^{\alpha\beta}\left(\frac{1}{\lambda^2} - \delta_{\lambda\mu} A^{\lambda\mu}\right)\right] \frac{\partial \hat{W}}{\partial I_2} \tag{18.23}$$

and $t^{\alpha 3} = t^{33} = 0$.

Membrane elements With the kinematics now defined in terms of the displacements of the middle surface, we can proceed to develop equilibrium equations for finite elements of elastic membranes. Isolating a typical element e with N_e nodes, we represent the local displacement field, as usual, by

$$u_i = \psi_N(x^1, x^2) u_i^N \tag{18.24}$$

where u_i are the components of displacement of a particle x^1, x^2 on the middle surface and $N = 1, 2, \ldots, N_e$; $i = 1, 2, 3$. Then, according to (18.15),

$$\begin{aligned} 2\gamma_{\alpha\beta} &= \psi_{N,\alpha} u_\beta^N + \psi_{N,\beta} u_\alpha^N + \psi_{N,\alpha} \psi_{M,\beta} u_k^N u_k^M \\ 2\gamma_{33} &= \lambda^2 - 1 \qquad \gamma_{\alpha 3} = 0 \end{aligned} \tag{18.25}$$

For compressible materials, λ^2 can also be approximated over the element by

$$\lambda^2 = \psi_N(x_1, x_2) \mu^N \tag{18.26}$$

where $\mu^N \equiv (\lambda^2)^N$ is the value of λ^2 at node N. For incompressible materials, λ^2 is obtained in terms of the membrane strains $\gamma_{\alpha\beta}$ by introducing (18.11) into (18.20):

$$\lambda^2 = (1 + 2\gamma_\alpha^\alpha + 2\epsilon^{\alpha\lambda}\epsilon^{\beta\mu}\gamma_{\alpha\beta}\gamma_{\lambda\mu})^{-1} \tag{18.27}$$

Since $t^{33} \approx 0$, the equations of equilibrium of a thin finite membrane element are

$$p_{Ni} = \int_{v_{0(e)}} \frac{\partial W}{\partial u_i^N} dv_0 = \int_{v_{0(e)}} \frac{\partial W}{\partial \gamma_{\alpha\beta}} \frac{\partial \gamma_{\alpha\beta}}{\partial u_i^N} dv_0 \tag{18.28}$$

Assuming that the initial thickness d_0 is constant over the undeformed element and denoting the undeformed middle-surface area by A_0^*, (18.28) becomes

$$p_{Ni} = d_0 \int_{A^*_{0(e)}} t^{\alpha\beta} \psi_{N,\alpha}(\delta_{\beta i} + \psi_{M,\beta} u_i^M)\, dA_0^* \tag{18.29}$$

The generalized nodal forces p_{Ni} are given by (16.93), which, for transversely applied loads, is more conveniently written

$$p_{N\alpha} = \int_{A^*_{0(e)}} \hat{F}^*_\alpha \psi_N(x^1,x^2)\, dA^*_0 + \int_{A^*_{0(e)}} S^{*j}(\delta_{\alpha j} + \psi_{M,\alpha} u_j^M)\psi_N(x^1,x^2)\, dA^*_0 \tag{18.30a}$$

$$p_{N3} = \int_{A^*_{0(e)}} \hat{F}^*_3 \psi_N\, dA^*_0 + \int_{A^*_{0(e)}} S^{*3}\psi_N\, dA^*_0 \tag{18.30b}$$

where $\hat{F}^*_i(x^1,x^2)$ are the components of body forces per unit undeformed middle-surface area and S^{*j} are the surface tractions per unit undeformed middle-surface area referred to the convected lines x^i [see (16.96)]. For applied tangential forces on the curve Γ describing the boundary of the element on its middle surface, we use a two-dimensional version of (16.98). However, in all subsequent applications, we prefer to use the approximate generalized forces of the type described by (16.113), (16.117), and (16.118).

In the case of compressible materials, specific forms of the stiffness relations for elements of isotropic membranes are obtained by introducing (18.18*a*) and (18.26) into (18.29). We find

$$2d_0 \int_{A^*_{0(e)}} \left[\left(\frac{\partial W}{\partial I_1} + \psi_R \mu^R \frac{\partial W}{\partial I_2}\right)\delta^{\alpha\beta} + \left(\frac{\partial W}{\partial I_2} + \psi_R \mu^R \frac{\partial W}{\partial I_3}\right) f^{\alpha\beta}\right] \psi_{N,\alpha} \times (\delta_{\beta i} + \psi_{M,\beta} u_i^M)\, dA^*_0 = p_{Ni} \tag{18.31}$$

where

$$\begin{aligned} f^{\alpha\beta} &= AA^{\alpha\beta} \\ &= \delta^{\alpha\beta} + 2\epsilon^{\alpha\lambda}\epsilon^{\beta\mu}\gamma_{\lambda\mu} \\ &= \delta^{\alpha\beta} + \epsilon^{\alpha\lambda}\epsilon^{\beta\mu}(\psi_{N,\mu} u_\lambda^N + \psi_{N,\lambda} u_\mu^N + \psi_{N,\lambda}\psi_{M,\mu} u_k^N u_k^M) \end{aligned} \tag{18.32}$$

The N_e values $\mu^R = (\lambda^2)^R$ are determined from the condition that t^{33} must vanish in an average sense over the element. Then, in view of (18.18*c*), we have at each node N,

$$\int_{A^*_0(e)} \left(\frac{\partial W}{\partial I_1} + \frac{\partial W}{\partial I_2}\delta_{\alpha\beta} f^{\alpha\beta} + \frac{\partial W}{\partial I_3} A\right)\psi_N\, dA^*_0 = 0 \tag{18.33}$$

The $3N_e$ equations (18.31) together with the N_e equations (18.33) constitute $4N_e$ equations in the $4N_e$ unknowns u_i^N and μ^N. If λ^2 is assumed to be uniform over each element, say $\lambda^2 = \mu_0$, we simply replace $\psi_R \mu^R$ in (18.31) by μ_0 and add to the $3N_e$ equilibrium equations the single condition $\int_{A^*_0} t^{33}\, dA^*_0 = 0$, which follows from (18.33) by replacing ψ_N by unity.

In the case of isotropic incompressible materials, we have, instead of (18.31),

$$2d_0 \int_{A_0^*} \left\{ (\delta^{\alpha\beta} - \lambda^4 f^{\alpha\beta}) \frac{\partial \hat{W}}{\partial I_1} + [\lambda^2 \delta^{\alpha\beta} + f^{\alpha\beta}(1 - 2\lambda^4 - 2\lambda^4 \gamma^\mu_\mu)] \frac{\partial \hat{W}}{\partial I_2} \right\} \psi_{N,\alpha} \times (\delta_{\beta i} + \psi_{M,\beta} u_i^M)\, dA_0^* = p_{Ni} \quad (18.34)$$

in which $f^{\alpha\beta}$ is as defined in (18.32) and, according to (18.27),

$$\lambda^2 = [1 + 2\psi_{N,\alpha}(\delta_{\alpha k} + \tfrac{1}{2}\psi_{M,\alpha} u_k^M) u_k^N + \tfrac{1}{2}\epsilon^{\alpha\lambda}\epsilon^{\beta\mu}(\psi_{N,\alpha} u_\beta^N + \psi_{N,\beta} u_\alpha^N + \psi_{N,\alpha}\psi_{M,\beta} u_k^N u_k^M)(\psi_{R,\lambda} u_\mu^R + \psi_{R,\mu} u_\lambda^R + \psi_{R,\lambda}\psi_{P,\mu} u_j^R u_j^P)]^{-1} \quad (18.35)$$

Here $i, j, k = 1, 2, 3$; $\alpha, \beta, \lambda, \mu = 1, 2$,; $N, M, P, R = 1, 2, \ldots, N_e$.

We shall now consider special forms of (18.34) and discuss numerical results obtained by applying these to specific problems.

Stretching of an elastic sheet We first consider the relatively simple problem of prescribed stretching of a thin elastic sheet. A thin rectangular body is clamped along opposite edges and stretched in its plane as shown in Fig. 18.2. The nodal forces along the edges $x^2 = \pm b/2$ are zero, and the nodal displacement components u_1^N of nodes on the boundary $x^1 = \pm a/2$ are prescribed as $\pm a(\epsilon - 1)/2$, where ϵ is the extension ratio of the x^1-coordinate line. This problem corresponds to the so-called biaxial strip test commonly used to characterize and to determine ultimate properties of materials such as synthetic rubbers, polymers, solid propellant fuels, and binders. Despite its simple appearance,

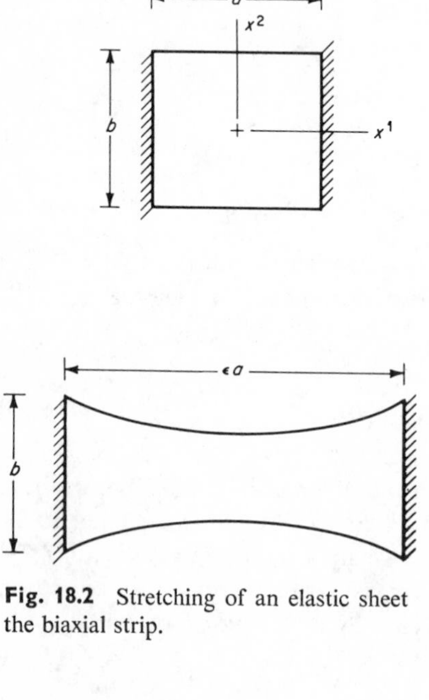

Fig. 18.2 Stretching of an elastic sheet the biaxial strip.

no exact solutions to the biaxial strip problem for finite elastic deformations appear to be available. Computationally, however, the finite-element analysis of the problem is particularly straightforward owing to the fact that nonzero displacements rather than forces are prescribed at the boundaries. Thus, the problem of handling changes in geometry of loading surfaces does not exist.

We first consider a simplex approximation of the local displacement fields corresponding to a network of thin triangular elements. Then

$$\psi_N(x^1,x^2) = a_N + b_{N\alpha}x^\alpha \tag{18.36}$$

and

$$\begin{aligned} u_\alpha &= \psi_N u_\alpha^N = a_N u_\alpha^N + b_{N\beta} x^\beta u_\alpha^N \\ u_3 &= u_3^N = 0 \end{aligned} \tag{18.37}$$

Here $N = 1, 2, 3$; $\alpha, \beta = 1, 2$; and the arrays a_N, $b_{N\alpha}$ are as defined in (10.106).

Assuming that the strip is incompressible, we obtain the governing equilibrium equations for a typical element by introducing (18.37) into (18.34):

$$\begin{aligned} 2v_{0(e)}\Bigg(&(\delta^{\alpha\beta} - \lambda^4 f^{\alpha\beta}) \frac{\partial \hat{W}}{\partial I_1} + \{\lambda^2\delta^{\alpha\beta} + f^{\alpha\beta}[1 - 2\lambda^4 - 2\lambda^4 b_{R\mu}(\delta_{\mu\lambda} \\ &+ \tfrac{1}{2} b_{M\mu} u_\lambda^M)]u_\lambda^R\} \frac{\partial \hat{W}}{\partial I_2}\Bigg) b_{N\alpha}(\delta_{\beta\rho} + b_{P\beta}u_\rho^P) = p_{N\rho} \end{aligned} \tag{18.38a}$$

$$\begin{aligned} \lambda^2 = [1 &+ 2b_{N\alpha}(\delta_{\alpha\beta} + b_{M\alpha}u_\beta^M)u_\beta^N + \tfrac{1}{2}\epsilon^{\alpha\lambda}\epsilon^{\beta\mu}(b_{N\alpha}u_\beta^N + b_{N\beta}u_\alpha^N \\ &+ b_{N\alpha}b_{M\beta}u_\lambda^N u_\lambda^M)(b_{R\lambda}u_\mu^R + b_{R\mu}u_\lambda^R + b_{R\lambda}b_{P\mu}u_\rho^R u_\rho^P)]^{-1} \end{aligned} \tag{18.38b}$$

where $v_{0(e)} = d_0 A_{0(e)}^*$ is the initial volume of the element and $\alpha, \beta, \lambda, \mu, \rho = 1, 2$; $M, N, P, R = 1, 2, 3$. These equations correspond to simplex elements of any type of incompressible, isotropic, hyperelastic material. In the case of Mooney material, $\hat{W}(I_1,I_2)$ is given by (15.62) and we set

$$\frac{\partial \hat{W}}{\partial I_1} = C_1 \qquad \frac{\partial \hat{W}}{\partial I_2} = C_2 \tag{18.39}$$

whereas setting $C_2 = 0$ yields a form of $\hat{W}$ corresponding to so-called neo-hookean materials. Most of the numerical results reported here pertain to Mooney materials.

In Fig. 18.3, the numerical results obtained by Oden and Sato [1967*a*]† by applying (18.38) to a specific biaxial strip problem are reproduced. In this example, an 8.0-in.-square rubber sheet, 0.05 in. thick, is stretched to twice its original length ($\epsilon = 2$). The material is assumed to be of the Mooney type, with material constants C_1 and C_2 of 24.0 and 1.5 psi, respectively. The deformed shapes for various finite-element networks are illustrated in Fig. 18.3. In this particular application, the Newton-Raphson method was used to solve the system of nonlinear equations generated in the analysis. Initial test points were obtained by first analyzing a very coarse representation of the sheet using a small number of Newton-Raphson iterations. These results were then used as starting values for a more refined network, the starting values of the displacements at added nodal points being obtained by linear interpolation. The net horizontal boundary force F required to produce an extension $\epsilon = 2$ is of interest in experimental studies. The value of F corresponding to each finite-element network is obtained by simply summing the x^1 components of the generalized forces at the boundary. In this example, the net force

† Finite-element analyses of finite deformations of thin elastic membranes were considered earlier by Oden [1966*a*], and finite plane deformations of elastic sheets were studied by Becker [1966] using finite-element models. See Figs. 18.6 and 18.7.

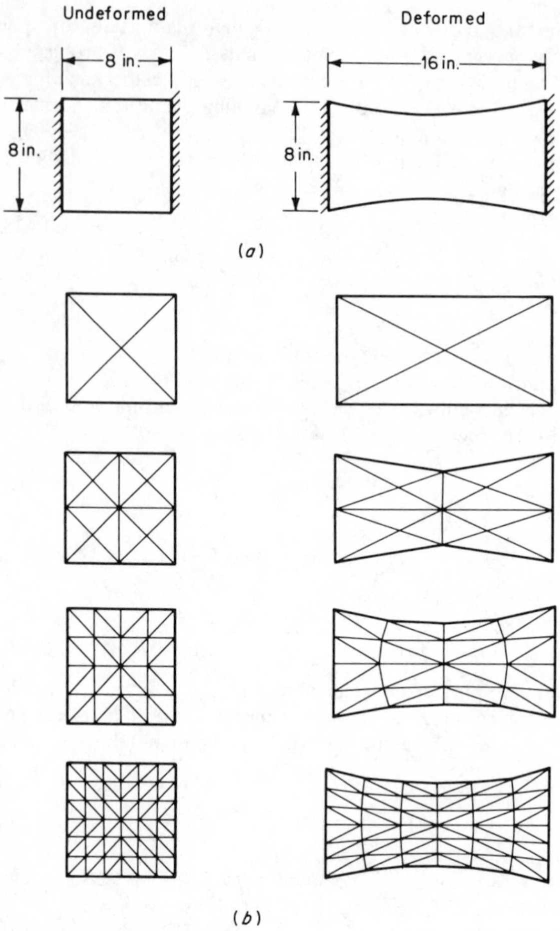

Fig. 18.3 Finite-element solutions of the biaxial-strip problem. (*a*) Continuous membrane; (*b*) various finite-element representations.

converged monotonically, as indicated in Fig. 18.4, to approximately 36.0 lb as the network was refined.

A different approach to the finite-element analysis of the biaxial strip problem was used by Becker [1966], who employed a bilinear approximation of the local displacement field, that is,

$$\psi_N(x^1,x^2) = a_N + b_{N\alpha}x^\alpha + \bar{c}_N x^1 x^2 \qquad (18.40)$$
$$u_\alpha = \psi_N u_\alpha^N = a_N u_\alpha^N + b_{N\beta}x^\beta u_\alpha^N + \bar{c}_N x^1 x^2 u_\alpha^N$$

Here $N = 1, 2, 3, 4$ and the arrays a_N, $b_{N\alpha}$, and $\bar{c}_N$ are as given in (10.118). The direct substitution of (18.40) into (18.35), however, leads to a very complicated expression for λ^2, which, in view of (18.34), must be integrated over the undeformed middle surface. To avoid such complications, a simplified form of the incompressibility condition $\lambda^2 A = I_3 = 1$ can be derived by following essentially the same procedure as that used to obtain (16.77). Considering the typical rectangular element of dimensions $a_0 \times b_0$ and thickness

d_0 in Fig. 18.5, the deformed middle-surface area A^* can be written in terms of the nodal displacements as the sum of the triangular areas 124 and 234:

$$A^* = \frac{1}{2}\det\begin{vmatrix} u_1^1 & a_0 + u_1^2 & u_1^4 \\ u_2^1 & u_2^2 & b_0 + u_2^4 \\ 1 & 1 & 1 \end{vmatrix} + \frac{1}{2}\det\begin{vmatrix} a_0 + u_1^2 & a_0 + u_1^1 & u_1^4 \\ u_2^2 & b_0 + u_2^3 & b_0 + u_2^4 \\ 1 & 1 & 1 \end{vmatrix} \tag{18.41}$$

Since $\lambda d_0 = d$, the condition of incompressibility is simply

$$A_0^* = a_0 b_0 = \lambda A^* \tag{18.42}$$

Hence, instead of (18.35) we use

$$\lambda = a_0 b_0 / A^* \tag{18.43}$$

The remaining terms in (18.34) are simple polynomials in x^1 and x^2 for the case in which u_α are defined locally by (18.40).

Rather than derive explicit nonlinear stiffness relations for such quadrilateral elements, Becker chose to view the problem as one of finding values of the nodal displacements which minimize the total potential energy $\Pi(\mathbf{U}^\Delta)$ computed using the approximation (18.40). This approach, described in Sec. 17.1, can be used in conjunction with any of the functional minimization schemes discussed in Art. 17. The solutions quoted here were obtained using a modified relaxation (univariant) scheme of the type outlined in Sec. 17.3. Becker's numerical results are reproduced in Figs. 18.6 and 18.7. Figure 18.6 shows a quarter of an initially square sheet of Mooney material subjected to an extension $\epsilon = 2.0$ for various finite-element representations. Nondimensional Mooney constants of $C_1 = 8$, $C_2 = 1$ were used and were found to yield results in close agreement with experiments performed on rubber sheets. Figure 18.7 contains contours of axial and lateral stresses

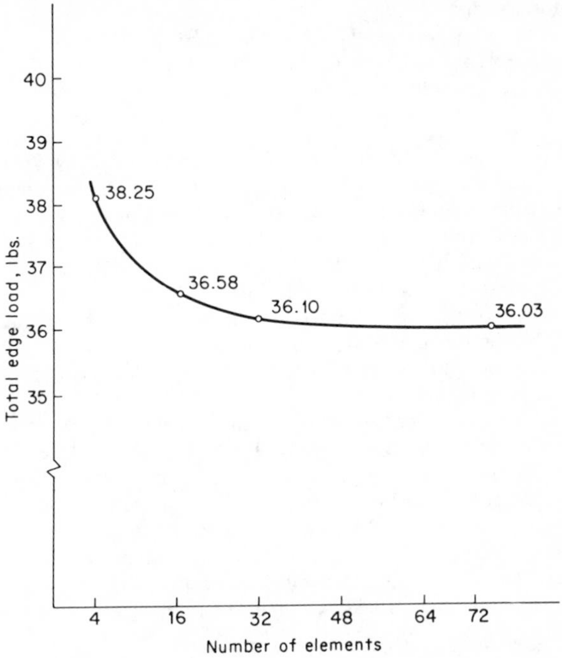

Fig. 18.4 Total edge load versus number of elements for stretching of a square membrane.

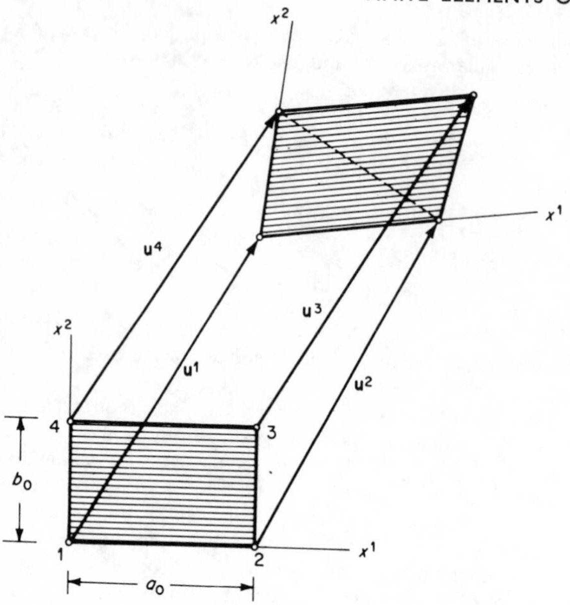

Fig. 18.5 Deformation of a quadrilateral element.

for the case of a square sheet stretched to 1.5 times its original length ($\epsilon = 1.5$). These contours depart significantly from those predicted by the infinitesimal theory of incompressible solids,† which, unlike the theory of finite deformations of Mooney materials, depicts the ratio of lateral to axial stress as being independent of the extension ratio ϵ and the material constants.

Sheet with a circular hole Rivlin and Thomas [1951] obtained experimental data on the behavior of a thin circular sheet of rubber containing a small, centrally located circular hole and subjected to finite axisymmetric stretching in the plane of the sheet. The experiments of Rivlin and Saunders [1951] on the same material indicate that it can be adequately characterized by a Mooney form of the strain-energy function with material constants $C_1 = C_2/0.08 = 18.35$ psi. The experiments of Rivlin and Thomas concerned a specimen of this material, which, in its undeformed state, was 5.0 in. in diameter and 0.0625 in. thick and which contained a 1.0-in.-diameter hole. Radial loads were applied at the outside edge of the sheet, and the displacement profiles for various radial extension ratios were measured.

We consider here a finite-element analysis of the same problem‡ in which the simplex model of (18.38) is employed. Figure 18.8 shows the complete finite-element model used in the analysis; owing to symmetry, only one octant of the model was actually analyzed. After generating the connected model, generalized forces at the boundary nodes of the interior circle were equated to zero and those acting at boundary nodes on the exterior circle were prescribed so as to represent a uniformly applied radial load. The nonlinear equilibrium equations generated in the analysis were solved by the method of incremental loading so as to provide a complete history of the behavior of the sheet.

† Williams and Schapery [1962].

‡ The finite-element solutions presented here were obtained by Oden and Key [1971*a*].

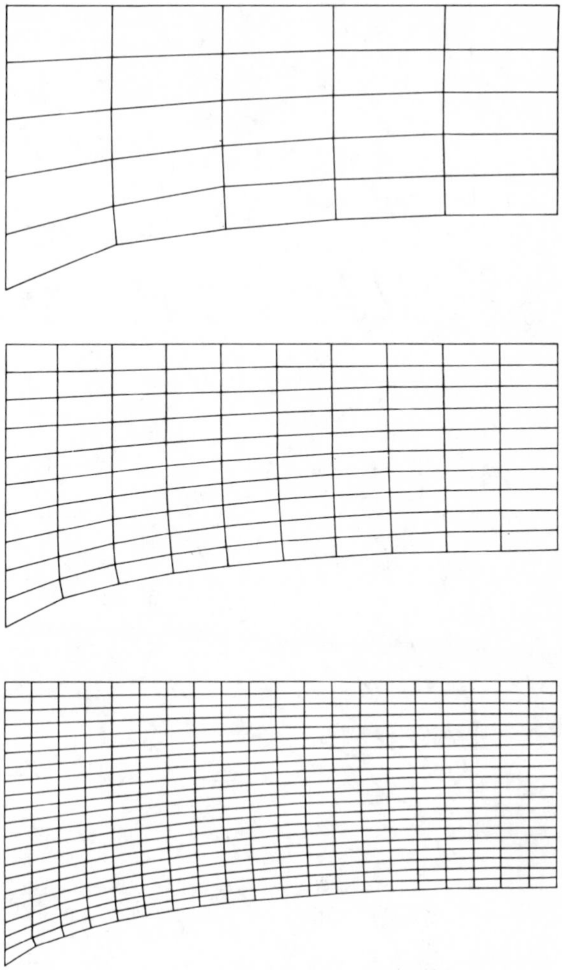

Fig. 18.6 Finite-element representations of a quarter of a deformed elastic sheet; $\epsilon = 2.0$, and $a_0/b_0 = 1.0$.

Figure 18.9 indicates the computed variation in the displacements of the interior and exterior nodes with the applied edge force P acting on one-eighth of the exterior boundary. We observe that the material softens with increasing load; under small loads, the material appears to be rather stiff, the slope S of the interior curve being 77 lb/in., while at larger strains the response is almost linear and $S \approx 22$ lb/in. We note that the use of a linearized theory in this analysis would lead to a displacement of the exterior node which is only 28 percent of the displacement predicted by the nonlinear theory.

Figure 18.10 contains a comparison of the computed displacement profiles with those determined experimentally by Rivlin and Thomas for various values of the radial extension ratio λ at $r = \infty$. The dimensionless ordinates of the curves are the ratios of the deformed

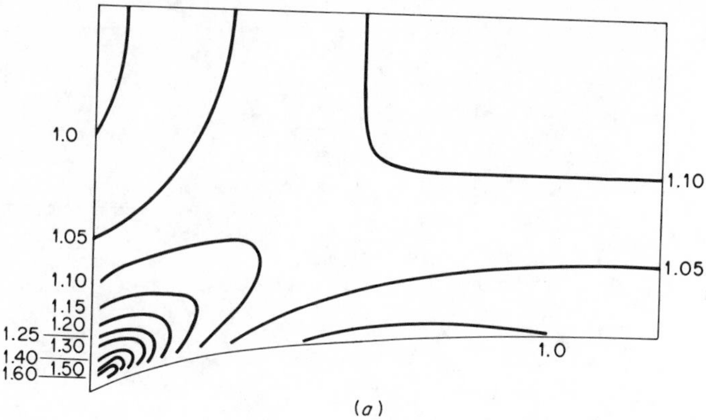

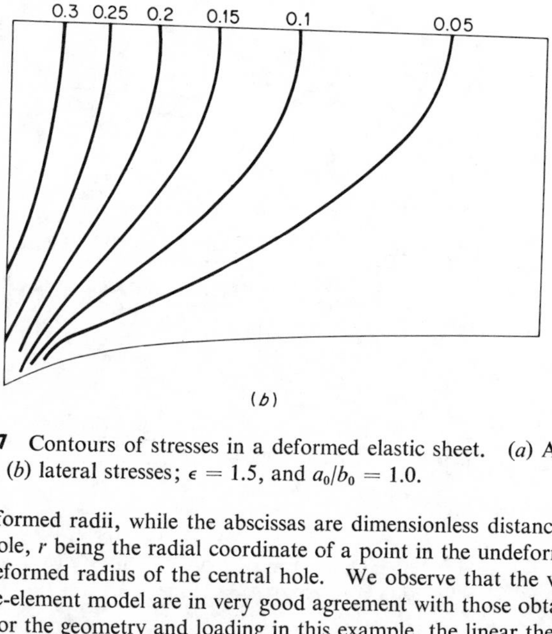

Fig. 18.7 Contours of stresses in a deformed elastic sheet. (*a*) Axial stresses; (*b*) lateral stresses; $\epsilon = 1.5$, and $a_0/b_0 = 1.0$.

to undeformed radii, while the abscissas are dimensionless distances r/a from the center of the hole, r being the radial coordinate of a point in the undeformed body and a being the undeformed radius of the central hole. We observe that the values computed using the finite-element model are in very good agreement with those obtained experimentally.

For the geometry and loading in this example, the linear theory of elasticity yields a stress-concentration factor of 2 [that is, $\sigma^{11}(\infty)/\sigma^{22}(a) = 2$]. In the general case considered here, however, the stress-concentration factor is dependent on the deformation and the character of the material. One factor contributing to the change in stress concentration is, of course, the significant decrease in thickness of the sheet when subjected to large, extensional strains; for example, when radial strains reach approximately 150 percent, the average thickness of the sheet is only around one-fourth of the undeformed thickness. Figure 18.11 shows the computed stress-concentration factor k versus the radial extension ratio for the radially loaded sheet of Mooney material considered here. For small strains, we see that the value of $k = 2$ corresponding to the linear theory is obtained. However, for increases in extension ratios, k also increases, reaching a value

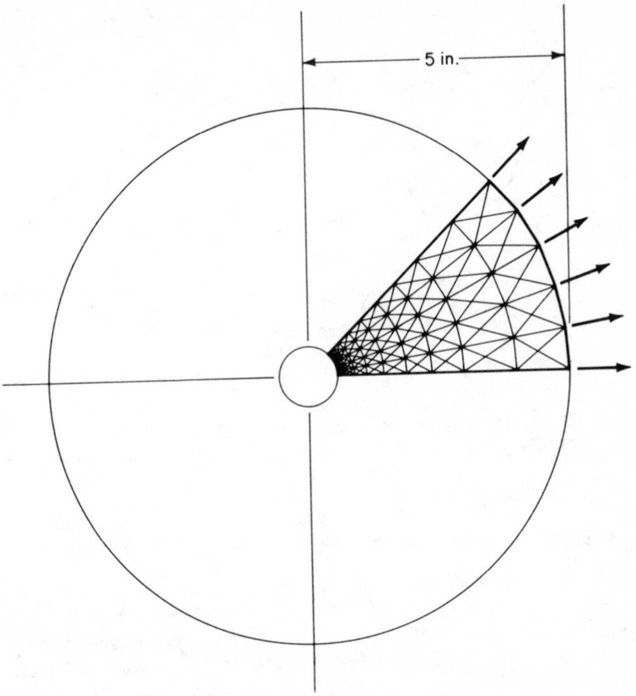

Fig. 18.8 Finite-element model of thin sheet containing a circular hole.

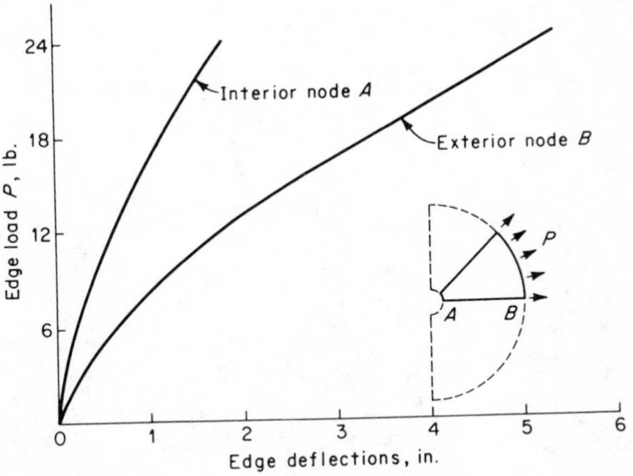

Fig. 18.9 Variation in nodal displacements with applied edge force.

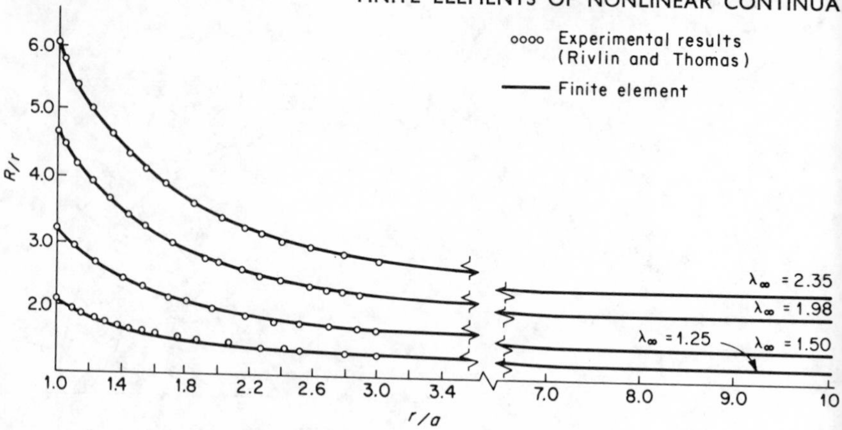

Fig. 18.10 Comparison of computed displacement fields with those determined experimentally by Rivlin and Thomas [1951] for various radial extension ratios.

of approximately 6 when $\lambda_\infty = 2.0$. This result is in agreement with the conclusions of Yang [1967] on stress concentrations in Mooney materials.

Uniaxial stretching of a sheet with a circular hole We consider the problem of uniform stretching of a thin, initially square, homogeneous sheet containing a centrally located circular hole. The origin of an x^1, x^2 material coordinate system is established at the center of the hole, with the x^1 and x^2 axes parallel to the sides of the sheet, and the sheet is stretched in successive stages in the x^1 direction while its length in the x^2 direction is held constant.

This problem is motivated by the experiments of Segal and Klosner [1970] on elastomeric sheets for which the strain-energy function deviated substantially from the Mooney form. After a series of tests on samples of a certain incompressible elastomer,† they proposed the form of the strain-energy function, cubic in I_2, given by (15.65). Thus,

† Klosner and Segal [1969].

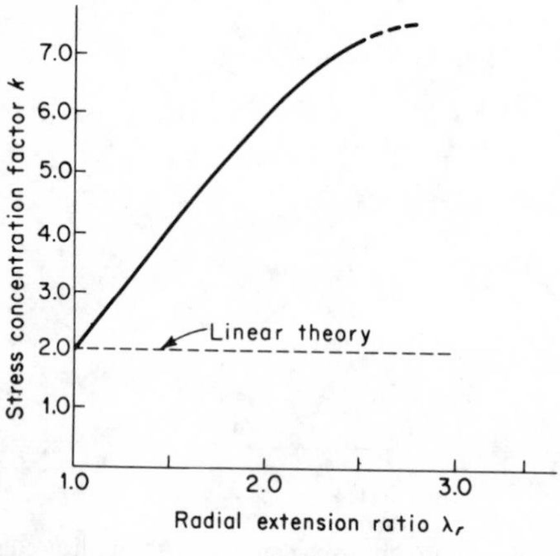

Fig. 18.11 Stress-concentration factor as a function of deformation.

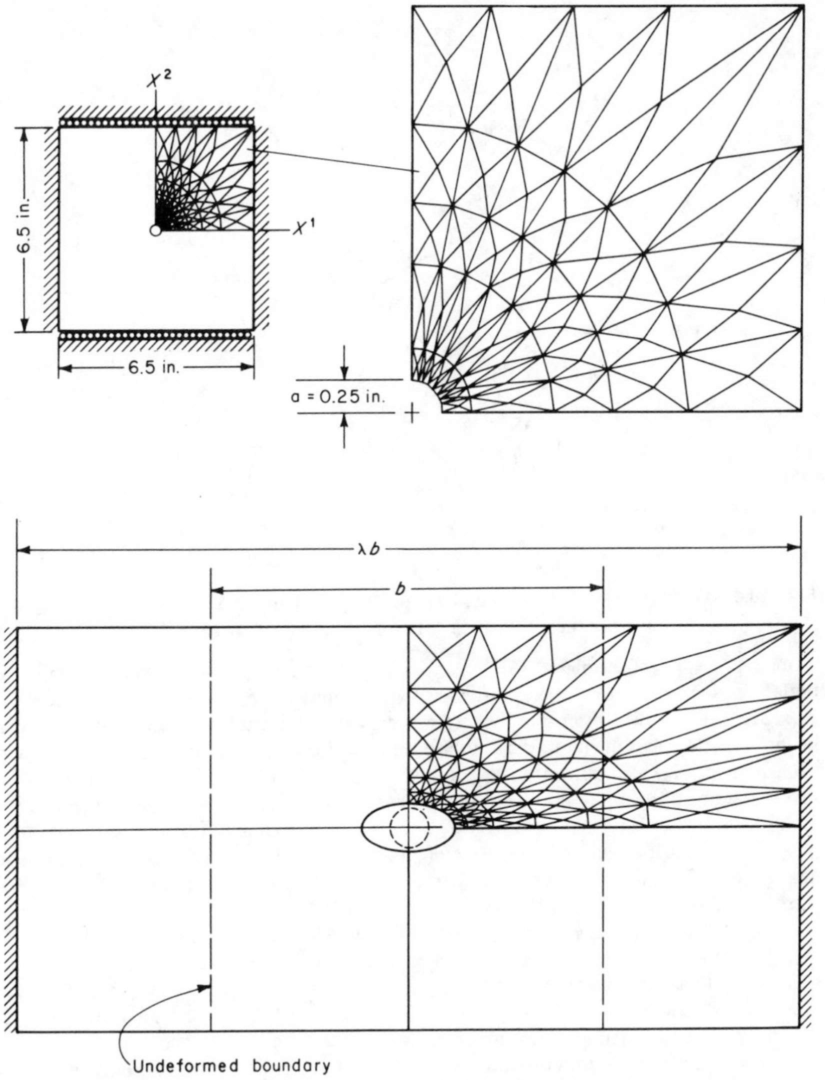

Fig. 18.12 Undeformed and deformed finite-element model of finite uniaxial stretching of a square sheet containing a circular hole.

the strain-energy function involves four material constants. Approximate values of these constants—$C_1 = 20.28$ psi, $C_2 = 5.808$ psi, $C_3 = -0.7200$ psi, and $C_4 = 0.04596$ psi—were determined from experiments on samples of a natural rubber. Tests were then run to determine the displacement field and the deformed shape of a 6.5-in.-square specimen of this material, 0.079 in. thick and containing a 0.5-in.-diameter circular hole. The specimen was subjected to the program of uniaxial extension described above.

We now discuss briefly the results of a finite-element analysis of the problem.† Figure 18.12*a* shows a finite-element model of half the sheet, which consists of 192 simplex

† Oden and Key [1971*a*].

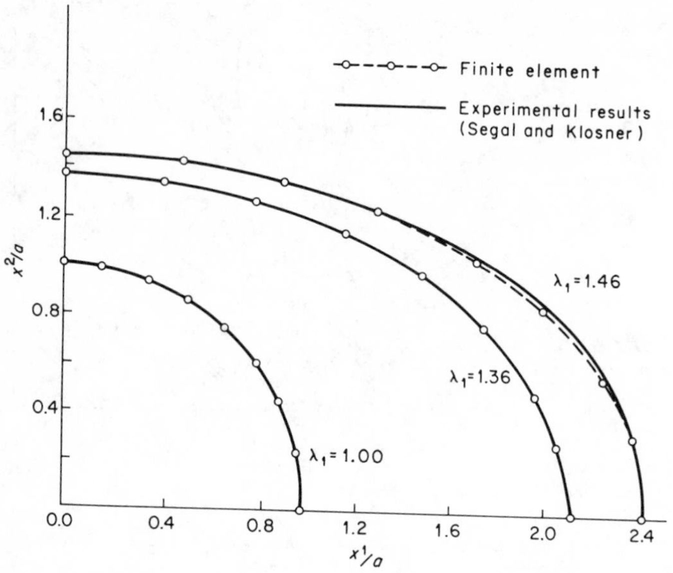

Fig. 18.13 Comparison of computed shapes of deformed hole with shapes determined experimentally by Segal and Klosner [1970].

elements. Application of (18.38) leads to a system of 198 simultaneous equations in the nodal displacement components after applying appropriate boundary conditions. The presence of the $(I_2 - 3)^3$ term in the strain-energy function leads to terms of twelfth degree in the nodal displacements in the equilibrium equations of each element. The transverse extension ratio λ, also appearing raised to the fourth power in the equilibrium equations, involves terms of degree -4 in the displacements. Nodal displacements in the x^1 direction along $x^1 = \pm 3.25$ were set equal to one another, and stretching of the sheet was accomplished by prescribing uniform load increments in the x^1 direction of the boundary nodes. The system of nonlinear equilibrium equations was thus solved using the method of incremental loading described in Sec. 17.5.

The deformed finite-element network calculated at $\lambda_1 = 2$ is shown in Fig. 18.12*b*. We observe that the circular hole is stretched in an elliptical shape and that lines, originally radial from the hole in the first octant of the sheet, flatten and acquire a noticeable curvature. An enlarged view of the deformed shape of a quarter of the hole is shown in Fig. 18.13, in which the calculated profiles are compared with those determined experimentally by Segal and Klosner for extension ratios $\lambda_{x_1} = \lambda_1$ of 1.0, 1.36, and 1.46. Agreement between computed and measured deformations is excellent. Figure 18.14 contains a comparison of the calculated variation in displacements along the x^1 and x^2 axes with the variations measured by Segal and Klosner for various values of the longitudinal extension ratio λ_1. Again, good agreement is obtained. Some differences occur between the calculated and measured vertical displacements of nodes on the x^2 axis at low values of λ_1. This is due chiefly to the fact that the load increments used in the calculations did not lead to an extension ratio exactly equal to that prescribed in the experiments; also, at low extension ratios, less accuracy in experimentally determined displacements can be expected.

Effect of form of the strain-energy function It is interesting to compare numerical solutions of this same problem (Fig. 18.12) obtained using different forms of the strain-energy function W. For purposes of illustration, we will consider the problem of uniaxial stretching of a square sheet containing a circular hole assuming seven different forms for the strain energy function: (1) the neo-hookean form (15.63), (2) the Mooney form (15.62), (3) the Klosner-Segal form (15.65) (considered previously), (4) the IHT-(Isihara, Hashitsume, Tatibana) form (15.66), (5) the Biderman form (15.67), (6) the Hart-Smith exponential form (15.68), and (7) the Alexander exponential form (15.70).

To establish a basis for comparison, the experimental data of Treloar† on an 8 percent sulfur rubber are used. These yield values of the material constants of the first five materials of $C = C_1 = 27.02$ psi, $C_2 = 1.42$ psi, $C_3 = -0.959$ psi, $C_4 = 0.0612$ psi, $B_1 = -0.27$ psi, $B_2 = 0.00654$ psi. Coefficients of all like terms in the polynomial forms

† Treloar [1944, 1958].

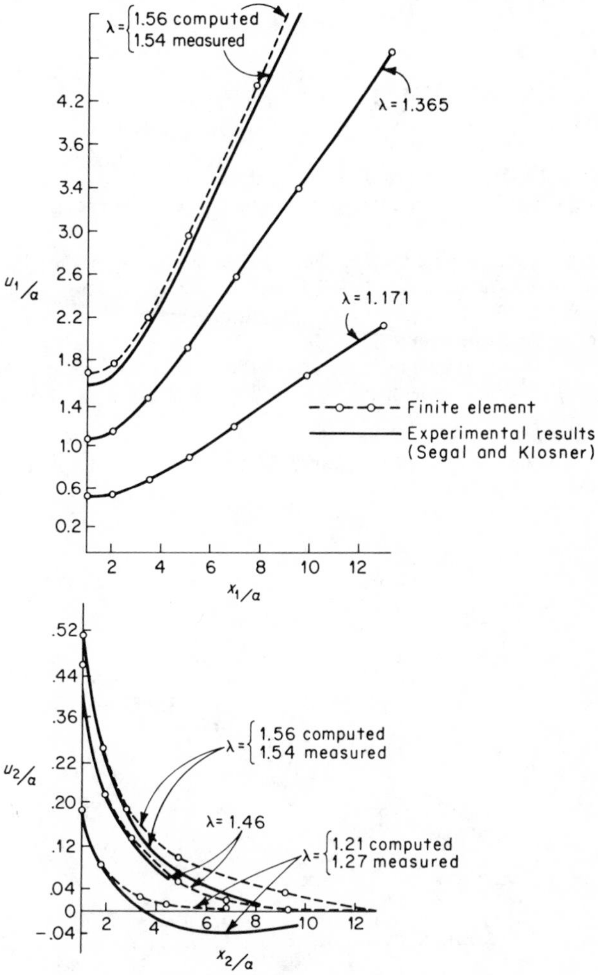

Fig. 18.14 Comparison of computed displacements with those determined by Segal and Klosner [1970].

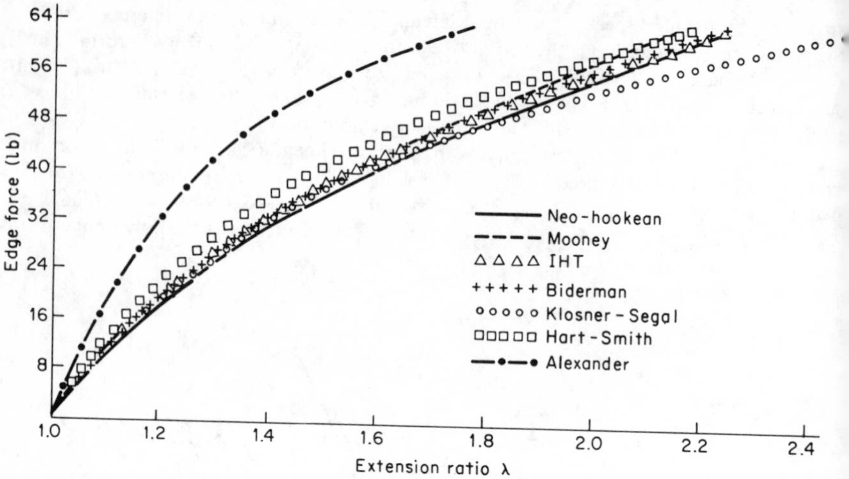

Fig. 18.15 Total edge force in x_1 direction along boundaries $x_1 = \pm a$ versus extension ratio λ at $x_1 = \pm\infty$ for various forms of the strain-energy function.

are thus made to coincide. Treloar's data yield values for the Hart-Smith form (15.68) of $G = 22.75$ psi, $k_1 = 0.00028$, and $k_2 = 1.20$. By requiring that like terms of (15.68) and (15.70) coincide and adjusting, in proportion, the experimental results of Alexander† on neoprene film, we obtain, for (15.70), $C_2 = 1.34$ psi, $C_3 = 26.55$ psi.

The computed variations‡ of the total applied force in the x_1-direction at the $x_1 = \pm a/2$ boundaries with the x_1-extension ratio λ at $x_1 = \pm\infty$ (see Fig. 18.12) for various

† Alexander [1968].

‡ This discussion and the numerical results given in Figs. 18.15–18.18 were presented by Oden and Key [1971*c*].

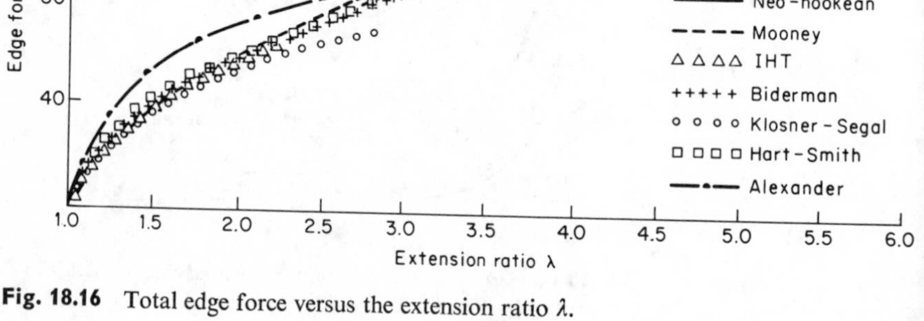

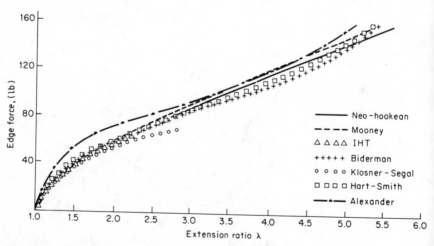

Fig. 18.16 Total edge force versus the extension ratio λ.

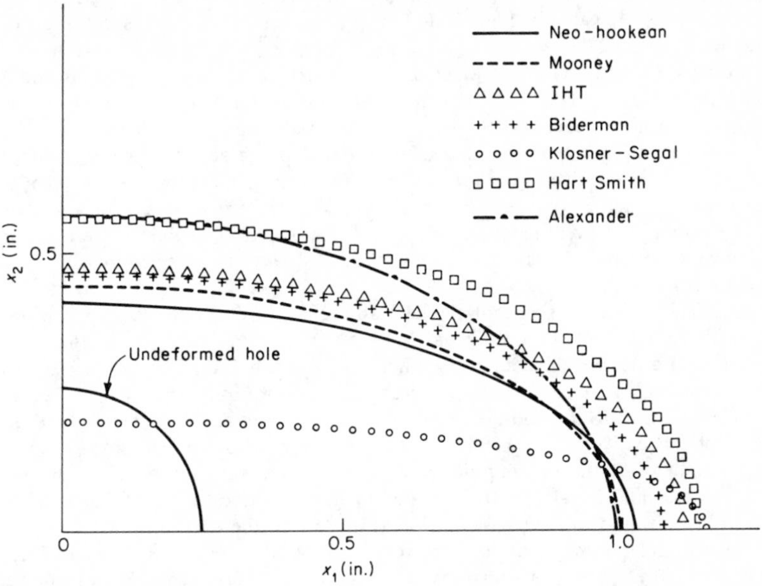

Fig. 18.17 Deformed shape of quarter of central hole produced by an applied edge force of 64.0 lb.

forms of the strain-energy function are shown in Fig. 18.15. Owing to a lack of experimental data, it was impossible to determine precise material constants for the Alexander form for the rubber used in Treloar's experiments. This may explain the difference between the initial slope of the response of the Alexander material and the other materials in Fig. 18.15. For values of the extension ratio λ of 1.0 – 2.0, the response of the

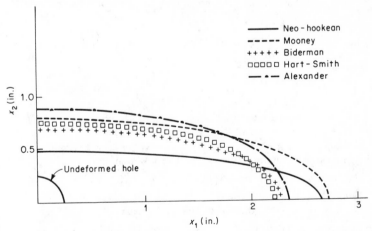

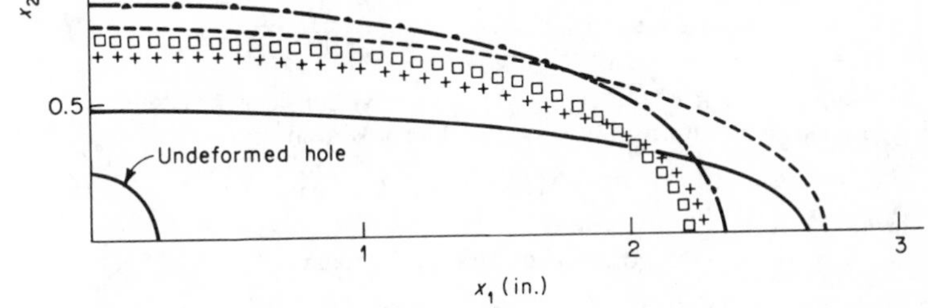

Fig. 18.18 Deformed shape of quarter of central hole produced by an applied edge force of 160.0 lb.

Alexander material appears to deviate from that of the other materials considered. However, as indicated in Fig. 18.16, for larger extension ratios this deviation is less pronounced. For $2.0 \leq \lambda < 5.5$, the uniaxial response of the neo-hookean and Mooney materials is practically linear, whereas the exponential forms of Hart-Smith and Alexander and the quadratic IHT form depict a stiffening of the material. The responses of the Klosner-Segal material and the Biderman material were not obtained for $\lambda > 2.0$.

It is important to note that the uniaxial response depicted in Figs. 18.15 and 18.16 is relatively insensitive to the form of the strain-energy function. Consequently, uniaxial experiments of the type described here would likely be inadequate for an accurate characterization of a given material, unless other features of the behavior were also measured. Similar observations were made by Alexander [1968] in connection with uniaxial tests on thin strips.

The deformed shape of the centrally located hole, however, appears to be more sensitive to the form of the strain energy function. Figures 18.17 and 18.18 show computed profiles of a quarter of the deformed hole for edge forces P of 64 and 160 lb, respectively. These forces, of course, produce different extension ratios λ at $x_1 = \pm\infty$. At $P = 64$ lb, the central hole in the Klosner-Segal material is seen to assume a much flatter shape than that of other materials. At smaller loads (corresponding to extension ratios λ of 1.0 to approximately 2.0), our computed profiles are in excellent agreement with those measured by Klosner and Segal [1969]; however, their report does not contain experimental data for higher values of λ. At $P = 64$ lb (Fig. 18.17), the shapes of the holes in the neo-hookean and Mooney materials are very similar; at higher loads (Fig. 18.18), however, significant differences in the transverse displacements at $x_2 = d/2$ are observed. Conversely, differences between profiles of the Hart-Smith and Biderman materials are less pronounced at $P = 160$ lb than at $P = 64$ lb. Whereas the Alexander material considered here is slightly stiffer than the Hart-Smith material, hole profiles are apparently similar in shape for a wide range of extension ratios. The profile of the hole in the Biderman material is similar in shape to that of the IHT material at $P = 64$ lb and the Hart-Smith material at $P = 160$ lb.

Inflation of elastic membranes† We now consider applications of (18.34) to the more general problem of finite inflation of membranes subjected to pressures normal to the middle surface. Employing once again a simplex representation, we have locally

$$u_i = a_N u_i^N + b_{N\beta} x^\beta u_i^N \tag{18.44}$$

where a_N and $b_{N\beta}$ are the same arrays as those given in (18.37) [and (10.106)]. Now, however, u_i and u_i^N are not zero, and instead of (18.38*a*), we have locally

$$2v_{0(e)}\left((\delta^{\alpha\beta} - \lambda^4 f^{\alpha\beta})\frac{\partial \hat{W}}{\partial I_1} + \left\{\lambda^2\delta^{\alpha\beta} + f^{\alpha\beta}\left[1 - 2\lambda^4 - 2\lambda^4 b_{R\mu} \times \left(\delta_{\mu k} + \frac{1}{2} b_{M\mu} u_k^M\right) u_k^R\right]\right\}\frac{\partial \hat{W}}{\partial I_2}\right) b_{N\alpha}(\delta_{\beta i} + b_{P\beta} u_i^P) = p_{N_i} \tag{18.45}$$

where $\alpha,\beta,\mu = 1, 2$; $N,M,R,P,i,k = 1, 2, 3$. We consider the case in which each element is subjected to a uniform transverse pressure q_0 so that, in accordance with (16.113),

$$p_{Nk} = \tfrac{1}{6} q_0 \epsilon_{ijk}(x_i^2 + u_i^2 - x_i^1 - u_i^1)(x_j^3 + u_j^3 - x_j^1 - u_j^1) \tag{18.46}$$

where, at least locally, $x_3^N = 0$.

Experimental and theoretical data on the behavior of an initially flat circular rubber

† Finite-element solutions of problems of inflation and finite deformation of thin elastic membranes were presented by Oden [1966*a*], Oden and Sato [1967*a*, 1967*b*], and Oden and Kubitza [1967]. See also Oden [1971*c*].

membrane, 50.8 mm in diameter and 0.2 mm thick and subjected to a uniform pressure of 0.097 kg/mm², were reported by Hart-Smith and Crisp [1967]. On the basis of experimental evidence and the molecular theory of Gent and Thomas [1958],† Hart-Smith and Crisp proposed a free-energy function of the type given in (15.68) for which

$$\frac{\partial \hat{W}}{\partial I_1} = C \exp\,[k_1(I_1 - 3)^2] \quad \text{and} \quad \frac{\partial \hat{W}}{\partial I_2} = \frac{Ck_2}{I_2} \tag{18.47}$$

Here C, k_1, and k_2 are material constants. Since k_1 ranges from 0 to 0.0004 and k_2 ranges from 0 to 1.5 for rubberlike materials, the functions in (18.47) approximate Mooney constants C_1 and C_2 in regions in which the values of I_1 and I_2 are uniform over each element, and Mooney constants can be calculated for each element using (18.47) and the experimental data of Hart-Smith and Crisp. The finite-element model shown in Fig. 18.19

† See also Hart-Smith [1966].

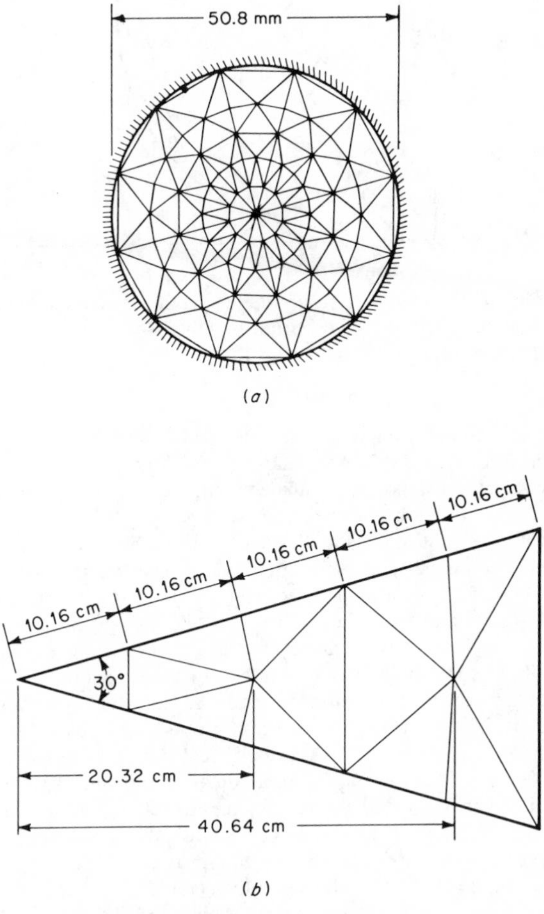

Fig. 18.19 (*a*) Finite-element representation of a circular membrane. (*b*) A typical 30° segment.

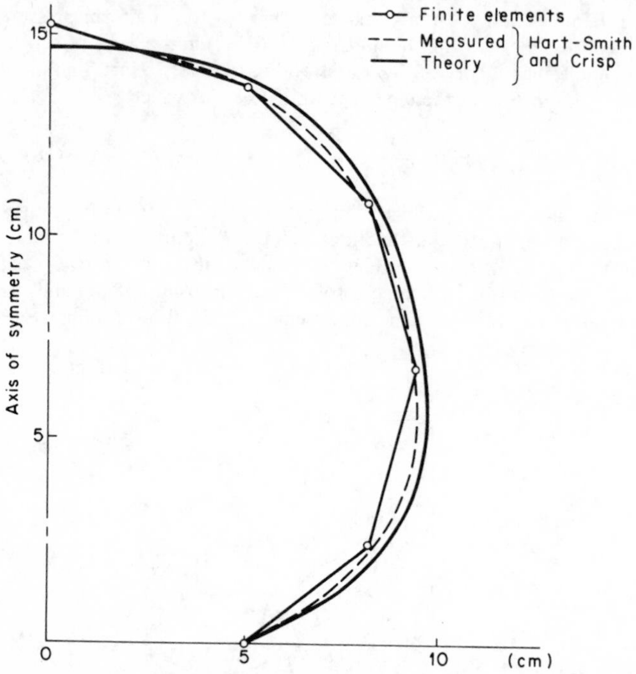

Fig. 18.20 Comparison of inflated profile calculated by the finite-element method with profiles obtained experimentally and theoretically.

was used in the analysis, and numerical results obtained by solving the system of nonlinear equations by the Newton-Raphson method are shown in Fig. 18.20.† Here we see the computed profile of the inflated membrane compared with that given by Hart-Smith and Crisp. Note that a rather coarse network of elements yields results in good agreement with experimental data, the maximum difference between computed and measured displacements being around 6 percent. Except for minor variations in the vertical displacement of the central node, essentially the same results were reproduced using averaged Mooney constants of $C_1 = 9.5C_2 = 1.75$ kg/cm² throughout the model.

The highly nonlinear character of problems of this type is worth noting. Figure 18.21 shows the variation of internal pressure with the tangential polar-extension ratio λ_p at the crown of an inflated sheet. These results were obtained experimentally by Oden and Kubitza [1967] by performing tests on circular disks of pure-gum natural rubber, 0.0173 cm thick and 38.1 cm in diameter. The disks were subjected to successive stages of internal pressure up to approximately 100 mm of water. A finite-element analysis of this behavior based on the assumption of a Mooney material with $C_1 = 1.14$ kg/cm², $C_2 = 0.14$ kg/cm² yielded deformed profiles in good agreement with the experimental data. We observe that for a pressure of approximately 90 mm of water multiple polar-extension ratios are possible, indicating an instability (in tension). This aspect of the behavior can lead to considerable computational difficulties. Successful finite-element analyses of

† These results were obtained by Oden and Kubitza [1967].

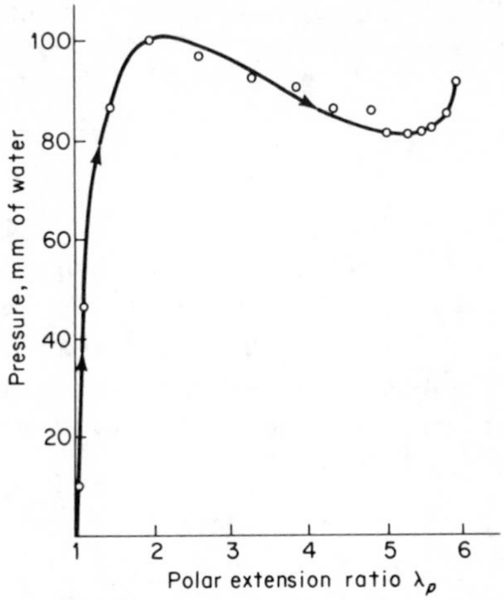

Fig. 18.21 Variation in polar-extension ratio with internal pressure.

this problem have been conducted, however, using the Newton-Raphson method and a judicious choice of starting values (initial test points) for given internal pressures.†

18.2 INCOMPRESSIBLE ELASTIC SOLIDS OF REVOLUTION‡

Consider an elastic solid of revolution of arbitrary cross-sectional shape, and suppose that the locations of material particles in a reference configuration of the body are given by the convected (intrinsic) cylindrical coordinates $\xi^1 = r$, $\xi^2 = z$, and $\xi^3 = \theta$. The deformation of the body is determined by the displacement field $\mathbf{u} = \mathbf{u}(r,z,\theta)$ and its gradient $\mathbf{u}_{,i} \equiv \partial\mathbf{u}/\partial\xi^i$. In the following, we confine our attention to the case of purely axisymmetric deformations, for which $\mathbf{u} = \mathbf{u}(r,z)$ and the displacement field is determined by radial and axial components u_1 and u_2. In this case, the strain-displacement relations reduce to

$$2\gamma_{\alpha\beta} = u_{\alpha,\beta} + u_{\beta,\alpha} + u_{\mu,\alpha}u^{\mu}_{,\beta} \tag{18.48a}$$

$$\gamma_{\alpha 3} = 0 \tag{18.48b}$$

$$\gamma_{33} = \frac{r^2}{2}(\lambda^2 - 1) \tag{18.48c}$$

† Oden [1971*c*].

‡ Applications of the finite-element method to the linear problem of symmetric infinitesimal deformations of hookean solids of revolution were presented by Rashid [1964, 1966] and Clough and Rashid [1965]. Wilson [1965] considered the linear problem of general infinitesimal deformations of axisymmetric elastic solids, and Becker and Brisbane [1965] developed finite-element models for the analysis of infinitesimal deformations of incompressible solids of revolution. Applications to the general problem of finite axisymmetric deformations of incompressible solids of revolution were presented in the paper by Oden and Key [1970], on which the present section is based.

where $\gamma_{\alpha\beta}(\alpha,\beta,\mu = 1,2)$ are the covariant components of the strain tensor, $u_\mu = u^\mu$, and

$$\lambda = 1 + \frac{u_1}{r} \tag{18.49}$$

The function $\lambda = \lambda(r,z)$ is the extension ratio in the circumferential direction; that is, λ is the ratio of the length of a circumferential fiber in the deformed body to its original length in the reference configuration.

Assuming, as before, that the body is isotropic and hyperelastic, the strain-energy function W can be expressed as a function of the principal invariants of the deformation tensor. For the type of deformations under consideration,

$$\begin{aligned} I_1 &= 2(1 + \gamma_\alpha^\alpha) + \lambda^2 \\ I_2 &= 2\lambda^2(1 + \gamma_\alpha^\alpha) + \varphi \\ I_3 &= \lambda^2\varphi \end{aligned} \tag{18.50}$$

where

$$\varphi = 1 + 2\gamma_\alpha^\alpha + 2\epsilon^{\alpha\beta}\epsilon^{\lambda\mu}\gamma_{\alpha\lambda}\gamma_{\beta\mu} \tag{18.51}$$

and, as before, $\epsilon^{\alpha\beta}$ is the two-dimensional permutation symbol. Consequently,

$$t^{\alpha\beta} = 2\frac{\partial W}{\partial I_1}\delta^{\alpha\beta} + 2\frac{\partial W}{\partial I_2}[\delta^{\alpha\beta}(1 + \lambda^2) + 2\epsilon^{\alpha\lambda}\epsilon^{\beta\mu}\gamma_{\lambda\mu}] + 2\frac{\partial W}{\partial I_3}\lambda^2(\delta^{\alpha\beta} + 2\epsilon^{\alpha\lambda}\epsilon^{\beta\mu}\gamma_{\lambda\mu}) \tag{18.52a}$$

$$t^{\alpha 3} = 0 \tag{18.52b}$$

$$r^2 t^{33} = 2\frac{\partial W}{\partial I_1} + 4(1 + \gamma_\alpha^\alpha)\frac{\partial W}{\partial I_2} + 2\frac{\partial W}{\partial I_3}\varphi \tag{18.52c}$$

In the case of incompressible materials, $W = \hat{W}(I_1,I_2)$, $I_3 = 1$, and

$$t^{\alpha\beta} = 2\frac{\partial \hat{W}}{\partial I_1}\delta^{\alpha\beta} + 2\frac{\partial \hat{W}}{\partial I_2}[\delta^{\alpha\beta}(1 + \lambda^2) + 2\epsilon^{\alpha\lambda}\epsilon^{\beta\mu}\gamma_{\lambda\mu} + 2h\lambda^2(\delta^{\alpha\beta} + 2\epsilon^{\alpha\lambda}\epsilon^{\beta\mu}\gamma_{\lambda\mu})] \tag{18.53a}$$

$$t^{\alpha 3} = 0 \tag{18.53b}$$

$$r^2 t^{33} = 2\frac{\partial \hat{W}}{\partial I_1} + 4(1 + \gamma_\alpha^\alpha)\frac{\partial \hat{W}}{\partial I_2} + 2h\varphi \tag{18.53c}$$

where h is the hydrostatic pressure.

Finite-element approximations We now set out to construct a discrete model of the body by representing it as a collection of a number E of finite

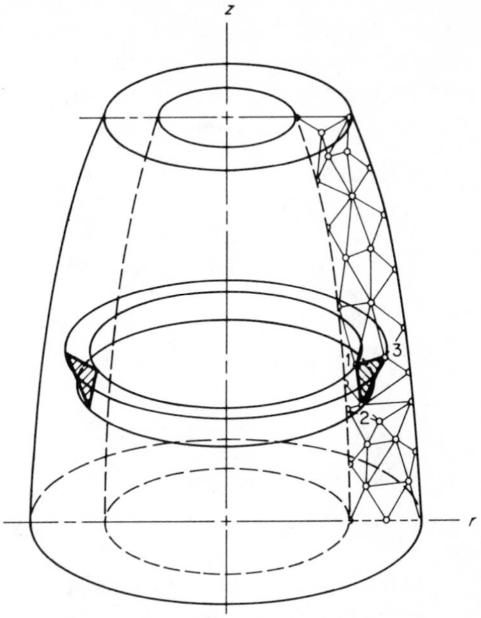

Fig. 18.22 Finite-element model of a solid of revolution.

elements of revolution, as indicated in Fig. 18.22. Following the usual procedure, we isolate a typical finite element e and approximate the local displacement field $u_\alpha(\xi^1,\xi^2)$ over the element by functions of the form $u_\alpha = \psi_N(\xi^1,\xi^2)u_\alpha^N$, where $\xi^1 = r$ and $\xi^2 = z$. The covariant components of the metric tensor in the undeformed body are simply $g_{11} = g_{22} = 1$, $g_{33} = r^2$, and $g_{ij} = 0$ $(i \neq j)$, and the Riemann-Christoffel symbols are $\Gamma^3_{13} = \Gamma^3_{31} = 1/r$, $\Gamma^1_{33} = -r$, all other $\Gamma^i_{jk} = 0$. Since $u_3 = \partial u_i/\partial \xi^3 = 0$, we have $u^\alpha_{;\beta} = u^\alpha_{,\beta} = u_{\alpha,\beta}$. Hence, upon introducing these quantities together with (18.53) into (16.192) and simplifying, we arrive at the following equilibrium equations for an isotropic incompressible finite element of revolution:

$$\int_{v_{0(e)}} \left\{ 2\frac{\partial \hat{W}}{\partial I_1}\frac{\partial \gamma_\beta^\beta}{\partial u_\alpha^N} + \frac{\partial \hat{W}}{\partial I_2}\frac{\partial \varphi}{\partial u_\alpha^N} + \left[\frac{\partial \hat{W}}{\partial I_1} + 2(1+\gamma_\beta^\beta)\frac{\partial \hat{W}}{\partial I_2}\right]\frac{\partial \lambda^2}{\partial u_\alpha^N} \right.$$
$$\left. + 2\frac{\partial \hat{W}}{\partial I_2}\frac{\partial \gamma_\beta^\beta}{\partial u_\alpha^N}\lambda^2 + h\left(\lambda^2\frac{\partial \varphi}{\partial u_\alpha^N} + \varphi\frac{\partial \lambda^2}{\partial u_\alpha^N}\right)\right\} dv_0 = p_{N\alpha} \quad (18.54)$$

where $h = \psi_N h^N$ is the element hydrostatic pressure and $dv_0 = 2\pi r\, dr\, dz$. We must also impose the discrete incompressibility conditions of the form (16.54) or (16.62).

Simplified forms In various applications to be discussed subsequently, we shall use a simplified version of (18.54) arising from the use of simplex elements together with an acceptable approximation of λ^2. In the case of simplex elements, $N_e = 3$, the hydrostatic pressure is assumed to be uniform over each element and, as usual, $u_\alpha = a_N u_\alpha^N + b_{N1} r u_\alpha^N + b_{N2} z u_\alpha^N$, with a_N and $b_{N\alpha}$ given by (10.106). Then

$$2\gamma_{\alpha\beta} = b_{N\alpha} u_\beta^N + b_{N\beta} u_\alpha^N + b_{N\beta} b_{M\alpha} u^{N\lambda} u_\lambda^M \tag{18.55a}$$

$$2\gamma_{33} = r^2(\lambda^2 - 1) \tag{18.55b}$$

$$\lambda = 1 + \frac{1}{r}(a_N u_1^N + b_{N\beta} x^\beta u_1^N) \tag{18.55c}$$

and

$$\frac{\partial \gamma_\beta^\beta}{\partial u_\alpha^N} = b_{N\beta}(\delta_\alpha^\beta + b_{M\beta} u_\alpha^M) \tag{18.56a}$$

$$\frac{\partial \lambda^2}{\partial u_\alpha^N} = 2\frac{\lambda}{r}\delta_{1\alpha}(a_N + b_{N\beta} x^\beta) \tag{18.56b}$$

$$\frac{\partial \varphi}{\partial u_\alpha^N} = 2b_{N\lambda}(\delta_\alpha^\beta + b_{M\beta} u_\alpha^M)[\delta_\lambda^\beta + \gamma_{\rho\mu}(\epsilon^{\beta\rho}\epsilon^{\lambda\mu} + \epsilon^{\rho\lambda}\epsilon^{\mu\beta})] \tag{18.56c}$$

Unfortunately, the integration of terms in (18.54) which involve the circumferential extension ratio λ of (18.55*c*) leads to extremely complex logarithmic forms. To avoid these complications, we shall use instead of (18.55*c*) an approximate λ which is calculated using the average radial displacement over the element and which converges to the exact λ as the dimensions of the element are made arbitrarily small:

$$\lambda = 1 + \frac{\bar{u}_1}{\bar{r}} \tag{18.57}$$

where $\bar{u}_1$ and $\bar{r}$ are the average radial displacement of the nodes and the average radial coordinate of the nodes:

$$\bar{u}_1 = \tfrac{1}{3}(u_1^1 + u_1^2 + u_1^3) \qquad \bar{r} = \tfrac{1}{3}(r^1 + r^2 + r^3) \tag{18.58}$$

The extension ratio λ is now treated as being uniform over the element, and instead of (18.56*b*), we use

$$\frac{\partial \lambda^2}{\partial u_\alpha^N} = 2\frac{\lambda}{\bar{r}}(\delta^{N1} + \delta^{N2} + \delta^{N3})\delta^{1\alpha} \tag{18.59}$$

With these simplifications, (18.54) becomes

$$\begin{aligned} 2v_0\Bigg\{&\left(\frac{\partial \hat{W}}{\partial I_1} + \lambda^2\frac{\partial \hat{W}}{\partial I_2}\right) b_{N\beta}(\delta_\alpha^\beta + b_{M\beta} u_\alpha^M) + \left[\frac{\partial \hat{W}}{\partial I_1} + 2\frac{\partial \hat{W}}{\partial I_2}(1 + \gamma_\beta^\beta)\right. \\ &\left. + h\varphi\right]\frac{\lambda}{\bar{r}}(\delta^{N1} + \delta^{N2} + \delta^{N3})\delta^{1\alpha} + \left(\frac{\partial \hat{W}}{\partial I_2} + h\lambda^2\right) b_{N\lambda}(\delta_\alpha^\beta \\ &+ b_{M\beta} u_\alpha^M)[\delta_\lambda^\beta + \gamma_{\rho\mu}(\epsilon^{\beta\rho}\epsilon^{\lambda\mu} + \epsilon^{\rho\lambda}\epsilon^{\mu\beta})]\Bigg\} = p_{N\alpha} \end{aligned} \tag{18.60}$$

where $v_0 = 2\pi\bar{r}A_0$, A_0 being the undeformed cross-sectional area of the element. The nodal forces $p_{N\alpha}$ are calculated using procedures outlined in Sec. 16.3. For example, in the case of a uniform traction of intensity q normal to an element boundary, we use, in accordance with (16.118),

$$p_{1\alpha} = p_{2\alpha} = (-1)^{(\alpha)}\epsilon_{\beta\alpha}\tfrac{1}{2}q\pi(\xi_1^2 + \xi_1^1 + u_1^1 + u_1^2)\epsilon_{\alpha\beta}L_*^\beta \tag{18.61}$$

where $L_*^\beta = L_\beta^*$ are the projections of the line connecting nodes 1 and 2 on the ξ_β axis after deformation ($\xi_1 = r$; $\xi_2 = z$).

Since (18.60) represents six equations in the seven unknowns $h_{(e)}$ and u_α^N, we must also construct the incompressibility condition for the element. Thus we obtain by simply comparing deformed and undeformed element volumes:

$$v(\mathbf{u}^N) - v_0 = 0 \tag{18.62}$$

Here

$$v_0 = 2\pi\bar{r}A_0 \qquad \text{and} \qquad v = 2\pi(\bar{r} + \bar{u}_1)A \tag{18.63}$$

where $\bar{u}_1$ and $\bar{r}$ are as defined in (18.58) and A and A_0 are the deformed and undeformed cross-sectional areas of the ring elements:

$$A_0 = \left| \frac{1}{2}\sum_{P=1}^{3} \epsilon_{MNP}\epsilon^{\alpha\beta 3}x_\alpha^M x_\beta^N \right| \tag{18.64a}$$

$$A = \left| \frac{1}{2}\sum_{P=1}^{3} \epsilon_{MNP}\epsilon^{\alpha\beta 3}(x_\alpha^M + u_\alpha^M)(x_\beta^N + u_\beta^N) \right| \tag{18.64b}$$

The infinite-cylinder problem We first consider the special case of finite axisymmetric deformations of an infinitely long thick-walled cylinder subjected to internal pressure. This problem is of special interest because it is one of the few cases for which results can be compared with known exact solutions† and it is one-dimensional, a fact which enables us to reduce the nonlinear stiffness relations to particularly simple forms.

The triangular finite elements of revolution developed previously can be used to portray axisymmetric (radial) deformations by constructing a finite-element model of a thin disk, as shown in Fig. 18.23*a*. Although the problem can be greatly simplified by equating the radial displacements of vertically opposed nodes I and I', the finite-element characterization is subjected to a more severe test by allowing all nodes to displace freely in the radial direction. Then a model with E finite elements leads to $2E + 2$ nonlinear equations in the $E + 2$ unknown nodal displacements and E element hydrostatic pressures.

As a first example, a hollow cylinder, 7.00 in. internal radius and 18.625 in. external radius and of Mooney material with $C_1 = 80$ psi, $C_2 = 20$ psi, is considered. The cylinder is subjected to an internal pressure of $p = 128.2$ psi.‡ Displacement and hydrostatic-pressure profiles for the case $E = 10$ (22 unknowns) are shown in Fig. 18.24. We see that for this rather crude representation, slight differences occur between nodal values of radial displacement of vertically opposed nodes. For the 10-element case, these differences

† Green and Zerna [1968].

‡ Exact solutions for a cylinder with similar properties were obtained by Baltrukonis and Vaishnav [1965].

reach as much as 5 percent, but the average values of top and bottom nodes differ from the exact by only 2 percent. Hydrostatic pressures in the elements represent only rough averages for this coarse finite-element mesh. The values indicated in Fig. 18.24 are obtained by averaging the elemental hydrostatic pressures of adjacent (upper and lower) elements and assigning these values to points which are radially midway between nodes. The method of incremental loading, discussed in Sec. 17.2, was used to solve the system of nonlinear equations generated in this example. Element stresses, obtained by averaging the mean stresses in vertically adjacent elements, for the case of 20 elements are shown in Fig. 18.25. It is seen that very good agreement with the exact solution is obtained.

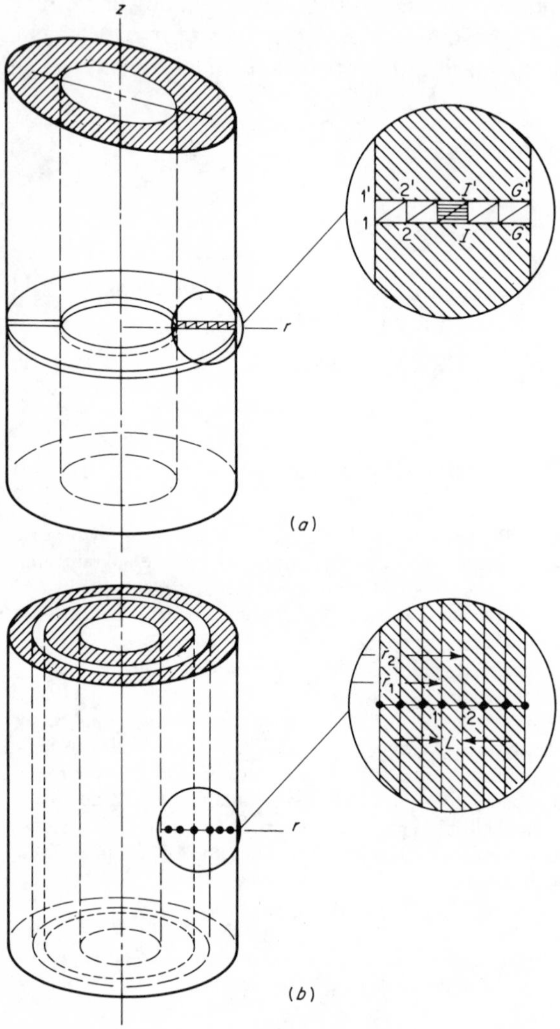

Fig. 18.23 Finite-element representations of an infinite cylinder.

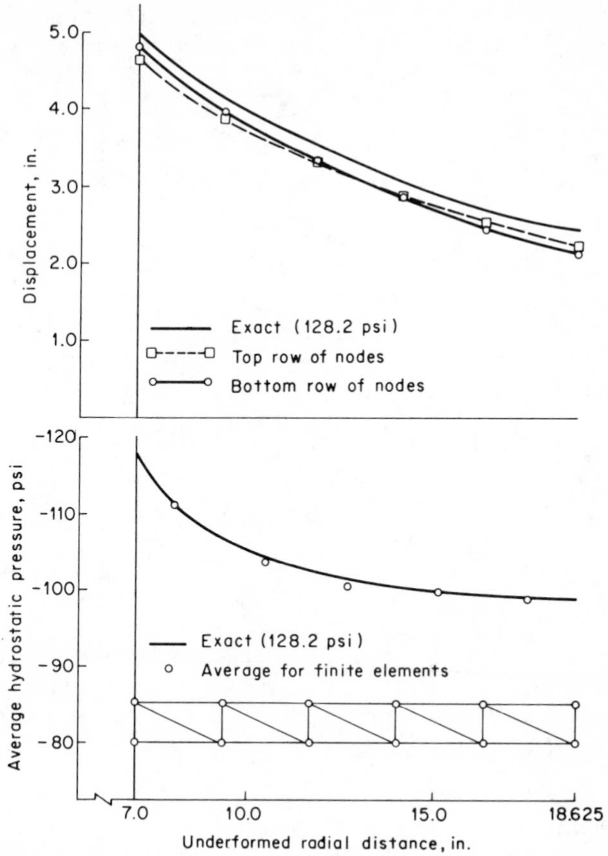

Fig. 18.24 Displacement and hydrostatic-pressure profiles.

An alternate finite-element representation of the infinite-cylinder problem is obtained by using finite thin cylindrical elements as indicated in Fig. 18.23*b*. Although numerical results obtained using this representation are practically the same as those obtained using models of the type in Fig. 18.23*a*, the stiffness equations derived from the purely one-dimensional kinematic relations are significantly simpler than those obtained from two-dimensional elements.

For a one-dimensional element of unit height constructed of a Mooney material, we find that the nonlinear stiffness relations are

$$p_N^1 = 2\pi(r_2^2 - r_1^2)\left\{\frac{\Delta}{L_0}(\epsilon^{1N} + \epsilon^{2N})[C_1 + C_2(1 + \lambda^2) + \lambda^2 h] + \frac{\lambda}{\bar{r}}[C_1 + C_2(1 + \Delta^2) + \Delta^2 h]\right\} \tag{18.65}$$

where

$$\bar{r} = \tfrac{1}{2}(r_1 + r_2) \qquad L_0 = r_2 - r_1 \tag{18.66}$$

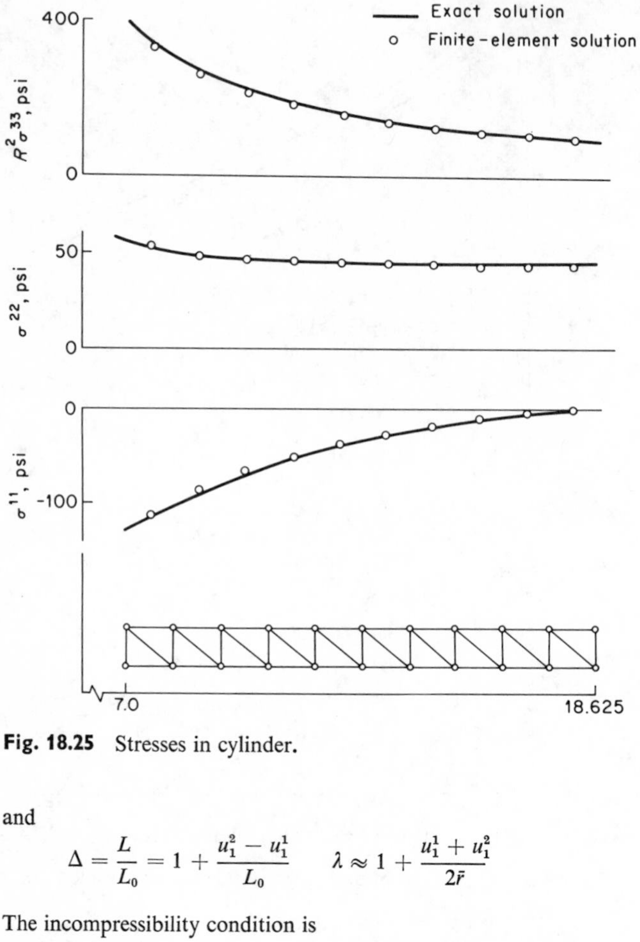

Fig. 18.25 Stresses in cylinder.

and

$$\Delta = \frac{L}{L_0} = 1 + \frac{u_1^2 - u_1^1}{L_0} \qquad \lambda \approx 1 + \frac{u_1^1 + u_1^2}{2\bar{r}} \tag{18.67}$$

The incompressibility condition is

$$1 = \lambda^2\Delta^2 \tag{18.68}$$

and the generalized force at the interior (or exterior) node due to internal (external) pressure p is

$$p_N^1 = \pm 2\pi(r_N + u_1^N)p \tag{18.69}$$

where the positive or negative sign is used if p is an internal pressure (N is the interior node) or an external pressure (N is the exterior node), respectively.

Figures 18.26 to 18.30 contain numerical results obtained using Eqs. (18.65) to (18.69) to solve the thick-walled cylinder problem described previously (that is, $r_{\text{int}} =$ 7.00 in., $r_{\text{out}} =$ 18.625 in., $C_1 = 80$ psi, and $C_2 = 20$ psi). Solutions for a variety of internal pressures were obtained, ranging from 0 to 150 psi. For this material, an applied internal pressure of 150 psi corresponds to strains on the order of 150 percent, so that the behavior falls well outside of that capable of being predicted by the classical, infinitesimal theory. The solution of the system of nonlinear equations generated in the analysis was obtained using the method of incremental loading described in Sec. 17.5. Figure 18.26 shows the variation of internal pressure with the radial displacement of the interior node obtained using 10, 20, and 40 increments and nine finite elements. For pressures less than 100 psi, relatively little error appears to be propagated for the 40-increment solution.

As an indication of the rate of convergence of the finite-element solutions, Fig. 18.27 shows the ratios of the displacement of the interior wall to the exact displacement plus the ratio of the hydrostatic pressure in the first interior element to the average of the exact hydrostatic pressure over the element, plotted versus the number of finite elements. As expected, the finite-element model is inherently stiffer than the actual cylinder. The displacement of the interior node converges quickly and monotonically from below to the exact value. Hydrostatic pressures, of course, converge less rapidly, but for only five elements, they differ only 1.5 percent from the exact values. The rather fast, monotonic convergence of the hydrostatic pressure indicated in Fig. 18.27 cannot be interpreted as an indication of the convergence of element stresses, even though the influence of hydrostatic pressure on the final stress values may be quite significant. The maximum difference between the exact stresses and those calculated from the finite-element solution occurs in the first (innermost) element. Figure 18.28 shows the variation of the ratio of the stress in this interior element to the exact average over the element versus the number of finite elements. We observe that convergence is not monotonic from below for all components, nor is it as rapid for all components as in the case of hydrostatic pressure. Nevertheless, for the nine-element representation, Figs. 18.29 and 18.30 show that stress profiles predicted by the finite-element analysis are in close agreement with the exact profiles.

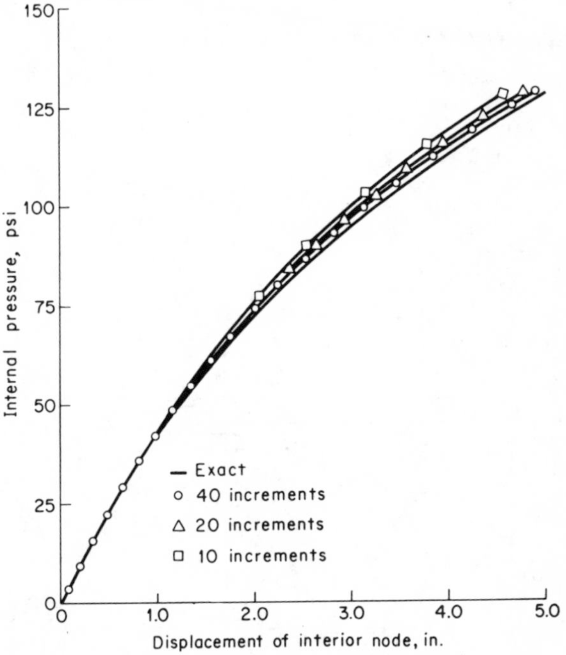

Fig. 18.26 Pressure-displacement curves obtained by incremental loading.

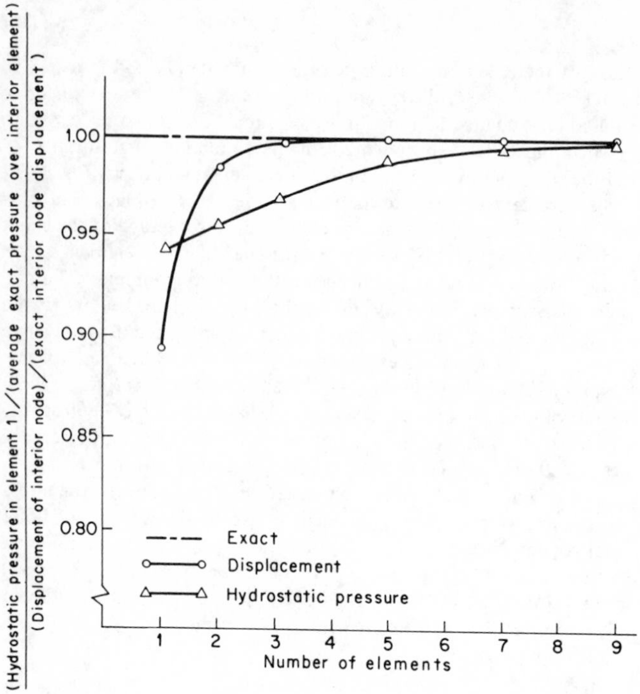

Fig. 18.27 Convergence of finite-element solutions.

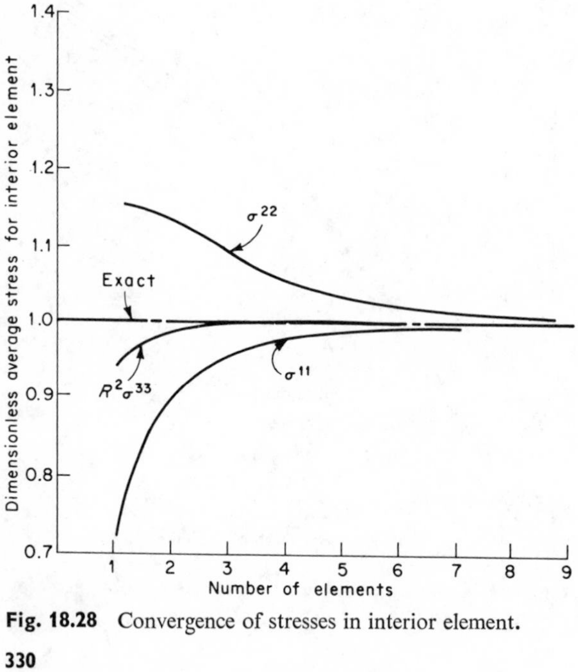

Fig. 18.28 Convergence of stresses in interior element.

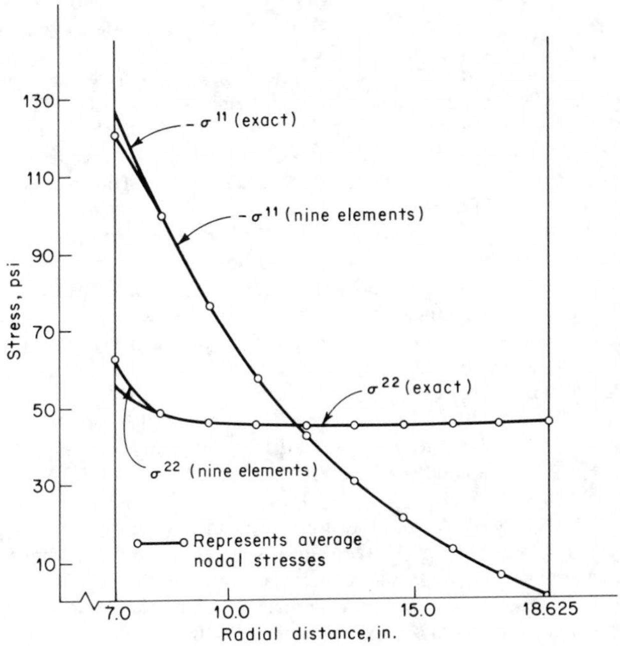

Fig. 18.29 Stress profiles σ^{11} and σ^{22}.

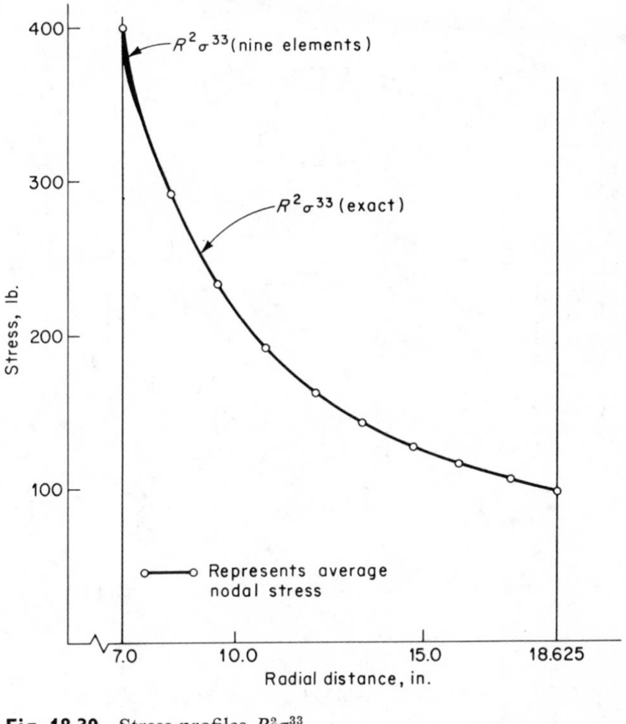

Fig. 18.30 Stress profiles $R^2\sigma^{33}$.

Thick-walled container We now consider finite axisymmetric deformations of the thick-walled incompressible elastic container shown in Fig. 18.31. Again, it is specified that the material be of the Mooney type with material constants $C_1 = 80$ psi and $C_2 =$ 20 psi. The body is subjected to a uniform internal pressure of 190 psi along the interior boundary *BC*. No forces are applied along *AB*. Dimensions of the undeformed body are indicated in the figure.

Because of symmetry, only one quadrant (half of the container) need be considered. Thus, the finite-element representation involves 48 finite elements connected together at 35 nodal points. Upon applying boundary conditions, this corresponds to 113 unknowns: 48 element hydrostatic pressures and 65 components of nodal displacement. The particular finite-element model used in this analysis leads to nonlinear stiffness equations for each element which are polynomials of sixth degree in the unknown nodal displacements and hydrostatic pressures.

The method of incremental loading was used to solve the system of nonlinear equations, and nineteen 10-psi load increments were employed. Approximate gradients $[\partial F_i/\partial X_j]$, computed by finite differences with a specified $\Delta X_i = 0.0001$, were used in the recurrence formulas.

The deformed and undeformed geometries of the assemblage of finite elements are shown to scale in Fig. 18.32. Since volumes of the finite elements are conserved, a large increase in radius creates a significant decrease in the cross-sectional areas. Straight radial lines in the undeformed body are seen to be considerably distorted in the deformed body. Maximum strains and stresses occur in the interior elements near the junction of the cylindrical and toroidal portions of the container, as expected. Maximum circumferential

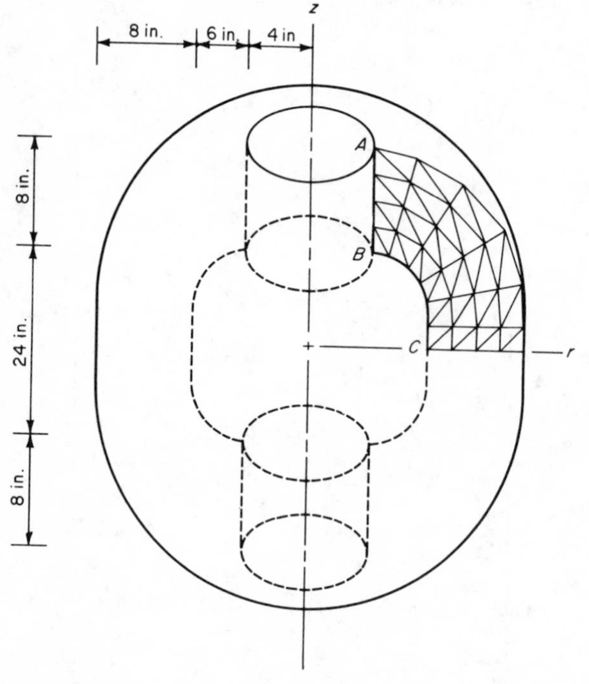

Fig. 18.31 Thick-walled highly elastic container subjected to internal pressure.

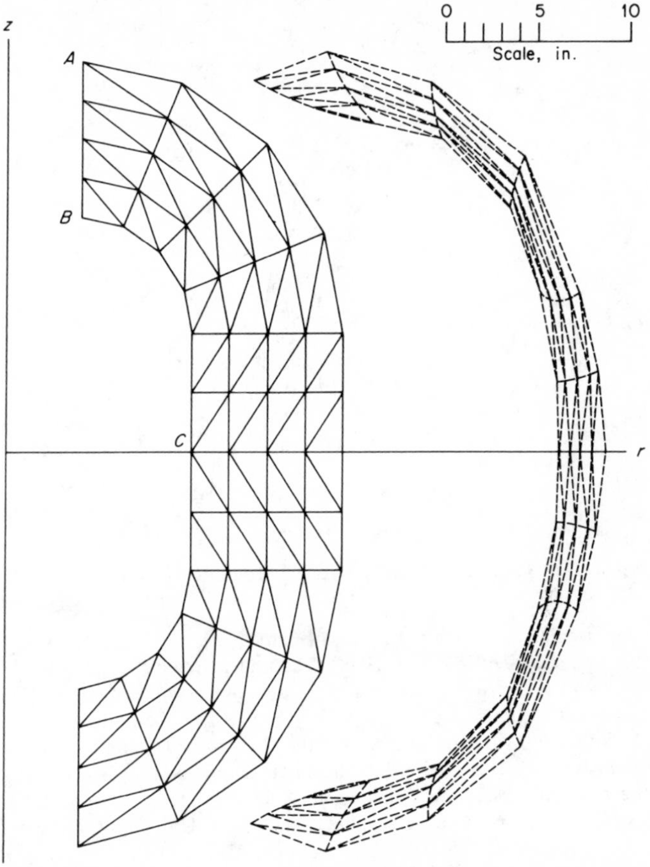

Fig. 18.32 Undeformed and deformed cross sections of container (to scale).

strains on the order of 400 percent are developed. Values for element stresses are obtained as described in Sec. 16.3. Stress contours for components t^{11} and R^2t^{33} are shown in Fig. 18.33.

Figure 18.34 shows the deformed profile of the body due to 14, 17, and 19 load increments (corresponding to 140-, 170-, 190-psi internal pressures). To investigate the behavior for higher internal pressures, 11 additional 10-psi increments after the 190-psi level were added to bring the total internal pressure up to 300 psi. The geometry changes radically for each increment after 190 psi, indicating a pronounced decrease in stiffness with an increase in load. For 300 psi, deformations are unrealistically large, some element areas being reduced to nearly zero. The results indicate that smaller load increments are required after the 190-psi level in order to obtain realistic deformations for higher pressures.

Bending and inflation of a circular plate† An interesting application of the finite-element formulation for axisymmetric solids concerns the quasi-static behavior of a simply

† We reproduce here the results obtained by Oden and Key [1971*a*].

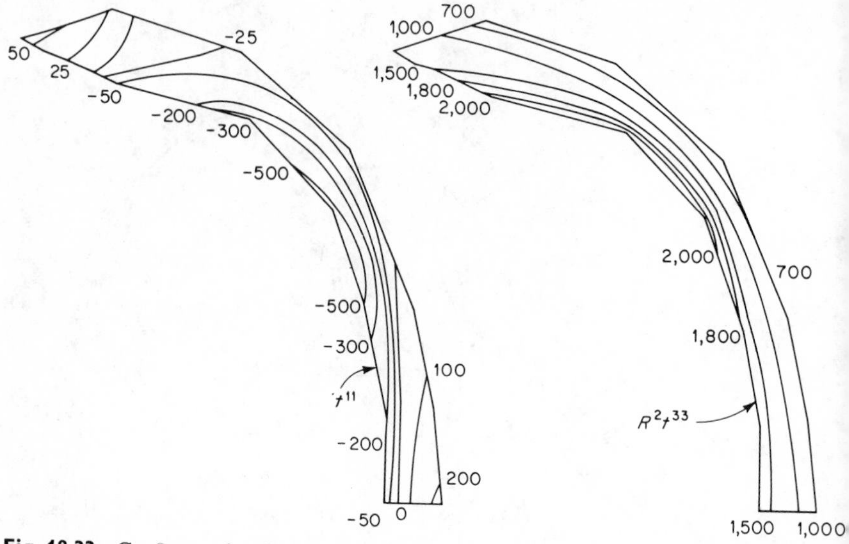

Fig. 18.33 Contours of radial stress t^{11} and circumferential stress R^2t^{33} (in psi and lb).

supported flat circular plate subjected to a piecewise-linear varying external pressure of the type indicated in Fig. 18.35*a*. The undeformed plate is 15 in. in diameter and 0.5 in. thick and is constructed of an isotropic incompressible material of the Mooney type with material constants of $C_1 = 80$ psi, $C_2 = 20$ psi. The finite-element model of the plate is shown in Fig. 18.35*b*; note that the thickness is exaggerated in this figure for clarity. The analysis was performed using the incremental loading procedure described previously; the external pressure was applied in increments corresponding to one unit of time in Fig. 18.35*a*, from 0 psi to a peak value of 43.7 psi. Unloading of the plate followed the same

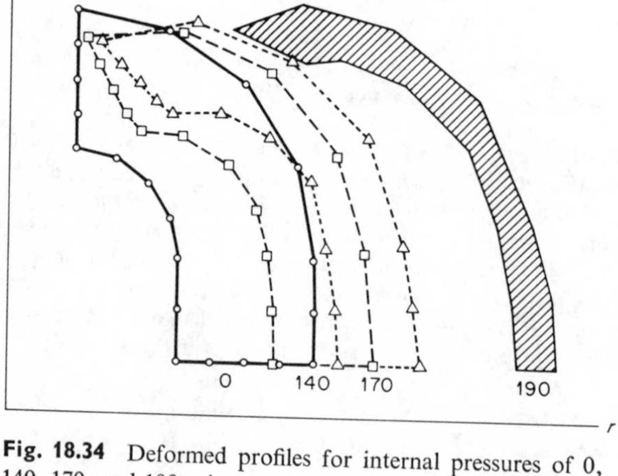

Fig. 18.34 Deformed profiles for internal pressures of 0, 140, 170, and 190 psi.

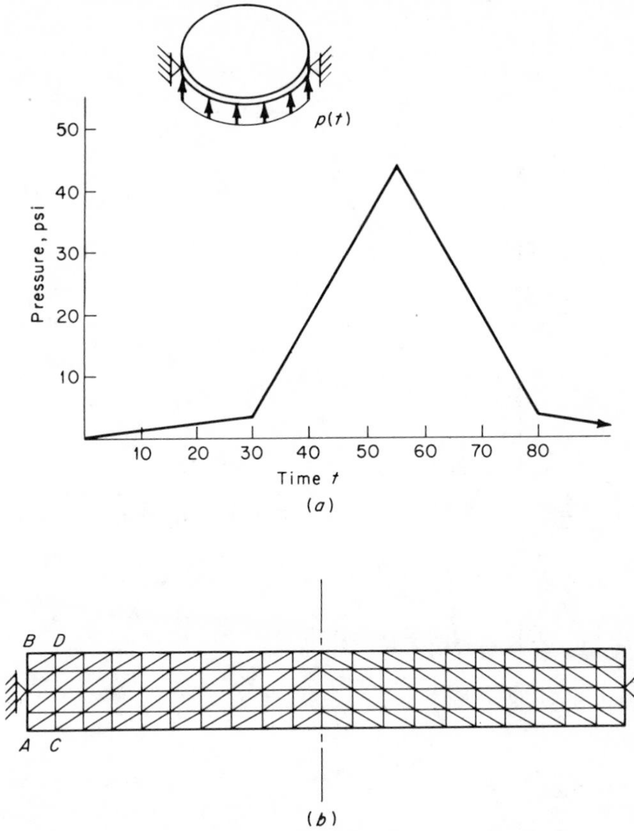

Fig. 18.35 (*a*) Load history of an initially flat circular plate. (*b*) A finite-element model of the plate.

loading history in reverse. We remark that the applied loads in this problem are nonconservative; element loading surfaces change markedly in magnitude and orientation with increases in applied pressure.

A number of preliminary analyses of the problem revealed an extreme sensivitity of the predicted response to the choice of load-increment size. This is illustrated in Fig. 18.36, in which the variation of the transverse deflection of the center of the plate with internal pressure is shown. We observe that the response quickly departs from that predicted by the linear theory after the pressure reaches only 1 psi; the curve then swings upward, being approximately linear from 3 to 28 psi, and then acquires a gradually decreasing slope for increases in applied pressure. Since it is difficult, if not impossible, to anticipate this type of behavior prior to selecting a load-increment size, several trial solutions using coarse finite-element models were obtained using various constant load-increment sizes. The results of these analyses are indicated by the dashed lines in Fig. 18.36; typically, the results of too large a load increment are characterized by a large initial displacement followed by essentially no increase in the displacements of certain nodes with an indefinitely large increase in pressure. Also, local equilibrium of element forces is violated and the deformed

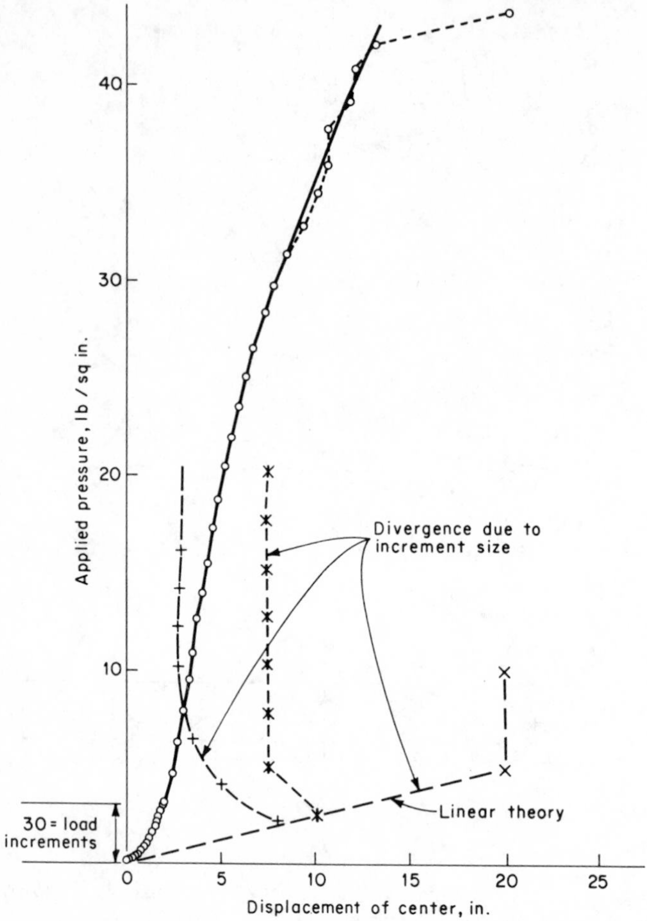

Fig. 18.36 Variation in transverse displacement of the center of the plate with applied pressure.

finite-element network assumes unrealistic distortions. In this particular problem, it was necessary to use at least 30 load increments to depict the response corresponding to an external pressure of only 3 psi; an additional external pressure of 40 psi was then applied in 25 increments.

It was impossible to predict the response of the plate beyond 42 psi without modifying the finite-element model. This was due to numerical instabilities encountered in the incremental loading process and due to the incompressibility of the elements; i.e., at strains on the order of 500 percent, the cross-sectional areas of certain finite elements shrink to nearly zero in order to maintain a constant volume of the ring element, and this results in poor conditioning of the jacobian matrices. To overcome this difficulty, it is necessary to stop the incremental loading process prior to such instabilities and to construct a new, refined finite-element model of the deformed body. Starting values of the nodal displacements of the new model can be obtained from the deformed shape of the initial model using linear or quadratic interpolation. If such a procedure is employed, it is advisable to perform several cycles of Newton-Raphson iteration to further correct nodal displacements and hydrostatic pressures prior to returning to the incremental loading process.

The assumed incompressibility of the material also leads to other computational difficulties in problems of this type. For example, suppose edge $\overline{AB}$ in Fig. 18.35*b* is clamped rather than simply supported and that the plate is subjected to essentially the same program of external pressure as that indicated in Fig. 18.35*a*. Under small increases in pressure, the nodes along, say, $\overline{CD}$ will displace approximately vertically, so that all elements in the rectangle $\overline{ABCD}$ will be approximately in a state of homogeneous, simple shear. This being the case, *any* value of the vertical displacement of nodes along $\overline{CD}$ will lead to isochoric (volume-preserving) deformations. Consequently, the system of

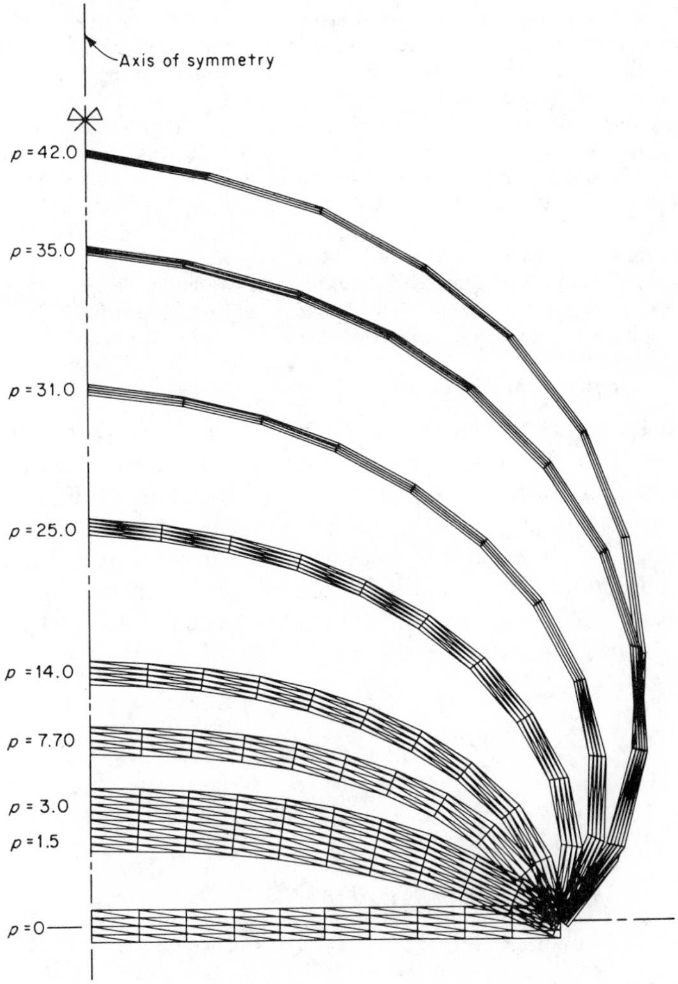

Fig. 18.37 Deformed shapes of the finite-element model of a circular plate for various values of external pressure.

nonlinear equations (equilibrium and incompressibility conditions) will then not possess a unique solution. While it is possible to reformulate the equations in such cases so as to obtain a determinate system, use of an alternate finite-element network, typically of an irregular pattern at the boundaries, may avoid such singularities.

The computed deformed shapes of the finite-element model of the plate for various values of applied pressure are shown in Fig. 18.37. In this particular problem, linear theory appears to be adequate for only about eight 0.1-psi load increments. At pressures of approximately 0.8 to 1.0, extensional strains are still small but rotations begin to be appreciable. Strains appear to remain small (less than 5 percent) until the external pressure reaches a value of approximately 4.0 psi. At this level of pressure, rotations of certain line elements reach values of 50°. Shear deformations and transverse strains at this level of loading are not appreciable; straight lines originally normal to the undeformed middle surface are still approximately straight and normal to the deformed middle surface, i.e., the Kirchhoff-Love hypothesis appears to hold for relatively large deflections. Results indicate that nonlinear plate theories of the Von Karman type (i.e., nonlinear theories based on the Kirchhoff hypothesis which assume moderate rotations but infinitesimal strains) should be adequate for pressures up to approximately 4.0 psi. After 4.0 psi rotations become quite large, and at 6.0 to 8.0 psi, extensional strains of the middle surface attain values of approximately 20 percent. At this load level, membrane action begins to take a dominant role in the behavior of the plate; i.e., the plate begins to inflate rather than bend. At 25 psi, transverse extensional strains reach −20 percent and the rotations of certain line elements exceed 90°. At a peak pressure of 43.7 psi, strains of the order of 500 percent are obtained, line elements near the support have suffered rotations of 128°, the thickness of the plate has shrunk from 0.5 in. to approximately 0.15 in., and the plate assumes the considerably distorted shape indicated in the figure.

18.3 FINITE-PLANE STRAIN

The theory of finite-plane strain of isotropic hyperelastic bodies can be regarded as a special case of that for axisymmetric solids discussed in the previous section. Consider a homogeneous solid subjected to plane deformations parallel to the x^1, x^2 plane. In addition, the body may be subjected to a uniform extension parallel to the x^3 axis defined by a prescribed constant extension ratio λ. Then $\gamma_{\alpha\beta}$ is given by (18.48a), $\gamma_{\alpha 3} = 0$, and $\gamma_{33} = \frac{1}{2}(\lambda^2 - 1)$, where λ is now a known constant. The principal invariants of the deformation tensor are

$$
\begin{aligned}
I_1 &= 2(1 + \gamma^\alpha_\alpha) + \lambda^2 \\
I_2 &= 1 + 2\lambda^2 + 2(1 + \lambda^2)\gamma^\alpha_\alpha + \hat{\varphi} \\
I_3 &= \lambda^2(1 + 2\gamma^\alpha_\alpha + \hat{\varphi})
\end{aligned}
\qquad (18.70)
$$

in which $\gamma^\alpha_\alpha = \gamma_{\alpha\alpha} = \gamma_{11} + \gamma_{22}$ and

$$\hat{\varphi} = 4(\gamma_{11}\gamma_{22} - \gamma_{12}\gamma_{21}) = 2\epsilon^{\alpha\lambda}\epsilon^{\beta\mu}\gamma_{\alpha\beta}\gamma_{\lambda\mu} \qquad (18.71)$$

For isochoric deformations, $I_3 = 1$ and we can rewrite (18.70) in the form

$$
\begin{aligned}
I_1 &= c - \hat{\varphi} \\
I_2 &= c - \lambda^2\hat{\varphi}
\end{aligned}
\qquad (18.72)
$$

where

$$c = 1 + \lambda^2 + \frac{1}{\lambda^2} = \text{const} \tag{18.73}$$

Consequently, for isotropic incompressible solids in finite-plane strain,

$$t^{\alpha\beta} = 2\frac{\partial \hat{W}}{\partial I_1}\delta^{\alpha\beta} + 2\frac{\partial \hat{W}}{\partial I_2}\left[\delta^{\alpha\beta}\left(\lambda^2 + \frac{1}{\lambda^2}\right) + \frac{1}{2}\frac{\partial \hat{\varphi}}{\partial \gamma_{\alpha\beta}}\right] + h\left(\delta^{\alpha\beta} + \frac{1}{2}\frac{\partial \hat{\varphi}}{\partial \gamma_{\alpha\beta}}\right) \tag{18.74a}$$

$$t^{\alpha 3} = 0 \tag{18.74b}$$

$$t^{33} = 2\lambda^2\frac{\partial \hat{W}}{\partial I_1} + 4\lambda^2\frac{\partial \hat{W}}{\partial I_2}(1 + \gamma_\alpha^\alpha) + h \tag{18.74c}$$

Finite-element approximations† Following the usual procedure, we obtain for the equilibrium equations of an isotropic incompressible finite element in plane strain

$$2\int_{v_{0(e)}} (\delta_{\beta\alpha} + \psi_{M,\beta}u_\alpha^M)\left\{\psi_{N,\beta}\left[\frac{\partial \hat{W}}{\partial I_1} + (1 + \lambda^2)\frac{\partial \hat{W}}{\partial I_2} + \lambda^2 h\right] + 2\left(\frac{\partial \hat{W}}{\partial I_2} + \lambda^2 h\right)\epsilon^{\nu\mu}\epsilon^{\lambda\beta}\psi_{N,\mu}\gamma_{\nu\lambda}\right\} dv_0 = p_{N\alpha} \tag{18.75}$$

In the case of simplex approximations, (18.75) reduces to

$$2v_{0(e)}(\delta_{\beta\alpha} + b_{M\beta}u_\alpha^M)\left\{b_{N\beta}\left[\frac{\partial \hat{W}}{\partial I_1} + (1 + \lambda^2)\frac{\partial \hat{W}}{\partial I_2} + \lambda^2 h\right] + 2\left(\frac{\partial \hat{W}}{\partial I_2} + \lambda^2 h\right)\epsilon^{\nu\mu}\epsilon^{\lambda\beta}b_{N\mu}\gamma_{\nu\lambda}\right\} = p_{N\alpha} \tag{18.76}$$

The incompressibility condition is then simply

$$A_0 = A \tag{18.77}$$

where A_0 and A are as defined in (18.64). Here we assume that λ is prescribed prior to subjecting the body to plane deformations. The generalized forces are evaluated using procedures outlined in Sec. 16.3. In the case of uniform normal tractions, $p_{N\alpha}$ is given by (18.61).

Simple and generalized shear We first consider applications of (18.76) to some rather simple problems in finite-plane strain for which exact solutions are known.

For convenience, the material is assumed to have a strain-energy function of the Mooney form. The classical problem of simple shear involves finite-plane deformations of a rectangular section of a material in which straight-line elements parallel to, say, the x_1 axis are displaced relative to one another in the y_1 direction but remain straight and parallel in the deformed body. A cubic specimen in simple shear is shown in Fig. 18.38.

Let A, B, C, and D denote corner points in the specimen, and let Δ denote the total

† We follow here the work of Oden [1968*b*].

horizontal displacement of A and B relative to C and D. Assuming that $\lambda = 1$ and employing the finite-element network shown in Fig. 18.38*b*, it is found that for the kth element on the line AC (shaded in the figure),

$$[b_{N\alpha}] = n \begin{bmatrix} 0 & 1 \\ -1 & -1 \\ 1 & 0 \end{bmatrix} \tag{18.78}$$

and

$$\begin{aligned} u^1_{1(k)} &= \frac{k\Delta}{n} \qquad u^2_{1(k)} = u^3_{1(k)} = \frac{(k-1)\,\Delta}{n} \\ u^1_{2(k)} &= u^2_{2(k)} = u^3_{3(k)} = 0 \end{aligned} \tag{18.79}$$

We find for the strains $\gamma_{11} = \gamma_{23} = \gamma_{33} = 0$ and $\gamma_{12} = \gamma_{21} = \Delta/2, \gamma_{22} = (\frac{1}{2})\Delta^2$. Thus, the strain components are independent of n (and k), and the same result is obtained for all elements in the network, regardless of the number of elements used. This follows immediately, of course, from the fact that in simple shear all elements experience pure homogeneous strain, which is precisely the mode of deformation assumed in deriving stiffness relations

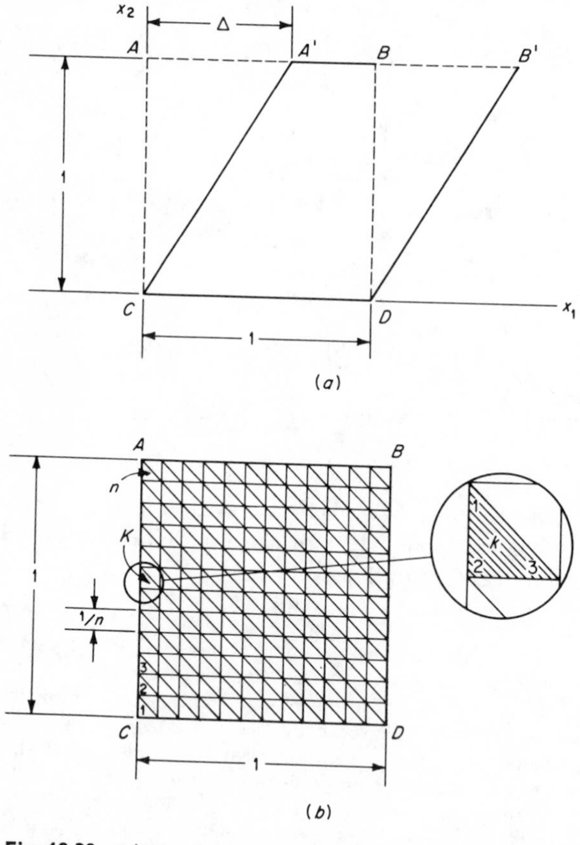

Fig. 18.38 (*a*) Simple shear deformation. (*b*) Finite-element representation of square specimen.

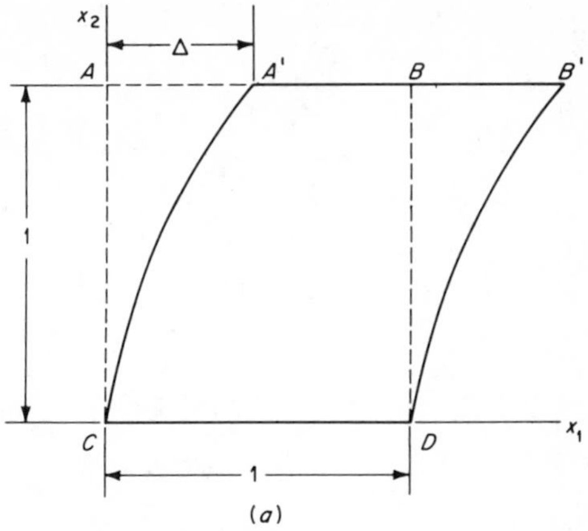

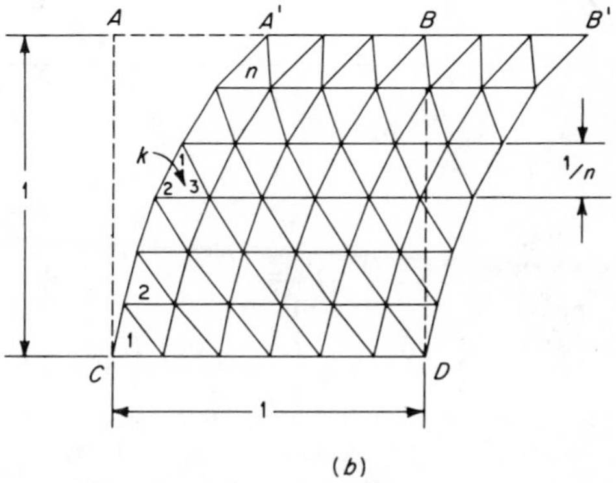

Fig. 18.39 (*a*) Generalized shear deformations. (*b*) Deformed finite-element network.

for finite elements. Thus, *the simplex finite-element solution is exact for the case of simple shear.*

For a Mooney material, we find that

$$
\begin{aligned}
t^{11} &= 2C_1 + 2C_2(2 + \Delta^2) + p(1 + \Delta^2) \\
t^{12} &= -(2C_2 + p)\Delta \\
t^{33} &= 2C_1 + 4C_2 + p
\end{aligned}
\tag{18.80}
$$

The incompressibility condition is automatically satisfied (isochoric deformation), and the stresses are in agreement with exact solutions.†

In the case of generalized shear, sides AC and BC of the specimen are allowed to deform in a general manner (Fig. 18.39). For an incompressible isotropic solid with strain energy of a Mooney form, Green and Adkins [1960, p. 122] have shown that the curve AC (and BC) in the deformed state takes the shape of a quadric parabola. Interesting

† Green and Adkins [1960].

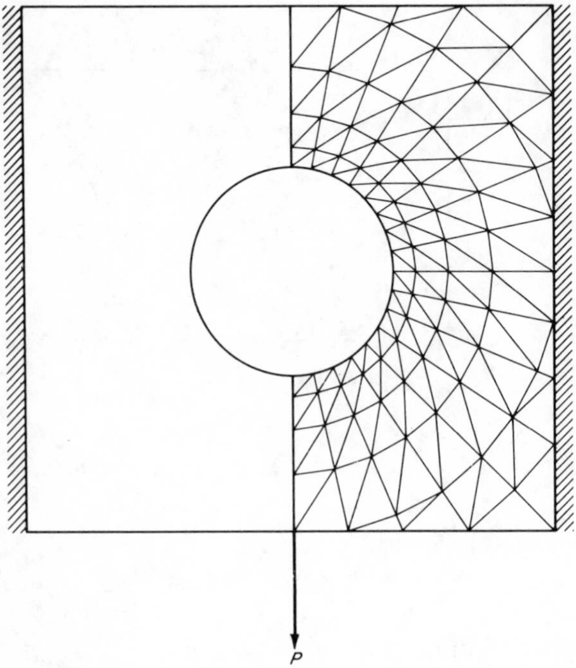

Fig. 18.40 Incompressible plane body with central hole subjected to concentrated edge force P, with finite-element model of the body.

results are obtained in a finite-element analysis of this problem in which the displacements of node points along lines parallel to AC are prescribed to fit a simple parabola. The deformed finite-element network for $2n^2$ elements (n divisions along each side) is shown in Fig. 18.39*b*.

For the kth element,

$$u^1_{1(k)} = \Delta\left(\frac{k}{n}\right)^2 \qquad u^2_{1(k)} = u^3_{1(k)} = \Delta\left(\frac{k-1}{n}\right)^2$$

$$\gamma_{12} = \frac{\Delta}{2}\left(\frac{2k}{n} - \frac{1}{n}\right) \tag{18.81}$$

$$\gamma_{22} = \frac{\Delta^2}{2}\left[\left(\frac{2k}{n}\right)^2 - \frac{2k}{n^2} + \frac{1}{n^2}\right]$$

the remaining components being zero. Stresses are calculated as before. For example, we find for the component t^{11} in element k,

$$t^{11} = 2C_1 + 4C_2 + p + (2C_2 + p)\frac{\Delta^2}{n^2}(2k-1)^2 \tag{18.82}$$

Plane body with circular hole As a final example, we describe briefly numerical results obtained from a finite-element analysis† of the problem of finite-plane strain of an

† See Oden and Key [1971*a*].

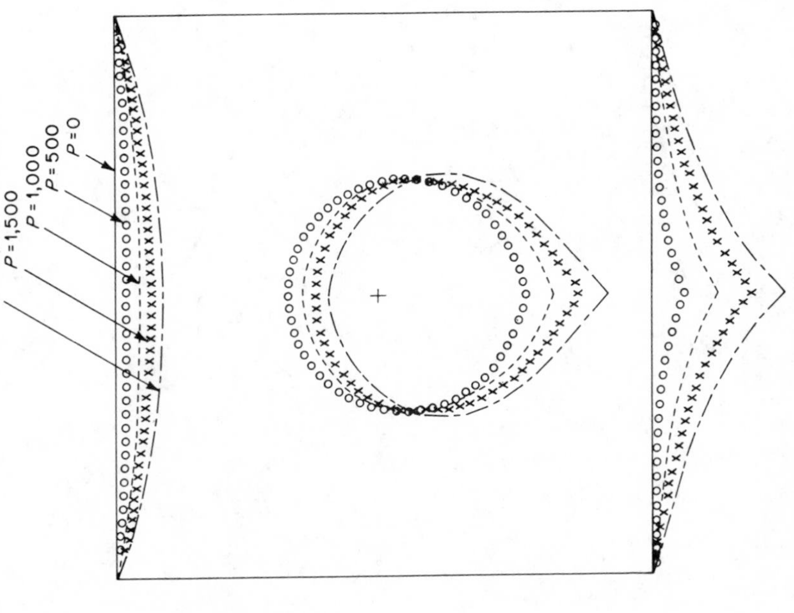

Fig. 18.41 Computed deformed profiles for various values of applied force.

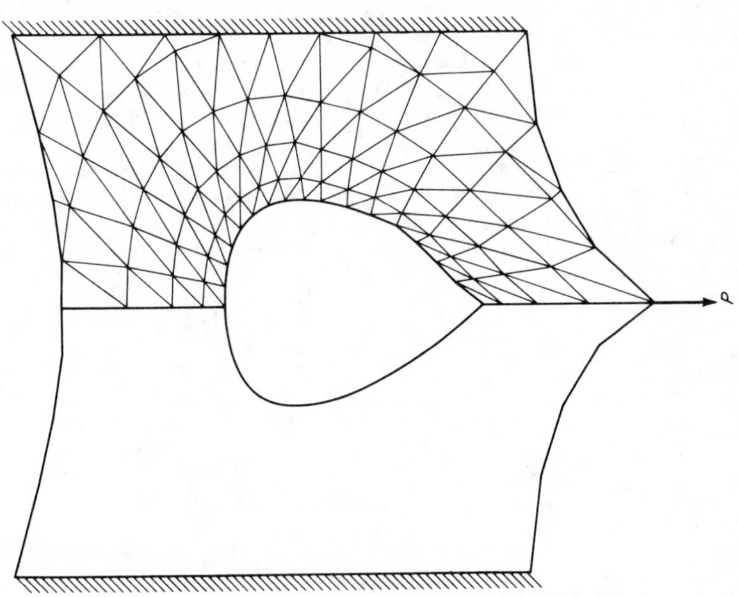

Fig. 18.42 Deformed shape of the finite-element model at a load P of 2,000 lb.

incompressible square specimen of Mooney-type rubber containing a centrally located circular hole. The specimen is clamped along two opposite edges, the remaining edges being free, and is subjected to a concentrated edge force at the center of the bottom edge, as indicated in Fig. 18.40. Dimensions of the undeformed plate are 10 × 10 in., the central hole being initially 4.0 in. in diameter, and the Mooney constants of the plate were assumed to be $C_1 = 80$ psi, $C_2 = 20$ psi. The finite-element model indicated in Fig. 18.40 was used in the analysis, and an external load P of 2,000 lb was applied in twenty 100 lb increments.

The computed deformed profiles of the plate at various levels of external load are shown in Fig. 18.41, and the deformed finite-element network at $P = 2{,}000$ lb is shown in Fig. 18.42. As expected, strain singularities at the point of application of the load are evident. Strains reached an order of 50 percent and the principal "diameter" of the distorted, heart-shaped central hole reached 5 in.

V
Nonlinear Thermomechanical Behavior

In this chapter, we apply the general equations of motion and heat conduction developed in Chap. III for finite elements of general continua to specific nonlinear problems in the study of thermomechanical behavior of solids. While, in principle, the generation of such discrete models is accomplished by merely introducing into the general equations appropriate constitutive equations for the material at hand, the character and complexity as well as the quantitative solution of the resulting systems of equations can be appreciated only when specific cases are considered. In the pages to follow, we first examine the theory of simple materials with memory and then consider special forms of the constitutive functionals consistent with that theory. We conclude our study by constructing finite-element models of such materials and by applying these to the analysis of representative nonlinear problems.

19 THERMOMECHANICS OF MATERIALS WITH MEMORY

It is common experience that mechanical working of a solid body generates heat and that, conversely, heating a body produces mechanical work. It is also common experience that the properties of many materials may vary with time, temperature, and the rate at which they are deformed. Plastics, polymers, biological tissue, and many metals, for example, may continue to deform under constant applied load, and the rate of continued deformation may be highly sensitive to changes in temperature. The behavior of materials of this type may depend not only upon the current deformation and temperature but also upon their histories. In general, their response is also characterized by a high dissipation of energy, a feature which distinguishes such materials markedly from the hyperelastic materials discussed previously. The thermomechanical behavior of a large class of dissipative materials is encompassed by the thermodynamical theory of simple materials; we examine highlights of this theory in the following sections.

19.1 THERMODYNAMICS OF SIMPLE MATERIALS†

We recall from Art. 14 [see, for example, (14.27)] that thermomechanically simple materials are characterized by a collection of four constitutive equations giving the stress, heat flux, entropy density, and free energy as functionals of the histories of the deformation, the temperature, and possibly the temperature gradients. In the present discussion, we assume, for convenience, that the material is homogeneous, i.e., that a homogeneous configuration C_0 has been taken as the reference configuration. Then the constitutive functionals and the density $\rho = \rho_0$ are independent of the material particle $\mathbf{x}$. Further, making use of the fact that $\gamma_{ij} = (\frac{1}{2})(z_{m,i}z_{m,j} - \delta_{ij})$ and of the local-action principle, we choose to take as the dependent constitutive variables the strain histories $\boldsymbol{\gamma}^t(\mathbf{x},s) = \boldsymbol{\gamma}(\mathbf{x}, t - s)$, $\boldsymbol{\gamma}(\mathbf{x},t)$ being the strain tensor at $\mathbf{x}$ at time t, the temperature history $\theta^t(\mathbf{x},s) = \theta(\mathbf{x}, t - s)$, and the current temperature gradient $\mathbf{g}(\mathbf{x},t) = \boldsymbol{\nabla}\theta(\mathbf{x},t)$; that is, we consider a class of simple materials described by constitutive equations of the form

$$
\begin{aligned}
\psi &= \mathop{\Psi}_{s=0}^{\infty} [\boldsymbol{\gamma}^t(s),\theta^t(s);\mathbf{g}(t)] \\
\mathbf{T} &= \mathop{\mathfrak{T}}_{s=0}^{\infty} [\boldsymbol{\gamma}^t(s),\theta^t(s);\mathbf{g}(t)] \\
\eta &= \mathop{\mathcal{N}}_{s=0}^{\infty} [\boldsymbol{\gamma}^t(s),\theta^t(s);\mathbf{g}(t)] \\
\mathbf{q} &= \mathop{\mathfrak{Q}}_{s=0}^{\infty} [\boldsymbol{\gamma}^t(s),\theta^t(s);\mathbf{g}(t)]
\end{aligned}
\tag{19.1}
$$

† The mechanical theory of simple materials was formulated by Noll [1958]; it was extended to thermomechanical phenomena by Coleman [1964] and was further generalized by Coleman and his collaborators in subsequent studies. See, in particular, the writings of Coleman, Gurtin, and Herrera [1965], Coleman and Gurtin [1965*a*, 1965*b*], and Coleman and Mizel [1967], on which our present development is based.

where ψ and η are the free energy and entropy densities, $\mathbf{T} = \sigma^{ij}\mathbf{G}_i \otimes \mathbf{G}_j$ is the stress tensor, $\mathbf{q}$ is the heat flux, and $\Psi^{\infty}_{s=0}, \ldots, \mathfrak{Q}^{\infty}_{s=0}$ are the corresponding constitutive functionals. We have added limits to the functionals in (19.1) to emphasize that the past-time parameters range from 0, corresponding to the current time, to ∞. The dependence of ψ, $\mathbf{T}$, η, and $\mathbf{q}$ on $\mathbf{x}$ and t and of $\boldsymbol{\gamma}^t(s)$, $\theta^t(s)$, and $\mathbf{g}(t)$ on $\mathbf{x}$ is understood. Note that the forms of the functionals $\Psi^{\infty}_{s=0}, \ldots, \mathfrak{Q}^{\infty}_{s=0}$ may differ from those in, say, (14.28) owing to our present choice of slightly different measures of the deformation. Note also that the same collection of variables appear as arguments of each constitutive functional in accordance with the equipresence rule discussed in Art. 14.2.

To facilitate subsequent calculations, it is convenient to introduce the notion of past histories. The idea amounts to simply distinguishing the current value of a function from all its values in the past. For example, let $u^t(s)$ be a real-valued function on $s \in [0,\infty)$ for fixed t; $u^t(s) = u(t - s)$ is, as before, referred to as the total history of u at time t. The restriction of $u^t(s)$ to the open interval $(0,\infty)$ is called the past history of $u^t(s)$ and is denoted $u^t_r(s)$; that is,

$$u^t_r(s) = u(t - s) \qquad 0 < s < \infty \tag{19.2}$$

Thus, the total history $u^t(s)$ can be described by the pair $[u^t_r(s), u^t(0)]$.

Returning to (19.1), we now choose to express the constitutive functionals in terms of the past histories $\boldsymbol{\gamma}^t_r(s)$, $\theta^t_r(s)$ rather than the total histories $\boldsymbol{\gamma}^t(s)$, $\theta^t(s)$. Introducing new functionals $\bar{\Psi}^{\infty}_{s=0}, \ldots, \bar{\mathfrak{Q}}^{\infty}_{s=0}$ of the type

$$\mathop{\bar{\Psi}}_{s=0}^{\infty} [\boldsymbol{\gamma}^t_r(s), \theta^t_r(s); \boldsymbol{\gamma}, \theta, \mathbf{g}] = \mathop{\Psi}_{s=0}^{\infty} [\boldsymbol{\gamma}^t(s), \theta^t(s); \mathbf{g}] \tag{19.3}$$

etc., where $\boldsymbol{\gamma}$, θ, and $\mathbf{g}$ are functions of $\mathbf{x}$ and t, we then have, instead of (19.1),

$$\begin{aligned}
\psi &= \mathop{\bar{\Psi}}_{s=0}^{\infty} [\boldsymbol{\gamma}^t_r(s), \theta^t_r(s); \boldsymbol{\gamma}, \theta, \mathbf{g}] \\
\mathbf{T} &= \mathop{\bar{\mathfrak{T}}}_{s=0}^{\infty} [\boldsymbol{\gamma}^t_r(s), \theta^t_r(s); \boldsymbol{\gamma}, \theta, \mathbf{g}] \\
\eta &= \mathop{\bar{\mathcal{N}}}_{s=0}^{\infty} [\boldsymbol{\gamma}^t_r(s), \theta^t_r(s); \boldsymbol{\gamma}, \theta, \mathbf{g}] \\
\mathbf{q} &= \mathop{\bar{\mathfrak{Q}}}_{s=0}^{\infty} [\boldsymbol{\gamma}^t_r(s), \theta^t_r(s); \boldsymbol{\gamma}, \theta, \mathbf{g}]
\end{aligned} \tag{19.4}$$

In order to be physically admissible, the functionals in (19.4) must be such that the basic physical laws governing thermomechanical behavior are not violated (rule 1, Art. 14.2). In particular, we require that the local forms of the law of conservation of energy and the Clausius-Duhem inequality be satisfied;† for example,

$$\rho\dot{\psi} = \sigma^{ij}\dot{\gamma}_{ij} - \rho\eta\dot{\theta} - \delta$$
$$\rho\theta\dot{\eta} - \nabla \cdot \mathbf{q} + \frac{1}{\theta}\mathbf{q} \cdot \mathbf{g} - \rho r \geq 0 \tag{19.5}$$

where δ is the internal-dissipation density.

To test the suitability of (19.4) by introducing the constitutive equations into (19.5), however, requires that we compute time rates of change of certain of the constitutive functionals. This necessitates the use of Frechet differentiation, as described in Sec. 10.1 [see (10.19)]. We restate and extend pertinent ideas below.

Smoothness properties of functionals Let $\mathcal{H}$ be a real normed linear space, the elements of which are functions u defined on the open interval $(0,\infty)$ and which have norms $\|u\|$. Let $\mathcal{H}^\dagger$ denote the set of real-valued functions u defined on $[0,\infty)$ such that the restriction u_r of u to $(0,\infty)$ belongs to $\mathcal{H}$. Then $\mathcal{H}^\dagger$ is also a normed linear space on which the norm is defined by

$$\|u\|^\dagger = \|u_r(s)\| + |u(0)| \tag{19.6}$$

Let $K_{s=0}^{\infty}$ be a functional defined on each $u \in \mathcal{H}^\dagger$ which is Frechet differentiable in the sense of (10.19). Then, for arbitrary $h \in \mathcal{H}^\dagger$, we may write

$$\underset{s=0}{\overset{\infty}{K}}(u+h) = \underset{s=0}{\overset{\infty}{K}}(u) + \delta^\dagger \underset{s=0}{\overset{\infty}{K}}(u \mid h) + \omega(u,h) \tag{19.7}$$

where $\delta^\dagger K_{s=0}^{\infty}(u \mid h)$ is the first Frechet differential of $K_{s=0}^{\infty}$ in $\mathcal{H}^\dagger$ and the remainder $\omega(u,h)$ has the property

$$\lim_{\|h\|^\dagger \to 0} \frac{1}{\|h\|^\dagger} \|\omega(u,h)\|^\dagger = 0 \tag{19.8}$$

We assume that $\delta^\dagger K_{s=0}^{\infty}(u \mid h)$ is jointly continuous in u and h, and the vertical stroke in $\delta^\dagger K_{s=0}^{\infty}(u \mid h)$ is introduced to emphasize that the differential is linear in h. Repeating the operation (19.7) N times, we may also write

$$\underset{s=0}{\overset{\infty}{K}}(u+h) = \underset{s=0}{\overset{\infty}{K}}(u) + \sum_{n=1}^{N} \frac{1}{n!} \delta^{\dagger n} \underset{s=0}{\overset{\infty}{K}}(u \mid h, h, \ldots, h) + \omega^N(u,h) \tag{19.9}$$

where $\delta^{\dagger n} K_{s=0}^{\infty}(u \mid h, h, \ldots, h)$ is an N-linear form in h, called the Nth Frechet differential of $K_{s=0}^{\infty}$ at u, and

$$\lim_{\|h\|^\dagger \to 0} \frac{1}{\|h\|^{\dagger N}} \|\omega^N(u,h)\|^\dagger \to 0 \tag{19.10}$$

Here $\delta^{\dagger n} K_{s=0}^{\infty}(u \mid h, h, \ldots, h)$ is a symmetric n-linear mapping of $\mathcal{H}^\dagger \times \mathcal{H}^\dagger \times \cdots \times \mathcal{H}^\dagger$ into the reals.

It is convenient to recast $\delta^\dagger K_{s=0}^{\infty}(u \mid h)$ into a form which differentiates u from its restriction u_r on $(0,\infty)$. This is accomplished by using, instead of $K_{s=0}^{\infty}(u)$, another

† See Eqs. (12.31), (12.33), and (12.35).

functional $\bar{K}$ defined by

$$\mathop{\bar{K}}_{s=0}^{\infty}[v(s);w] = \mathop{K}_{s=0}^{\infty}(u) \tag{19.11}$$

where $v(s) = u_r(s)$ is the restriction of u to $(0,\infty)$ and $w = u(0)$. Then, the *total* Frechet differential of $K_{s=0}^{\infty}(u)$ is

$$\delta^{\dagger} \mathop{K}_{s=0}^{\infty}(u \mid h) = \delta \mathop{\bar{K}}_{s=0}^{\infty}[v(s);w \mid h_r(s)] + D \mathop{\bar{K}}_{s=0}^{\infty}[v(s);w]h(0) \tag{19.12}$$

where

$$\delta \mathop{\bar{K}}_{s=0}^{\infty}[v(s);w \mid h_r(s)] = \mathop{\bar{K}}_{s=0}^{\infty}[v + h_r;w] - \mathop{\bar{K}}_{s=0}^{\infty}(v;w) - \omega(v,h_r) \tag{19.13a}$$

$\omega(v,h_r) = o(\|h_r\|)$, and in view of (10.21),

$$D \mathop{\bar{K}}_{s=0}^{\infty}[v(s);w] = \frac{\partial}{\partial w} \mathop{\bar{K}}_{s=0}^{\infty}[v(s);w] \tag{19.13b}$$

Clearly, $D\,\bar{K}_{s=0}^{\infty}[\ ;\ ]h(0)$ is a linear function of $h(0)$ and $\delta\,\bar{K}_{s=0}^{\infty}[\ ;\ \mid h_r]$ is a linear functional of $h_r(s)$.

It is not too difficult to extend the above ideas to more general spaces. To cite one such extension of particular importance in derivations to be presented shortly, let $\mathbf{\Gamma}$ denote the ordered pair $(\mathbf{T},\beta)$, where $\mathbf{T}$ is an arbitrary second-order tensor on $\mathscr{E}^{33}$ and β is a scalar. Further, let $\mathscr{A}$ denote the set of all such quantities, and suppose that we introduce binary operations of addition and scalar multiplication on pairs of elements of $\mathscr{A}$ by the definitions

$$\begin{aligned} \mathbf{\Gamma}_1 + \mathbf{\Gamma}_2 &= (\mathbf{T}_1 + \mathbf{T}_2, \beta_1 + \beta_2) \\ \alpha \cdot \mathbf{\Gamma} &= (\alpha\mathbf{T},\alpha\beta) \end{aligned} \tag{19.14}$$

where $\mathbf{\Gamma}_1 = (\mathbf{T}_1,\beta_1)$, $\mathbf{\Gamma}_2 = (\mathbf{T}_2,\beta_2) \in \mathscr{A}$, and α is a scalar. The algebraic system $(\mathscr{A},+,\cdot)$, denoted $\mathscr{V}^{10}$, constitutes a 10-dimensional linear space with zero element $\mathbf{O} = (0,0)$. It is also possible to introduce an inner product into $\mathscr{V}^{10}$—defined, for example, by

$$\mathbf{\Gamma}_1 \cdot \mathbf{\Gamma}_2 \equiv \langle \mathbf{\Gamma}_1,\mathbf{\Gamma}_2 \rangle = \operatorname{tr} \mathbf{T}_1\mathbf{T}_2^T + \beta_1\beta_2 \tag{19.15}$$

where tr denotes the trace and $\mathbf{T}_2^T$ is the transpose of $\mathbf{T}_2$. The natural norm is then denoted

$$\|\mathbf{\Gamma}\|_0 = \sqrt{\langle \mathbf{\Gamma},\mathbf{\Gamma} \rangle} \tag{19.16}$$

The set C^{10} consisting of elements $\mathbf{\Gamma} = (\mathbf{T},\beta)$ such that $\det \mathbf{T} \neq 0$ and $\beta > 0$ forms a *cone* in $\mathscr{V}^{10}$.

We now consider the Banach space $\mathscr{H}$ of past histories formed by Lebesque measurable functions from $(0,\infty)$ into $\mathscr{V}^{10}$, with norm $\|\mathbf{\Gamma}\|$, called the *norm on past histories*, which has the following properties:†

1. If $\mathbf{\Gamma}_1 \in \mathscr{H}$ and $\mathbf{\Gamma}$: $(0,\infty) \to \mathscr{V}^{10}$ and if $\|\mathbf{\Gamma}\|_0 \le \|\mathbf{\Gamma}_1\|_0$ almost everywhere in $(0,\infty)$, then $\|\mathbf{\Gamma}\| \le \|\mathbf{\Gamma}_1\|$; partial ordering of norms is preserved under the mapping. Moreover, if $\{\mathbf{\Gamma}_n\} \in \mathscr{H}$ and $\|\mathbf{\Gamma}_n\| \le K$, where K is finite, and if $\|\mathbf{\Gamma}_n(s)\|_0 \to \|\mathbf{\Lambda}(s)\|_0$, $\mathbf{\Lambda}$ being measurable, then $\mathbf{\Lambda} \in \mathscr{H}$ and $\lim_{n\to\infty} \|\mathbf{\Gamma}_n\| = \|\mathbf{\Lambda}\| \le K$.

† Cf. Coleman and Mizel [1967].

2. Let $\mathbf{\Omega} \in \mathscr{V}^{10}$ be constant. Then there exists in $\mathscr{H}$ an element $\boldsymbol{\omega}$ such that $\boldsymbol{\omega}(s) = \mathbf{\Omega}$ for every $s \in (0,\infty)$.
3. Let $C^{10} \subset \mathscr{V}^{10}$ be the cone $C^{10} = \{\mathbf{\Gamma} = (\mathbf{T},\beta) \mid \mathbf{\Gamma} \in \mathscr{V}^{10}; \det \mathbf{T} \neq 0, \beta > 0\}$. Then C^{10} is the image of a cone $\mathscr{C} \subset \mathscr{H}$, the absolutely continuous functions of which, whose derivatives belong to $\mathscr{H}$, are dense in $\mathscr{C}$.
4. If $\mathbf{\Gamma} \in \mathscr{H}$, then, for every $\epsilon \geq 0$, $S^\epsilon \mathbf{\Gamma}(s)$ and $S_\epsilon \mathbf{\Gamma}(s) \in \mathscr{H}$ where $S^\epsilon \mathbf{\Gamma}(s) = \mathbf{\Gamma}(s - \epsilon)$ if $s \in (\epsilon,\infty)$ or $= \mathbf{0}$ if $s \in (0,\epsilon]$ and $S_\epsilon \mathbf{\Gamma}(s) = \mathbf{\Gamma}(s + \epsilon)$ for $s \in (0,\infty)$.

In addition to $\mathscr{H}$, we may define the Banach space $\mathscr{H}^\dagger$ of total histories, which consists of the set of Lebesgue-measurable functions $\mathbf{\Gamma}^t$ which map $[0,\infty)$ into $\mathscr{V}^{10}$ and which have finite norm: $\|\mathbf{\Gamma}_r^t\| < \infty$. Here $\mathbf{\Gamma}_r^t$ is the equivalence class in $\mathscr{H}^\dagger$ corresponding to the restriction of $\mathbf{\Gamma}^t$ to $(0,\infty)$. The norm in $\mathscr{H}^\dagger$, called the *norm on total histories*, is defined by

$$\|\mathbf{\Gamma}^t\|^\dagger = \|\mathbf{\Gamma}_r^t\| + \|\mathbf{\Gamma}^t(0)\|_0 \tag{19.17}$$

The elements in $\mathscr{H}^\dagger$ corresponding to functions $\mathbf{\Gamma}^t$ mapping $[0,\infty)$ into the cone C^{10} form a cone $\mathscr{C}^\dagger$ in $\mathscr{H}^\dagger$.

Now let $K_{s=0}^{\infty}$ be a continuous functional defined on $\mathscr{C}^\dagger \oplus \mathscr{V}^3$, where $\mathscr{V}^3$ is a three-dimensional vector space. Then

$$\underset{s=0}{\overset{\infty}{K}}[\mathbf{\Gamma}^t;\mathbf{v}] = \underset{s=0}{\overset{\infty}{\bar{K}}}[\mathbf{\Gamma}_r^t;\mathbf{\Gamma},\mathbf{v}] \tag{19.18}$$

where $\mathbf{v} \in \mathscr{V}^3$. Assuming that $\bar{K}_{s=0}^{\infty}$ is continuously Frechet differentiable, we may write

$$\underset{s=0}{\overset{\infty}{\bar{K}}}[\mathbf{\Gamma}_r^t + \mathbf{\Phi}; \mathbf{\Gamma} + \mathbf{\Omega}, \mathbf{v} + \mathbf{w}] = \underset{s=0}{\overset{\infty}{\bar{K}}}[\mathbf{\Gamma}_r^t;\mathbf{\Gamma},\mathbf{v}] + \delta \underset{s=0}{\overset{\infty}{\bar{K}}}[\mathbf{\Gamma}_r^t;\mathbf{\Gamma},\mathbf{v} \mid \mathbf{\Phi}]$$
$$+ D_{\mathbf{\Gamma}} \underset{s=0}{\overset{\infty}{\bar{K}}}[\mathbf{\Gamma}_r^t;\mathbf{\Gamma},\mathbf{v}] \cdot \mathbf{\Omega} + \partial_{\mathbf{v}} \underset{s=0}{\overset{\infty}{\bar{K}}}[\mathbf{\Gamma}_r^t;\mathbf{\Gamma},\mathbf{v}] \cdot \mathbf{w} + \omega(\mathbf{\Gamma}_r^t;\mathbf{\Gamma},\mathbf{v},\mathbf{\Phi},\mathbf{\Omega},\mathbf{w}) \tag{19.19}$$

where $\mathbf{\Phi} \in \mathscr{H}$, $\mathbf{w} \in \mathscr{V}^3$, each "partial" Frechet differential is jointly continuous in its arguments, $\delta \bar{K}_{s=0}^{\infty}[\ ;\ ,\ \mid\ ]$ denotes the Frechet differential with respect to $\mathscr{H}$, $\partial_{\mathbf{v}}$ is ordinary partial differentiation with respect to $\mathbf{v}$,

$$D_{\mathbf{\Gamma}} \underset{s=0}{\overset{\infty}{\bar{K}}}[\mathbf{\Gamma}_r^t;\mathbf{\Gamma},\mathbf{v}] \cdot \mathbf{\Omega} = \underset{s=0}{\overset{\infty}{\bar{K}}}[\mathbf{\Gamma}_r^t; \mathbf{\Gamma} + \mathbf{\Omega}, \mathbf{v}] - \underset{s=0}{\overset{\infty}{\bar{K}}}[\mathbf{\Gamma}_r^t;\mathbf{\Gamma},\mathbf{v}] - \bar{\omega}(\mathbf{\Gamma}_r^t;\mathbf{\Gamma},\mathbf{v},\mathbf{\Omega}) \tag{19.20}$$

$\bar{\omega} = o(\|\mathbf{\Omega}\|_0)$, and

$$\lim_{\|\mathbf{\Phi}\|+\|\mathbf{\Omega}\|_0+|\mathbf{w}|\to 0} \left[\frac{\|\omega(\mathbf{\Gamma}_r^t;\mathbf{\Gamma},\mathbf{v},\mathbf{\Phi},\mathbf{\Omega},\mathbf{w})\|}{\|\mathbf{\Phi}\| + \|\mathbf{\Omega}\|_0 + |\mathbf{w}|}\right] = 0 \tag{19.21}$$

that is, $\omega = o(\|\Phi\| + \|\Omega\|_0 + |\mathbf{w}|)$. If we regard $\bar{K}_{s=0}^{\infty}[\ ;\ ,\ ]$ as a function of the parameter t, then rates of change of $\bar{K}_{s=0}^{\infty}[\ ;\ ,\ ]$ with respect to t can be computed with the aid of (19.19) and property (4) of $\mathscr{H}$. Denoting $d\,\bar{K}_{s=0}^{\infty}/dt$ by $\dot{\bar{K}}(t)$ we have†

$$\dot{\bar{K}}(t) = \delta \underset{s=0}{\overset{\infty}{\bar{K}}}[\mathbf{\Gamma}_r^t;\mathbf{\Gamma},\mathbf{v} \mid \dot{\mathbf{\Gamma}}_r^t] + D_{\mathbf{\Gamma}} \underset{s=0}{\overset{\infty}{\bar{K}}}[\mathbf{\Gamma}_r^t;\mathbf{\Gamma},\mathbf{v}] \cdot \dot{\mathbf{\Gamma}}(t) + \partial_{\mathbf{v}} \underset{s=0}{\overset{\infty}{\bar{K}}}[\mathbf{\Gamma}_r^t;\mathbf{\Gamma},\mathbf{v}] \cdot \dot{\mathbf{v}}(t) \tag{19.22}$$

where

$$\dot{\mathbf{\Gamma}}_r^t = \frac{d}{d\tau}\mathbf{\Gamma}(\tau)\big|_{\tau=t-s} = -\frac{d}{ds}\mathbf{\Gamma}_r^t(s) \tag{19.23}$$

for almost all $s \in (0,\infty)$.

† Ibid., p. 262.

Fading memory While in most of the subsequent developments the character of the norm $\|\cdot\|$ in (19.6) et seq. need not be specified, it is meaningful in many important applications to construct the norm of $\mathcal{H}$ so as to exhibit a fading memory. In the physical world, a material has a fading memory if its current response is not appreciably affected by its response at times distant in the past. Mathematically, we can model such properties by regarding the normed linear space $\mathcal{H}$, described previously, as a Hilbert space $\mathcal{H}_{(i)}$ and introducing a fixed monotone decreasing square integrable *influence function* $i(s)$ which is continuous on $[0,\infty)$ and which has the property $\lim_{s\to 0} i(s) = 0$. Then we define for the inner product of any two elements $u, v \in \mathcal{H}_{(i)}$,

$$\langle u,v\rangle = \int_0^\infty u(s)v(s)i^2(s)\,ds \tag{19.24}$$

The norm corresponding to (19.6) is

$$\|u\| = \left[\int_0^\infty u^2(s)i^2(s)\,ds\right]^{\frac{1}{2}} \tag{19.25}$$

In the case of the more general space $\mathcal{H}$ containing elements $\mathbf{\Gamma} = (\mathbf{T},\beta)$, we may, for example, use as a "fading memory" norm corresponding to (19.17):

$$\|\mathbf{\Gamma}\| = \left[\int_0^\infty \operatorname{tr}\mathbf{T}(s)\mathbf{T}^T(s)i^2(s)\,ds\right]^{\frac{1}{2}} + \left[\int_0^\infty \beta^2(s)i^2(s)\,ds\right]^{\frac{1}{2}} + [\operatorname{tr}\mathbf{T}(0)\mathbf{T}^T(0)]^{\frac{1}{2}} + |\beta(0)| \tag{19.26}$$

Some basic properties of thermomechanically simple materials To apply the above concepts of differentiation on normed spaces to the constitutive functionals (19.4), the structure of the associated norm spaces should be identified. Toward this end, we temporarily adopt a more concise notation: Let $\mathbf{\Lambda}(t)$ denote the ordered pair $[\boldsymbol{\gamma}(t),\theta(t)]$; i.e., the first member of $\mathbf{\Lambda}$ is the strain, a second-order tensor, and the second is the absolute temperature θ, a scalar. That is, with the appropriate units, $\mathbf{\Lambda}$ is a member of the space $\mathscr{V}^{10}$ discussed previously. Then $\mathbf{\Lambda}^t(s)$ denotes the pair of histories $[\boldsymbol{\gamma}^t(s),\theta^t(s)]$ and $\mathbf{\Lambda}_r^t(s)$ the restriction of $\mathbf{\Lambda}^t(s)$ to $(0,\infty)$. Equations (19.4) can then be written

$$\psi = \overset{\infty}{\underset{s=0}{\overline{\Psi}}}\,[\mathbf{\Lambda}_r^t(s);\mathbf{\Lambda},\mathbf{g}] \qquad \mathbf{T} = \overset{\infty}{\underset{s=0}{\overline{\mathfrak{T}}}}\,[\mathbf{\Lambda}_r^t(s);\mathbf{\Lambda},\mathbf{g}]$$
$$\eta = \overset{\infty}{\underset{s=0}{\overline{\mathscr{N}}}}\,[\mathbf{\Lambda}_r^t(s);\mathbf{\Lambda},\mathbf{g}] \qquad \mathbf{q} = \overset{\infty}{\underset{s=0}{\overline{\mathfrak{Q}}}}\,[\mathbf{\Lambda}_r^t(s);\mathbf{\Lambda},\mathbf{g}] \tag{19.27}$$

It is clear from (19.18) that $\overline{\Psi}_{s=0}^{\infty}, \ldots, \overline{\mathfrak{Q}}_{s=0}^{\infty}$ represent functionals defined on the direct sum of the space of total histories $\mathcal{H}^\dagger$ and $\mathscr{V}^{(3)}$, the space to which $\mathbf{g}(t)$ belongs. Thus, the time rate of change of, say, $\overline{\Psi}^{\infty}_{s=0}$ can be obtained with the aid of (19.22):

$$\dot{\psi} = \delta \overset{\infty}{\underset{s=0}{\overline{\Psi}}} [\mathbf{\Lambda}_r^t;\mathbf{\Lambda},\mathbf{g} \mid \dot{\mathbf{\Lambda}}_r^t] + D_{\Lambda} \overset{\infty}{\underset{s=0}{\overline{\Psi}}} [\mathbf{\Lambda}_r;\mathbf{\Lambda},\mathbf{g}] \cdot \dot{\mathbf{\Lambda}}(t) + \partial_{\mathbf{g}} \overset{\infty}{\underset{s=0}{\overline{\Psi}}} [\mathbf{\Lambda}_r^t;\mathbf{\Lambda},\mathbf{g}] \cdot \dot{\mathbf{g}}(t) \quad (19.28)$$

Here $\delta\, \overline{\Psi}^{\infty}_{s=0}\ [\ ;\ ,\ |\ ]$ is the Frechet differential in $\mathcal{H}$, D_{Λ} is the operator defined in (19.20), and $\partial_{\mathbf{g}}$ is the ordinary partial differential operator. By decomposing $\mathbf{\Lambda}$ into its components $\boldsymbol{\gamma}$ and θ, we can also write (19.28) in the expanded form:

$$\dot{\psi} = \delta_{\gamma} \overset{\infty}{\underset{s=0}{\overline{\Psi}}} [- \mid \dot{\boldsymbol{\gamma}}_r^t] + \delta_{\theta} \overset{\infty}{\underset{s=0}{\overline{\Psi}}} [- \mid \dot{\theta}_r^t] + D_{\gamma} \overset{\infty}{\underset{s=0}{\overline{\Psi}}} [-] \cdot \dot{\boldsymbol{\gamma}} + D_{\theta} \overset{\infty}{\underset{s=0}{\overline{\Psi}}} [-]\dot{\theta} + \partial_{\mathbf{g}} \overset{\infty}{\underset{s=0}{\overline{\Psi}}} [-] \cdot \dot{\mathbf{g}} \quad (19.29)$$

where, for simplicity in notation, we have used $[-] = [\boldsymbol{\gamma}_r^t,\theta_r^t;\boldsymbol{\gamma},\theta,\mathbf{g}]$. The operations δ_{γ}, δ_{θ}, D_{γ}, and D_{θ} are obviously defined analogously to δ and D_{Λ} of (19.28); computationally, it is sometimes convenient to use the following definitions:

$$\delta_{\gamma} \overset{\infty}{\underset{s=0}{\overline{\Psi}}} [- \mid \mathbf{h}_r] = \lim_{\alpha\to 0} \frac{\partial}{\partial \alpha} \left\{ \overset{\infty}{\underset{s=0}{\overline{\Psi}}} [\boldsymbol{\gamma}_r^t + \alpha \mathbf{h}_r,\ \theta_r^t;\ \boldsymbol{\gamma},\ \theta,\ \mathbf{g}] \right\} \quad (19.30a)$$

$$\delta_{\theta} \overset{\infty}{\underset{s=0}{\overline{\Psi}}} [- \mid \beta_r] = \lim_{\alpha\to 0} \frac{\partial}{\partial \alpha} \left\{ \overset{\infty}{\underset{s=0}{\overline{\Psi}}} [\boldsymbol{\gamma}_r^t,\ \theta_r^t + \alpha\beta_r;\ \boldsymbol{\gamma},\ \theta,\ \mathbf{g}] \right\} \quad (19.30b)$$

$$D_{\gamma} \overset{\infty}{\underset{s=0}{\overline{\Psi}}} [-] \cdot \mathbf{h} = \lim_{\alpha\to 0} \frac{\partial}{\partial \alpha} \left\{ \overset{\infty}{\underset{s=0}{\overline{\Psi}}} [\boldsymbol{\gamma}_r^t,\ \theta_r^t;\ \boldsymbol{\gamma} + \alpha \mathbf{h},\ \theta,\ \mathbf{g}] \right\} \quad (19.30c)$$

$$D_{\theta} \overset{\infty}{\underset{s=0}{\overline{\Psi}}} [-] \cdot \beta = \lim_{\alpha\to 0} \frac{\partial}{\partial \alpha} \left\{ \overset{\infty}{\underset{s=0}{\overline{\Psi}}} [\boldsymbol{\gamma}_r^t,\ \theta_r^t;\ \boldsymbol{\gamma},\ \theta + \alpha\beta,\ \mathbf{g}] \right\} \quad (19.30d)$$

Thus, in general, ordinary partial differentiation with respect to γ_{ij} and θ is implied by (19.30*c*) and (19.30*d*).

Returning now to the first member of (19.5), and recalling (12.35), we rewrite the Clausius-Duhem inequality in the form

$$\sigma^{ij}\dot{\gamma}_{ij} - \rho(\dot{\psi} + \eta\dot{\theta}) + \frac{1}{\theta}\mathbf{q}\cdot\mathbf{g} \geq 0 \quad (19.31)$$

The constitutive functionals (19.27) must be such that (19.31) is not violated for any thermodynamical process we wish to perform on the material. Consequently, converting to indicial notation and introducing (19.27) and

(19.29) into (19.31), we arrive at the requirement

$$\left(\overset{\infty}{\underset{s=0}{\mathfrak{T}^{ij}}}[-] - \rho D_{\gamma_{ij}} \overset{\infty}{\underset{s=0}{\Psi}}[-]\right) \dot{\gamma}_{ij} - \rho\left(D_\theta \overset{\infty}{\underset{s=0}{\Psi}}[-] + \overset{\infty}{\underset{s=0}{\mathcal{N}}}[-]\right)\dot{\theta}$$
$$- \rho\delta_{\gamma_{ij}} \overset{\infty}{\underset{s=0}{\Psi}}[- \mid \dot{\gamma}^t_{ij(r)}] - \rho\delta_\theta \overset{\infty}{\underset{s=0}{\Psi}}[- \mid \dot{\theta}^t_r]$$
$$- \rho\partial_{g_i} \overset{\infty}{\underset{s=0}{\Psi}}[-]\dot{g}_i + \frac{1}{\theta}\overset{\infty}{\underset{s=0}{\mathfrak{Q}^i}}[-]g_i \geq 0 \quad (19.32)$$

Now for fixed past histories $\gamma^t_{ij(r)}$, θ^t_r and fixed values γ_{ij}, θ, and g_i, we may allow $\dot{\gamma}_{ij}$, $\dot{\theta}$, and $\dot{g}_i$ to vary arbitrarily. In order that (19.32) hold for such arbitrary variations, it is necessary that the coefficients of $\dot{\gamma}_{ij}$, $\dot{\theta}$, and $\dot{g}_i$ vanish and that the remaining terms be positive at all times; that is,

$$\partial_{g_i} \overset{\infty}{\underset{s=0}{\Psi}}[-] = 0 \quad (19.33a)$$

$$\overset{\infty}{\underset{s=0}{\mathfrak{T}^{ij}}}[-] = \rho D_{\gamma_{ij}} \overset{\infty}{\underset{s=0}{\Psi}}[-] \quad (19.33b)$$

$$\overset{\infty}{\underset{s=0}{\mathcal{N}}}[-] = -D_\theta \overset{\infty}{\underset{s=0}{\Psi}}[-] \quad (19.33c)$$

$$-\rho\left\{\delta_{\gamma_{ij}} \overset{\infty}{\underset{s=0}{\Psi}}[- \mid \dot{\gamma}^t_{ij(r)}] + \delta_\theta \overset{\infty}{\underset{s=0}{\Psi}}[- \mid \dot{\theta}^t_r]\right\} + \frac{1}{\theta}\overset{\infty}{\underset{s=0}{\mathfrak{Q}^i}}[-]g_i \geq 0 \quad (19.33d)$$

Also, in view of these results and (12.33), the internal dissipation is given by

$$\delta = -\rho\left\{\delta_{\gamma_{ij}} \overset{\infty}{\underset{s=0}{\Psi}}[- \mid \dot{\gamma}^t_{ij(r)}] + \delta_\theta \overset{\infty}{\underset{s=0}{\Psi}}[- \mid \dot{\theta}^t_r]\right\} \quad (19.34)$$

With the development of (19.33) and (19.34), we have established the following *fundamental properties of thermomechanically simple materials:*†

1. *The free energy functional* $\overset{\infty}{\underset{s=0}{\Psi}}[-]$ *is independent of the temperature gradient* **g** [as is implied by (19.33*a*)].
2. *The constitutive functionals* $\overset{\infty}{\underset{s=0}{\mathfrak{T}^{ij}}}[-]$ and $\overset{\infty}{\underset{s=0}{\mathcal{N}}}[-]$ *for stress and entropy are determined from the free-energy functional; consequently, both are also independent of* **g** [as is implied by (19.33*b*) and (19.33*c*)].
3. *The internal dissipation* δ *is determined from the free-energy functional by* (19.34) *and satisfies the general dissipation inequality* (19.33*d*).

† Coleman [1964].

It follows from properties 1 and 2 that *a simple material is characterized by only two constitutive functionals, one describing the free energy and the other the heat flux;* for example,

$$\begin{aligned} \psi &= \overset{\infty}{\underset{s=0}{\hat{\Psi}}} [\gamma_r^t, \theta_r^t; \gamma, \theta] \\ Q^i &= \overset{\infty}{\underset{s=0}{\hat{\mathfrak{Q}}^i}} [\gamma_r^t; \theta_r^t; \gamma, \theta, \mathbf{g}] \end{aligned} \tag{19.35}$$

The stress, entropy, internal dissipation, and internal energy are determined by the functionals

$$\sigma^{ij} = \overset{\infty}{\underset{s=0}{\hat{\mathfrak{T}}^{ij}}} [\gamma_r^t, \theta_r^t; \gamma, \theta] \equiv \rho D_{\gamma_{ij}} \overset{\infty}{\underset{s=0}{\hat{\Psi}}} [\gamma_r^t, \theta_r^t; \gamma, \theta] \tag{19.36a}$$

$$\eta = \overset{\infty}{\underset{s=0}{\hat{\mathscr{N}}}} [\gamma_r^t, \theta_r^t; \gamma, \theta] \equiv -D_\theta \overset{\infty}{\underset{s=0}{\hat{\Psi}}} [\gamma_r^t, \theta_r^t; \gamma, \theta] \tag{19.36b}$$

$$\begin{aligned} \delta = \overset{\infty}{\underset{s=0}{\hat{\mathscr{D}}}} [\gamma_r^t, \theta_r^t; \gamma, \theta] \equiv -\rho \Big\{ & \delta_{\gamma_{ij}} \overset{\infty}{\underset{s=0}{\hat{\Psi}}} [\gamma_r^t, \theta_r^t; \gamma, \theta \mid \dot{\gamma}^t_{ij(r)}] \\ & + \delta_\theta \overset{\infty}{\underset{s=0}{\hat{\Psi}}} [\gamma_r^t, \theta_r^t; \gamma, \theta \mid \dot{\theta}_r^t] \Big\} \end{aligned} \tag{19.36c}$$

$$\epsilon = \psi + \eta\theta = \overset{\infty}{\underset{s=0}{\mathfrak{e}}} [\gamma_r^t, \theta_r^t; \gamma, \theta] \equiv (1 - \theta D_\theta) \overset{\infty}{\underset{s=0}{\hat{\Psi}}} [\gamma_r^t, \theta_r^t; \gamma, \theta] \tag{19.36d}$$

In addition to (19.36), the constitutive functionals are also required to obey the principle of material frame indifference, discussed in Sec. 14.4, and to be form-invariant under appropriate unimodular transformations belonging to the isotropy group of the material. We remark that in most applications of this theory to homogeneous solids, we prefer to use, instead of ψ, σ^{ij}, η, Q^i, and δ, the quantities $\varphi = \rho_0 \psi$, $t^{ij} = \sqrt{G}\, \sigma^{ij}$, $\xi = \rho_0 \eta$, $q_i = \delta_{im} \sqrt{G}\, Q^m$, and $\sigma = \delta \sqrt{G}$. The basic conclusions and, with minor modifications, the forms of the equations are, of course unaltered. In the following section we consider briefly various specific forms of the constitutive functionals for specific materials.

19.2 SPECIAL FORMS OF THE CONSTITUTIVE EQUATIONS FOR SIMPLE MATERIALS

In searching for forms of the constitutive functionals appropriate for specific materials, it is natural to first consider possible expansions of the free-energy functional† in much the same spirit as expansions of the strain-energy function were examined in Art. 15. While experimental data justifying such expansions for specific nonlinear materials are incomplete, various linearizations of the constitutive functionals so obtained do agree with those of classical linear viscoelasticity and thermoelasticity. Insofar as our present discussion is concerned, it suffices to cite only representative examples of possible forms of these functionals.

The tensor-valued and scalar-valued functions $\mathbf{0}^t(s)$ and $0^t(s)$, defined by

$$\mathbf{0}^t(s) = \mathbf{0} \qquad 0^t(s) = 0 \qquad 0 \leq s \leq \infty \tag{19.37}$$

are referred to as zero histories. If $\mathscr{F}_{s=0}^{\infty}\,[\boldsymbol{\gamma}^t,\theta^t]$ is a functional of the total histories $\boldsymbol{\gamma}^t(s)$, $\theta^t(s)$, we can, with the aid of (19.9) and subject to appropriate conditions of smoothness on $\mathscr{F}_{s=0}^{\infty}\,[-]$, construct a generalized Taylor expansion of $\mathscr{F}_{s=0}^{\infty}\,[-]$ about the zero histories (19.37) of the form

$$\begin{aligned}\mathop{\mathscr{F}}_{s=0}^{\infty}\,[\mathbf{0}^t(s) + \boldsymbol{\gamma}^t(s),\, 0^t(s) + \theta^t(s)] &= \mathop{\mathscr{F}}_{s=0}^{\infty}\,[\mathbf{0}^*] + \delta_{\boldsymbol{\gamma}} \mathop{\mathscr{F}}_{s=0}^{\infty}\,[\mathbf{0}^* \mid \boldsymbol{\gamma}^t(s)] \\ &\quad + \delta_\theta \mathop{\mathscr{F}}_{s=0}^{\infty}\,[\mathbf{0}^* \mid \theta^t(s)] + \frac{1}{2!}\,\delta_{\boldsymbol{\gamma}}^2 \mathop{\mathscr{F}}_{s=0}^{\infty}\,[\mathbf{0}^* \mid \boldsymbol{\gamma}^t(s),\boldsymbol{\gamma}^t(s)] \\ &\quad + \delta_{\boldsymbol{\gamma}}\delta_\theta \mathop{\mathscr{F}}_{s=0}^{\infty}\,[\mathbf{0}^* \mid \boldsymbol{\gamma}^t(s),\theta^t(s)] + \frac{1}{2!}\,\delta_\theta^2 \mathop{\mathscr{F}}_{s=0}^{\infty}\,[\mathbf{0}^* \mid \theta^t(s),\theta^t(s)] + \cdots\end{aligned} \tag{19.38}$$

where $\mathbf{0}^* = [\mathbf{0}^t(s),0^t(s)]$. Alternatively, we can express $\mathscr{F}_{s=0}^{\infty}[-]$ in terms of the past histories $\boldsymbol{\gamma}_r^t(s)$, $\theta_r^t(s)$ via the new functional

$$\mathop{\bar{\bar{\mathscr{F}}}}_{s=0}^{\infty}\,[\boldsymbol{\gamma}_r^t,\theta_r^t;\boldsymbol{\gamma},\theta] = \mathop{\mathscr{F}}_{s=0}^{\infty}\,[\boldsymbol{\gamma}^t,\theta^t] \tag{19.39}$$

† The idea of representing various constitutive functionals in Frechet-type expansions, the terms of which are generally multiple integrals in the strain and/or temperature histories, has been explored by a number of authors. Isothermal cases, for example, were discussed by Green and Rivlin [1957, 1960], Green, Rivlin, and Spencer [1959], Coleman and Noll [1960], Chacon and Rivlin [1964], Pipkin [1964, 1966], and others. While the field is still relatively new, several authors (e.g., Ward and Onat [1963]) have attempted to use truncated expansions in the experimental characterization of certain viscoelastic and thermoviscoelastic materials. A different basis for defining the free energy has been used by Schapery [1968]. See also Lianis [1965] and McGuirt and Lianis [1969].

and write, instead of (19.38),

$$
\begin{aligned}
\overline{\mathscr{F}}_{s=0}^{\infty}[\mathbf{0}_r^t + \boldsymbol{\gamma}_r^t, 0_r^t + \theta_r^t; \mathbf{0} + \boldsymbol{\gamma}, 0 + \theta] &= \overline{\mathscr{F}}_{s=0}^{\infty}[\bar{\mathbf{0}}^*] + \delta_{\boldsymbol{\gamma}} \overline{\mathscr{F}}_{s=0}^{\infty}[\bar{\mathbf{0}}^* \mid \boldsymbol{\gamma}_r^t] \\
&+ \delta_\theta \overline{\mathscr{F}}_{s=0}^{\infty}[\bar{\mathbf{0}}^* \mid \theta_r^t] + D_{\boldsymbol{\gamma}} \overline{\mathscr{F}}_{s=0}^{\infty}[\bar{\mathbf{0}}^*] \cdot \boldsymbol{\gamma} + D_\theta \overline{\mathscr{F}}_{s=0}^{\infty}[\bar{\mathbf{0}}^*]\,\theta \\
&+ \frac{1}{2!}\delta_{\boldsymbol{\gamma}}^2 \overline{\mathscr{F}}_{s=0}^{\infty}[\bar{\mathbf{0}}^* \mid \boldsymbol{\gamma}_r^t, \boldsymbol{\gamma}_r^t] + \delta_{\boldsymbol{\gamma}}\delta_\theta \overline{\mathscr{F}}_{s=0}^{\infty}[\bar{\mathbf{0}}^* \mid \boldsymbol{\gamma}_r^t, \theta_r^t] \\
&+ \delta_{\boldsymbol{\gamma}} D_{\boldsymbol{\gamma}} \overline{\mathscr{F}}_{s=0}^{\infty}[\bar{\mathbf{0}}^* \mid \boldsymbol{\gamma}_r^t] \cdot \boldsymbol{\gamma} + \delta_{\boldsymbol{\gamma}} D_\theta \overline{\mathscr{F}}_{s=0}^{\infty}[\bar{\mathbf{0}}^* \mid \boldsymbol{\gamma}_r^t]\theta \\
&+ \frac{1}{2!} D_{\boldsymbol{\gamma}}^2 \overline{\mathscr{F}}_{s=0}^{\infty}[\bar{\mathbf{0}}^*] \cdot \boldsymbol{\gamma}^2 + \frac{1}{2!} D_\theta^2 \overline{\mathscr{F}}_{s=0}^{\infty}[\bar{\mathbf{0}}^*]\theta^2 + \frac{1}{2!}\delta_\theta^2 \overline{\mathscr{F}}_{s=0}^{\infty}[\bar{\mathbf{0}}^* \mid \theta_r^t, \theta_r^t] + \cdots
\end{aligned}
\tag{19.40}
$$

where $\bar{\mathbf{0}}^* = (\mathbf{0}_r^t, 0_r^t; \mathbf{0}, 0)$.

If we set $\overline{\mathscr{F}}_{s=0}^{\infty}[-] = \Psi_{s=0}^{\infty}[-]$, where $\Psi_{s=0}^{\infty}[-]$ is the free-energy functional, then, by truncating the expansion (19.40) after various terms, various approximations of the constitutive functionals for specific materials can be obtained. For example, Achenbach, Vogel, and Herrmann [1968, p. 13]† proposed the following free-energy functional obtained from (19.40) by retaining only cubic terms in $\boldsymbol{\gamma}$, θ, and their histories:

$$
\begin{aligned}
\Phi_{s=0}^{\infty} = A_0 + A_1^{ij}\gamma_{ij} + A_2^{ij}\gamma_{ij}T + A_3^{ij}\gamma_{ij}T^2 + A_4^{ijmn}\gamma_{ij}\gamma_{mn} \\
+ A_5^{ijmn}\gamma_{ij}\gamma_{mn}T + B_1 T + B_2 T^2 + B_3 T^3
\end{aligned}
\tag{19.41}
$$

Here $\Phi_{s=0}^{\infty}$ is the free-energy functional per unit undeformed volume, $T = \theta - T_0$ is the change in temperature from a uniform constant reference temperature T_0, and‡

$$
\begin{aligned}
A_0 = \Phi_{s=0}^{\infty}[-] &+ \tfrac{1}{2}\delta_T^2 \Phi_{s=0}^{\infty}[- \mid T_r^t, T_r^t] + \tfrac{1}{2}\delta_{\boldsymbol{\gamma}}^2 \Phi_{s=0}^{\infty}[- \mid \boldsymbol{\gamma}_r^t, \boldsymbol{\gamma}_r^t] \\
&+ \delta_{\boldsymbol{\gamma}}\delta_T \Phi_{s=0}^{\infty}[- \mid \boldsymbol{\gamma}_r^t, T_r^t] + \tfrac{1}{6}\delta_T^3 \Phi_{s=0}^{\infty}[- \mid T_r^t, T_r^t, T_r^t] \\
&+ \tfrac{1}{2}\delta_{\boldsymbol{\gamma}}^2\delta_T \Phi_{s=0}^{\infty}[- \mid \boldsymbol{\gamma}_r^t, \boldsymbol{\gamma}_r^t, T_r^t] + \tfrac{1}{2}\delta_{\boldsymbol{\gamma}}\delta_T^2 \Phi_{s=0}^{\infty}[- \mid \boldsymbol{\gamma}_r^t, T_r^t, T_r^t]
\end{aligned}
\tag{19.42a}
$$

$$
\begin{aligned}
A_1^{ij} = D_{\gamma_{ij}} \Big\{ \Phi_{s=0}^{\infty}[-] &+ \delta_{\boldsymbol{\gamma}} \Phi_{s=0}^{\infty}[- \mid \boldsymbol{\gamma}_r^t] + \delta_T \Phi_{s=0}^{\infty}[- \mid T_r^t] \\
&+ \tfrac{1}{2}\delta_T^2 \Phi_{s=0}^{\infty}[- \mid T_r^t, T_r^t] + \delta_{\boldsymbol{\gamma}}\delta_T \Phi_{s=0}^{\infty}[- \mid \boldsymbol{\gamma}_r^t, T_r^t] \Big\}
\end{aligned}
\tag{19.42b}
$$

† This study was confined to the one-dimensional case.

‡ Note that the terms $\delta_{\boldsymbol{\gamma}}\Phi[- \mid \boldsymbol{\gamma}_r^t]$ and $\delta_\theta\Phi[- \mid \theta_r^t]$ do not enter (19.42) since these functionals vanish for constant strain and temperature histories.

$$A_2^{ij} = D_{\gamma_{ij}} D_T \left\{ \mathop{\Phi}_{s=0}^{\infty} [-] + \delta_T \mathop{\Phi}_{s=0}^{\infty} [- \mid T_r^t] + \delta_\Upsilon \mathop{\Phi}_{s=0}^{\infty} [- \mid \Upsilon_r^t] \right\} \tag{19.42c}$$

$$A_3^{ij} = \tfrac{1}{2} D_{\gamma_{ij}} D_T^2 \mathop{\Phi}_{s=0}^{\infty} [-] \qquad A_4^{ijmn} = \tfrac{1}{2} D_{\gamma_{ij}} D_{\gamma_{mn}} \left\{ \mathop{\Phi}_{s=0}^{\infty} [-] + \delta_T \mathop{\Phi}_{s=0}^{\infty} [- \mid T_r^t] \right\} \tag{19.42d,e}$$

$$A_5^{ijmn} = \tfrac{1}{2} D_{\gamma_{ij}} D_{\gamma_{mn}} D_T \mathop{\Phi}_{s=0}^{\infty} [-] \tag{19.42f}$$

$$\begin{aligned} B_1 = D_T \Big\{ & \mathop{\Phi}_{s=0}^{\infty} [-] + \delta_T \mathop{\Phi}_{s=0}^{\infty} [- \mid T_r^t] + \delta_\Upsilon \mathop{\Phi}_{s=0}^{\infty} [- \mid \Upsilon_r^t] \\ & + \tfrac{1}{2} \delta_T^2 \mathop{\Phi}_{s=0}^{\infty} [- \mid T_r^t, T_r^t] + \tfrac{1}{2} \delta_\Upsilon^2 \mathop{\Phi}_{s=0}^{\infty} [- \mid \Upsilon_r^t, \Upsilon_r^t] + \delta_\Upsilon \delta_T \mathop{\Phi}_{s=0}^{\infty} [- \mid \Upsilon_r^t, T_r^t] \end{aligned} \tag{19.42g}$$

$$B_2 = \tfrac{1}{2} D_T^2 \left\{ \mathop{\Phi}_{s=0}^{\infty} [-] + \delta_T \mathop{\Phi}_{s=0}^{\infty} [- \mid T_r^t] + \delta_\Upsilon \mathop{\Phi}_{s=0}^{\infty} [- \mid \Upsilon_r^t] \right\} \tag{19.42h}$$

$$B_3 = \tfrac{1}{6} D_T^3 \mathop{\Phi}_{s=0}^{\infty} [-] \tag{19.42i}$$

where $\mathop{\Phi}_{s=0}^{\infty} [- \mid \cdot\,\cdot] \equiv \mathop{\Phi}_{s=0}^{\infty} [\mathbf{0}_r^t, 0_r^t; \mathbf{0}, 0 \mid \cdot\,\cdot]$, $\delta_T = \delta_{\theta - T_0} = \delta_\theta$, and $D_T = D_{\theta - T_0} \equiv D_\theta$. More specifically, we may introduce the strain- and temperature-dependent integral coefficients†

$$\begin{aligned} A_0 = {} & a + \mathbf{R}_{11} * \Upsilon * \Upsilon' + \mathbf{R}_{12} * \Upsilon * T' + \mathbf{R}_{22} * T * T' \\ & + \mathbf{R}_{112} * \Upsilon * \Upsilon' * T' + \mathbf{R}_{222} * T * T' * T' + \mathbf{R}_{122} * \Upsilon * T' * T' \end{aligned} \tag{19.43a}$$

$$A_1^{ij} = E^{ij} + G_1'^{ijmn} * \gamma_{mn} + M_1'^{ij} * T + M_2'^{ij} * T * T' + G_2'^{ijmn} * \gamma_{mn} * T' \tag{19.43b}$$

$$A_2^{ij} = M_1^{ij}(0) + K_1'^{ijmn} * \gamma_{mn} + K_2'^{ij} * T \qquad A_3^{ij} = K_2^{ij}(0) \tag{19.43c,d}$$

$$A_4^{ijmn} = \tfrac{1}{2} G_1^{ijmn}(0) + N'^{ijmn} * T \qquad A_5^{ijmn} = K_1^{ijmn}(0) \tag{19.43e,f}$$

$$B_1 = H + L_1' * T + L_2'^{ij} * \gamma_{ij} + L_{11}' * T * T' + L_{22}'^{ijmn} * \gamma_{ij} * \gamma_{mn}' + L_{12}'^{ij} * \gamma_{ij} * T' \tag{19.43g}$$

$$B_2 = \tfrac{1}{2} [L_1(0) + F_1' * T + F_2'^{ij} * \gamma_{ij}] \qquad B_3 = \tfrac{1}{3} F_1(0) \tag{19.43h}$$

Here $\mathbf{R}_{11}, \mathbf{R}_{12}, \ldots, G^{ijmn}, M_1^{ij}, \ldots, L_{11}, L_{22}^{ijmn}, \ldots, F_1$ are material kernels

† Achanbach, Vogel, and Herrman [1968, p. 5].

and $*$ denotes a generalized convolution operator which, with an appropriate change of variables, indicates integrals of the form

$$\mathbf{R}_{11} * \boldsymbol{\gamma} * \boldsymbol{\gamma}' = \int_0^t \int_0^t R_{11}^{ijmn}(t - s_1, t - s_2)\gamma_{ij}(s_1) \frac{d\gamma_{mn}(s_2)}{ds_2} \, ds_1 \, ds_2 \tag{19.44a}$$

$$\mathbf{R}_{112} * \boldsymbol{\gamma} * \boldsymbol{\gamma}' * T' = \int_0^t \int_0^t \int_0^t R_{112}^{ijmn}(t - s_1, t - s_2, t - s_3)\gamma_{ij}(s_1) \times \frac{d\gamma_{mn}(s_2)}{ds_2} \frac{dT(s_3)}{ds_3} \, ds_1 \, ds_2 \, ds_3 \tag{19.44b}$$

. . . .

$$M_1'^{ij} * T = \int_0^t \frac{dM^{ij}(t - s)}{d(t - s)} T(s) \, ds \tag{19.44c}$$

. . . .

The lower limit of integration is assumed to be the time at which the deformation of the body is initiated. Introducing (19.41) into (19.36*a*) and (19.36*b*) and recalling that $t^{ij} = \sqrt{G}\,\sigma^{ij} = \rho\sqrt{G}\, D_{\gamma}\, \bar{\Psi}_{s=0}^{\infty}\,[-] = \rho_0 D_{\gamma}\, \bar{\Psi}_{s=0}^{\infty}\,[-] = D_{\gamma}\, \Phi_{s=0}^{\infty}\,[-]$ and $\xi = \rho_0\eta = -\rho_0 D_{\theta}\, \bar{\Psi}_{s=0}^{\infty}\,[-] = -D_T\, \Phi_{s=0}^{\infty}\,[-]$, we arrive at the following constitutive equations for stress and entropy:

$$t^{ij} = A_1^{ij} + A_2^{ij}T + A_3^{ij}T^2 + 2A_4^{ijmn}\gamma_{mn} + 2A_5^{ijmn}\gamma_{mn}T \tag{19.45a}$$

$$-\xi = B_1 + 2B_2T + 3B_3T^2 + A_2^{ij}\gamma_{ij} + 2A_3^{ij}\gamma_{ij}T + A_5^{ijmn}\gamma_{ij}\gamma_{mn} \tag{19.45b}$$

The internal-dissipation functional is obtained in a similar manner using (19.36*c*). Obviously, theories of higher order are obtained in precisely the same manner by simply retaining additional terms in the expansion (19.40). It is interesting to note that by equating to zero $\theta_r^t(s)$ and $\theta(t)$ in (19.45*a*), we obtain the classical constitutive equation for linear isothermal viscoelasticity. Alternatively, if the integral coefficients are replaced by constants, (19.45*a*) reduces to the constitutive equation for a class of thermoelastic materials with temperature-dependent material properties. We shall examine this latter class of materials in more detail later.

To complete the characterization of the material, we must add to (19.45) a constitutive equation for the heat flux $\mathbf{q}$. Among a variety of forms that might be assumed for the functional $\bar{\mathfrak{Q}}_{s=0}^{\infty}\,[-]$, we mention the linear law proposed by Christensen and Naghdi [1967]:

$$Q^i = \int_0^\infty \kappa^{ij}(t - s) \frac{\partial T_{,j}(s)}{\partial s} \, ds \tag{19.46}$$

where $\kappa^{ij}(t - s)$ is a thermal conductivity kernel. Alternatively, we may use the generalized Fourier law

$$Q^i = \underset{s=0}{\overset{\infty}{\kappa^{ij}}}[\Upsilon_r^t, T_r^t; \Upsilon, T]T_{,j} \tag{19.47}$$

where $\underset{s=0}{\overset{\infty}{\kappa^{ij}}}[-]$ is a functional of the indicated arguments. If $\underset{s=0}{\overset{\infty}{\kappa^{ij}}}[-]$ is replaced by a constant κ^{ij}, (19.47) reduces to the classical Fourier law of heat conduction.

Thermorheologically simple materials Experimental evidence obtained from tests on a large class of viscoelastic materials has led to the identification of an important subclass of materials with memory, commonly referred to as *thermorheologically simple materials*.† This classification arose from the observation that, among certain amorphous high polymers which approximately obey established linear and nonlinear viscoelastic laws at uniform temperatures are a group which exhibits a simple property with a change of temperature, namely, a translational shift of various material properties when plotted against the logarithm of time at different uniform temperatures. This shift phenomenon is the basic characteristic of all thermorheologically simple materials and makes it possible to establish an equivalence relation between temperature and $\ln t$.

To fix ideas, note that in the case of isothermal deformations of an isotropic linearly viscoelastic solid, (19.45*a*) reduces to a constitutive law that can be written in the form

$$t^{ij} = \int_0^t J(t - s)\frac{\partial \gamma'_{ij}}{\partial s}\,ds + \frac{1}{3}\delta^{ij}\int_0^t K(t - s)\frac{\partial \gamma}{\partial s}\,ds \tag{19.48}$$

where γ'_{ij} and γ are the deviatoric and dilatational components of γ_{ij} and $J(t - s)$ and $K(t - s)$ are relaxation moduli. Now, in the case of a thermorheologically simple material, we choose to write the relaxation moduli at uniform temperature T, for example, $J(t - s)$, as a function of $\ln t$, denoted $E_T(\ln t)$. Then the shift property is apparent if

$$E_T(\ln t) = E_{T_0}[\ln t + f(T)] \tag{19.49}$$

where $f(T)$ is a shift function relative to T_0 such that $df/dT > 0$ and $f(T_0) = 0$. With this property, the modulus of the relaxation curve will shift toward shorter times with an increase in T, as indicated in Fig. 19.1. If we introduce a *shift factor* $b(T)$, defined by

$$b(T) = \exp f(T) \tag{19.50}$$

† See Schwarzl and Staverman [1952] or Williams, Landel, and Ferry [1955]. A summary discussion has been given by Cost [1969]. See also Lianis [1968].

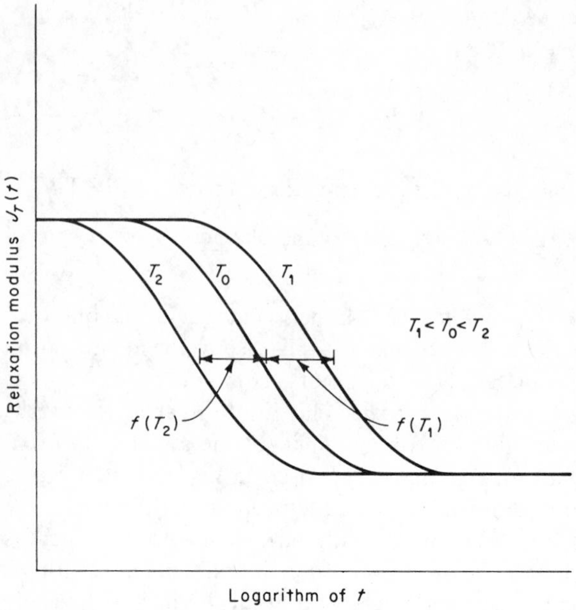

Fig. 19.1 Shift phenomenon in thermorheologically simple material.

then the relaxation modulus $J_T(t)$ at a uniform temperature T has the property

$$\begin{aligned} J_T(t) &= E_T(\ln t) = E_{T_0}[\ln t + \ln b(T)] \\ &= E_{T_0}\{\ln [tb(T)]\} \\ &= J_{T_0}(\zeta) \end{aligned} \tag{19.51}$$

where ζ is a *reduced time*, defined by

$$\zeta = tb(T) \tag{19.52}$$

If it is possible to invert (19.52) so as to obtain $t = g(\zeta)$, then (19.48) can be written in the form

$$\hat{t}^{ij} = \int_0^\zeta J(\zeta - \hat{s}) \frac{\partial \hat{\gamma}'_{ij}}{\partial \hat{s}} \, d\hat{s} + \frac{1}{3} \delta^{ij} \int_0^\zeta K(\zeta - \hat{s}) \frac{\partial \hat{\gamma}}{\partial \hat{s}} \, d\hat{s} \tag{19.53}$$

where $\hat{t}^{ij}(\mathbf{x},\zeta) = t^{ij}[\mathbf{x},g(\zeta)]$, $\hat{\gamma}'_{ij}(\mathbf{x},\zeta) = \gamma'_{ij}[\mathbf{x},g(\zeta)]$, $\hat{\gamma}(\mathbf{x},\zeta) = \gamma[\mathbf{x},g(\zeta)]$, and $\hat{s} = f(s)$. That is, when transformed to reduced time, the constitutive equations for a thermorheologically simple material assume the same form as the constitutive law for the isothermal case.

While the above development pertains to the very special case of linear response of a body at various uniform temperatures, the basic ideas can be extended to finite deformations of bodies subjected to transient, nonhomogeneous temperatures.† For example, if it is assumed that the relaxation properties at a particle $\mathbf{x}$ depend only on the current temperature $T(\mathbf{x},t)$ at $\mathbf{x}$, we may further postulate that an increment $\Delta\zeta$ in reduced time is related to an increment Δt in real time according to $\Delta\zeta = b[T(\mathbf{x},t)]\,\Delta t$. Then, instead of (19.52), we have

$$\zeta(\mathbf{x},t) = \int_0^t b[T(\mathbf{x},s)]\,ds \tag{19.54}$$

Clearly, in the case of uniform, constant temperatures, (19.54) reduces to (19.52). As another possibility, Taylor, Pister, and Goudreau [1970] suggested a reduced time ζ dependent upon the histories of both T and the temperature rates:

$$\hat{\zeta}(\mathbf{x},t) = \int_0^t \hat{b}\left[T(\mathbf{x},s), \frac{\partial T(\mathbf{x},s)}{\partial s}\right] ds \tag{19.55}$$

Still other generalizations are possible.

As a representative example of a constitutive functional proposed for the free energy of a thermorheologically simple material, we cite the functional suggested by Cost [1969, p. 55]:‡

$$\begin{aligned}\Phi[-] &= \Phi_0 + \int_0^\zeta D^{ij}(\zeta - \hat{s})\frac{\partial \hat{\gamma}_{ij}}{\partial \hat{s}}\,d\hat{s} - \int_0^\zeta f(\zeta - \hat{s})\frac{\partial \hat{T}}{\partial \hat{s}}\,d\hat{s}\\ &+ \frac{1}{6}\delta^{ij}\delta^{mn}\int_0^\zeta\int_0^\zeta [3K(\zeta - \hat{s}_1, \zeta - \hat{s}_2) - 2G(\zeta - \hat{s}_1, \zeta - \hat{s}_2)]\\ &\times \frac{\partial \hat{\gamma}_{ij}}{\partial \hat{s}_1}\frac{\partial \hat{\gamma}_{mn}}{\partial \hat{s}_2}\,d\hat{s}_1\,d\hat{s}_2 + \frac{1}{4}(\delta^{ij}\delta^{mn} + \delta^{im}\delta^{jn})\int_0^\zeta\int_0^\zeta 2G\\ &\times (\zeta - \hat{s}_1, \zeta - \hat{s}_2)\frac{\partial \hat{\gamma}_{ij}}{\partial \hat{s}_1}\frac{\partial \hat{\gamma}_{mn}}{\partial \hat{s}_2}\,d\hat{s}_1\,d\hat{s}_2 - \delta^{ij}\int_0^\zeta\int_0^\zeta 3\alpha K(\zeta - \hat{s}_1, \zeta - \hat{s}_2)\\ &\times \frac{\partial \hat{\gamma}_{ij}}{\partial \hat{s}_1}\frac{\partial \hat{T}}{\partial \hat{s}_2}\,d\hat{s}_1\,d\hat{s}_2 - \frac{1}{2}\int_0^\zeta\int_0^\zeta \mathscr{M}(\zeta - \hat{s}_1, \zeta - \hat{s}_2)\frac{\partial \hat{T}}{\partial \hat{s}_1}\frac{\partial \hat{T}}{\partial \hat{s}_2}\,d\hat{s}_1\,d\hat{s}_2\end{aligned} \tag{19.56}$$

where, for the sake of compactness, all quantities have been transformed to reduced time; that is, $\hat{\gamma}_{ij}(\mathbf{x},\zeta) = \gamma_{ij}[\mathbf{x},g(\zeta)]$, $\hat{T}(\mathbf{x},\zeta) = T[\mathbf{x},g(\zeta)]$, $\hat{\varphi}(\mathbf{x},\zeta) = \varphi[\mathbf{x},g(\zeta)]$, where $t = g(\zeta)$, $\zeta = f(t)$ is given by (19.54), and $\hat{s}_1 = f(s_1)$, $\hat{s}_2 = f(s_2)$. In (19.56), $D^{ij}(\ )$, $K(\ ,\ )$, $G(\ ,\ )$, and $\mathscr{M}(\ ,\ )$ are material kernels

† Generalizations of the theory of thermorheologically simple materials to nonlinear continua were explored by Lianis [1968] and were applied successfully to the characterization of finite deformations of styrene-butadiene rubber in the experiments of McGuirt and Lianis [1969].

‡ This particular form is a generalization of one proposed by Christensen and Naghdi [1967].

and α is a material constant. Converting (19.56) to real time and ignoring the contribution due to $\partial T(\mathbf{x},t)/\partial x_i$, we find upon introducing (19.56) into (19.36), using the measures t^{ij}, ξ, and σ and performing the indicated differentiations, that

$$t^{ij} = D^{ij}(0) + \delta^{im}\delta^{jn}\int_0^t 2G[f(t) - \hat{s}]\frac{\partial\gamma_{mn}}{\partial\hat{s}}\,d\hat{s} + \frac{1}{3}\,\delta^{ij}\int_0^t M[f(t) - \hat{s}] \times \frac{\partial\gamma_{kk}}{\partial\hat{s}}\,d\hat{s} - \delta^{ij}\int_0^t \alpha 3K[f(t) - \hat{s}]\frac{\partial T}{\partial\hat{s}}\,d\hat{s} \quad (19.57a)$$

$$\xi = f(0) + 3\alpha K(0)\gamma_{ii} + \mathcal{M}(0)T(t) - \int_0^t \frac{\partial}{\partial\hat{s}}\,[3\alpha K(\hat{s})]\gamma_{ii}(f(t) - \hat{s})\,d\hat{s} - \int_0^t \frac{\partial}{\partial\hat{s}}\,[\mathcal{M}(\hat{s})]T(f(t) - \hat{s})\,d\hat{s} \quad (19.57b)$$

$$\begin{aligned}\sigma = &-\frac{1}{6}\int_0^t\int_0^t \frac{\partial}{\partial t}\,\{M[f(t) - \hat{s}_1, f(t) - \hat{s}_2]\}\frac{\partial\gamma_{ii}}{\partial\hat{s}_1}\frac{\partial\gamma_{jj}}{\partial\hat{s}_2}\,d\hat{s}_1\,d\hat{s}_2 \\ &-\frac{1}{2}\int_0^t\int_0^t \frac{\partial}{\partial t}\,\{2G[f(t) - \hat{s}_1, f(t) - \hat{s}_2]\}\frac{\partial\gamma_{ij}}{\partial\hat{s}_1}\frac{\partial\gamma_{ij}}{\partial\hat{s}_2}\,d\hat{s}_1\,d\hat{s}_2 \\ &+\int_0^t\int_0^t \frac{\partial}{\partial t}\,\{3\alpha K[f(t) - \hat{s}_1, f(t) - \hat{s}_2]\}\frac{\partial\gamma_{ii}}{\partial\hat{s}_1}\frac{\partial T}{\partial\hat{s}_2}\,d\hat{s}_1\,d\hat{s}_2 \\ &+\frac{1}{2}\int_0^t\int_0^t \frac{\partial}{\partial t}\,\{\mathcal{M}[f(t) - \hat{s}_1, f(t) - \hat{s}_2]\}\frac{\partial T}{\partial\hat{s}_1}\frac{\partial T}{\partial\hat{s}_2}\,d\hat{s}_1\,d\hat{s}_2 \end{aligned} \quad (19.57c)$$

In these equations, $K(s) = K(0,s)$, $G(s) = G(0,s)$, $M(s) = M(0,s) \equiv 3K(0,s) - 2G(0,s)$, and $\mathcal{M}(s) = \mathcal{M}(0,s)$. Clearly, $D^{ij}(0)$ is the initial stress in the material. The material kernels $K(s)$ and $G(s)$ are relaxation bulk and shear moduli, α is the coefficient of thermal expansion, and $\mathcal{M}(0)T$ can be associated with the specific heat of the material. While (19.57) is expected to apply to infinitesimal deformations of solids for which temperature gradients are small, we note that upon introducing $\zeta = \hat{f}(t)$, we generally obtain a highly nonlinear constitutive equation in $T(\mathbf{x},t)$.

Thermoelasticity An important special class of simple materials is characterized by free energies which depend only upon the current values of γ_{ij} and T and not upon their histories. We refer to such materials as *thermoelastic materials*.† Consider, for example, the case in which the free

† See (15.22) and (15.23). The theory of thermoelasticity dates back over a century to the work of Duhamel [1837]. The linear theory is treated in some detail in various books and survey articles, e.g., Boley and Weiner [1960], Chadwick [1960], Nowacki [1962], Parkus [1964, 1968], and Johns [1965].

energy per unit undeformed volume in an isotropic thermoelastic solid is a differentiable function of the current strain $\gamma_{ij}(\mathbf{x},t)$ and temperature $T(\mathbf{x},t) = \theta(\mathbf{x},t) - T_0$:

$$\varphi(\mathbf{x},t) = \Phi(\gamma_{ij},T) \tag{19.58}$$

Then, in accordance with (19.36*a*) and (19.36*b*),†

$$t^{ij} = \frac{\partial\Phi(\gamma_{ij},T)}{\partial\gamma_{ij}} \qquad \xi = -\frac{\partial\Phi(\gamma_{ij},T)}{\partial T} \tag{19.59a,b}$$

while (19.36*c*) reveals that for thermoelastic solids

$$\sigma = 0 \tag{19.59c}$$

Following the procedure used to obtain (19.42), we now consider a homogeneous isotropic solid for which $\Phi(\ )$ is an analytic function of T and certain invariants of γ_{ij}; for example,

$$\varphi = \Phi(J_1,J_2,J_3,T) \tag{19.60}$$

where

$$\begin{aligned} J_1 &= \gamma_{ii} \\ J_2 &= \tfrac{1}{2}(\gamma_{ii}\gamma_{jj} - \gamma_{ij}\gamma_{ji}) \\ J_3 &= \det \gamma_{ij} \end{aligned} \tag{19.61}$$

Then, expanding (19.60) in a power series in T and the strain invariants, we obtain‡

$$\begin{aligned} \Phi = a_0 &+ a_1J_1 + a_2J_2 + a_3J_3 + a_4T + a_5J_1^2 + a_6J_1^3 \\ &+ a_7T^2 + a_8J_1T + a_9J_1J_2 + a_{10}J_1T^2 + a_{11}J_2T \\ &+ a_{12}T^3 + a_{13}J_1^2T + a_{14}J_1^4 + a_{15}J_2^2 + a_{16}J_1J_3 + \cdots \end{aligned} \tag{19.62}$$

where $a_0, a_1, \ldots$ are material constants. For simplicity, we shall confine our attention to materials which are stress-free in the undeformed state and for which the free energy is a function of no terms of order higher than fourth in the strains and third in the temperature increments. Then a_0 and $a_{17}, a_{18}, a_{19}, \ldots$ are zero. Note that no assumptions about magnitudes of strains or temperature are required in specifying such materials; rather, we merely identify a class of thermoelastic materials which are characterized by free-energy functions of this form.

† Since we assume the existence of a differentiable potential function $\Phi(\gamma,T)$, a more accurate description of the material (19.59) would be a "hyperthermoelastic" solid.

‡ Such expansions were used by Dillon [1962] to obtain a nonlinear theory of thermoelasticity. Other nonlinear stress-strain laws for thermoelastic solids were proposed by Reiner [1958] and Jindra [1959]. We follow here the discussion of Oden [1969*d*]. See also Oden and Poe [1970].

In the case of finite thermoelasticity, manageable theories can be developed for only relatively simple forms of the free energy. For example, a certain incompressible thermoelastic material of the Mooney type is described by a free-energy function of the form

$$\Phi = a_1 J_1 + a_2 J_2 + a_7 T^2 + a_8 J_1 T \tag{19.63}$$

where

$$a_1 = 2(C_1 + 2C_2) \qquad a_2 = 4C_2 \tag{19.64}$$

and C_1 and C_2 are the usual Mooney constants for the isothermal case. Also, the following incompressibility condition must be satisfied:

$$J_3 + \tfrac{1}{4}J_1 + \tfrac{1}{2}J_2 = 0 \tag{19.65}$$

According to (19.59*a*,*b*) and (15.58), the stress and entropy are given by

$$\begin{aligned} t^{ij} = a_1\delta^{ij} + a_2(J_1\delta^{ij} - \delta^{im}\delta^{jn}\gamma_{mn}) + \delta^{ij}a_8 T + h[2\delta^{ij}(1 + 4J_2) \\ + 4J_1(\delta^{ij} - 2\delta^{im}\delta^{jn}\gamma_{mn}) + 4\delta^{im}\delta^{jn}\gamma_{mk}(2\gamma_{nk} - \delta_{nk})] \end{aligned} \tag{19.66a}$$

$$\xi = -2a_7 T - a_8\gamma_{ii} \tag{19.66b}$$

where h is an arbitrary hydrostatic pressure.

If it is assumed that the displacement gradients $|u_{i,j}| \ll 1$, the classical strain-displacement relations $2\gamma_{ij} = u_{i,j} + u_{j,i}$ hold and we may use

$$\begin{aligned} \Phi = a_2 J_2 + a_3 J_3 + a_5 J_1^2 + a_6 J_1^3 + a_7 T^2 + a_8 J_1 T + a_9 J_1 J_2 \\ + a_{10} J_1 T^2 + a_{11} J_2 T + a_{12} T^3 + a_{13} J_1^2 T + a_{14} J_1^4 \\ + a_{15} J_2^2 + a_{16} J_1 J_3 \end{aligned} \tag{19.67}$$

as a characterization of a materially nonlinear thermoelastic solid. Here it is assumed that terms of order higher than fourth do not contribute substantially to the free energy. Then

$$\begin{aligned} t^{ij} = \delta^{ij}[a_2 J_1 + a_3 J_2 + 2a_5 J_1 + 3a_6 J_1^2 + a_8 T + a_9 J_2 + a_9 J_1^2 \\ + a_{10} T^2 + a_{11} T J_1 + 2a_{13} J_1 T + 4a_{14} J_1^3 + 2a_{15} J_1 J_2 \\ + a_{16} J_3 + a_{16} J_2 J_1] - \gamma_{ij}(a_2 + a_3 J_1 + a_9 J_1 + a_{11} T \\ + 2a_{15} J_2 + a_{16} J_1^2) + \gamma_{ik}\gamma_{jk}(a_3 + a_{16} J_1) \end{aligned} \tag{19.68a}$$

$$-\xi = 2a_7 T + a_8 J_1 + 2a_{10} J_1 T + a_{11} J_2 + 3a_{12} T^2 + a_{13} J_1^2 \tag{19.68b}$$

Dillon [1962]†, noting that the constitutive equations (19.68) are too complicated to be manageable in most stress-analysis problems, proposed a

† Our results here, i.e., (19.71) and (19.72), represent a slight generalization of the form proposed by Dillon.

simplified version of (19.67) based on the following assumptions and observations:

1. It is assumed that the linearized versions of (19.68) coincide with the classical equations describing a linear isotropic thermoelastic solid. Then

$$a_2 = -2\mu \qquad a_5 = \tfrac{1}{2}(\lambda + 2\mu) \qquad a_8 = -\alpha(3\lambda + 2\mu) \qquad -2Ta_7 = c \tag{19.69}$$

where λ and μ are the Lamé constants, α is the coefficient of thermal expansion, and c is the specific heat at constant deformation [observe that $a_7 = -c/2T$ is inconsistent with (19.68*b*), which was derived assuming that a_7 is independent of T; however, this discrepancy is also found in the linear theory, where c is introduced as a constant for which $\nabla \cdot \mathbf{q} = c\dot{T}$ for constant deformation].

2. The specific heat is

$$c_d = c - 2(T + T_0)(3a_{12}T + a_{10}J_1) \tag{19.70}$$

We assume that $c_d = c =$ constant, so that $a_{10} = a_{12} = 0$.

3. If the heating of the material in compression equals the cooling in tension, it is required that $a_{13} = 0$.

4. Experiments on most metallic-type materials reveal that the relationship between dilatational components of stress and strain is linear to a larger extent than that between the deviatoric components. Thus, we omit terms of fourth order and higher in the strains by setting $a_{14} = a_{15} = a_{16} = 0$ but retain $a_{11}J_2T$ since it represents a nonlinear contribution to the deviatoric strain.

With these simplifications, $\Phi(\ )$ reduces to

$$\Phi = -2\mu J_2 + a_3J_3 + \tfrac{1}{2}(\lambda + 2\mu)J_1^2 + a_6J_1^3 + a_7T^2 - \alpha(3\lambda + 2\mu)J_1T + a_9J_1J_2 + a_{11}J_2T \tag{19.71}$$

Thus

$$t^{ij} = \delta^{ij}[\lambda J_1 a_3 J_2 + 3a_6J_1^2 - \alpha(3\lambda + 2\mu)T + a_9(J_2 + J_1^2) + a_{11}J_1T] + \gamma_{ij}(2\mu - a_3J_1 - a_9J_1 - a_{11}T) + a_3\gamma_{ik}\gamma_{jk} \tag{19.72a}$$

$$\xi = -2a_7T + \alpha(3\lambda + 2\mu)J_1 - a_{11}J_2 \tag{19.72b}$$

Equations (19.71) and (19.72) are recorded only as examples of one possible reduced form of the constitutive equations for nonlinear thermoelastic solids. A number of other forms of the free energy which lead to nonlinear constitutive equations for stress and entropy can be obtained by deleting or adding appropriate terms in the power series expansion in (19.62). Which form is appropriate for a given material can be determined only through experimentation. Equations (19.71) and (19.72) reduce to the classical equations of linear isotropic thermoelasticity upon deleting nonlinear terms. By setting $a_3 = a_6 = a_9 = 0$, the material described by Dillon [1962] which is nonlinear in the deviatoric strains is obtained. If either of

a_3, a_6, or $a_9 \neq 0$, a material which is also mildly nonlinear in dilatation i obtained. For purely dilatational strains, ξ is identical to that of th classical theory.

20 FINITE-ELEMENT MODELS OF NONLINEAR THERMOMECHANICAL BEHAVIOR

In this article, we develop finite-element models of the nonlinear thermo mechanical behavior of solids, and we examine applications of these t representative problems in nonlinear thermoviscoelasticity and thermo elasticity.

20.1 FINITE ELEMENTS OF MATERIALS WITH MEMORY†

Let $u_i^{(e)}(\mathbf{x},t) = \psi_N^{(e)}(\mathbf{x})u_{i(e)}^N(t)$ and $T^{(e)}(\mathbf{x},t) = \varphi_N^{(e)}(\mathbf{x})T_{(e)}^N(t)$ denote loca approximations of the displacement and incremental temperature fields ove a typical element e of a finite-element model of a thermomechanically simpl body, and suppose that $\underset{s=0}{\overset{\infty}{\varphi^{(e)}}}[-]$, $\underset{s=0}{\overset{\infty}{\mathfrak{T}^{ij}_{(e)}}}[-]$, $\underset{s=0}{\overset{\infty}{\mathscr{N}^{(e)}}}[-]$, and $\underset{s=0}{\overset{\infty}{\mathfrak{Q}^{i}_{(e)}}}[-]$ are th constitutive functionals for the free energy per unit initial volume, the stres tensor, the entropy per unit initial volume, and the heat flux for the elemen Following the arguments of the previous article, we choose to write $\underset{s=0}{\overset{\infty}{\varphi^{(e)}}}[-$ in terms of the past histories $\boldsymbol{\gamma}^t_{r(e)}$, $\theta^t_{r(e)}$ of the strain and temperature and c

† Finite-element formulations of problems of viscoelasticity, creep, thermoelasticity, an thermoviscoelasticity have been investigated by a number of authors. Among the firs studies of creep and thermoelasticity were those of Padlog, Huff, and Holloway [1960] Turner, Dill, Martin, and Melosh [1960], Gallagher [1961], Gallagher, Quinn, and Padlo [1961], Gallagher, Padlog, and Bijlaard [1962], and Warren [1962]. See also the survey o Warren, Castle, and Gloria [1962]. Finite-element analyses of problems of linear visco elasticity were presented by King [1965], Chang [1966], White [1969], and others, and th formulation of finite-element models of isothermal nonlinearly viscoelastic bodies wa discussed by Oden [1967*a*, 1970*a*] and Dong, Pister, and Dunham [1968]. The linea heat-conduction problem was first treated using finite elements by Wilson and Nicke [1966] and Becker and Parr [1967]; a nonlinear heat-conduction problem was solved b Aguirre-Ramirez and Oden [1969]. Transient linear thermoelastic behavior of solids unde prescribed temperature fields was investigated by Visser [1966], while Fujino and Ohsak [1969] examined heat conduction and thermal analysis of solids by the finite-elemen method; see also Besseling [1966]. Transient linear coupled thermoelasticity problem were explored by Nickell and Sackman [1968] and Oden and Kross [1969], and finite element solutions of nonlinear problems in dynamic coupled thermoelasticity wer presented by Oden [1969*d*] and Oden and Poe [1970]. The development of general finite element models of thermomechanical behavior of simple materials was discussed by Ode and Aguirre-Ramirez [1969] and Oden [1970*b*]. Finite-element analyses of heat con duction and motion of thermoviscoelastic materials were presented by Taylor and Chan [1966], Taylor, Pister, and Goudreau [1970], Cost [1969], Oden and Armstrong [1971] and Oden [1971*d*].

their current values $\boldsymbol{\gamma}^t_{(e)}(0)$, $\theta^t_{(e)}(0)$. Temporarily omitting the element identification label e for simplicity, we note that the total strain and temperature histories for a finite element may be written in the form

$$\gamma^t_{ij}(s) = \tfrac{1}{2}[\psi_{N,i}u_j^{N(t)}(s) + \psi_{N,j}u_i^{N(t)}(s) + \psi_{N,i}\psi_{M,j}u_k^{N(t)}(s)u_k^{M(t)}(s)] \tag{20.1}$$

$$\theta^t(s) = T_0 + \varphi_N T^{N(t)}(s) \tag{20.2}$$

where it is understood that $\gamma^t_{ij}(s)$, $\psi_{N,i}$, $\theta^t(s)$, and φ_N are functions of $\mathbf{x}$. In (20.1) and (20.2), $u_i^{N(t)}(s)$ and $T^{N(t)}(s)$ are histories of the nodal values of the displacement components and the change in temperature (from a uniform, constant reference temperature T_0) at node N of the element. Consequently, the constitutive equations for the element can be expressed as functionals of the histories of u_i^N and T^N in accordance with (19.36):

$$\varphi^{(e)} = \underset{s=0}{\overset{\infty}{\Upsilon^{(e)}}}[\mathbf{u}_r^{N(t)}, T_r^{N(t)}; \mathbf{u}^N, T^N] \tag{20.3a}$$

$$t^{ij}_{(e)} = \underset{s=0}{\overset{\infty}{\mathfrak{S}^{ij}_{(e)}}}[\mathbf{u}_r^{N(t)}, T_r^{N(t)}; \mathbf{u}^N, T^N] \tag{20.3b}$$

$$\xi_{(e)} = \underset{s=0}{\overset{\infty}{\mathfrak{H}_{(e)}}}[\mathbf{u}_r^{N(t)}, T_r^{N(t)}; \mathbf{u}^N, T^N] \tag{20.3c}$$

$$q^i_{(e)} = \underset{s=0}{\overset{\infty}{\mathcal{Q}^i_{(e)}}}[\mathbf{u}_r^{N(t)}, T_r^{N(t)}; \mathbf{u}^N, T^N] \tag{20.3d}$$

Here $\mathbf{u}_r^{N(t)}$, $T_r^{N(t)}$ are the restrictions of $\mathbf{u}^N(t-s)$, $T^N(t-s)$ to $(0,\infty)$,

$$\underset{s=0}{\overset{\infty}{\Upsilon^{(e)}}}[\mathbf{u}^{N(t)}, T^{N(t)}] = \underset{s=0}{\overset{\infty}{\Phi^{(e)}}}\{\boldsymbol{\gamma}^t_r(\mathbf{x},\mathbf{u}_r^{N(t)}), \theta^t_r(\mathbf{x}, T_r^{N(t)}); \boldsymbol{\gamma}[\mathbf{x},\mathbf{u}^{N(t)}(0)], \theta[\mathbf{x}, T^{N(t)}(0)]\} \tag{20.4a}$$

$$\underset{s=0}{\overset{\infty}{\mathfrak{S}^{ij}_{(e)}}}[\mathbf{u}^{N(t)}, T^{N(t)}] = D_{\gamma_{ij}} \underset{s=0}{\overset{\infty}{\Phi^{(e)}}}\{\boldsymbol{\gamma}^t_r(\mathbf{x},\mathbf{u}_r^{N(t)}), \theta^t_r(\mathbf{x}, T_r^{N(t)}); \boldsymbol{\gamma}[\mathbf{x},\mathbf{u}^{N(t)}(0)], \theta[\mathbf{x}, T^{N(t)}(0)]\} \tag{20.4b}$$

$$\underset{s=0}{\overset{\infty}{\mathfrak{H}_{(e)}}}[\mathbf{u}^{N(t)}, T^{N(t)}] = -D_\theta \underset{s=0}{\overset{\infty}{\Phi^{(e)}}}\{\boldsymbol{\gamma}^t_r(\mathbf{x},\mathbf{u}_r^{N(t)}), \theta^t_r(\mathbf{x}, T_r^{N(t)}); \boldsymbol{\gamma}[\mathbf{x},\mathbf{u}^{N(t)}(0)], \theta[\mathbf{x}, T^{N(t)}(0)]\} \tag{20.4c}$$

$$\underset{s=0}{\overset{\infty}{\mathcal{Q}^i_{(e)}}}[\mathbf{u}^{N(t)}, T^{N(t)}] = \underset{s=0}{\overset{\infty}{\mathfrak{Q}^i_{(e)}}}\{\boldsymbol{\gamma}^t_r(\mathbf{x},\mathbf{u}_r^{N(t)}), \theta^t_r(\mathbf{x}, T_r^{N(t)}); \boldsymbol{\gamma}[\mathbf{x},\mathbf{u}^{N(t)}(0)], \theta[\mathbf{x}, T^{N(t)}(0)], T^N(t)\boldsymbol{\nabla}\varphi_N(\mathbf{x})\} \tag{20.4d}$$

and, for simplicity, we have denoted $[\mathbf{u}_r^{N(t)}, T_r^{N(t)}; \mathbf{u}^N(t), T^N(t)] = [\mathbf{u}^{N(t)}, T^{N(t)}]$. Likewise, the internal dissipation for the element is given by

$$\sigma_{(e)} = \underset{s=0}{\overset{\infty}{\mathfrak{D}}}{}_{(e)}[\mathbf{u}_r^{N(t)}, T_r^{N(t)}; \mathbf{u}^N, T^N] \tag{20.5}$$

where

$$-\underset{s=0}{\overset{\infty}{\mathfrak{D}}}{}_{(e)}[-] = \delta_{\boldsymbol{\gamma}} \underset{s=0}{\overset{\infty}{\Phi}}{}^{(e)}[+ \mid \dot{\mathbf{u}}_r^{N(t)}] + \delta_\theta \underset{s=0}{\overset{\infty}{\Phi}}{}^{(e)}[+ \mid \dot{T}_r^{N(t)}] \tag{20.6}$$

Here we have used the notation

$$[+ \mid \dot{\mathbf{u}}_r^{N(t)}] = \{\boldsymbol{\gamma}_r^t(\mathbf{x}, \mathbf{u}_r^{N(t)}), \theta_r^t(\mathbf{x}, T_r^{N(t)}); \boldsymbol{\gamma}[\mathbf{x}, \mathbf{u}_r^{N(t)}(0)], \theta[\mathbf{x}, T_r^{N(t)}(0)] \mid f_{Nij}^{k(t)}(\mathbf{x}, s)\dot{\mathbf{u}}_{k(r)}^{N(t)}\} \tag{20.7a}$$

where $s \in (0, \infty)$,

$$f_{Nij}^{k(t)}(\mathbf{x}, s) = \frac{1}{2} \psi_{N,m} z_{k,n}^t(s)(\delta_{jn}\delta_{mi} + \delta_{in}\delta_{jm}) \tag{20.7b}$$

and $z_{k,n}^t(s)$ are the histories of the deformation gradients:

$$z_{k,n}^t(s) = \delta_{nk} + \psi_{M,n} u_k^{M(t)}(s) \tag{20.7c}$$

The array in (20.7b) provides for the transformation

$$\dot{\gamma}_{kj(r)}^t = -\frac{d}{ds} \gamma_{ij(r)}^t(s) = -\frac{\partial \gamma_{ij}}{\partial u_k^N} \frac{du_{k(r)}^{N(t)}}{ds} = -f_{Nij}^{k(t)} \dot{u}_{k(r)}^{N(t)}(s) \tag{20.8}$$

Introducing (20.3) and (20.5) into (13.60) and (13.117), we arrive at the general equations of motion and heat conduction of finite elements of thermomechanically simple media:

$$m_{NM}\ddot{u}_i^M + \int_{v_{0(e)}} \underset{s=0}{\overset{\infty}{\mathfrak{S}}}{}_{(e)}^{mj}[\mathbf{u}^{N(t)}, T^{N(t)}]\psi_{N,m}(\delta_{ji} + \psi_{M,j}u_i^M)\, dv_0 = p_{Ni} \tag{20.9}$$

$$\int_{v_{0(e)}} \{\varphi_N(T_0 + \varphi_M T^M) \frac{d}{dt} \underset{s=0}{\overset{\infty}{\mathfrak{H}}}{}_{(e)} [\mathbf{u}^{N(t)}, T^{N(t)}] + \varphi_{N,i} \underset{s=0}{\overset{\infty}{\mathcal{Q}}}{}^i_{(e)}[\mathbf{u}^{N(t)}, T^{N(t)}]\}\, dv_0 = q_N + \sigma_N \tag{20.10}$$

where p_{Ni} and q_N are the generalized forces and normal heat-flux components at node N, as defined by (13.54) and (13.115); and, in this case, the generalized nodal dissipation σ_N of (13.116) is given by

$$\sigma_N = \int_{v_{0(e)}} \underset{s=0}{\overset{\infty}{\mathfrak{D}}}{}_{(e)}[\mathbf{u}^{N(t)}, T^{N(t)}]\varphi_N\, dv_0 \tag{20.11}$$

Since global forms of these equations are obtained by direct application of (13.74) to (13.77) and (13.118) to (13.120), we do not reproduce them here.

It is interesting to note that (20.9) can also be written in the form

$$p_{Ni} = m_{NM}\ddot{u}_i^M + \int_{v_{0(e)}} D_{u_i^N} \overset{\infty}{\underset{s=0}{\Phi}}{}^{(e)}[\boldsymbol{\gamma}^t,\theta^t]\, dv_0 \tag{20.12}$$

in analogy with (16.131), where

$$D_{u_k^N} \overset{\infty}{\underset{s=0}{\Phi}}{}^{(e)}[\boldsymbol{\gamma}^t,\theta^t] = D_{\gamma_{ij}} \overset{\infty}{\underset{s=0}{\Phi}}{}_{(e)}[\boldsymbol{\gamma}_r^t,\theta_r^t;\boldsymbol{\gamma},\theta]\frac{\partial \gamma_{ij}}{\partial u_k^N} = \overset{\infty}{\underset{s=0}{\mathfrak{S}}}{}^{ij}_{(e)}[\mathbf{u}^{N(t)},T^{N(t)}]\frac{\partial \gamma_{ij}}{\partial u_k^N} \tag{20.13}$$

Similarly, in (20.10), we may use instead of $d \overset{\infty}{\underset{s=0}{\mathfrak{H}}}{}_{(e)} [\mathbf{u}^{N(t)},T^{N(t)}]/dt$,

$$\frac{d}{dt} D_{T^N} \overset{\infty}{\underset{s=0}{\Phi}}{}^{(e)}[\boldsymbol{\gamma}^t,\theta^t] = D_\theta \overset{\infty}{\underset{s=0}{\Phi}}{}_{(e)}[\boldsymbol{\gamma}_r^t,\theta_r^t;\boldsymbol{\gamma},\theta]\frac{\partial\theta}{\partial T^N}\dot{T}^N \tag{20.14}$$

In (20.11) to (20.13) it is understood that $\boldsymbol{\gamma}^t$ and θ^t are to be expressed in terms of (or determined by) the histories of the nodal quantities $\mathbf{u}^{N(t)}$ and $T^{N(t)}$. Thus, specific forms of the equations of motion and heat conduction for an element can be obtained directly from the free-energy functional and heat-flux functional corresponding to a specific material. In the following articles, we examine applications of these equations to specific problems in thermoviscoelasticity and transient thermoelasticity.

20.2 APPLICATION TO THERMORHEOLOGICALLY SIMPLE MATERIALS

In the case of a thermorheologically simple material of the type defined by (19.56), (19.57), and, say, (19.46), it is convenient to transform various field quantities to reduced time ζ. Thus, if $f(\mathbf{x},t)$ is any function of real time t (and $\mathbf{x}$), $g(\zeta) = t$ exists, and $\hat{b}[\hat{\theta}(\mathbf{x},\zeta)] = b[\theta(\mathbf{x},t)]$ is the shift factor of (19.50) (with $\theta = T_0 + T$), then

$$\dot{f}(\mathbf{x},t) = \mathring{\hat{f}}(\mathbf{x},\zeta)\hat{b}[\hat{\theta}(\mathbf{x},\zeta)] \tag{20.15}$$

where $\mathring{\hat{f}} = d\hat{f}/d\zeta$. It follows that

$$\begin{aligned} \dot{u}_i^M &= \hat{b}(\hat{\theta})\mathring{\hat{u}}_i^M \\ \ddot{u}_i^M &= \hat{b}^2(\hat{\theta})\ddot{\hat{u}}_i^M + \hat{b}(\hat{\theta})\mathring{\hat{b}}(\hat{\theta})\mathring{\hat{u}}_i^M \\ \dot{\theta} &= \hat{b}(\hat{\theta})\mathring{\hat{\theta}} \end{aligned} \tag{20.16}$$

etc., wherein dependence of the quantities $\mathring{\hat{u}}_i^M$, $\hat{\theta}$, etc., on ζ is understood. Similar relations hold for ξ, Q^i, and σ.

To cite a specific example, consider the material defined by (19.46), (19.56), and (19.57). Suppose $\rho\theta\dot{\xi} \approx \rho T_0\dot{\xi}$ and $2\gamma_{ij} = u_{i,j} + u_{j,i}$. Then, transforming the resulting equations into reduced time with the aid of (20.15) and (20.16), we obtain the following equations of motion and heat conduction of a thermorheologically simple element in reduced time:

$$m_{NM}^{(1)}\ddot{\hat{u}}_k^M + m_{NM}^{(2)}\dot{\hat{u}}_k^M + a_{NMk}^i \int_0^\zeta 2G(\zeta - \hat{s}) \frac{\partial \hat{u}_i^M}{\partial \hat{s}} d\hat{s}$$

$$+ b_{NMk}^i \int_0^\zeta [3K(\zeta - \hat{s}) - 2G(\zeta - \hat{s})] \frac{\partial \hat{u}_i^M}{\partial \hat{s}} d\hat{s}$$

$$-c_{NMk} \int_0^\zeta 3K(\zeta - \hat{s}) \frac{\partial \hat{T}^M}{\partial \hat{s}} d\hat{s} = \hat{p}_{Nk}(\zeta) \tag{20.17}$$

$$h_{MN}^{(1)}\dot{\hat{T}}^N + h_{MN}^{(2)i}\dot{\hat{u}}_i^N + h_{MN}^{(3)} \int_0^\zeta \frac{\partial}{\partial \zeta} [\mathscr{M}(\zeta - \hat{s})] \frac{\partial \hat{T}^N}{\partial \hat{s}} d\hat{s}$$

$$+ h_{MN}^{(4)i} \int_0^\zeta \frac{\partial}{\partial \zeta} [3\acute{K}(\zeta - \hat{s})] \frac{\partial \hat{u}_i^N}{\partial \hat{s}} d\hat{s}$$

$$+ k_{MN} \int_0^\zeta \kappa(\zeta - \hat{s}) \frac{\partial \hat{T}^N}{\partial \hat{s}} d\hat{s} + d_{MNL}^{(1)ij} \int_0^\zeta \int_0^\zeta \frac{\partial}{\partial \zeta} [3K(\zeta - \hat{s}_1,$$

$$\zeta - \hat{s}_2) - 2G(\zeta - \hat{s}_1, \zeta - \hat{s}_2)] \frac{\partial \hat{u}_i^N}{\partial \hat{s}_1} \frac{\partial \hat{u}_j^L}{\partial \hat{s}_2} d\hat{s}_1 \, d\hat{s}_2$$

$$+ d_{MNL}^{(2)ij} \int_0^\zeta \int_0^\zeta \frac{\partial}{\partial \zeta} [2G(\zeta - \hat{s}_1, \zeta - \hat{s}_2)] \frac{\partial \hat{u}_i^N}{\partial \hat{s}_1} \frac{\partial \hat{u}_j^L}{\partial \hat{s}_2} d\hat{s}_1 \, d\hat{s}_2$$

$$-d_{MNL}^{(3)i} \int_0^\zeta \int_0^\zeta \frac{\partial}{\partial \zeta} [3K(\zeta - \hat{s}_1, \zeta - \hat{s}_2)] \frac{\partial \hat{u}_i^N}{\partial \hat{s}_1} \frac{\partial \hat{T}^L}{\partial \hat{s}_2} d\hat{s}_1 \, d\hat{s}_2$$

$$- d_{MNL}^{(4)} \int_0^\zeta \int_0^\zeta \frac{\partial}{\partial \zeta} [\mathscr{M}(\zeta - \hat{s}_1, \zeta - \hat{s}_2)] \frac{\partial \hat{T}^N}{\partial \hat{s}_1} \frac{\partial \hat{T}^L}{\partial \hat{s}_2} d\hat{s}_1 \, d\hat{s}_2 = \hat{q}_M(\zeta) \tag{20.18}$$

Here $\hat{s} = \hat{f}(s)$, $\hat{f}$ being the function of (20.15), and

$$m_{NM}^{(1)} = \int_{v_{0(e)}} \rho_0 \hat{b}^2 \psi_N \psi_M \, dv_0 \qquad m_{NM}^{(2)} = \int_{v_{0(e)}} \rho_0 \hat{b}\dot{\hat{b}} \psi_N \psi_M \, dv_0$$

$$a_{NMk}^{i} = \frac{1}{2} \int_{v_{0(e)}} \hat{b}(\psi_{M,j}\delta_{ik} + \psi_{M,i}\delta_{jk})\psi_{N,j} \, dv_0$$

$$b_{NMk}^{i} = \frac{1}{3} \int_{v_{0(e)}} \hat{b}\psi_{M,k}\psi_{N,i} \, dv_0 \qquad c_{NMk} = \alpha \int_{v_{0(e)}} \hat{b}\psi_{M,k}\varphi_N \, dv_0$$

$$h_{MN}^{(1)} = T_0 m(0) \int_{v_{0(e)}} \varphi_M \varphi_N \, dv_0 = m(0) h_{MN}^{(3)}$$

$$h_{MN}^{(2)i} = 3\alpha T_0 K(0) \int_{v_{0(e)}} \varphi_M \psi_{N,i} \, dv_0 = 3K(0) h_{MN}^{(4)i}$$

$$k_{MN} = \int_{v_{0(e)}} \varphi_{M,i} \varphi_{N,i} \, dv_0$$

$$d_{MNL}^{(1)ij} = \frac{1}{6} \int_{v_{0(e)}} \varphi_M \psi_{N,i} \psi_{L,j} \, dv_0$$

$$d_{MNL}^{(2)ij} = \frac{3}{2} d_{MLN}^{(1)ij} + \frac{1}{4} \delta^{ij} \int_{v_{0(e)}} \varphi_M \psi_{N,m} \psi_{L,m} \, dv_0$$

$$d_{MNL}^{(3)i} = \int_{v_{0(e)}} \alpha \varphi_M \psi_{N,i} \varphi_L \, dv_0$$

$$d_{MNL}^{(4)} = \frac{1}{2} \int_{v_{0(e)}} \varphi_M \varphi_N \varphi_L \, dv_0 \tag{20.19}$$

To obtain quantitative solutions of (20.17) and (20.18), we shall assume that the kernels appearing in the integrals in these equations pertain to materials with fading memory; i.e., the current response is not appreciably affected by strains and temperatures at times distant in the past. This property is often assured by considering each kernel to be representable in the form of a decaying Prony series; for example, a typical material kernel $K(\zeta)$ is defined by a series of the form

$$K(\zeta) = \sum_{i=1}^{n} K_i \exp\left(\frac{-\zeta}{\nu_i}\right) \tag{20.20}$$

where K_i and ν_i are experimentally determined constants, $\zeta \geq 0$; and $\nu_i > 0$. Kernels of the type $G(\zeta,\zeta')$ are often assumed to have the property

$$G(\zeta - \zeta', \zeta - \zeta'') = \bar{G}(2\zeta - \zeta' - \zeta'') \tag{20.21}$$

so that, in accordance with (20.20), we can write

$$G(\zeta - \zeta', \zeta - \zeta'') = \sum_{i=1}^{n} G_i \exp\left(\frac{-2\zeta + \zeta' + \zeta''}{\lambda_i}\right)$$
$$\frac{\partial G}{\partial \zeta'}(\zeta - \zeta', \zeta - \zeta'') = \sum_{i=1}^{n} \frac{G_i}{\lambda_i} \exp\left(\frac{-2\zeta + \zeta' + \zeta''}{\lambda_i}\right) \tag{20.22}$$

etc., λ_i being a constant > 0.

Now consider a typical integral appearing in one of the constitutive equations cited earlier; e.g., define

$$IG_{Nj} \equiv \int_0^{\zeta} 2G(\zeta - \zeta') \frac{\partial \hat{u}_j^N}{\partial \zeta'} d\zeta' \tag{20.23}$$

We may obtain an alternate representation of (20.23) which is easily adapted to numerical approximation by dividing the reduced time interval $[0,\zeta]$ into r subintervals $[\zeta_m, \zeta_{m+1}]$, $m = 1, 2, \ldots, r + 1$. Then, using the representation (20.20) for the kernel in (20.23), we have

$$IG_{Nj} = \sum_{i=1}^{r} G_i e^{-\zeta/\lambda_i} \sum_{m=1}^{r} \int_{\zeta_m}^{\zeta_{m+1}} e^{+\zeta'/\lambda_i} \frac{\partial \hat{u}_j^N}{\partial \zeta'} d\zeta' \tag{20.24}$$

If $h_m = \zeta_{m+1} - \zeta_m$ is the time step, we may use Simpson's rule to approximate the integrals in (20.24):

$$\int_{\zeta_m}^{\zeta_{m+1}} \frac{\partial \hat{u}_j^N}{\partial \zeta'} \exp\left(\frac{\zeta'}{\lambda_i}\right) d\zeta' \approx \frac{h_m}{6}\left\{\frac{\partial \hat{u}_j^N(\zeta_m)}{\partial \zeta}\right.$$
$$\times \exp\left(\frac{\zeta_m}{\lambda_i}\right) + 4\,\frac{\partial \hat{u}_j^N(\zeta_m + h_m/2)}{\partial \zeta} \exp\left[\frac{(\zeta_m + h_m/2)}{\lambda_i}\right] + \frac{\partial \hat{u}_j^N(\zeta_m + h_m)}{\partial \zeta}$$
$$\left. \times \exp\left(\frac{\zeta_m + h_m}{\lambda_i}\right)\right\} \tag{20.25}$$

To proceed further, the rates $\partial \hat{u}_j^N/\partial \zeta$ must be replaced by difference approximations. In the present study, we assume a quadratic variation of $\hat{u}_i^N$ between nodes $m - 1$, m, $m + 1$ which leads to approximations of $O(h^2)$ of the rates. Introducing these into (20.25) and rearranging terms, we arrive at the following form of (20.23):

$$IG_{Nj} = \sum_i G_i J_{ji}^N(u,\lambda) \tag{20.26}$$

wherein

$$J_{ji}^N(u,\lambda) = H(\hat{u}_j^N,\lambda_i) \exp\left[\frac{-(\zeta_r + h_r)}{\lambda_i}\right] + C(\hat{u}_j^N,\lambda_i) \tag{20.27a}$$

$$H(\hat{u}_j^N,\lambda_i) = \sum_{m=1}^{r-1} \left\{ \tfrac{1}{12}(\hat{u}_{j(m+1)}^N - \hat{u}_{j(m-1)}^N) \exp\left(\frac{\zeta_m}{\lambda_i}\right) + \tfrac{2}{3}(\hat{u}_{j(m+1)}^N - \hat{u}_{j(m)}^N) \right.$$
$$\times \exp\left[\frac{(\zeta_m + h_m/2)}{\lambda_i}\right] + \tfrac{1}{12}(3\hat{u}_{j(m+1)}^N) - 4\hat{u}_{j(m)}^N$$
$$\left. + \hat{u}_{j(m-1)}^N) \exp\left(\frac{\zeta_m + h_m}{\lambda_i}\right) \right\} \tag{20.27b}$$

$$C(\hat{u}_j^N,\lambda_i) = \tfrac{1}{12}(\hat{u}_{j(r+1)}^N - \hat{u}_{j(r-1)}^N) \exp\left(\frac{-\zeta_r}{\lambda_i}\right) + \tfrac{2}{3}(\hat{u}_{j(r+1)}^N - \hat{u}_{j(r)}^N)$$
$$\times \exp\left[\frac{-(\zeta_r + h_r/2)}{\lambda_i}\right] + \tfrac{1}{12}(3\hat{u}_{j(r+1)}^N - 4\hat{u}_{j(r)}^N + \hat{u}_{j(r-1)}^N)$$
$$\times \exp\left[\frac{-(\zeta_r + h_r)}{\lambda_i}\right] \tag{20.27c}$$

In (20.27*a*) we have separated the histories $H(\hat{u}_j^N,\lambda_i)$ from the current values $C(\hat{u}_j^N,\lambda_i)$; that is, the value of the integral at the current time $\zeta_r + h_r$ is obtained by adding to the accumulated sum at ζ_r the contribution between ζ_r and $\zeta_r + h_r$. By the notation $\hat{u}_{j(m)}^N, \hat{u}_{j(m-1)}^N$, etc., we mean $\hat{u}_j^N(\zeta_m)$, $\hat{u}_j^N(\zeta_m - h_m)$, etc. Using the notation of (20.27), the equations of motion and heat conduction for a thermorheologically simple finite element then assume the forms

$$\hat{p}_{Nj} = m_{MN}^{(1)}\ddot{\hat{u}}_j^M + m_{MN}^{(2)}\dot{\hat{u}}_j^M + (a_{MNj}^i - b_{MNj}^i) \sum_k G_k J_{ik}^M(u,\lambda)$$
$$+ b_{MNj}^i \sum_k K_k J_{ik}^M(u,\nu) - c_{MNj} \sum_k K_k J_k^M(T,\nu) \tag{20.28a}$$

$$\hat{q}_N = h_{MN}^{(1)}\dot{\hat{T}}^M + h_{MN}^{(2)i}\dot{\hat{u}}_i^M - 3\alpha^2 h_{MN}^{(3)} \sum_k \frac{K_k}{\nu_k} J_k^M(T,\nu)$$
$$- h_{MN}^{(4)i} \sum_k \frac{K_k}{\nu_k} J_{ik}^M(u,\nu) - 2d_{MNL}^{(1)ij} \sum_k \frac{K_k}{\nu_k} J_{ik}^M(u,\nu) J_{jk}^L(u,\nu)$$
$$+ 2(d_{NML}^{(1)ij} - d_{NML}^{(2)ij}) \sum_k \frac{G_k}{\lambda_k} J_{ik}^M(u,\lambda) J_{jk}^L(u,\lambda) + 2d_{NML}^{(3)i}$$
$$\times \sum_k \frac{K_k}{\nu_k} J_{ik}^M(u,\lambda) J_k^L(T,\nu) + 6\alpha^2\, d_{NML}^{(4)} \sum_k \frac{K_k}{\nu_k} J_k^M(T,\nu) J_k^L(T,\nu)$$
$$+ k_{NM} \sum_k E_k J_k^M(T,\epsilon) \tag{20.28b}$$

Thus, upon connecting elements together, we obtain a system of highly nonlinear differential equations in the nodal values $\hat{u}_i^N$, $\hat{T}^N$ and their histories.

An example—transient response of a thermoviscoelastic thick-walled cylinder†

For illustration purposes, we now consider applications of (20.28) to the problem of an infinite, thick-walled, circular cylindrical tube subjected to prescribed mechanical and thermal conditions at its inner and outer boundaries. For the local finite-element approximations of the displacement components and the temperature, we use the simplex approximations $\psi_N(r) = a_N + b_N r$, $n = 1, 2$, a_N and b_N being constants dependent only on the length of a radial element [$a_1 = -a_2 = -1/(r_2 - r_1)$; $b_1 = r_2/(r_2 - r_1)$, $b_2 = r_1/(r_2 - r_1)$ for an element between nodes at radii r_1 and r_2]. By introducing this form of the interpolation functions into the arrays in (20.19) we obtain all properties of the equations of the discrete model that are independent of specific material properties.

To proceed further, we must identify the form of the shift function $b(\theta)$. The following empirical expression for $b(\theta)$ has been proposed for a wide variety of polymers, polymer solutions, organic and inorganic glasses [Williams, Landel, and Ferry, 1955]

$$\log_{10} b(\theta) = \frac{\alpha(\theta - T_0)}{\beta + \theta - T_0}$$

Here α and β are constants. Experimental data [Eringen, 1967] show that for $|\theta - T_0| \leq 50°\text{C}$, $\alpha \approx 9$ and $\beta \approx 100$, so that if $T_0 = 0°\text{C}$,

$$b(\theta) = 10^{9T/(100+T)} \tag{20.29}$$

where, for the finite element, $T = \varphi_N(r)T^N = (a_N + b_N r)T^N$. Transforming to reduced time, we find that

$$\hat{b}(\hat{\theta}) = 10^{\hat{\tau}} \qquad \dot{\hat{b}} = (\hat{b} \ln (10))\dot{\hat{\tau}} \tag{20.30a, b}$$

wherein

$$\hat{\tau} = \frac{9\varphi_N(r)\hat{T}^N}{100 + \varphi_L(r)\hat{T}^L} \tag{20.30c}$$

and $\dot{\hat{T}}^N$ is represented by the quadratic difference scheme described earlier.

In the numerical integration scheme, we shall treat the current value of $\hat{b}(\hat{\theta})$ at time ζ as a constant and allow $\hat{b}$ and $\hat{\tau}$ to lag the current values by the amount for the current step size h_r. The values of $\hat{b}$ and $\hat{\tau}$ are recomputed at the end of each time step and are carried forward into the next time interval to determine nodal displacements and temperatures. This procedure makes it possible to express the mass matrices $m_{MN}^{(1)}$ and $m_{MN}^{(2)}$ in the particularly simple forms

$$m_{MN}^{(1)} = 2\pi M_{NM} \qquad m_{MN}^{(2)} = 2\pi \ln (10)\dot{\hat{\tau}}\, M_{NM} \tag{20.31a, b}$$

$$M_{NM} = \rho_0\left[\frac{a_N a_M(r_2^4 - r_1^4)}{4} + \frac{(a_N b_M + a_M b_N)(r_2^3 - r_1^3)}{3} + \frac{b_N b_M(r_2^2 - r_1^2)}{2}\right] \tag{20.31c}$$

† The example was presented by Oden and Armstrong [1971]. See also Oden, Chung, and Key [1971].

and we have set $\varphi_N(r) \approx a_N + b_N(r_1 + r_2)/2$. Similar procedures are used to evaluate the remaining integrals in the equations of motion and heat conduction.

Neglecting body forces and internal heat sources, the generalized forces and heat fluxes become

$$\hat{p}_N = r_N(a_N + b_N r_N)S_N(\zeta) \qquad \hat{q}_N = r_N(a_N + b_N r_N)Q_N(\zeta) \tag{20.32a, b}$$

(no sum on N), where $S_N(\zeta)$ and $Q_N(\zeta)$ are prescribed functions of reduced time. Alternately, we can also prescribe arbitrary time-dependent nodal displacements and temperatures. For convective heat transfer problems, we set

$$Q_N = f_N(\hat{T} - T_0^*) \tag{20.33}$$

where f_N is the film coefficient at node N and T_0^* is the temperature of a medium in contact with the cylinder at a boundary.†

To complete the statement of the problem, we consider a hypothetical material for which $\rho_0 = 10^{-3}$ lb-sec²/in.⁴, $\alpha = 10^{-5}$ in./in.-°C, the dilatational relaxation kernel $K(\zeta) = 10 \times 10^5 \exp(-\zeta/10^{-5}) + 5 \times 10^5 \exp(-\zeta/\infty)$ lb/in.², the shear relaxation kernel $G(\zeta) = 1 \times 10^5 \exp(-\zeta/10^{-5}) + 0.1 \times 10^5 \exp(-\zeta/\infty)$ lb/in.², and the thermal conductivity kernel of the heredity-type heat-flux law (19.46) of $\kappa(\zeta) = 0.10 \exp(-\zeta/10^{-5}) + 0.01 \exp(-\zeta/\infty)$ in lb/(in.)(°C)(sec).

We now apply the above equations to the analysis of a cylinder with an inner radius of 10.0 in. and an outer radius of 11.0 in. for which the 1.0-in. wall is divided into twenty 0.05-in. elements. The system of nonlinear differential equations governing this model is integrated using the explicit quadratic scheme mentioned earlier; however, owing to the nonlinear heredity terms, we encounter at each time step a system of nonlinear algebraic equations in the current values of the nodal displacements and temperatures. In the examples to be cited below, these equations are solved by Newton-Raphson iteration. Starting values for the Newton-Raphson process are obtained from the initial conditions [e.g., if $\hat{u}^N(\zeta_0) = \dot{\hat{u}}^N(\zeta_0) = \hat{T}^N(\zeta_0) = \dot{\hat{T}}^N(\zeta_0) = 0$, then central difference approximations of $\hat{u}^N$ and $\hat{T}^N$ lead directly to expressions for $\dot{\hat{u}}^N$, $\ddot{\hat{u}}^N$, $\dot{\hat{T}}^N$, and $\ddot{\hat{T}}^N$ at time $\zeta_0 + h$ in terms of $\hat{u}^N$ and $\hat{T}^N$ at $\zeta_0 + h$]. Initial values so computed are introduced directly into (20.28) to obtain displacements and temperatures at $\zeta_0 + h$ and the value of $\hat{\tau}$ [see (20.30*c*)] to be used in the second time step. For the second time step, we again use the quadratic approximation $\dot{u}_{m+1} \approx (3u_{m+1} - 4u_m + u_{m-1})/2h$ for various rates. This results in adequate data to initiate the Newton-Raphson iteration process. Once convergence of a specified degree is obtained, another reduced time increment is added and the process is continued. Element stresses are computed at the end of each time step by introducing nodal displacements and temperatures into the stress constitutive equation (19.57*a*). All quantities (nodal displacements, temperatures, and element stresses) are computed as functions of reduced time and are then transformed to real time by using the relation $\Delta t = \Delta\zeta/\hat{b}[\theta(\zeta)]$, where $\Delta\zeta = h$ is the reduced time increment and $\hat{b}(\theta)$ is given by (20.30*a*). The real time at step $r + 1$ is then $t_r + h/\hat{b}[\theta(\zeta)]$. Note, however, that different values of real time should be considered for $\hat{u}^N$ and $\hat{T}^N$ than for element stresses; more specifically, for the nodal values $\hat{u}^N$, $\hat{T}^N$ the shift factor is associated with a temperature change at a single node, whereas the stress tensor is associated with a shift factor that involves the entire element temperature field and, consequently, depends upon the temperature changes at every point in the element. In the former case we simply set the

† For a more detailed discussion of models of various boundary conditions, see Oden and Kross [1968].

real time increment Δt^{Γ} associated with nodal quantities $\hat{u}^{\Gamma}$, $\hat{T}^{\Gamma}$ equal to $h/\hat{b}_{\Gamma}$ where $\hat{b}_{\Gamma}$ is the (global) value of $\hat{b}(\theta)$ at node Γ. In the latter case (transformation of stresses), we set the real time increment in element e, $\Delta t_{(e)}$, equal to $h/\hat{b}_{(e)}$ wherein $\hat{b}_{(e)}$ is an averaged shift factor for element e.

We shall consider as an example the dynamic response of the cylinder due to purely mechanical pressure, step loading $p = 10h(t)$ psi [$h(t)$ being the unit step function] applied at the inner boundary, while the thermal boundary conditions correspond to convective heat transfer. Film coefficients of 0.1 and 1.0 in.-lb/(in.^{2})(°C)(sec), respectively, are assumed for the inner and outer boundaries. The heat supply from internal sources is assumed to be zero and the initial temperature of the cylinder and the surrounding media are taken to be 300°K. Heat is thus generated through thermomechanical coupling. Since energy is dissipated in the system, the internal temperature increases during the transient part of the motion while heat is expelled at different rates at each boundary. For the material parameters and the loading considered, a steady-state condition is reached at approximately 10^{-3} sec. We can also obtain, for purposes of comparison, the solution of the isothermal case by ignoring all thermal effects, setting $\hat{b} = 1$, $\dot{\hat{b}} = 0$, and by treating the integration variable as real time.

Nondimensional plots of the computed circumferential and radial stress waves at various real times are shown for the isothermal and nonisothermal cases in Figs. 20.1 and 20.2; the displacement profiles are shown in Fig. 20.3. While stress and displacement profiles are obtained at different times for the isothermal and nonisothermal (cases owing to the necessity of transforming all computed quantities to real time in the nonisothermal case), the results do indicate a lag in the nonisothermal stress waves and a noticeable decrease in amplitude. Circumferential stresses are initially compressive but become tensile almost everywhere in the cylinder at $t = 10^{-3}$ sec. The circumferential stress wave requires approximately 2.5×10^{-5} sec to traverse the 1.0-in. thickness of the cylinder in the isothermal case

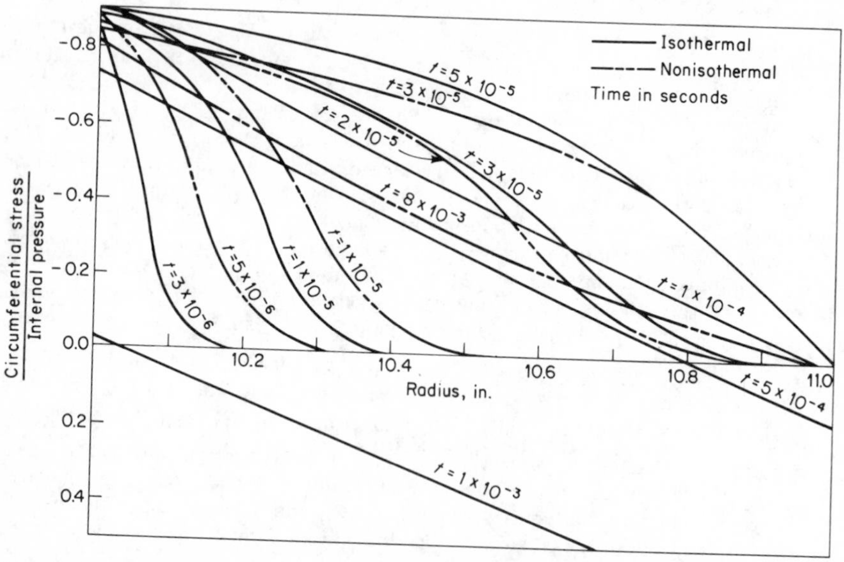

Fig. 20.1 Stress propagation in a thermorheologically simple cylinder—nondimensional circumferential stress versus the radial distance for various real times.

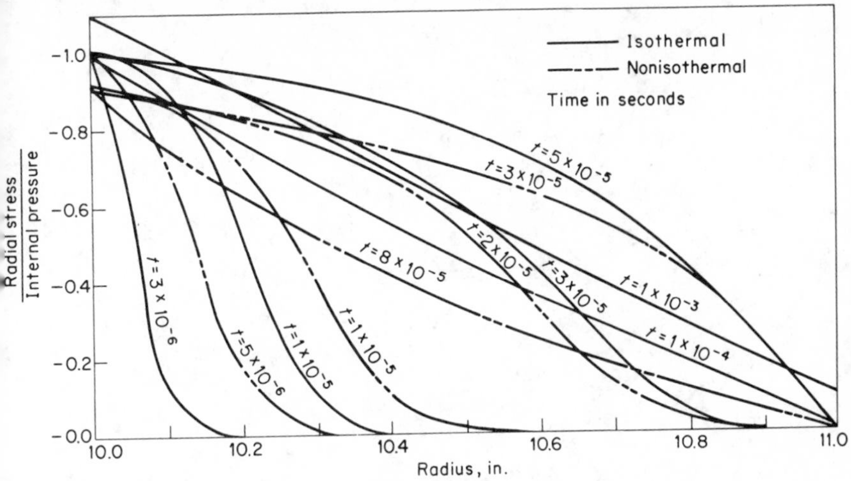

Fig. 20.2 Stress propagation in a thermorheologically simple cylinder—nondimensional radial stress versus radial distance for various real times.

and approximately 4.0×10^{-5} sec in the nonisothermal case. Displacement profiles are qualitatively the same for the isothermal and nonisothermal cases. The non-isothermal displacements are higher than those predicted by the isothermal analysis at the outer boundary, but they are lower on the inner boundary. This is due chiefly to the relative magnitudes of the film coefficients at these boundaries. We remark that the results indicated were computed using a reduced time increment of 10^{-6} sec; rather rapid convergence of the Newton-Raphson scheme was observed at each time step

Figure 20.4 indicates the computed temperature change over the thickness of the cylinder at various values of real time. Note that the mechanically loaded inner boundary is the first to experience an increase in temperature. The maximum temperature is reached at approximately $t = 4 \times 10^{-5}$ sec, and the temperature appears to decrease thereafter.

20.3 APPLICATIONS IN NONLINEAR COUPLED THERMOELASTICITY

In the case of thermoelastic solids, the internal dissipation is zero, the free energy is assumed to be a differentiable function of the current strain and temperature, and the stress and entropy are determined from the free energy by means of (19.59). The equations of motion and heat conduction for a typical thermoelastic element are then

$$m_{NM}\ddot{u}_k^M + \int_{v_{0(e)}} \frac{\partial \Phi(\gamma_{ij},T)}{\partial \gamma_{ij}} (\delta_{ik} + \psi_{M,k}u_i^M)\psi_{N,j}\, dv_0 = p_{Nk} \tag{20.34}$$

$$-\int_{v_{0(e)}} [\varphi_N(T_0 + \varphi_M T^M)\frac{d}{dt}\frac{\partial \Phi(\gamma_{ij},T)}{\partial T} - \varphi_{N,i}Q^i(\gamma_{ij},T,\nabla T)]\, dv_0 = q_N \tag{20.35}$$

Specific forms of (20.34) and (20.35) are obtained, of course, by identifying the form of $\Phi(\gamma_{ij},T)$ and $Q^i(\gamma_{ij},T,\nabla T)$ [as, for example, in (19.46), (19.63), (19.67), and (19.71)] for the material under consideration.

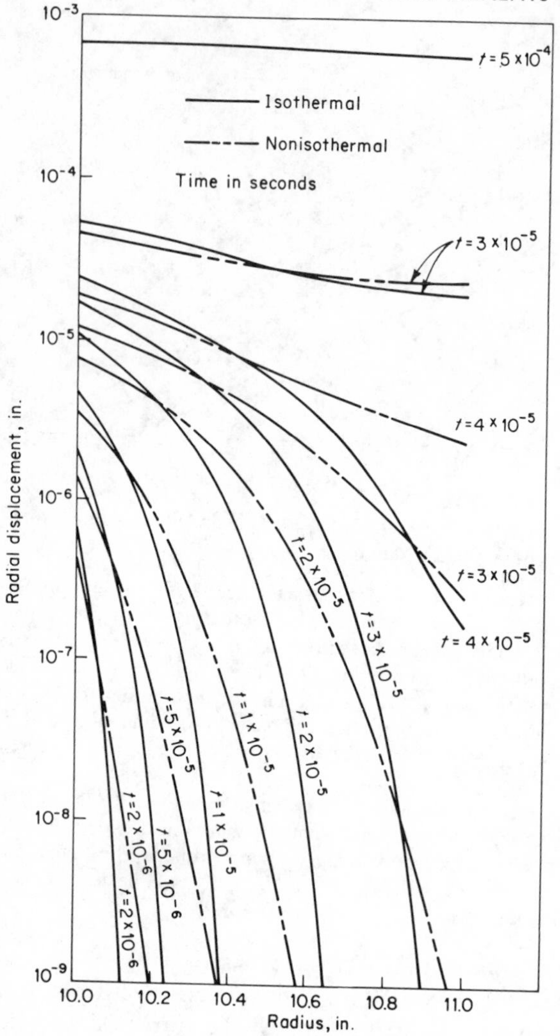

Fig. 20.3 Radial displacement profiles for various real times.

An example—A nonlinearly thermoelastic half-space As a specific example of an application of (20.34) and (20.35) to a transient problem in nonlinear thermoelasticity, we consider here a nonlinear version of the Danilovskaya problem,† i.e., the transient response of a thermoelastic half-space subjected to time-dependent heating at its boundary. We assume that while strains are infinitesimal, the response

† The linear uncoupled problem was first treated by its namesake, Danilovskaya [1952]. Closed-form solutions of the coupled linear problem were obtained by Sternberg and Chakravorty [1959], and finite-element solutions of the linear problem were among those presented by Nickell and Sackman [1968] and Oden and Kross [1969]. The nonlinear generalization discussed here was presented by Oden and Poe [1970].

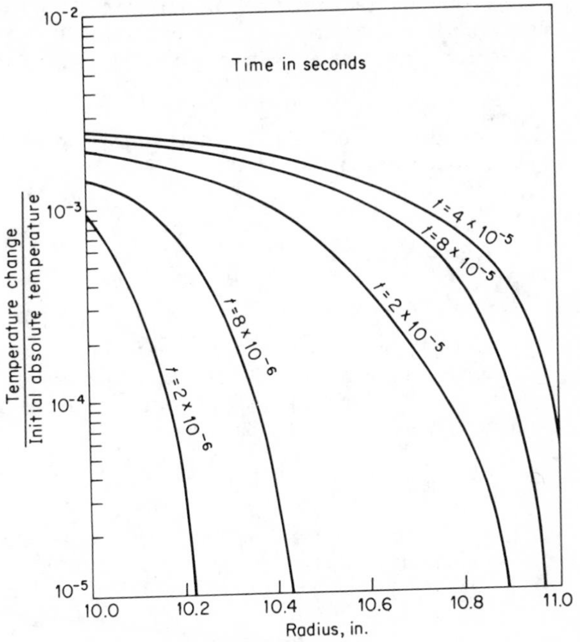

Fig. 20.4 Nondimensional temperature distributions at various real times.

of the material is governed by a nonlinear constitutive law of the type described in (19.71), (19.72*a*), and (19.72*b*) [for the sake of discussion; we also retain the a_{13} term of (19.67)], while the heat flux obeys the nonlinear Fourier law

$$q_i = \kappa_0(1 + \epsilon T)T_{,i} \tag{20.36}$$

Here κ_0 is the thermal conductivity at $T = 0$ and ϵ is a material parameter. Then (20.34) and (20.35) assume the forms

$$\begin{aligned} m_{NM}\ddot{u}_k^M + \int_{v_{0(e)}} &\{[(\lambda + 3a_6\gamma_{rr} + a_9\gamma_{rr} + a_{11}T + 2a_{13}T)\gamma_{ss} \\ &+ \tfrac{1}{2}(a_3 + a_9)(\gamma_{rr}\gamma_{ss} - \gamma_{rs}\gamma_{rs}) - \alpha(3\lambda + 2\mu)T]\delta_{ij} + (2\mu - a_3\gamma_{rr} \\ &- a_9\gamma_{rr} - a_{11}T)\gamma_{ij} + a_3\gamma_{ir}\gamma_{ir}\}\delta_{ik}\psi_{N,j}\, dv_0 = p_{Nk} \end{aligned} \tag{20.37}$$

$$\begin{aligned} T_0\int_{v_{0(e)}} &\varphi_N(x)[2a_7\dot{T} - \alpha(3\lambda + 2\mu)\dot{\gamma}_{rr} + a_{11}(\gamma_{rr}\dot{\gamma}_{ss} - \gamma_{rs}\dot{\gamma}_{rs}) \\ &+ 6a_{12}T\dot{T} + 2a_{13}\gamma_{rr}\dot{\gamma}_{ss}]\, dv_0 - \int_{v_{0(e)}} \kappa_0(1 + \epsilon T)\delta_{ij}\varphi_{N,i}T_{,j}\, dv_0 = q_N \end{aligned} \tag{20.38}$$

where λ and μ are the Lamé constants; α is the coefficient of thermal expansion; a_3, a_7, a_9, a_{11}, a_{12}, and a_{13} are material constants; and $\gamma_{ij} = (\psi_{N,i}u_j^N + \psi_{N,j}u_i^N)/2$, $T = \varphi_N T^N$.

Now consider an isotropic materially nonlinear thermoelastic half-space ($x \geq 0$), composed of the material corresponding to (20.37) and (20.38) and constrained to only uniaxial motion characterized by the displacement field

$$u_1 = u(x,t) \qquad u_2 = u_3 = 0$$

The bounding surface at $x_1 = 0$ is assumed to be stress-free and is subjected to a uniformly distributed ramp heating of the form

$$T_1 = 0 \qquad -\infty < t \leq 0$$

$$T_1 = \frac{T_f}{T_0} t \qquad 0 \leq t \leq t_0$$

$$T_1 = T_f \qquad t_0 \leq t < \infty$$

where T_1 is the initial surface temperature, T_f is the final surface temperature, and t_0 is the rise time of the boundary temperature. Since the body is assumed to be initially at rest, the displacements and stresses resulting from the temperature field T_1 are governed by the initial conditions

$$u_1(x,0) = 0 \qquad \frac{\partial u_1(x,0)}{\partial t} = 0 \qquad 0 \leq x < \infty$$

For simplicity, we use simplex approximations of element displacement and temperature fields, so that

$$u_1 = \psi_N(x) u_1^N \qquad T = \psi_N(x) T^N$$

where

$$\psi_N(x) = a_N + b_{N1} x \qquad N = 1, 2$$

With these selections, the equations of motion and heat conduction for a typical finite element become

$$p_1 = \frac{\rho_0 L}{6}(2\ddot{u}_1 + \ddot{u}_2) + \frac{\lambda + 2\mu}{L}(u_1 - u_2) + \frac{\alpha}{2}(3\lambda + 2\mu)(T_1 + T_2) - \frac{3}{L^2} a_6 (u_1 - u_2)^2 + \frac{a_{13}}{L}(T_1 + T_2)(\dot{u}_1 - \dot{u}_2) \qquad (20.39a)$$

$$p_2 = \frac{\rho_0 L}{6}(\ddot{u}_1 + 2\ddot{u}_2) - \frac{1}{L}(\lambda + 2\mu)(u_1 - u_2) - \frac{\alpha}{2}(3\lambda + 2\mu)(T_1 + T_2) + \frac{3}{L^2} a_6 (u_1 - u_2)^2 - \frac{a_{13}}{L}(T_1 + T_2)(\dot{u}_1 - \dot{u}_2) \qquad (20.39b)$$

$$q_1 = \frac{\rho_0 c_D(T) T_0}{6}(2\dot{T}_1 + \dot{T}_2) + \frac{\kappa_0}{L}\left[1 + \frac{\epsilon}{2L}(T_1 + T_2)\right](T_1 - T_2) - \frac{T_0}{2}\alpha(3\lambda + 2\mu)(\dot{u}_1 - \dot{u}_2) - \frac{T_0}{L} a_{13}(u_1 - u_2)(\dot{u}_1 - \dot{u}_2) \qquad (20.40a)$$

$$q_2 = \frac{\rho_0 c_D(T) T_0}{6}(\dot{T}_1 + 2\dot{T}_2) - \frac{\kappa_0}{L}\left[1 + \frac{\epsilon}{2L}(T_1 + T_2)\right](T_1 - T_2) - \frac{T_0}{2}\alpha(3\lambda + 2\mu)(\dot{u}_1 - \dot{u}_2) - \frac{T_0}{L} a_{13}(u_1 - u_2)(\dot{u}_1 - \dot{u}_2) \qquad (20.40b)$$

where L is the length of the element. We introduce dimensionless variables as follows:

$$l = \frac{a}{\kappa} x_1 \qquad \zeta = \frac{a^2}{\kappa} t$$

$$\bar{\theta} = \frac{T}{T_0} \qquad \bar{U} = \frac{a(\lambda + 2\mu)}{\kappa\beta T_0} U \tag{20.41a}$$

where

$$\kappa = \frac{\kappa_0}{\rho_0 c_D} \qquad a^2 = \frac{\lambda + 2\mu}{\rho}$$

$$\beta = \bar{\alpha}(3\lambda + 2\mu) \qquad \delta = \frac{\beta^2 T_0}{\rho c_D(\lambda + 2\mu)} \tag{20.41b}$$

In the above relations, l is a characteristic length, t is the real time, κ_0 is the initial coefficient of thermal conductivity, $c_D(T)$ is the temperature-dependent specific heat at constant deformation, λ and μ are Lamé constants, α is the linear coefficient of thermal expansion, and the quantity δ is the thermomechanical coupling parameter.

Numerical results showing the influence of thermomechanical coupling, material nonlinearities, and temperature-dependent specific heat and thermal conductivity in the solution of the half-space problem are presented in Figs. 20.5 to 20.12. Solutions of the finite-element differential equations were obtained by a Runge-Kutta-Gill integration scheme. Figures 20.5 and 20.6 contain numerical results for the linearized material as well as the exact solutions [Sternberg and Chakravorty, 1959]; here we find the dimensionless temperature $\bar{\theta}$ and displacement $\bar{U}$ at $l = 1.0$,

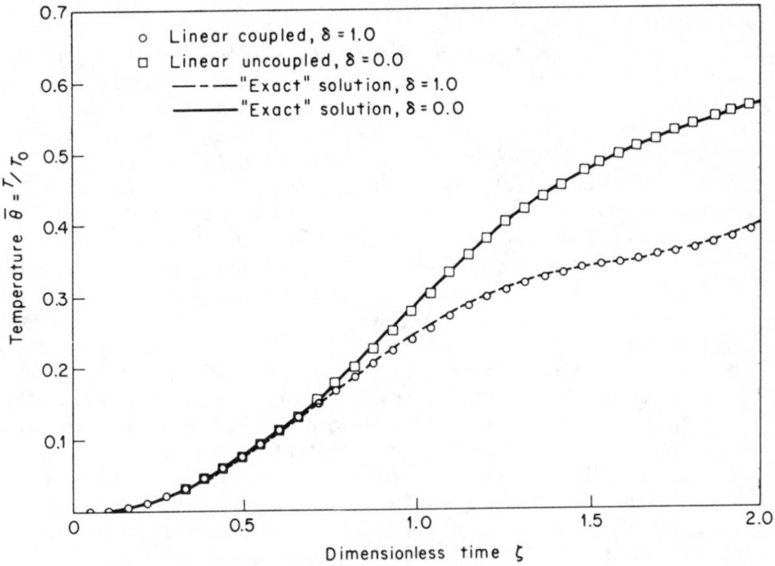

Fig. 20.5 Temperature at $l = 1.0$ linear coupled and uncoupled half-space with $\zeta_0 = 1.0$.

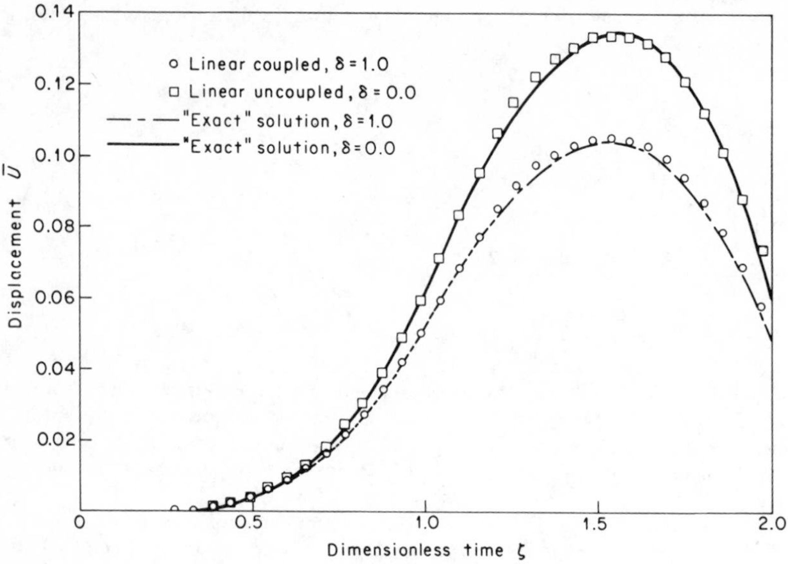

Fig. 20.6 Displacement at $l = 1.0$ in linear coupled and uncoupled half-space with $\zeta_0 = 1.0$.

with the thermomechanical coupling parameter $\delta = 0.0$ and $\delta = 1.0$ as a function of dimensionless time ζ for the cases $\zeta_0 = 1.0$ and $\zeta_0 = 0.25$, respectively. These results were obtained using a 50-element model having 10 elements between the bounding surface and $l = 1.0$.

A qualitative description of the effect of material nonlinearities in the half-space is illustrated in Figs. 20.7 and 20.8. These data were generated using the same 50-element model for a material having $\epsilon = 0.0$, $\delta = 1.0$ for the case $\zeta_0 = 1.0$. We note that a coupling parameter of the magnitude used represents a high degree of thermomechanical coupling for metallic materials. The influence of the nonlinearities introduced by the a_6 and a_{13} terms on the heat-conduction equation must be transmitted through this coupling term. The magnitudes of the material constants a_6 and a_{13} in (20.39) and (20.40), when nondimensionalized according to (20.41), are denoted A_6 and A_{13}. Figure 20.7 shows the temperature at $l = 1.0$ as a function of ζ for the four cases: $A_6 = 0.05$, $A_{13} = 0.0$; $A_6 = 0.25$, $A_{13} = 0.0$; $A_6 = 0.0$, $A_{13} = 0.05$; $A_6 = 0.0$, $A_{13} = 0.05$. Figure 20.8 depicts the variation in displacement at $l = 1.0$ versus time for the same four cases. Very little variation in the temperature is observed for $A_6 = A_{13} = 0.05$. For the case $A_6 = 0.25$, no significant departure from the linear theory occurs until $\zeta \geq 1.2$. This is due to the fact that this nonlinearity manifests itself only in terms of second order in the displacements and apparently requires a rather large value for A_6 in order to influence the heat-conduction equations. The effect of the A_{13} is much more pronounced in the temperature variations due to the fact that it appears in the elemental heat-conduction equations as well as in the equations of motion. These effects become apparent at $\zeta \geq 0.8$ for $A_{13} \geq 0.25$. Noticeable deviations in the displacements also occur at $\zeta = 1.0$, and these increase significantly with time. For values of A_6 and A_{13} greater than 0.25, the influence of material nonlinearities is quite pronounced. We note that the effects of shear do not appear in this example problem.

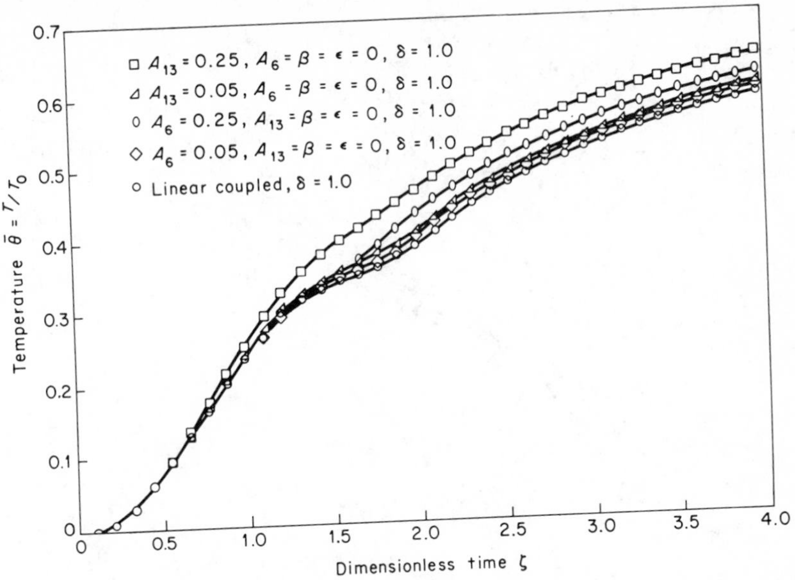

Fig. 20.7 Temperature at $l = 1.0$ in nonlinear half-space with $\zeta_0 = 1.0$ for various values of nonlinear material constants.

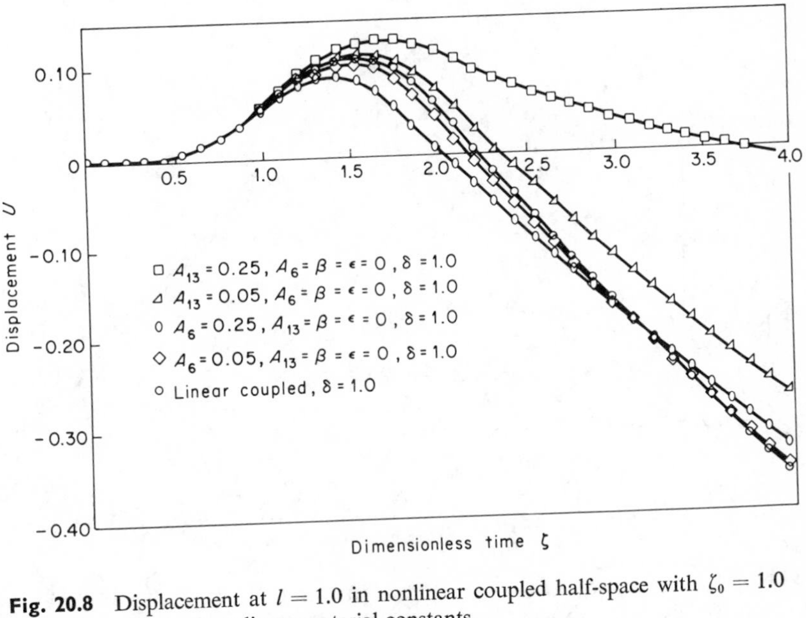

Fig. 20.8 Displacement at $l = 1.0$ in nonlinear coupled half-space with $\zeta_0 = 1.0$ for various values of nonlinear material constants.

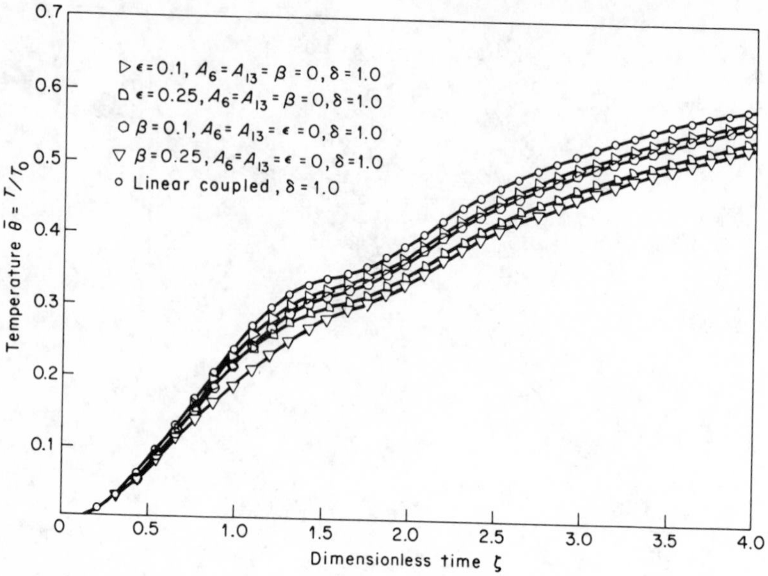

Fig. 20.9 Temperature at $l = 1.0$ in coupled half-space having temperature-dependent specific heat and thermal conductivity with $\zeta_0 = 1.0$.

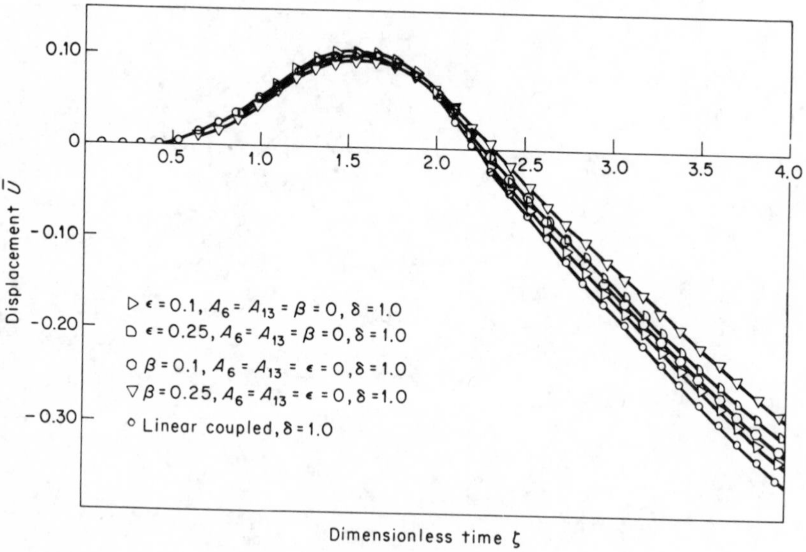

Fig. 20.10 Displacement at $l = 1.0$ in coupled half-space having temperature-dependent specific heat and thermal conductivity with $\zeta_0 = 1.0$.

Figures 20.9 and 20.10 display the effects of temperature-dependent coefficients of specific heat and thermal conductivity for materials with $\delta = 1.0$ and $A_6 = A_{13} = 0$ subjected to a ramp heating corresponding to $\zeta_0 = 1.0$. For simplicity, we assume that the specific heat varies only with temperature changes at each individual node and is of the form

$$c_D = c_0 + \beta T \tag{20.42}$$

where c_0 is the conventionally used constant specific heat and β is the term governing the rate of change in specific heat with temperature. Although this form of the specific heat was not used in assessing the magnitude of the incremental temperature as described in the derivation of (20.37) and (20.38), it is included for the purpose of illustrating quantitatively the effect of a temperature-dependent specific heat on the response of the material. According to Hodgman, Weast, and Selby [1956], the relation (20.42) accurately describes the variation in specific heat with temperature obtained experimentally for both iron and aluminum in the range of 0 to 400°C. Using the value of c_0 for iron at 0°C, the corresponding value calculated in (20.42) for β had an insignificant effect on the temperatures or displacements at $l = 1.0$. The results displaying the influence of β shown in Figs. 20.9 and 20.10 were obtained using the specific heat of iron at 0°C with $\beta = 0.25$ and $\beta = 0.1$ for $\epsilon = 0.0$. For this case, $\beta = 0.1$ approximately doubles the magnitude of c_D, and $\beta = 0.25$ increases c_D by approximately 250 percent for a change in temperature of $T = 1.0$. Similarly, the value ϵ calculated from experimental data [Jakob, 1962] for iron has essentially no effect on the temperatures or displacements at $l = 1.0$ for the case for the linearized material with $\beta = 0$. The other cases shown in Figs. 20.9 and 20.10 correspond to $\epsilon = -0.1$ and -0.25, which represent a 10 and 25 percent decrease in thermal

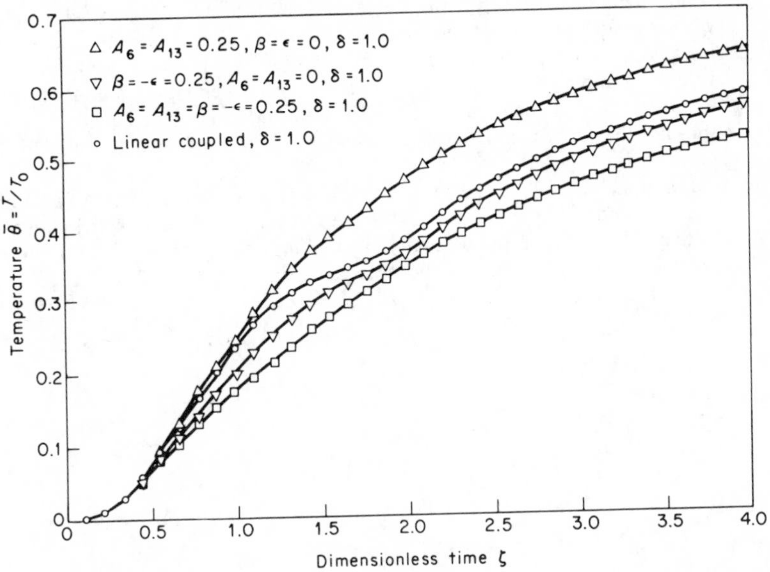

Fig. 20.11 Temperature at $l = 1.0$ in coupled half-space having both material nonlinearities and temperature-dependent specific heat and thermal conductivity with $\zeta_0 = 1.0$.

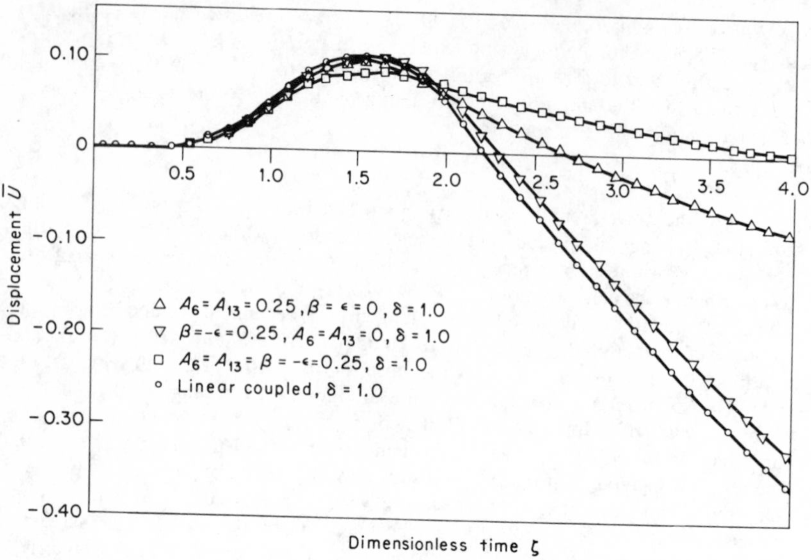

Fig. 20.12 Displacement at $l = 1.0$ in coupled half-space having both material nonlinearities and temperature-dependent specific heat and thermal conductivity with $\zeta_0 = 1.0$.

conductivity, respectively, for $T = 1$. Such decreases in thermal conductivity with increases in temperature are consistent with experimental data on certain metallic materials such as iron [Jakob, 1962]. The results indicate that effective values of ϵ and β produce a decrease in the temperatures in the material with increases in time. These temperature effects are transmitted into the equations of motion through the thermomechanical coupling parameter and tend to dampen the displacements at $l = 1.0$.

Figures 20.11 and 20.12 display the combined quantitative effects of the nonlinear terms A_6 and A_{13} and the temperature-dependent thermal conductivity and specific heat on the response of the material to ramp heating with $\zeta_0 = 1.0$. The cases shown correspond to values of the material constants of $A_6 = A_{13} = 0.25$; $\beta = \epsilon = 0$; $A_6 = A_{13} = 0$, $\beta = 0.25$, $\epsilon = -0.25$; $A_6 = A_{13} = \beta = 0.25$, $\epsilon = -0.25$.

References

The following list is limited to works consulted in writing this book. Numbers in parentheses following each reference indicate the article and/or section in which the work is cited.

Achenbach, J. D., S. M. Vogel, and G. Herrmann [1968]: On Stress Waves in Viscoelastic Media Conducting Heat, in H. Parkus and L. I. Sedov (eds.), "Irreversible Aspects of Continuum Mechanics and Transfer of Physical Characteristics in Moving Fluids," pp. 1–15, Springer-Verlag New York Inc., New York. (19.2)

Aguirre-Ramirez, G., and J. T. Oden [1969]: The Finite Element Technique Applied to Heat Conduction with Temperature Dependent Thermal Conductivity, Heat Transfer Division, ASME Winter Meeting, Nov. 16–20, *ASME Paper* 69-WA/HT-34. (13.8, 20.1)

Aguirre-Ramirez, G., J. T. Oden, and S. T. Wu [1970]: A Numerical Solution of the Boltzmann Equation by the Finite Element Technique, *Proc. 7th Inter. Symp. Rarefied Gas Dyn.*, Universitâ Degli Studi Di Pisa, Pisa, June 29–July 3, Academic Press, Inc., New York. (11.4)

Ahlberg, J. H., E. N. Nilson, and J. L. Walsh [1967]: "The Theory of Splines and Their Applications," Academic Press, Inc., New York. (10)

Ahmad, S., B. M. Irons, and O. C. Zienkiewicz [1969]: Curved Thick Shell and Membrane Elements with Particular Reference to Axisymmetric Problems, *Proc. 2d Conf. Matrix Methods Struct. Mech.*, AFFDL-TR-68-150 (Oct. 15–17, 1968), pp. 539–572, Wright-Patterson AFB, Ohio. (7.5, 10.4)

Alexander, H. [1968]: A Constitutive Relation for Rubber-like Materials, *Inter. J. Eng. Sci.*, vol. 6, no. 9, pp. 549–563. (15.5, 18.1)

Allen, D. N. de G. [1954]: "Relaxation Methods," McGraw-Hill Book Company, New York. (17.3)

Ames, W. F. [1965]: "Nonlinear Partial Differential Equations in Engineering," Academic Press, Inc., New York. (10.3)

Arantes e Oliveira, E. R. de [1968]: Theoretical Foundations of the Finite Element Method, *Inter. J. Solids Struct.*, vol. 4, pp. 929–952. (10, 10.2)

Arantes e Oliveira, E. R. de [1969]: Completeness and Convergence in the Finite Element Method, *Proc. 2d Conf. Matrix Methods Struct. Mech.*, AFFDL-TR-68-150 (Oct. 15–17, 1968), pp. 1061–1090, Wright-Patterson AFB, Ohio. (10)

Arantes e Oliveira, E. R. de [1971]: The Convergence Theorems and Their Role in the Theory of Structures, *IUTAM Colloq. High Speed Computing Elastic Struct.*, Liege. (10)

Archer, J. S. [1963]: Consistent Mass Matrix for Distributed Mass Systems, *J. Struct. Div., ASCE*, vol. 89, pp. 161–178. (9.7, 13.2)

Archer, J. S. [1965]: Consistent Matrix Formulation for Structural Analysis Using Finite Element Techniques, *AIAA J.*, vol. 3, pp. 1910–1918. (9.7, 13.2)

Argyris, J. H. [1954]: Energy Theorems and Structural Analysis, *Aircraft Eng.*, vol. 26, pp. 347–356 (Oct.), 383–387, 394 (Nov.). (2.1, 6.1)

Argyris, J. H. [1955]: Energy Theorems and Structural Analysis, *Aircraft Eng.*, vol. 27, pp. 42–58 (Feb.), 80–94 (March), 125–134 (April), 145–158 (May). (2.1)

Argyris, J. H. [1956]: The Matrix Analysis of Structures with Cut-outs and Modifications, *Proc. 9th Inter. Congr. Appl. Mech., sec. II, Mechanics of Solids*, September. (2.1)

Argyris, J. H. [1957]: The Matrix Theory of Statics, *Ingr. Arch.*, vol. 25, pp. 174–192. (2.1)

Argyris, J. H. [1958]: On the Analysis of Complex Structures, *Appl. Mech. Rev.*, vol. 11, pp. 331–338. (2.1)

Argyris, J. H. [1959]: Recent Developments of Matrix Theory of Structures, *Tenth Meeting of the AGARD Structures and Materials Panel*, Aachen, Germany. (16.5)

Argyris, J. H. [1965*a*]: Matrix Analysis of Three-dimensional Elastic Media, Small and Large Deflections, *AIAA J.*, vol. 3, pp. 45–51. (10.4, 16.5)

Argyris, J. H. [1965*b*]: Tetrahedron Elements with Linearly Varying Strain for the Matrix Displacement Method, *J. Roy. Aeron. Soc.*, vol. 69, pp. 877–880. (10.4)

Argyris, J. H. [1965*c*]: Triangular Elements with Linearly Varying Strain for the Matrix Displacement Method, *J. Roy. Aeron. Soc.*, vol. 69, pp. 711–713. (10.4)

Argyris, J. H. [1966*a*]: Continua and Discontinua, *Proc. Conf. Matrix Methods Struct. Mech.*, AFFDL-TR-66-80 (Oct. 26–28, 1965), Wright-Patterson AFB, Ohio. (2.1, 10.4)

Argyris, J. H. [1966*b*]: Matrix Displacement Analysis of Plates and Shells, *Ingr. Arch.*, vol. 35, pp. 102–142. (2.1)

Argyris, J. H., and S. Kelsey [1956]: Structural Analysis by the Matrix Force Method with Applications to Aircraft Wings, *Wiss. Ges. Luftfahrt Jahrb.*, pp. 78–98. (2.1)

Argyris, J. H., and S. Kelsey [1959]: The Analysis of Fuselages of Arbitrary Cross-section and Taper, *Aircraft Eng.*, vol. 31, pp. 62–74 (March), 101–112 (April), 133–143 (May), 169–180 (June), 192–203 (July), 224–256 (Aug.), 272–283 (Sept.). (2.1)

Argyris, J. H., and S. Kelsey [1960]: "Energy Theorems and Structural Analysis," Butterworth and Co. (Publishers), Ltd., London. (2.1)

Argyris, J. H., and S. Kelsey [1961]: The Analysis of Fuselages of Arbitrary Cross-section and Taper, *Aircraft Eng.*, vol. 33, pp. 34 (Feb.), 71–83 (March), 103–113 (April), 164–174 (June), 193–200 (July), 227–238 (Aug.). (2.1)

Argyris, J. H., and S. Kelsey [1963]: "Modern Fuselage Analysis and the Elastic Aircraft," Butterworth and Co. (Publishers), Ltd., London. (2.1)

Argyris, J. H., S. Kelsey, and H. Kamel [1964]: Matrix Methods of Structural Analysis: A Précis of Recent Developments, in B. Fraeijs de Veubeke (ed.), "Matrix Methods of Structural Analysis," pp. 1–164, AGARDograph 72, Pergamon Press Ltd., Oxford. Also published in 1964 as Recent Advances in Matrix Methods of Structural Analysis, in "Progress in the Aeronautical Sciences," vol. 4, Pergamon Press, Ltd., Oxford. (2.1, 16.5)

Argyris, J. H., and K. E. Buck [1968]: A Sequel to Technical Note 14 on the TUBA Family of Plate Elements, *Aeron. J. Roy. Aeron. Soc.*, vol. 72, no. 675, pp. 977–983. (10.4)

Argyris, J. H., and I. Fried [1968]: The LUMINA Element for the Matrix Displacement Method, *Aeron. J. Roy. Aeron. Soc.*, vol. 72, pp. 514–517. (2, 10.4)

Argyris, J. H., I. Fried, and D. W. Scharpf [1968*a*]: The HERMES 8 Element for the Matrix Displacement Method, *Aeron. J. Roy. Aeron. Soc.*, vol. 72, pp. 613–617. (10.4)

Argyris, J. H., I. Fried, and D. W. Scharpf [1968*b*]: The TET 20 and TEA 8 Elements for the Matrix Displacement Method, *Aeron. J. Roy. Aeron. Soc.*, vol. 72, pp. 618–623. (10.4)

Argyris, J. H., I. Fried, and D. W. Scharpf [1968*c*]: The TUBA Family of Plate Elements for the Matrix Displacement Method, *Aeron. J. Roy. Aeron. Soc.*, vol. 72, pp. 702–709. (10.4)

Argyris, J. H., and D. W. Scharpf [1968]: The SHEBA Family of Shell Elements for the Displacement Method, *Aeron. J. Roy. Aeron. Soc.*, vol. 72, no. 694, pp. 873–883. (10.4)

Argyris, J. H., K. E. Buck, D. W. Scharpf, H. M. Hilber, and G. Mareczek [1969]: Some New Elements for the Matrix Displacement Method, *Proc. 2d Conf. Matrix Methods Struct. Mech.*, AFFDL-TR-68-150 (Oct. 15–17, 1968), pp. 333–366, Wright-Patterson AFB, Ohio. (10.4)

Argyris, J. H., and D. W. Scharpf [1969]: Finite Elements in Space and Time, *Aeron. J. Roy. Aeron. Soc.*, vol. 73, no. 708, pp. 1041–1044. Also published in *Nucl. Eng. Design*, vol. 10, no. 4, pp. 456–464. (11.1)

Avila, J. [1970]: Continuation Methods for Nonlinear Equations, doctoral dissertation, University of Maryland, College Park. (17.5)

Babuška, I. [1970]: Approximation by Hill Functions, *Tech. Note* BN-648, University of Maryland, College Park. (10)

Bubuška, I. [1971]: The Rate of Convergence for the Finite Element Method, *SIAM Journal on Numerical Analysis*, vol. 8, no. 2, pp. 304–315. (10.2)

Baltrukonis, J. H., and R. N. Vaishnav [1965]: Finite Deformations under Pressurization in an Infinitely Long, Thick-walled Elastic Container Ideally Bonded to a Thin Elastic Case, *Trans. Soc. Rheol.*, vol. 9, pp. 273–291. (18.2)

Barsoum, R. [1970]: A Finite Element Formulation for the General Stability Analysis of Thin-walled Members, doctoral dissertation, Cornell University, Ithaca, N.Y. (16.5)

Bazeley, G. P., Y. K. Cheung, B. M. Irons, and O. C. Zienkiewicz [1966]: Triangular Elements in Plate Bending—Conforming and Non-conforming Solutions, *Proc. Conf. Matrix Methods Struct. Mech.*, AFFDL-TR-66-80 (Oct. 26–28, 1965), pp. 547–576, Wright-Patterson AFB, Ohio. (10.2, 10.4)

Becker, E. B. [1966]: A Numerical Solution of a Class of Problems of Finite Elastic Deformation, doctoral dissertation, University of California, Berkeley. (16, 16.2, 17.1, 18.1)

Becker, E. B., and J. J. Brisbane [1965]: Application of the Finite Element Method to the Stress Analysis of Solid Propellant Rocket Grains, *Spec. Rep.* S-67, vol. I, Rohm and Haas Co., Huntsville, Ala. (10.4, 16.2, 18.2)

Becker, E. B., and C. H. Parr [1967]: Application of the Finite Element Method to Heat Conduction in Solids, *Tech. Rep.* S-117, Rohm and Haas Redstone Research Laboratories, Huntsville, Ala. (13.8, 20.1)

Bell, K. [1969]: A Refined Triangular Plate Bending Finite Element, *Int. J. Numerical Methods Eng.*, vol. 1, no. 1, pp. 101–122. (10.4)

Berke, L., R. M. Bader, W. J. Mykytow, J. S. Przemieniecki, and M. H. Shirk (eds.) [1969]: *Proc. 2d Conf. Matrix Methods Struct. Mech.*, AFFDL-TR-68-150 (Oct. 15–17, 1968), Wright-Patterson AFB, Ohio. (2.1)

Besseling, J. R. [1966]: Matrix Analysis of Creep and Plasticity Problems, *Proc. Conf. Matrix Methods Struct. Mech.*, AFFDL-TR-66-80 (Oct. 26–28, 1965), pp. 665–678, Wright-Patterson AFB, Ohio. (20.1)

Best, G. C. [1963]: A Formula for Certain Types of Stiffness Matrices of Structural Elements, *AIAA J.*, vol. 1, pp. 478–479. (10.4)

Best, G. C., and J. T. Oden [1963]: Analysis of Shell-type Structures, *Eng. Res. Rep.* 157, General Dynamics, Fort Worth, Texas, December. (10.4)

Biderman, V. L. [1958]: Calculation of Rubber Parts (in Russian), *Rascheti na Prochnost*, Moscow. (15.5, 18.1)

Biot, M. A. [1955]: Variational Principles in Irreversible Thermodynamics with Application to Viscoelasticity, *Phys. Rev.*, vol. 97, no. 6, pp. 1463–1469. (12)

Biot, M. A. [1956]: Thermoelasticity and Irreversible Thermodynamics, *J. Appl. Phys.*, vol. 27, no. 3, pp. 240–257. (12)

Biot, M. A. [1965]: "Mechanics of Incremental Deformation," John Wiley & Sons, Inc., New York. (16.5)

Birkhoff, G., and S. MacLane [1941]: "A Survey of Modern Algebra," The Macmillian Co., New York. (7.1)

Birkhoff, G., and H. L. Garabedian [1960]: Smooth Surface Interpolation, *J. Math. Phys.*, vol. 39, pp. 258–268. (10)

Birkhoff, G., and C. de Boor [1965]: Piecewise Polynomial Interpolation and Approximation, in H. L. Garabedian (ed.), "Approximation of Functions," pp. 164–190, Elsevier Publishing Co., Amsterdam. (8.1, 10)

Bisplinghoff, R. L., H. Ashley, and R. L. Halfman [1955]: "Aeroelasticity," Addison-Wesley Publishing Company, Inc., Reading, Mass. (16.5)

Bogner, F. K. [1968]: Finite Deflection, Discrete Element Analysis of Shells, doctoral dissertation, Case Western Reserve University, Cleveland, Ohio. Also *Tech. Rep.* AFFDL-TR-67-185, Wright-Patterson AFB, Ohio, December. (17.1)

Bogner, F. K., R. H. Mallett, M. D. Minich, and L. A. Schmit, Jr. [1965]: Development and Evaluation of Energy Search Methods of Nonlinear Structural Analysis, *Tech. Rep.* AFFDL-TR-113, Wright-Patterson AFB, Ohio. (17.1)

Bogner, F. K., R. L. Fox, and L. A. Schmit, Jr. [1966]: The Generalization of Interelement, Compatible Stiffness and Mass Matrices by the Use of Interpolation Formulas, *Proc. Conf. Matrix Methods Struct. Mech.*, AFFDL-TR-66-80 (Oct. 26–28, 1965), pp. 397–443, Wright-Patterson AFB, Ohio. (7.5, 8.1, 10.4)

Bogner, F. K., R. L. Fox, and L. A. Schmit, Jr. [1967]: A Cylindrical Shell Discrete Element, *AIAA J.*, vol. 5, no. 4, pp. 745–750. (7.5, 10.4)

Boley, B. A., and J. H. Weiner [1960]: "Theory of Thermal Stresses," John Wiley & Sons, Inc., New York. (12, 19.2)

Bosshard, W. [1968]: Ein Neues Vollverträgliches Endliches Element für Plattenbiegung, *Abhandlungen IVBH*, Zurich, pp. 1–14. (10.4)

Box, M. J. [1966]: A Comparison of Several Current Optimization Methods, and the Use of Transformations in Constrained Problems, *Computer J.*, vol. 9, pp. 67–77. (17)

Bramble, J., and M. Zlamal [1970]: Triangular Elements in the Finite Element Method, *Math. Computers*, vol. 24, no. 112, pp. 809–820. (10, 10.4)

Bramlette, T. T., and R. H. Mallett [1970]: A Finite Element Solution Technique for the Boltzman Equation, *J. Fluid Mech.*, vol. 42, I, pp. 177–191. (11.4)

Brauchli, H. J., and J. T. Oden [1971]: Conjugate Approximation Functions in Finite Element Analysis, *Quart. Appl. Math.*, vol. 29, no. 3, pp. 65–90. (9, 13.2)

Brooks, S. H. [1958]: A Discussion of Random Methods for Seeking Maxima, *J. Operations Res.*, vol. 6, pp. 244–251. (17, 17.6)

Brooks, S. H. [1959]: A Comparison of Maximum Seeking Methods, *J. Operations Res.*, vol. 7, pp. 430–457. (17, 17.6)

Brown, J. E., J. M. Hutt, and A. E. Salama [1968]: Finite Element Solution to Dynamic Stability of Bars, *AIAA J.*, vol. 6, no. 7, pp. 1423–1425. (16.5)

Cauchy, A. L. [1829]: Sur l'équilibre et le mouvement intérieur des corps considérés comme des masses continues, *Ex. Math.*, vol. 4, pp. 293–319; *Oeuvres*, vol. 9, pp. 342–369. (16.5)

Cauchy, A. L. [1847]: Méthode générale pour la résolution des systèmes d'équations simultanées, *C.R. Acad. Sci. Paris*, vol. 25, pp. 536–538. (17.3)

Chacon, R. V. S., and R. S. Rivlin [1964]: Representation Theorems in the Mechanics of Materials with Memory, *Angew. Math. Phys.*, vol. 15, pp. 444–447. (19.2)

Chadwick, P. [1960]: Thermoelasticity. The Dynamical Theory, in I. N. Sneddon and R. Hill (eds.), "Progress in Solid Mechanics," vol. 1, North-Holland Publishing Co., Amsterdam. (19.2)

Chang, T. Y. [1966]: Approximate Solutions in Linear Viscoelasticity, doctoral dissertation, University of California, Berkeley. (20.1)

Christensen, R. M., and P. M. Naghdi [1967]: Linear Non-isothermal Viscoelastic Solids, *Acta Mechanica*, vol. III/1, pp. 1–12. (19.2)

Ciarlet, P. G., M. H. Schultz, and R. S. Varga [1967]: Numerical Methods of High-order Accuracy for Nonlinear Boundary Value Problems, *Numerische Math.*, vol. 9, pp. 394–430. (10)

Clough, R. W. [1960]: The Finite Element Method in Plane Stress Analysis, *J. Struct. Div., ASCE*, Proc. 2d Conf. Electronic Computation, pp. 345–378. (2.1)

Clough, R. W., and Y. Rashid [1965]: Finite Element Analysis of Axisymmetric Solids, *J. Eng. Mech. Div., ASCE*, vol. 91, no. EM1, pp. 71–85. (10.4, 18.2)

Clough, R. W., and J. L. Tocher [1966]: Finite Element Stiffness Matrices for Analysis of Plate Bending, *Proc. Conf. Matrix Methods in Struct. Mech.*, AFFDL-TR-66-80 (Oct. 26–28, 1965), pp. 515–546, Wright-Patterson AFB, Ohio. (10, 10.2, 10.4)

Coleman, B. D. [1964]: Thermodynamics of Materials with Memory, *Arch. Rational Mech. Anal.*, vol. 17, pp. 1–46. (12, 12.4, 14.3, 14.4, 19.1)

Coleman, B. D., and W. Noll [1960]: An Approximation Theorem for Functionals with Applications in Continuum Mechanics, *Arch. Rational Mech. Anal.*, vol. 6, pp. 355–370. (19.2)

Coleman, B. D., and W. Noll [1961]: Foundations of Linear Viscoelasticity, *Rev. Mod. Phys.*, vol. 33, pp. 239–249. (14.4)

Coleman, B. D., and W. Noll [1963]: The Thermodynamics of Elastic Materials with Heat Conduction and Viscosity, *Arch. Rational Mech. Anal.*, vol. 13, pp. 167–178. (12.3)

Coleman, B. D., and V. J. Mizel [1964]: Existence of Caloric Equations of State in Thermodynamics, *J. Chem. Phys.*, vol. 40, pp. 1116–1125. (12.3)

Coleman, B. D., and M. E. Gurtin [1965*a*]: Waves in Materials with Memory. III. Thermodynamic Influences on the Growth and Decay of One-dimensional Acceleration Waves, *Arch. Rational Mech. Anal.*, vol. 19, pp. 266–298. Also reprinted in C. Truesdell (ed.), "Wave Propagation in Dissipative Materials," Springer-Verlag, Berlin, 1965. (19.1)

Coleman, B. D., and M. E. Gurtin [1965*b*]: Waves in Materials with Memory. IV. Thermodynamics and the Velocity of General Acceleration Waves, *Arch. Rational Mech. Anal.*, vol. 19, pp. 317–338. Also reprinted in C. Truesdell (ed.), "Wave Propagation in Dissipative Materials," Springer-Verlag, Berlin, 1965. (19.1)

Coleman, B. D., M. E. Gurtin, and R. I. Herrera [1965]: Waves in Materials with Memory. I. The Velocity of One-dimensional Shock and Acceleration Waves, *Arch. Rational*

Mech. Anal., vol. 19, pp. 1–19. Also reprinted in C. Truesdell (ed.), "Wave Propagation in Dissipative Materials," Springer-Verlag, Berlin, 1965. (19.1)

Coleman, B. D., and V. J. Mizel [1967]: A General Theory of Dissipation in Materials with Memory, *Arch. Rational Mech. Anal.*, vol. 27, pp. 255–274. (19.1)

Collatz, L. [1966]: "The Numerical Treatment of Differential Equations," 3d ed., Springer-Verlag New York Inc., New York. (17.5)

Cost, T. L. [1969]: Thermomechanical Coupling Phenomena in Non-isothermal Viscoelastic Solids, doctoral dissertation, University of Alabama, University. (19.2, 20.1)

Courant, R. [1943]: Variational Methods for the Solution of Problems of Equilibrium and Vibrations, *Bull. Amer. Math. Soc.*, vol. 49, pp. 1–23. (2.1, 8.1, 10)

Courant, R., K. O. Fredrichs, and H. Lewy [1928]: Über die partiellen differenzengleichungen der mathematischen physik, *Math. Ann.*, vol. 100, pp. 32–74. (2.1, 11.1)

Crandall, S. H. [1956]: "Engineering Analysis: A Survey of Numerical Procedures," McGraw-Hill Book Co., New York. (10.3)

Dana, J. D., and C. S. Hurlbut [1959]: "Dana's Manual of Mineralogy," 17th ed., John Wiley & Sons, Inc., New York. (15.5)

Daniel, J. W. [1967]: The Conjugate Gradient Method for Linear and Nonlinear Operator Equations, *SIAM J. Numerical Anal.*, vol. 4, pp. 10–25. (17.3)

Danilovskaya, V. I. [1952]: On a Dynamical Problem of Thermoelasticity (in Russian), *Prikl. Mat. i Mekhan.*, vol. 16, no. 3, pp. 341–344. (20.3)

Davidenko, D. [1965]: An Application of the Method of Variation of Parameters to the Construction of Iterative Formulas of Increased Accuracy for Numerical Solutions of Nonlinear Integral Equations, *Dokl. Akad. Nauk SSSR*, vol. 162, pp. 499–502. (17.5)

Davidon, W. C. [1959]: Variable Metric Method for Minimization, *AEC Res. Develop. Rep.* ANL-5990 (rev.). (17.3)

Deak, A. L., and T. H. H. Pian [1967]: Application of the Smooth-surface Interpolation to the Finite Element Analysis, *AIAA J.*, vol. 5, no. 1, pp. 187–189. (8.1)

DeGroot, S. R., and P. Mazur [1962]: "Non-equilibrium Thermodynamics," North-Holland Publishing Co., Amsterdam. (12)

Dillon, O. W., Jr. [1962]: A Nonlinear Thermoelasticity Theory, *J. Mech. Phys. Solids*, vol. 10, pp. 123–131. (19.2)

Dong, R. G., K. S. Pister, and R. S. Dunham [1968]: Mechanical Characterization of Nonlinear Viscoelastic Solids for Iterative Solution of Boundary-value Problems, *Struct. Mater. Res. Rep.* 68-11, Structural Engineering Laboratory, University of California, Berkeley. (20.1)

Duhamel, J. M. C. [1837]: Second Mémoire sur les Phénomenés Thermoméchaniques, *J. Ecole Polytech.*, vol. 15, no. 25, p. 1. (19.2)

Dunne, P. C. [1968]: Complete Polynomial Displacement Fields for Finite Element Method, *Aeron. J. Roy. Aeron. Soc.*, vol. 72, pp. 245–246. (10, 10.4)

Eckart, C. [1948]: The Thermodynamics of Irreversible Processes. IV. The Theory of Elasticity and Anelasticity, *Phys. Rev.*, vol. 72, no. 2, pp. 373–382. (12)

Ergatoudis, I. [1966]: Quadrilateral Elements in Plane Analysis, M.S. thesis, University of Wales, Swansea. (10.4)

Ergatoudis, I., B. M. Irons, and O. C. Zienkiewicz [1968*a*]: Curved Isoparametric, "Quadrilateral" Elements for Finite Element Analysis, *Int. J. Solids Struct.*, vol. 4, pp. 31–42. (7.5, 10.4)

Ergatoudis, I., B. M. Irons, and O. C. Zienkiewicz [1968*b*]: Three Dimensional Analysis of Arch Dams and Their Foundations, *Proc. Symp. Arch Dams*, pp. 21–34. (7.5, 10.4)

Eringen, A. C. [1962]: "Nonlinear Theory of Continuous Media," McGraw-Hill Book Co., New York. (5, 12.2, 14)

Eringen, A. C. [1967]: "Mechanics of Continua," John Wiley & Sons, Inc., New York. (5, 12, 12.3, 13.8, 14, 14.4, 20.2)

Felippa, C. A. [1966]: Refined Finite Element Analysis of Linear and Nonlinear Two-dimensional Structures, doctoral dissertation, University of California, Berkeley. (2.1, 8.1, 10.4)

Felippa, C. A., and R. W. Clough [1968]: The Finite Element Method in Solid Mechanics, *Symp. Numerical Soln. Field Probl. Continuum Mech.*, Durham, North Carolina, April 5–6. (2.1, 10, 10.2, 10.4)

Ficken, F. [1951]: The Continuation Method for Nonlinear Functional Equations, *Commun. Pure Appl. Math.*, vol. 4, pp. 435–456. (17.5)

Finkbeiner, D. T., II [1966]: "Introduction to Matrices and Linear Transformations," 2d ed., W. H. Freeman and Co., San Francisco. (7.1)

Finlayson, B. A., and L. E. Scriven, [1966]: The Method of Weighted Residuals—A Review, *Appl. Mech. Rev.*, vol. 19, no. 9, pp. 735–748. (10.3)

Fix, G., and G. Strang [1969]: Fourier Analysis of the Finite-element Method in Ritz-Galerkin Theory, *Studies Appl. Math.*, vol. 48, pp. 265–273. (10)

Fix, G., and N. Nassif [1971]: The Accuracy of Finite Element Semi-discretizations of Parabolic Problems, *Numerische Math.* (to appear). (10.2)

Fletcher, R. [1965]: Function Minimization without Evaluating Derivatives, *Computer J.*, vol. 8, pp. 33–51. (17)

Fletcher, R., and M. J. D. Powell [1963]: A Rapidly Convergent Descent Method for Minimization, *Computer J.*, vol. 6, no. 2, pp. 163–188. (17.3)

Fletcher, R., and C. M. Reeves [1964]: Function Minimization by Conjugate Gradients, *Computer J.*, vol. 7, pp. 149–154. (17.3)

Fort, T. [1948]: "Finite Differences," Clarendon Press, Oxford. (8.1)

Fox, R. L., and E. L. Stanton [1968]: Developments in Structural Analysis by Direct Energy Minimization, *AIAA J.*, vol. 6, no. 6, pp. 1036–1042. (17.1)

Fraeijs de Veubeke, B. M. (ed.) [1964*a*]: "Matrix Methods of Structural Analysis," Pergamon Press, Ltd., Oxford. (2.1)

Fraeijs de Veubeke, B. M. [1964*b*]: Upper and Lower Bounds in Matrix Structural Analysis, in B. M. Fraeijs de Veubeke (ed.), "Matrix Methods of Structural Analysis," Pergamon Press, Ltd., Oxford. (10)

Fraeijs de Veubeke, B. M. [1965]: Displacements and Equilibrium Models in the Finite Element Method, in O. C. Zienkiewicz and G. S. Holister (eds.), "Stress Analysis," pp. 145–197, John Wiley & Sons, Inc., New York. (10, 10.4)

Fraeijs de Veubeke, B. M. [1966]: Bending and Stretching of Plates—Special Models for Upper and Lower Bounds, *Proc. Conf. Matrix Methods Struct. Mech.*, AFFDL-TR-66-80 (Oct. 26–28, 1965), pp. 863–886, Wright-Patterson AFB, Ohio. (10, 10.4)

Fraeijs de Veubeke, B. M. (ed.) [1971]: High-speed Computing of Elastic Structures, Proc. IUTAM Sym. High-speed Computing Elastic Struct., August, 1970, Liège. (2.1)

Fried, I. [1969*a*]: Some Aspects of the Natural Coordinate System in the Finite Element Method, *AIAA J.*, vol. 7, no. 7, pp. 1366–1368. (10.4)

Fried, I. [1969*b*]: Finite-element Analysis of Time-dependent Phenomena, *AIAA J.*, vol. 7, no. 6, pp. 1170–1172. (11.1)

Fried, I. [1971]: Discretization and Round-off Error in the Finite Element Analysis of Elliptic Boundary Value Problems and Eigenvalue Problems, doctoral dissertation, Massachusetts Institute of Technology, Cambridge. (10.2)

Friedrichs, K. O. [1934]: Spektraltheorie Halbbeschränkter Operatoren und Anwendung auf die Spektralzerlegung von Differentialoperatoren, *Math. Ann.*, vol. 109, pp. 465–487. (10.1)

Friedrichs, K. O., and H. B. Keller [1967]: A Finite Difference Scheme for Generalized Neumann Problems, in J. H. Bramble (ed.), "Numerical Solution of Partial Differential Equations," Academic Press, Inc., New York. (10)

Fujino, T., and K. Ohsaka [1969]: The Heat Conduction and Thermal Stress Analysis by the Finite Element Method, *Proc. 2d Conf. Matrix Methods Struct. Mech.*, AFFDL-TR-68-150 (Oct. 15–17, 1968), pp. 1121–1164, Wright-Patterson AFB Ohio. (20.1)

Fung, Y. C. [1965]: "Foundations of Solid Mechanics," Prentice-Hall, Inc., Englewood Cliffs, N.J. (12)

Galerkin, B. G. [1915]: Rods and Plates. Series Occurring in Various Questions Concerning the Elastic Equilibrium of Rods and Plates (in Russian), *Vestn. Inghenerov*, vol. 19, pp. 897–908. (10.1, 10.3)

Gallagher, R. H. [1961]: Matrix Structural Analysis of Heated Airframes, *Proc. ASD Symp. Aerothermoelasticity*, ASD-TR-61-645, November. (20.1)

Gallagher, R. H. [1964]: "A Correlation Study of Methods of Matrix Structural Analysis," Pergamon Press, Ltd., Oxford. (2.1)

Gallagher, R. H., J. F. Quinn, and J. Padlog [1961]: Deformational Response Determinations for Practical Heated Wing Structures, *Proc. AIR-ONR Symp. Struct. Dyn. High Speed Flight*, Los Angeles. (20.1)

Gallagher, R. H., J. Padlog, and P. O. Bijlaard [1962]: Stress Analysis of Heated Complex Shapes, *J. Amer. Rocket Soc.*, vol. 32, pp. 700–707. (10.4, 16.1, 20.1)

Gallagher, R. H., I. Rattinger, and J. S. Archer [1964]: "A Correlation Study of Methods of Matrix Structural Analysis," AGARDograph 69, Pergamon Press, Ltd., Oxford. (10.4)

Gallagher, R. H., Y. Yamada, and J. T. Oden (eds.) [1970]: "Recent Advances in Matrix Methods in Structural Analysis and Design," Proceedings of the U.S.-Japan Seminar on Matrix Methods in Structural Analysis and Design, Tokyo, 1969, University of Alabama Press, University. (2.1)

Gavurin, M. [1958]: Nonlinear Functional Equations and Continuous Analogues of Iterative Methods, *Izv. Vysshikh Ucebn Zavedenii Mat.*, vol. 6, pp. 18–31. (17.5)

Gent, A. N., and A. G. Thomas [1958]: Forms of the Stored (Strain) Energy Function for Vulcanized Rubber, *J. Polymer Sci.*, vol. 28, pp. 625–628. (15.5, 18.1)

Goël, J. J. [1968]: Construction of Basic Functions for Numerical Utilization of Ritz's Method, *Numerische Math.*, vol. 12, pp. 435–447. (10)

Goldstein, A. A. [1962]: Cauchy's Method of Minimization, *Numerische Math.*, vol. 4, pp. 146–150. (17.3)

Graves, L. M. [1956]: "The Theory of Functions of Real Variables," 2d ed. McGraw-Hill Book Co., New York. (10.4)

Green, A. E., and R. S. Rivlin [1957]: The Mechanics of Nonlinear Materials with Memory, Part I, *Arch. Rational Mech. Anal.*, vol. 1, pp. 1–21. (19.2)

Green, A. E., R. S. Rivlin, and A. J. M. Spencer [1959]: The Mechanics of Nonlinear Materials with Memory, Part II, *Arch. Rational Mech. Anal.*, vol. 3, pp. 82–90. (19.2)

Green, A. E., and J. E. Adkins [1960]: "Large Elastic Deformations (and Non-linear Continuum Mechanics)," Oxford University Press, London. (5.2, 15, 15.5, 18.3)

Green, A. E., and R. S. Rivlin [1960]: The Mechanics of Nonlinear Materials with Memory, Part III, *Arch. Rational Mech. Anal.*, vol. 4, pp. 387–404. (19.2)

Green, A. E., and W. Zerna [1968]: "Theoretical Elasticity," 2d ed., Oxford University Press, London. (4, 15, 18.2)

Greene, B. E., R. E. Jones, R. W. McLay, and D. R. Strome [1969]: On the Application of Generalized Variational Principles in the Finite Element Method, *AIAA J.*, vol. 7, no. 7, pp. 1254–1260. (10)

Greub, W. H. [1963]: "Linear Algebra," 2d ed., Academic Press, Inc., New York. (7.1, 10.4)

Haisler, W. E., J. A. Stricklin, and F. J. Stebbins [1971]: Development and Evaluation of Solution Procedures for Geometrically Nonlinear Structural Analysis by the Direct Stiffness Method, *Proc. AIAA/ASME 12th Struct., Struct. Dyn., Mater. Conf.*, Anaheim, Calif. (17.5)

Hart-Smith, L. J. [1966]: Elasticity Parameters for Finite Deformations of Rubber-like Materials, *Z. Angew. Math. Phys.*, vol. 17, pp. 608–625. (15.5, 18.1)

Hart-Smith, L. J., and J. D. C. Crisp [1967]: Large Elastic Deformations of Thin Rubber Membranes, *Int. J. Eng. Sci.*, vol. 5, no. 1, pp. 1–24. (15.5, 18.1)

Hatsopoulas, G. N., and J. H. Keenan [1965]: "Principles of General Thermodynamics," John Wiley & Sons, Inc., New York. (12)

Herrmann, L. R. [1965]: Elasticity Equations for Incompressible and Nearly Incompressible Materials by a Variational Theorem, *AIAA J.*, vol. 3, no. 10, pp. 1896–1900. (16.2)

Herrmann, L. R. [1966]: A Bending Analysis for Plates, *Proc. Conf. Matrix Methods Struct. Mech.*, AFFDL-TR-66-80 (Oct. 26–28, 1965), pp. 577–602, Wright-Patterson AFB, Ohio. (10.4)

Hestenes, M. R., and E. Stiefel [1952]: Method of Conjugate Gradients for Solving Linear Problems, *J. Res. Nat. Bur. Std. B*, vol. 49, pp. 409–436. (17.3)

Hibbit, H. D., P. V. Marcal, and J. R. Rice [1970]: A Finite Element Formulation for Problems of Large Strain and Large Displacement, *Int. J. Solids Struct.*, vol. 6, no. 8, pp. 1069–1086 (16.5)

Hildebrand, F. B. [1956]: "Introduction to Numerical Analysis," McGraw-Hill Book Co., New York. (8.1)

Hodgman, C. D., R. C. Weast, and S. M. Selby [1956]: "Handbook of Chemistry and Physics," Chemical Rubber Publishing Co., Cleveland. (20.3)

Hofmeister, L. D., G. Greenbaum, and D. A. Evensen [1970]: Large Strain, Elasto-plastic Finite Element Analysis, *Proc. 11th AIAA/ASME Struct. Structural Dyn. Mater. Conf.*, Fort Collins, Colo., April. (16)

Holand, I., and K. Bell (eds.) [1969]: "Finite Element Methods in Stress Analysis," Tapir Press, Trondheim, Norway. (2.1)

Hooke, R. [1678]: "Lectures de Potentia Restitutiva, or Of Spring, Explaining the Power of Springing Bodies," a pamphlet reproduced more recently by R. T. Gunther, "Early Science in Oxford," vol. 8, pp. 119–152, 1931. (15.1)

Hrennikoff, A. [1941]: Solution of Problems in Elasticity by the Framework Method, *J. Appl. Mech.*, vol. 8, pp. A169–A175. (2.1)

Huang, A. B., and D. P. Giddens [1967]: The Discrete Ordinate Method for the Linearized Boundary Value Problems in the Kinetic Theory of Gases, in "Rarified Gas Dynamics, 5th Symposium," vol. I, pp. 481–504, Academic Press, Inc., New York. (11.4)

Huang, K. [1963]: "Statistical Mechanics," John Wiley & Sons, Inc., New York. (11.4)

Hughes, T. J. R., and H. Allik [1969]: Finite Elements for Compressible and Incompressible Continua, *Proc. Symp. Civil Eng.*, Vanderbilt University, Nashville, Tenn., pp. 27–62. (10.4, 16.2)

Hutchinson, W. D., G. W. Becker, and R. F. Landel [1965]: Determination of the Stored Energy Function of Rubber-like Materials, *Bull. 4th Meeting Interagency Chem. Rocket Propulsion Group—Working Group Mech. Behavior*, *CPIA Publ.* 94U, vol. 1, pp. 141–152. (15.5)

Hutt, J. M. [1968]: Dynamic Stability of Plates by Finite Elements, doctoral dissertation, Oklahoma State University, Stillwater. (16.5)

Irons, B. [1966]: Engineering Application of Numerical Integration in Stiffness Method, *AIAA J.*, vol. 4, pp. 2035–2037. (10.4)

Irons, B., and K. J. Draper [1965]: Inadequacy of Nodal Connections in a Stiffness Solution for Plate Bending, *AIAA J.*, vol. 3, no. 5, p. 961. (10.2, 10.4)

Isaacson, E., and H. B. Keller [1966]: "Analysis of Numerical Methods," John Wiley & Sons, Inc., New York. (17, 17.2, 17.5)

Isihara, A., N. Hashitsume, and M. Tatibana [1951]: Statistical Theory of Rubber-like Elasticity-IV (Two-dimensional Stretching), *J. Chem. Phys.*, vol. 19, pp. 1508–1512. (15.5)

Jakob, M. [1962]: "Heat Transfer," vol. I, John Wiley & Sons, Inc., New York. (20.3)

Jaunzemis, W. [1967]: "Continuum Mechanics," The Macmillan Co., New York. (14)

Jindra, F. [1959]: Warmespannungen, bei einem Nichtlinearen Elastizatatsgeset, *Ing. Arch.*, vol. 38, pp. 109–116. (19.2)

John, F. [1958]: On Finite Deformations of Elastic Isotropic Material, *Inst. Math. Sci. Rep.* IMM-NYU 250, New York University. (15.5)

Johns, D. J. [1965]: "Thermal Stress Analysis," Pergamon Press, Ltd., Oxford. (19.2)

Johnson, M. W., Jr., and R. W. McLay [1968]: Convergence of the Finite Element Method in the Theory of Elasticity, *J. Appl. Mech.*, ser. E, vol. 35, no. 2, pp. 274–278. (10)

Kantorovich, L. V. [1933]: On a Direct Method for the Approximate Solution of the Problem of the Minimum of a Double Integral (in Russian), *Bull. Acad. Sci. USSR*, vol. 7, pp. 647–652. (11.4)

Kariappa, B., R. Somashekar, and C. G. Shah [1970]: Discrete Element Approach to Flutter of Skew Panels with In-plane Forces under Yawed Supersonic Flow, *AIAA J.*, vol. 8, no. 11, pp. 2017–2030. (16.5)

Kavanagh, K. T. [1969]: The Finite Element Analysis of Physically and Kinematically Non-linear Elastic Solids, doctoral dissertation, University of California, Berkeley. Also published as *Struct. Mater. Res. Rep.* 69-4, Structural Engineering Laboratory, University of California, Berkeley, April. (15.5, 16)

Kawai, T. [1970]: Finite Element Analysis of the Geometrically Nonlinear Problems, in R. H. Gallagher, Y. Yamuda, and J. T. Oden (eds.), "Recent Advances in Matrix Methods in Structural Analysis and Design," Proceedings of the U.S.-Japan Seminar on Matrix Methods in Structural Analysis and Design, Tokyo, 1969, University of Alabama Press, University. (16.5)

Key, S. W. [1966]: A Convergence Investigation of the Direct Stiffness Method, doctoral dissertation, University of Washington, Seattle. (10, 10.2, 10.3, 10.4)

Key, S. W. [1969]: A Variational Principle for Incompressible and Nearly Incompressible Anisotropic Elasticity, *Int. J. Solids Struct.*, vol. 5, no. 9, pp. 951–964. (16.2)

King, I. P. [1965]: On the Finite Element Analysis of Two-dimensional Stress Problems with Time Dependent Properties, doctoral dissertation, University of California, Berkeley. (20.1)

Kisner, W. [1964]: A Numerical Method for Funding Solutions to Nonlinear Problems, *SIAM J. Appl. Math.*, vol. 12, no. 2, pp. 424–428. (17.5)

Klosner, J. M., and A. Segal [1969]: Mechanical Characterization of a Natural Rubber, *PIBAL Rep.* 69-42, Polytechnic Institute of Brooklyn, New York. (15.5, 18.1)

Kolmogorov, A. N., and S. V. Fomin [1957]: "Elements of the Theory of Functions and Functional Analysis, vol. 1, Metric and Normed Spaces," translated from the 1954 Russian ed. by L. F. Baron, Graylock Press, Rochester, N.Y. (7.1, 10.1)

Kolmogorov, A. N., and S. V. Fomin [1961]: "Elements of the Theory of Functions and Functional Analysis, vol. 2, Measure. The Lebesque Integral. Hilbert Space," translated from the 1960 Russian ed. by H. Kamel and H. Komm, Graylock Press, Rochester, N.Y. (7.1, 7.4)

Korneev, V. [1967]: The Comparison of the Finite Element Method with the Variational-difference Method in the Theory of Elasticity (in Russian), *Izvestia Reports of the Whole-Union Scientific Investigation of the Hydro-Technical Institute*, vol. 83, pp. 286–307. (10.2)

Kowalik, J., and M. R. Osborne [1968]: "Methods for Unconstrained Optimization Problems," American Elsevier Publishing Co., Inc., New York. (17)

Krahula, J. L., and J. F. Polhemus [1968]: Use of Fourier Series in the Finite Element Method, *AIAA J.*, vol. 6, no. 4, pp. 726–728. (10.4)

Kron, G. [1939]: "Tensor Analysis of Networks," John Wiley & Sons, Inc., New York. (2.1, 6.1.)

Kron, G. [1944*a*]: Tensorial Analysis and Equivalent Circuits of Elastic Structures, *J. Franklin Inst.*, vol. 238, pp. 399–442. (2.1)

Kron, G. [1944*b*]: Equivalent Circuits of the Elastic Field, *J. Appl. Mech.*, vol. 66, pp. 149–167. (2.1)

Kron, G. [1953]: A Set of Principles to Interconnect the Solutions of Physical Systems, *J. Appl. Phys.*, vol. 24, pp. 965–980. (2.1)

Kron, G. [1954]: A Method to Solve Very Large Physical Systems in Easy Stages, *Proc. IRE*, vol. 42, pp. 680–686. (2.1)

Kron, G. [1955]: Solving Highly Complex Structures in Easy Stages, *J. Appl. Mech.*, vol. 22, pp. 235–244. (2.1)

Langhaar, H. L., and S. C. Chu [1968]: Piecewise Polynomials and the Partition Method for Ordinary Differential Equations, in "Developments in Theoretical and Applied Mechanics," vol. IV, Proceedings, Southeastern Conference on Theoretical and Applied Mechanics, pp. 553–564, Pergamon Press, Ltd., Oxford. (8.1, 10.4)

Leigh, D. C. [1968]: "Nonlinear Continuum Mechanics," McGraw-Hill Book Company, New York. (14)

Leonard, J. W., and T. T. Bramlette [1970]: Finite-element Solution of Differential Equations, *J. Struct. Div., ASCE*, vol. 96, no. EM6, pp. 1277–1284. (10.3)

Lianis, G. [1965]: Integral Constitutive Equations of Nonlinear Thermo-viscoelasticity, *School Aeron. Astronautics Eng. Sci. Rep.* AA&ES 65-1, Purdue University, Lafayette Ind. (19.2)

Lianis, G. [1968]: Nonlinear Thermorheologically Simple Materials, *5th Int. Congr. Rheol., Kyoto, Japan.* (19.2)

Liusternik, L. A., and V. J. Sobolev [1961]: "Elements of Functional Analysis," translated from the Russian ed. by A. E. Labarre, Jr., H. Izbicki, and H. W. Crowley, Frederick Ungar Publishing Co., New York. (7.1)

Mallett, R. H., and L. A. Schmit, Jr. [1967]: Nonlinear Structural Analysis by Energy Search, *J. Struct. Div. ASCE*, vol. 93, no. ST3, pp. 221–234. (17.1)

Malvern, L. E. [1970]: "An Introduction to the Mechanics of a Continuous Media," Prentice-Hall, Inc., Englewood Cliffs, N.J. (14)

Marcal, P. V. [1968]: Effect of Initial Displacements on Problems of Large Deflection and Stability, *Proc. ASCE Joint Specialty Conf. Optimization Nonlinear Probl.*, Chicago, April. (16.5)

Marcal, P. V. [1970]: Finite Element Analysis with Material Nonlinearities—Theory and Practice, in R. H. Gallagher, Y. Yamada, and J. T. Oden (eds.), "Recent Advances in Matrix Methods in Structural Analysis and Design," Proceedings of the U.S.-Japan Seminar on Matrix Methods in Structural Analysis and Design, Tokyo, 1969, University of Alabama Press, University. (2.1)

Martin, H. C. [1966*a*]: "Introduction to Matrix Methods of Structural Analysis," McGraw-Hill Book Company, New York. (2.1)

Martin, H. C. [1966*b*]: On the Derivation of Stiffness Matrices for the Analysis of Large Deflection and Stability Problems, *Proc. Conf. Matrix Methods Struct. Mech.*, AFFDL-TR-66-80 (Oct. 26–28, 1965), pp. 697–716, Wright-Patterson AFB, Ohio. (16.5)

Martin, H. C. [1970]: Finite Element Formulations of Geometrically Nonlinear Problems, in R. H. Gallagher, Y. Yamada, and J. T. Oden (eds.), "Recent Advances in Matrix Methods in Structural Analysis and Design," Proceedings of the U.S.-Japan Seminar on Matrix Methods in Structural Analysis and Design, Tokyo, 1969, University of Alabama Press, University. (2.1, 16.5)

McCoy, N. [1960]: "Introduction to Modern Algebra," Allyn and Bacon, Inc., Boston. (7.1)

McGuirt, C. W., and G. Lianis [1969]: Experimental Investigation of Non-linear Non-isothermal Viscoelasticity, *Int. J. Eng. Sci.*, vol. 7, pp. 579–599. (19.2)

McLay, R. W. [1963]: An Investigation into the Theory of the Displacement Method of Analysis for Linear Elasticity, doctoral dissertation, University of Wisconsin, Madison. (10)

McLay, R. W. [1967]: Completeness and Convergence Properties of Finite Element Displacement Functions—A General Treatment, *AIAA 5th Aerospace Sci. Meeting, AIAA Paper* 67-143, New York. (10)

McLay, R. W. [1971]: On Certain Approximations in the Finite Element Method, *J. Appl. Mech.*, vol. 38, no. 4, pp. 58–61. (10.2)

Meixner, J. [1943]: Zur Thermodynamik der Irreversiblen Prozesse, *Z. Angew. Phys. Chem., B*, vol. 53, pp. 235–263. (12)

Melosh, R. J. [1963*a*]: Basis of Derivation of Matrices for the Direct Stiffness Method, *AIAA J.*, vol. 1, pp. 1631–1637. (10, 10.2, 10.3)

Melosh, R. J. [1963*b*]: Structural Analysis of Solids, *J. Struct. Div. ASCE*, vol. 89, pp. 205–223. (10.4)

Melosh, R. J. [1966]: A Flat Triangular Shell Element Stiffness Matrix, *Proc. Conf. Matrix Methods Struct. Mech.*, AFFDL-TR-66-80 (Oct. 26–28, 1965), pp. 503–514, Wright-Patterson AFB, Ohio. (10.4)

Meyer, G. [1968]: On Solving Nonlinear Equations with a One-parameter Operator Imbedding, *SIAM J. Numerical Anal.*, vol. 5, pp. 739–752. (17.5)

Mikhlin, S. G. [1964]: "Variational Methods in Mathematical Physics," translated from the 1957 Russian ed. by T. Boddington, Pergamon Press, Ltd., Oxford. (10.1)

Mikhlin, S. G. [1965]: "The Problem of the Minimum of a Quadratic Functional," translated from the 1952 Russian ed. by A. Feinstein, Holden-Day, Inc., Publishers, San Francisco. (10.1, 10.3)

Mikhlin, S. G., and K. L. Smolitskiy [1967]: "Approximate Methods for the Solution of Differential and Integral Equations," translated from the Russian ed. by R. E. Kalaba, American Elsevier Publishing Company, Inc., New York. (10.1)

Mooney, M. [1940]: A Theory of Large Elastic Deformation, *J. Appl. Phys.*, vol. 11, pp. 582–592. (15.5)

Moore, J. T. [1962]: "Elements of Abstract Algebra," The Macmillan Company, New York. (7.1)

Morse, P. M., and H. Feshbach [1953]: "Methods of Theoretical Physics," part I, McGraw-Hill Book Company, New York. (11.2)

Mostow, G. D., J. H. Sampson, and J. P. Meyer [1963]: "Fundamental Structures of Algebra," McGraw-Hill Book Company, New York. (7.1)

Müller, I. [1967]: On the Entropy Inequality, *Arch. Rational Mech. Anal.*, vol. 26, pp. 118–141. (12.3)

Murray, D. W., and E. L. Wilson [1969*a*]: Finite Element Large Deflection Analysis of Plates, *J. Eng. Mech. Div. ASCE*, vol. 95, no. EM1, pp. 143–165. (16.5)

Murray, D. W., and E. L. Wilson [1969*b*]: Finite-element Postbuckling Analysis of Thin Elastic Plates, *AIAA J.*, vol. 7, no. 10, pp. 1915–1920. (16.5)

Murray, K. H. [1970]: Comments on the Convergence of Finite-element Solutions, *AIAA J.*, vol. 8, pp. 815–816. (10.2)

Nelder, J. A., and R. Mead [1964]: A Simplex Method for Function Minimization, *Computer Journal*, vol. 7, p. 308.

Nemat-Nasser, S., and H. D. Shatoff [1970]: A Consistent Numerical Method for the Solution of Nonlinear Elasticity Problems at Finite Strains, *Dep. Aerospace Mech. Eng. Sci. Tech. Rep.* 2, January. (16, 16.1)

Nickell, R. E., and J. J. Sackman [1968]: Approximate Solutions in Linear Coupled Thermoelasticity, *J. Appl. Mech.*, ser. E, vol. 35, no. 2, pp. 255–266. (11.1, 20.1, 20.2)

Noll, W. [1958]: A Mathematical Theory of the Mechanical Behavior of Continuous Media, *Arch. Rational Mech. Anal.*, vol. 2, pp. 197–226. (19.1)

Novozhilov, V. V. [1953]: "Foundations of Nonlinear Theory of Elasticity," translated from the 1948 Russian ed. by F. Bagemihl, H. Komm, and W. Seidel, Graylock Press, Rochester, N.Y. (4.2)

Nowacki, W. [1962]: "Thermoelasticity," Pergamon Press, Ltd., Oxford. (19.2)

Oden, J. T. [1966*a*]: Analysis of Large Deformations of Elastic Membranes by the Finite Element Method, *Proc. IASS Int. Congr. Large-span Shells, Leningrad.* (16, 18.1)

Oden, J. T. [1966*b*]: Calculation of Geometric Stiffness Matrices for Complex Structures, *AIAA J.*, vol. 4, no. 8, pp. 1480–1482. (16.5)

Oden, J. T. [1967*a*]: Numerical Formulation of a Class of Problems in Nonlinear Viscoelasticity, *Advan. Astronautical Sci.*, vol. 24, June. (6.1, 10.3, 10.4, 16.1, 20.1)

Oden, J. T. [1967*b*]: Numerical Formulation of Nonlinear Elasticity Problems, *J. Struct. Div. ASCE*, vol. 93, no. ST3, pp. 235–255. (6.1, 7, 10.4, 16, 16.1, 16.2)

Oden, J. T. [1967*c*]: Finite Element Applications in Linear and Nonlinear Thermoelasticity, in "The Discrete and Continuum Concepts in Micro and Macro Mechanics," *EDM Specialty Conf., ASCE*, Nov. 8–10, 1967, Raleigh, N.C., pp. 32–35. (10.3, 13.4)

Oden, J. T. [1968*a*]: Calculation of Stiffness Matrices for Thin Shells of Arbitrary Shape *AIAA J.*, vol. 6, no. 5, pp. 969–972. (7.5)

Oden, J. T. [1968*b*]: Finite Plane Strain of Incompressible Elastic Solids by the Finite Element Method, *Aeron. Quart.*, vol. 19, pp. 254–264. (13.1, 16, 16.2, 18.3)

Oden, J. T. [1969*a*]: A General Theory of Finite Elements. Part 1. Topological Considerations, *Int. J. Numerical Methods Eng.*, vol. 1, no. 2, pp. 205–221. (6.1, 7, 9.7, 10.4, 13.2)

Oden, J. T. [1969*b*]: A General Theory of Finite Elements. Part 2. Applications, *Int. J. Numerical Methods Eng.*, vol. 1, no. 3, pp. 247–259. (6.1, 7, 10.3, 11.1, 11.4, 13.4)

Oden, J. T. [1969*c*]: Finite Element Applications in Nonlinear Structural Analysis, *Proc. Symp. Appl. Finite Element Methods Civil Eng.*, Vanderbilt University, Nashville, Tenn., pp. 419–456. (2.1, 16, 16.5)

Oden, J. T. [1969*d*]: Finite Element Analysis of Nonlinear Problems in the Dynamical Theory of Coupled Thermoelasticity, *Nucl. Eng. Design*, vol. 10, no. 4, pp. 465–475. (13.8, 19.2, 20.1)

Oden, J. T. [1969*e*]: Discussion of "Finite Element Analysis of Nonlinear Structures," *J. Struct. Div. ASCE*, vol. 95, no. ST6, pp. 1379–1381. (16.5)

Oden, J. T. [1970*a*]: A Generalization of the Finite Element Concept and Its Application to a Class of Problems in Nonlinear Viscoelasticity, *Developments in Theoretical and Applied Mechanics*, IV, Edited by D. Friederick (Proceedings, 4th SECTAM, March 1968), Pergamon Press, Oxford, pp. 581–592. (13.1, 13.4, 13.6, 20.1)

Oden, J. T. [1970*b*]: Finite Element Formulation of Problems of Finite Deformation and Irreversible Thermodynamics of Nonlinear Continua, in R. H. Gallagher, Y. Yamada, and J. T. Oden (eds.), "Recent Advances in Matrix Methods in Structural Analysis and Design," Proceedings of the U.S.-Japan Seminar on Matrix Methods in Structural Analysis and Design, Tokyo, 1969, University of Alabama Press, University. (2.1, 13.4, 13.8, 16.5, 20.1)

Oden, J. T. [1970*c*]: Finite-element Approximation of a Class of Nonlinear Operators, *SIAM Soc. Indus. Appl. Math.*, Fall meeting, Boston, Mass. (10)

Oden, J. T. [1970*d*]: Note on an Approximate Method for Computing Nonconservative Generalized Forces on Finitely Deformed Finite Elements, *AIAA J.*, vol. 8, no. 11, pp. 2088–2090. (16.3)

Oden, J. T. [1971*a*]: The Theory of Conjugate Projections in Finite Element Analysis, *NATO Advanced Study Institute on Finite Element Methods in Continuum Mechanics*, Lisbon. (9.7)

Oden, J. T. [1971*b*]: Finite Element Formulation of Nonlinear Boundary-value Problems, *NATO Advanced Study Institute on Finite Element Methods in Continuum Mechanics*, Lisbon. (10, 10.2)

Oden, J. T. [1971*c*]: Finite Element Approximations in Nonlinear Elasticity, *NATO Advanced Study Institute on Finite Element Methods in Continuum Mechanics*, Lisbon. (16.1, 17, 18.1)

Oden, J. T. [1971*d*]: Finite Element Approximations in Nonlinear Thermoviscoelasticity, *NATO Advanced Study Institute on Finite Element Methods in Continuum Mechanics*, Lisbon. (20.1)

Oden, J. T., and W. K. Kubitza [1967]: Numerical Analysis of Nonlinear Pneumatic Structures, *Proc. Int. Colloq. Pneumatic Struct., Stuttgart*, May, pp. 82–107. (13.4, 16, 16.5, 18.1)

Oden, J. T., and T. Sato [1967*a*]: Finite Strains and Displacements of Elastic Membranes by the Finite Element Method, *Int. J. Solids Struct.*, vol. 1, pp. 471–488. (16, 16.2, 18.1)

Oden, J. T., and T. Sato [1967*b*]: Structural Analysis of Aerodynamic Deceleration Systems, *Advan. Astronautical Sci.*, vol. 24. (16, 18.1)

Oden, J. T., and G. A. Wempner [1967]: Numerical Analysis of Arbitrary Shell Structures under Arbitrary Static Loadings, *Univ. Alabama Res. Inst. Rep.* 47, November. (7.5, 10.4)

Oden, J. T., and G. Aguirre-Ramirez [1969]: Formulation of General Discrete Models of Thermomechanical Behavior of Materials with Memory, *Int. J. Solids Struct*, vol. 5, no. 10, pp. 1077–1093. (7, 10.4, 13.1, 13.2, 13.4, 13.6, 13.8, 20.1)

Oden, J. T., and D. A. Kross [1969]: Analysis of General Coupled Thermoelasticity Problems by the Finite Element Method, *Proc. 2d Conf. Matrix Methods Struct. Mech.*, AFFDL-TR-68-150 (Oct. 15–17, 1968), pp. 1091–1120, Wright-Patterson AFB, Ohio. (13.8, 20.1, 20.2, 20.3)

Oden, J. T., D. M. Rigsby, and D. Cornett [1969]: On the Numerical Solution of a Class of Problems in a Linear First Strain Gradient Theory of Elasticity, *Int. J. Numerical Methods Eng.*, vol. 1, no. 2, pp. 159–174. (10.4)

Oden, J. T., and J. E. Key [1970]: Numerical Analysis of Finite Axisymmetric Deformation of Incompressible Elastic Solids of Revolution, *Int. J. Solids Struct.*, vol. 6, pp. 497–518. (16, 16.2, 16.5, 18.2)

Oden, J. T., and J. Poe [1970]: On the Numerical Solution of a Class of Problems in Dynamic Coupled Thermoelasticity, *Developments in Theoretical and Applied Mechanics*, V (Proceedings, 5th SECTAM, April 1970), edited by D. Frederick, Pergamon Press, Oxford. (13.8, 19.2, 20.1, 20.3)

Oden, J. T., and H. J. Brauchli [1971*a*]: On the Calculation of Consistent Stress Distributions in Finite Element Applications, *Int. J. Numerical Methods Eng.*, vol. 3, no. 3, pp. 317–325. (9, 9.8, 16.3)

Oden, J. T., and H. J. Brauchli [1971*b*]: A Note on Accuracy and Convergence of Finite Element Approximations, *Int. J. Numerical Methods Eng.*, vol. 3, pp. 291–294. (10)

Oden, J. T., and J. E. Key [1971*a*]: Analysis of Finite Deformations of Elastic Solids by the Finite Element Method, *Proc. IUTAM Colloq. High Speed Computing Elastic Struct.*, Liège. (16, 16.2, 16.3, 18.1, 18.2, 18.3)

Oden, J. T., and J. E. Key [1971*b*]: On Some Generalizations of the Incremental Stiffness Relations for Finite Deformations of Compressible and Incompressible Finite Elements, *Nuc. Eng. Design*, vol. 15, pp. 121–134. (16.5)

Oden, J. T., and J. E. Key [1971*c*]: On the Effect of the Form of the Strain Energy Function on the Solution of a Boundary-value Problem in Finite Elasticity, *Computer and Structures*, forthcoming. (to be published). (18.1)

Oden, J. T., and W. H. Armstrong [1971]: Analysis of Nonlinear Dynamic Coupled Thermoviscoelasticity Problems by the Finite Element Method, *Intl. J. Comp. and Struct.* (to appear). (20.1, 20.2)

Oden, J. T., T. J. Chung, and J. E. Key [1971]: Analysis of Nonlinear Thermoelastic and Thermoplastic Behavior of Solids of Revolution by the Finite Element Method, *Proc.* 1*st Intern. Conf. Struct. Mech. Reactor Tech.*, Berlin. (20.2)

Oden, J. T., and J. N. Reddy [1971]: Mixed Conjugate Finite Element Approximations of Linear Operators, *Int. J. Struct. Mech.* (to appear). (9.7)

Ortega, J. M., and W. C. Rheinboldt [1970]: "Iterative Solution of Nonlinear Equations in Several Variables," Academic Press, Inc., New York. (17, 17.2, 17.5)

Padlog, J., R. D. Huff, and G. F. Holloway [1960]: Unelastic Behavior of Structures Subjected to Cyclic, Thermal, and Mechanical Stressing Conditions, *WADD-TR* 60-271, Wright-Patterson AFB, Ohio. (20.1)

Parkus, H. [1964]: Methods of Solution of Thermoelastic Boundary Value Problems, in A. M. Freudenthal, B. A. Boley, and H. Liebowitz (eds.), "High Temperature Structures and Materials," Proceedings of the Third Symposium on Naval Structural Mechanics, Pergamon Press, Ltd., Oxford. (19.2)

Parkus, H. [1968]: "Thermoelasticity," Blaisdell Publishing Company, Waltham, Mass. (19.2)

Parr, C. H. [1964]: The Application of Numerical Methods to the Solution of Structural Integrity Problems of Solid Propellant Rockets, II, *Solid Rocket Struct. Integrity Abstr.*, Solid Rocket Structural Integrity Information Center, University of Utah, Salt Lake City. (2.1)

Parr, C. H. [1967]: The Application of Numerical Methods to the Solution of Structural Integrity Problems of Solid Propellant Rockets, II, *Solid Rocket Struct. Integrity Abstr.*, Solid Rocket Structural Integrity Information Center, University of Utah, Salt Lake City. (2.1)

Percy, J. H. [1967]: Quadrilateral Finite Element Analysis in Elastic-Plastic Plane-stress Analysis, *AIAA J.*, vol. 5, no. 2, p. 367. (10.4)

Pestel, E. [1966]: Dynamic Stiffness Matrix Formulation by Means of Hermitian Polynomials, *Proc. Conf. Matrix Methods Struct. Mech.*, AFFDL-TR-66-80 (Oct. 26–28, 1965), pp. 479–502, Wright-Patterson AFB, Ohio. (8.1)

Pestel, E. C., and F. A. Leckie [1963]: "Matrix Methods in Elastomechanics," McGraw-Hill Book Company, New York. (2.1)

Peterson, F. E., D. M. Campbell, and L. R. Herrmann [1966]: Nonlinear Plane Stress Analysis Applicable to Solid Propellant Rocket Grains, *Bull. 5th Meeting Interagency Chem. Rocket Propellant Group—Working Group on Mech. Behavior, CPIA Publ.* 119, vol. 1, pp. 421–455. (16)

Pian, T. H. H. [1964*a*]: Derivation of Element Stiffness Matrices, *AIAA J.*, vol. 2, no. 3, pp. 576–577. (10.4)

Pian, T. H. H. [1964*b*]: Derivation of Element Stiffness Matrices by Assumed Stress Distributions, *AIAA J.*, vol. 2, pp. 1333–1336. (10.4)

Pian, T. H. H., and P. Tong [1969*a*]: Basis of Finite Element Method for Solid Continua, *Int. J. Numerical Methods Eng.*, vol. 1, pp. 3–28. (10.4)

Pian, T. H. H., and P. Tong [1969*b*]: Rationalization in Deriving Element Stiffness Matrix by Assumed Stress Approach, *Proc. 2d Conf. Matrix Methods Struct. Mech.*, AFFDL-TR-68-150 (Oct. 15–17, 1968), pp. 441–470, Wright-Patterson AFB, Ohio. (10.4)

Pipkin, A. C. [1964]: Small Finite Deformations of Viscoelastic Solids, *Rev. Mod. Phys.*, vol. 36, pp. 1034–1041. (19.2)

Pipkin, A. C. [1966]: Approximate Constitutive Equations, in S. Eskinazi (ed.), "Modern Developments in the Mechanics of Continua," pp. 89–108, Academic Press, Inc., New York. (19.2)

Polya, G. [1952]: Sur une Interpretation de la Méthode des Différences Finies Qui Peut Fournir des Bornes Supérieures ou Inférieures, *Compt. Rend.*, vol. 235, p. 995. (2.1)

Powell, M. J. D. [1969]: A Survey of Numerical Methods for Unconstrained Optimization, *SIAM Rev.*, vol. 12, no. 1, pp. 79–97. (17)

Prager, W., and J. L. Synge [1947]: Approximations in Elasticity Based on the Concept of Function Spaces, *Quart. Appl. Math.*, vol. 5, pp. 241–269. (2.1)

Prigogine, I. [1947]: "Etude Thermodynamique des Phénomènes Irréversibles," Desoor, Liège. (12)

Prigogine, I. [1961]: "Introduction to Thermodynamics of Irreversible Processes," 2d ed., Interscience Publishers, Inc., New York. (12)

Przemieniecki, J. S. [1968]: "Theory of Matrix Structural Analysis," McGraw-Hill Book Company, New York. (2.1)

Przemieniecki, J. S., R. M. Bader, W. F. Bozich, J. R. Johnson, and W. J. Mykytow (eds.) [1966]: *Proc. Conf. Matrix Methods Struct. Mech.*, AFFDL-TR-66-80 (Oct. 26–28, 1965), Wright-Patterson AFB, Ohio. (2.1)

Rall, L. B. [1969]: "Computational Solution of Nonlinear Operator Equations," John Wiley & Sons, Inc., New York. (17.2, 17.4)

Rashid, Y. R. [1964]: Solution of Elasto-static Boundary Value Problems by the Finite Element Method, doctoral dissertation, University of California, Berkeley. (10.4, 18.2)

Rashid, Y. R. [1966]: Analysis of Axisymmetric Composite Structures by the Finite Element Method, *Nucl. Eng. Design*, vol. 3, pp. 163–182. (10.4, 18.2)

Rayleigh, Lord (Sir John William Strutt) [1877]: "The Theory of Sound," 2d ed. rev. 1945, Dover Publications, Inc., New York. (10.3)

Reiner, M. [1958]: Rheology, in S. Flugge (ed.), "Encyclopedia of Physics," vol. VI, p. 507, Springer-Verlag OHG, Berlin. (19.2)

Ritz, W. [1908]: Uber Eine Neue Methode zur Lösung Gewisser Randwertaufgaben, *Gottingener Nachrichten, Mathematik und Physik*, Klasse, pp. 236–248. (10.3)

Ritz, W. [1909]: Uber Eine Neue Methode zur Lösung Gewisser Variationsprobleme der Mathematischen Physik, *J. Reine Angew. Math.*, vol. 135, no. 1, p. 1. (10.3)

Rivlin, R. S. [1948*a*]: Large Elastic Deformations of Isotropic Materials. I. Fundamental Concepts, *Phil. Trans. Roy. Soc.*, vol. A240, pp. 459–490. (15.5)

Rivlin, R. S. [1948*b*]: Large Elastic Deformations of Isotropic Materials. IV. Further Developments of the General Theory, *Phil. Trans. Roy. Soc.*, vol. A241, pp. 379–397; reprinted in "Problems of Non-linear Elasticity," International Science series, Gordon and Breach, Science Publishers, Inc., New York, 1965. (15.5)

Rivlin, R. S. [1949*a*]: Large Elastic Deformations of Isotropic Materials. V. The Problem of Flexure, *Proc. Roy. Soc.*, vol. A195, pp. 463–473; reprinted in "Problems of Non-linear Elasticity," International Science Series, Gordon and Breach, Science Publishers, Inc., New York, 1965. (15.5)

Rivlin, R. S. [1949*b*]: Large Elastic Deformations of Isotropic Materials. VI. Further Results in the Theory of Torsion, Shear, and Flexure, *Phil. Trans. Roy. Soc.*, vol. A242, pp. 173–195; reprinted in "Problems of Non-linear Elasticity," International Science Series, Gordon and Breach, Science Publishers, Inc., New York, 1965. (15.5)

Rivlin, R. S., and D. W. Saunders [1951]: Large Elastic Deformations of Isotropic Materials. VII. Experiments on the Deformation of Rubber, *Phil. Trans. Roy. Soc.*, vol. A243, pp. 251–288. (15.5, 18.1)

Rivlin, R. S., and A. G. Thomas [1951]: Large Elastic Deformations of Isotropic Materials. VIII. Strain Distribution around a Hole in a Sheet, *Phil. Trans. Roy. Soc.*, vol. A243, pp. 289–298. (18.1)

Rowan, W. H., Jr., and R. M. Hackett (eds.) [1969]: *Proc. Symp. Appl. Finite Element Methods Civil Eng.*, School of Engineering, Vanderbilt University, Nashville, Tenn. (2.1)

Rydzewski, J. R. (ed.) [1965]: "Theory of Arch Dams," Pergamon Press, Ltd., Oxford. (2.1)

Saaty, T. L. [1967]: "Modern Nonlinear Equations," McGraw-Hill Book Company, New York. (10.1, 17)

Saaty, T. L., and J. Bram [1964]: "Nonlinear Mathematics," McGraw-Hill Book Company, New York. (17, 17.3)

Saint Venant, A. J. C. B. de [1868]: Formules de L'elasticité des Corps Amorphes que des Compressions Permanentes et Inégales ont Rendus Hétérotropes, *J. Math. Pures Appl.*, vol. 13, pp. 242–254. (16.5)

Sakadi, Z. [1949]: On Elasticity Problems When the Second Order Terms of the Strain Are Taken into Account, II, *Mem. Fac. Eng. Nagoya Univ.*, vol. 1, pp. 95–107. (15.5)

San Miguel, A. [1965]: On the Characterization of Multiaxial Data in Terms of the Strain Energy Concept, *Bull. 4th Meeting Interagency Chem. Rocket Propellant Group—Working Group Mech. Behavior*, *CPIA Publ.* 94U, vol. 1, pp. 169–178. (15.5)

Schapery, R. A. [1968]: On the Application of a Thermodynamic Constitutive Equation to Various Nonlinear Materials, *School Aeron. Astronautics Eng. Sci. Rep.* AA&ES 68-4, Purdue University, Lafayette, Ind. (19.2)

Schmit, L. A., Jr., F. K. Bogner, and R. L. Fox [1968]: Finite Deflection Structural Analysis Using Plate and Shell Discrete Elements, *AIAA J.*, vol. 6, no. 5, pp. 781–791. (17.1)

Schmit, L. A., Jr., and G. R. Monforton [1969]: Finite Deflection Analysis of Sandwich Plates and Cylindrical Shells with Laminated Facings, *Proc. 10th AIAA/ASME Struct. Structural Dyn. Mater. Conf., New Orleans*, April. (17.1)

Schwarzl, F., and A. J. Staverman [1952]: Time-Temperature Dependence of Linear Viscoelastic Behavior, *J. Appl. Phys.*, vol. 23, pp. 838–843. (19.2)

Seeger, A. [1964]: Discussion to Paper by Foux, in M. Reiner and D. Abir (eds.), "Second-order Effects in Elasticity, Plasticity, and Fluid Dynamics," Proceedings of the International Symposium, Haifa, April 23–27, 1962, Pergamon Press, Ltd., Oxford (distributed by The Macmillan Company, New York). (15.5)

Seeger A., and O. Buck [1960]: Die Experimentalle Ermittlung der Elastichen Konstanten Höherer Ordung, *Z. Naturforsch*, vol. 15a, pp. 1056–1067. (15.5)

Segal, A., and J. M. Klosner [1970]: Stress Concentration in an Elastomeric Sheet Subject to Large Deformations, *PIBAL Rep.* 70-11, Polytechnic Institute of Brooklyn, March. (18.1)

Sheng, P. L. [1955]: Secondary Elasticity, *Chinese Assoc. Advan. Sci. Monograph*, ser. 1, vol. I, no. 1. (15.5)

Signorini, A. [1955]: Solidi Incomprimibili, *Ann. Mat. Pura Appl.*, pp. 147–201. (15.5)

Silvester, R. [1969]: Higher-order Polynomial Triangular Finite Elements for Potential Problems, *Int. J. Eng. Sci.*, vol. 7, no. 8, pp. 849–861. (10.4)

Singhal, A. C. [1969]: 775 Selected References on the Finite Element Method and Matrix Methods of Structural Analysis, *Civil Eng. Dept.*, *Rep.* S-12, Laval University, Quebec, January. (2.1)

Smith, G. F., and R. S. Rivlin [1958]: The Strain Energy Function for Anisotropic Elastic Materials, *Trans. AMS*, vol. 88, pp. 175–193. (15.5)

Southwell, R. V. [1940]: "Relaxation Methods in Engineering Science: A Treatise on Approximate Computation," Oxford University Press, London. (17.3)

Southwell, R. V. [1946]: "Relaxation Methods in Theoretical Physics: A Continuation of the Treatise," vol. I (vol. II, 1956), Oxford University Press, London. (17.3)

Spang, H. A., III [1962]: A Review of Minimization Techniques for Nonlinear Functions, *SIAM Rev.*, vol. 4, no. 4, pp. 343–364. (17, 17.6)

Sternberg, E., and J. G. Chakravorty [1959]: On Inertia Effects in a Transient Thermoelastic Problem, *J. Appl. Mech.*, vol. 26, no. 4, *Trans. ASME*, ser. E, vol. 81, pp. 503–509. (20.3)

Straneo, P. [1925]: Sull'Expressione die Fenomeni Ereditari, *Atti. Accad. Naz. Lincei Rend.*, ser. 6, vol. I, pp. 29–33. (14.4)

Strang, G. [1970]: The Finite Element Method and Approximation Theory, *Sym. Numerical Sol. Partial Differential Equations*, University of Maryland, College Park. (10)

Stricklin, J. A. [1968]: Integration of Area Coordinates in Matrix Structural Analysis, *AIAA J.*, vol. 6, no. 10, p. 2023. (10.4)

Synge, J. L. [1957]: "The Hypercircle Method in Mathematical Physics," Cambridge University Press, Cambridge. (2.1, 10, 10.4)

Szabo, B. A. [1969]: Principles of Discretization of Continuous Structures, doctoral dissertation, State University of New York, Buffalo. (10.3)

Szabo, B. A., and G. C. Lee [1969*a*]: Derivation of Stiffness Matrices for Problems in Plane Elasticity by Galerkin's Method, *Int. J. Numerical Methods Eng.*, vol. 1, no. 3, pp. 301–310. (10.3)

Szabo, B. A., and G. C. Lee [1969*b*]: Stiffness Matrix for Plates by Galerkin's Method, *J. Eng. Mech. Div. ASCE*, vol. 95, no. EM3, pp. 571–585. (10.3)

Tabandeh, N. [1970]: Convergence of the Finite Element Method, doctoral dissertation, Louisiana State University, Baton Rouge. (10)

Taig, I. C. [1961]: Structural Analysis by the Matrix Displacement Method, *English Elec. Aviation Ltd. Rep.* S-0-17. (10.4)

Taig, I. C., and R. I. Kerr [1964]: Some Problems in the Discrete Element Representation of Aircraft Structures, in B. Fraeijs de Veubeke (ed.), "Matrix Methods of Structural Analysis," AGARDograph 72, pp. 267–315, Pergamon Press, Ltd., Oxford. (10.4)

Taylor, A. E. [1958]: "Introduction to Functional Analysis," John Wiley & Sons, Inc., New York. (7.1)

Taylor, R. L., and T. Y. Chang [1966]: An Approximate Method for Thermoviscoelastic Stress Analysis, *Nucl. Eng. Design*, vol. 4, pp. 21–28. (20.1)

Taylor, R. L., K. S. Pister, and L. R. Herrmann [1968]: On a Variational Theorem for Incompressible and Nearly Incompressible Orthotropic Elasticity, *Int. J. Solids Struct.*, vol. 4, pp. 875–883. (16.2)

Taylor, R. L., K. S. Pister, and G. L. Goudreau [1970]: Thermomechanical Analysis of Viscoelastic Solids, *Int. J. Numerical Methods Eng.*, vol. 2, pp. 45–59. (19.2, 20.1)

Thomas, A. G. [1955]: The Departures from the Statistical Theory of Rubber Elasticity, *Trans. Faraday Soc.*, vol. 51, pp. 569–582. (15.5)

Tocher, J. L., and B. J. Hartz [1967]: Higher-order Finite Elements for Plane Stress, *J. Eng. Mech. Div. ASCE*, vol. 93, no. EM4, pp. 149–172. (8.1, 10.4)

Todhunter, I., and K. Pearson [1893]: "A History of the Theory of Elasticity and of the Strength of Materials from Galilei to Lord Kelvin," Cambridge University Press, Cambridge. (15.1)

Tong, P. [1969]: An Assumed Stress Hybrid Model for an Incompressible and Near-incompressible Material, *Int. J. Solids Struct.*, vol. 5, pp. 455–461. (16.2)

Tong, P., and T. H. H. Pian [1967]: The Convergence of Finite Element Method in Solving Linear Elastic Problems, *Int. J. Solids Struct.*, vol. 3, pp. 865–879. (10)

Treloar, L. R. G. [1944]: Stress-Strain Data for Vulcanized Rubber under Various Types of Deformation, *Trans. Faraday Soc.*, vol. 40, p. 59. (15.5, 18.1)

Treloar, L. R. G. [1958]: "The Physics of Rubber Elasticity," 2d ed., Oxford University Press, London. (15.5, 18.1)

Truesdell, C. [1952]: The Mechanical Foundations of Elasticity and Fluid Dynamics, *J. Rational Mech. Anal.*, vol. 1, pp. 125–300 (with corrections in vol. 2, pp. 595–616, 1953, vol. 3, p. 801, 1954); reprinted in "Continuum Mechanics I: The Mechanical Foundations of Elasticity and Fluid Dynamics," Gordon and Breach, Science Publishers, Inc., New York, 1966. (15.1)

Truesdell, C. [1966*a*]: "The Elements of Continuum Mechanics," Springer-Verlag New York Inc., New York. (4, 12, 12.3, 12.4, 14, 14.3)

Truesdell, C. [1966*b*]: Thermodynamics of Deformation, in R. J. Donnelly, R. Herman, and I. Prigogine (eds.), "Non-equilibrium Thermodynamics Variational Techniques and Stability," University of Chicago Press, Chicago. (12)

Truesdell, C. [1969]: "Rational Thermodynamics," McGraw-Hill Book Company, New York. (12, 12.3)

Truesdell, C., and R. Toupin [1960]: The Classical Field Theories, in "Encyclopedia of Physics," vol. III/1 ("Principles of Classical Mechanics and Field Theory"), Springer-Verlag New York Inc., New York. (4, 5, 13.1, 16.2)

Truesdell, C., and W. Noll [1965]: The Nonlinear Field Theories of Mechanics, in "Encyclopedia of Physics," vol. III/3, Springer-Verlag New York Inc., New York. (4, 5.2, 14, 14.3, 14.4, 15.2, 15.3, 15.5, 16.5)

Turner, M. J. [1959]: The Direct Stiffness Method of Structural Analysis, *10th Meeting AGARD Struct. Mater. Panel*, Aachen, Germany. (16.5)

Turner, M. J., R. W. Clough, H. C. Martin, and L. P. Topp [1956]: Stiffness and Deflection Analysis of Complex Structures, *J. Aeron. Sci.*, vol. 23, no. 9, pp. 805–823, 854. (2.1, 10.4)

Turner, M. J., E. H. Dill, H. C. Martin, and R. J. Melosh [1960]: Large Deflections of Structures Subjected to Heating and External Loads, *J. Aerospace Sci.*, vol. 27, pp. 97–102, 127. (16.5, 20.1)

Turner, M. J., H. C. Martin, and B. C. Weikel [1964]: Further Developments and Applications of the Stiffness Method, in B. M. Fraeijs de Veubeke (ed.), "Matrix Methods in Structural Analysis," AGARDograph 72, pp. 203–206, Pergamon Press, Ltd., Oxford. (16.5)

Utku, S. [1966]: Stiffness Matrices for Thin Triangular Elements of Nonzero Gaussian Curvature, *AIAA 4th Aerospace Sci. Meeting Paper* 66-530, Los Angeles, June 27–29. (10.4)

Utku, S., and R. J. Melosh [1967]: Behavior of Triangular Shell Element Stiffness Matrices Associated with Polyhedral Deflection Distributions, *AIAA 5th Aerospace Sci. Meeting Paper* 67-114, New York, January 23–27. (10.4)

Vainberg, M. M. [1964]: "Variational Methods for the Study of Nonlinear Operators," translated from the 1956 Russian ed. by A. Feinstein, Holden-Day, Inc., Publishers, San Francisco. (10.1)

Varga, R. S. [1967]: Hermite Interpolation-type Ritz Methods for Two-point Boundary Value Problems, in J. H. Bramble (ed.), "Numerical Solutions of Partial Differential Equations," pp. 365–373, Academic Press, Inc., New York. (10)

Visser, W. [1966]: A Finite-element Method for the Determination of Non-stationary Temperature Distributions and Thermal Deformations, *Proc. Conf. Matrix Methods Struct. Mech.*, AFFDL-TR-66-80 (Oct. 26–28, 1965), pp. 925–943, Wright-Patterson AFB, Ohio. (20.1)

Walz, J. E., R. E. Fulton, and N. J. Cyrus [1969]: Accuracy and Convergence of Finite Element Approximations, *Proc. 2d Conf. Matrix Methods Struct. Mech.*, AFFDL-TR-68-150 (Oct. 15–17, 1968), pp. 995–1027, Wright-Patterson AFB, Ohio. (10)

Ward, I. M., and E. T. Onat [1963]: Nonlinear Mechanical Behavior of Oriented Polypropylene, *J. Mech. Phys. Solids*, vol. 11, pp. 217–229. (19.2)

Warren, D. S. [1962]: A Matrix Method for the Analysis of Buckling of Structural Panels Subjected to Creep Environments, *AF Flight Dyn. Lab. Rep.* ASD-TR-62-740. (20.1)

Warren, D. S., R. A. Castle, and R. C. Gloria [1962]: An Evaluation of the State-of-the-art of Thermomechanical Analysis of Structures, *AF Flight Dyn. Lab. Rep.* WADD-TR-61-152. (2.1, 20.1)

Wempner, G. A. [1969]: Finite Elements, Finite Rotations, and Small Strains of Flexible Shells, *Int. J. Solids Struct.*, vol. 5, pp. 117–153. (7.5)

Wempner, G. A., J. T. Oden, and D. A. Kross [1968]: Finite Element Analysis of Thin Shells, *J. Eng. Mech. Div. ASCE*, vol. 94, no. EM6, pp. 1273–1294. (7.5, 10.4)

Weyl, H. [1946]: "The Classical Groups, Their Invariants and Representations" Princeton University Press, Princeton, N.J. (15.5)

Whaples, G. [1952]: Carathéordory's Temperature Equations, *J. Rational Mech. Anal.*, vol. 1, pp. 301–307. (12.1)

White, J. L. [1969]: Finite Elements in Linear Viscoelasticity, *Proc. 2d Conf. Matrix Methods Struct. Mech.*, AFFDL-TR-68-150 (Oct. 15–17, 1968), pp. 489–516, Wright-Patterson AFB, Ohio. (20.1)

Williams, M. L., F. R. Landel, and J. D. Ferry [1955]: The Temperature Dependence of Relaxation Mechanisms in Amorphous Polymers and Other Glass Forming Liquids, *J. ACS*, vol. 77, pp. 3701–3707. (19.2, 20.2)

Williams, M. L., and R. A. Schapery [1962]: Studies of Viscoelastic Media, *Calif. Inst. Tech. Rep.* ARL 62-366. (18.1)

Wilson, E. L. [1965]: Structural Analysis of Axisymmetric Solids, *AIAA J.*, vol. 3, pp. 2269–2274. (10.4, 18.2)

Wilson, E. L., and R. E. Nickell [1966]: Application of the Finite Element Method to Heat Conduction Analysis, *Nucl. Eng. Design*, vol. 4, no. 3, pp. 276–286. (13.8, 20.1)

Wissmann, J. W. [1962]: Numerical Analysis of Nonlinear Elastic Bodies, *Eng. Res. Rep.* ERR-AN-157, General Dynamics/Astronautics, September. (6.1, 10.4)

Wissmann, J. W. [1963]: Numerische Berechnung Nichtlinearer Elasticher Koerper, Dissertation, Hannover, (6.1, 10.4, 16.1)

Wissmann, J. W. [1966]: Nonlinear Structural Analysis; Tensor Formulation, *Proc. Conf. Matrix Methods Struct. Anal.*, AFFDL-TR-66-80 (Oct. 26–28, 1965), pp. 679–696, Wright-Patterson AFB, Ohio. (6.1, 16.1)

Withum, D. [1966]: Berechnung von Platten Nach Dim Ritz'schen Verfahren Mithilfe Drejeckförmiger Maschennetze, *Mitt. Inst. Statik Tech. Hochschule Hanover*, vol. 9. (10.4)

Yaghmai, S. [1968]: Incremental Analysis of Large Deformations in Mechanics of Solids with Applications to Axisymmetric Shells of Revolution, doctoral dissertation, University of California, Berkeley; also *Struct. Eng. Lab. Rep.* SESM 68-17, University of California, Berkeley. (16.5)

Yang, W. H. [1967]: Stress Concentration in a Rubber Sheet under Axially Symmetric Stretching, *J. Appl. Mech.*, vol. 34, pp. 942–946. (18.1)

Zienkiewicz, O. C. [1970]: The Finite Element Method: From Intuition to Generality, *Appl. Mech. Rev.*, vol. 23, no. 3, pp. 249–256. (2.1)

Zienkiewicz, O. C., and G. S. Holister (eds.) [1965]: "Stress Analysis," John Wiley & Sons, Inc., New York. (2.1)

Zienkiewicz, O. C., and Y. K. Cheung [1967]: "The Finite Element Method in Structural and Continuum Mechanics," McGraw-Hill Book Company, New York. (2.1, 10.4)

Zlamal, M. [1968]: On the Finite Element Method, *Numerische Math.*, vol. 12, pp. 394–409. (10, 10.4)

Zoutendijk, G. [1966]: Nonlinear Programming: A Numerical Survey, *SIAM J. Control*, vol. 4, pp. 194–210. (17)

Zudans, Z. [1969]: Survey of Advanced Structural Design Analysis Techniques, *Nucl. Eng. Design*, vol. 10, no. 4, pp. 400–440. (2.1)

INDEX

INDEX